Environmental Technologies to Treat Sulfur Pollution

Principles and Engineering – 2nd Edition

Environmental Technologies to Treat Sulfur Pollution

Principles and Engineering – 2nd Edition

Edited by
Piet N.L. Lens

PUBLISHING

Published by

IWA Publishing
Republic – Export Building, 1st Floor
2 Clove Crescent
London E14 2BE, UK
Telephone: +44 (0)20 7654 5500
Fax: +44 (0)20 7654 5555
Email: publications@iwap.co.uk
Web: www.iwapublishing.com

First published 2020

British Library Cataloguing in Publication Data
A CIP catalogue record for this book is available from the British Library

ISBN: 9781789060959 (Hardback)
ISBN: 9781789060966 (eBook)

This eBook was made Open Access in October 2020

Cover photographs (from left to right):

Top left: Chemically synthesized elemental sulfur lentils with the biofilm formed on its surface during autotrophic denitrification. Top right: Denitrifying biomass attached onto AnoxK™ 1 carriers from a moving bed biofilm reactor fed with biogenic sulfur. Bottom right: Denitrifying biomass attached onto AnoxK™ Z-200 carriers used in a moving bed biofilm reactor fed with biogenic sulfur. Central picture – left: Biogenic sulfur obtained from the incomplete oxidation of sulfide in gaseous streams under oxygen-limiting conditions by sulfur-oxidizing microorganisms. Central picture – central: Chemically synthesized elemental sulfur lentils and limestone particles with the biofilm formed on the surface of sulfur lentils during autotrophic denitrification. Central picture – right: Chemically synthesized elemental sulfur lentils and limestone particles with the biofilm formed on the surface of sulfur lentils during autotrophic denitritation. Photographs courtesy of Anastasiia Kostrytsia, Manuel Di Ruscio, Stefano Papirio and Giovanni Esposito (University of Cassino and Southern Lazio, Cassino, Italy).

Contents

Part II: The Sulfur Cycle

Preface

In recent decades, natural as well as man-made ecosystems have been increasingly affected by sulfur pollution. Adverse effects of sulfur pollution are well known, such as acid rain, odor nuisance from polluted rivers, landfills or treatment systems, corrosion of steel and concrete, and heavy metal and acidity release from oxygen exposed sediments and mineral ores. In contrast to the enormous efforts dedicated to the abatement of pollution by organic compounds, nitrogen and phosphorous, treatment of sulfur-induced pollution has received rather limited attention. Nevertheless, a whole spectrum of environmental technologies that rely on sulfur cycle conversions has been developed. It is the aim of this book to present these technologies and to inform the reader about their functioning and operation.

This book is the 2nd edition of the book published in 2000 with the same title, edited by Look Hulshoff Pol and myself. Since then, a lot of interesting research on the genetics, biochemistry, eco-physiology and technological applications of sulfur cycle bacteria has been performed. Sulfur cycle bacteria keep surprising with their unique metabolic features, e.g. sulfonates as an ecological currency for nutrient and energy exchange between microbial autotrophs and heterotrophs, bacterial giantism (e.g. *Thiomargarita*), substrate diversity, eventually in syntrophy with Archaea (e.g. alkanes as substrates), and long distance electron transfer (e.g. cable bacteria). This has allowed promising technological developments to solve environmental problems with polluted water, gases and solids, which are overviewed in the book. Interestingly, the shift from

Editor: Piet N.L. Lens
DOI: 10.2166/9781789060966_xix

end of pipe treatment towards technologies focusing on resource recovery also occurs in sulfur cycle based technologies, and some very exciting products that can be made using sulfur cycle bacteria have been discovered, e.g. sulfated biochemicals, organo-sulfur based intracellular storage products and nano-sized elemental sulfur, that have fine chemical and medical applications.

Our main objective in assembling the chapters of the 2nd edition has been to present major, up-to-date reviews of the different technological applications. Most chapters of this 2nd edition are updates of the chapters published in the 1st edition, capturing the gain in knowledge and developments since the book's first publication in 2000. In addition, some chapters on new areas that were not or hardly investigated in 2000 have been added: bioelectrical systems, use of the sulfate radical in advanced oxidation processes and nanosulfur (bio) manufacturing. For the reader's interest, chapter 1 introduces the different chapters in this 2nd edition book, and also refers to the relevant chapters of the 1st edition that have not been reproduced in this 2nd edition.

Each contributed chapter is presented on a stand-alone basis, so that the reader will find it helpful to consider only the theme of each chapter. There are nevertheless many connections between what may at first seem to be quite different fields of application. Like in the 1st edition, it was one of our purposes to draw out and emphasize these interdisciplinary linkages. For this reason, a comprehensive index is included to facilitate cross-referencing. We hope that the work described in this book will inspire those working in the field and will encourage those who are beginning to investigate this field.

I wish to thank all contributors of this book for their enthusiastic support, especially the authors who updated and revised the 20 year old chapters from the 1st edition. I also thank all past and present coworkers as well as all collaborators who joined in unraveling parts of the sulfur cycle and its application in environmental technology as described in this book, especially those at Wageningen University, UNESCO-IHE, Tampere University and National University Ireland Galway. In addition, the national and international granting agencies who supported this work on the sulfur cycle over the years are gratefully acknowledged, in particular the Science Foundation Ireland (SFI), who financially supported the open access publication of this book volume through the SFI Research Professorship Programme *Innovative Energy Technologies for Biofuels, Bioenergy and a Sustainable Irish Bioeconomy* (IETSBIO^3; grant number 15/RP/2763). I'm also grateful to Niall Cunniffe and Mark Hammond of IWA Publishing for their help and editorial support in realizing this book.

Galway, May 2020

Prof. Dr. Ir. Piet N.L. Lens
Editor

List of Contributors

Markéta Andreides
Department of Water Technology and Environmental Engineering, University of Chemistry and Technology Prague, Technická, 166 28 Prague, Czech Republic

Sergi Astals
Department of Chemical Engineering and Analytical Chemistry, University of Barcelona, Martí i Franquès 1, 08028 Barcelona, Spain

Jan Bartáček
Department of Water Technology and Environmental Engineering, University of Chemistry and Technology Prague, Technická, 166 28 Prague, Czech Republic

Nicolas Bernet
INRAE, Univ Montpellier, LBE, 102 Avenue des Etangs, 11100 Narbonne, France

Maria F. Carboni
Department of Microbiology, National University of Ireland, Galway, University Road, H91 TK33 Galway, Ireland

Rachel Costa
Department of Biochemistry and Chemical Technology, Institute of Chemistry, São Paulo State University. R. Francisco Degni, 55, 14800-060, Araraquara, SP, Brazil

Philippe M. Chazal
Laboratoire des Sciences de l'Eau et de l'Environnement, Faculté des Sciences, 123 Avenue A. Thomas, 87060 Limoges Cedex, France

Lirong Cheng
Engineering Research Center of Groundwater Pollution Control and Remediation, College of Water Sciences, Beijing Normal University, Beijing 100875, China

Christiane Dahl
Institut für Mikrobiologie & Biotechnologie, Rheinische Friedrich-Wilhelms-Universität Bonn, Meckenheimer Allee 168, 53115 Bonn, Germany

Aizhong Ding
Beijing Engineering Research Center of Petrochemical Wastewater

Reclamation, College of Water Sciences, Beijing Normal University, Beijing 100875, China

Ruth M. Fisher
School of Civil and Environmental Engineering, UNSW, Sydney, NSW 2052, Australia

Wenjuan Jia
College of Water Sciences, Beijing Normal University, Beijing 100875, China

James E. Hayes
School of Civil and Environmental Engineering, UNSW, Sydney, NSW 2052, Australia

Thorkild Hvitved-Jacobsen
Department of Chemistry and Bioscience, Aalborg University, Fredrik Bajers Vej 7H, 9220 Aalborg East, Denmark

Pavel Jeníček
Department of Water Technology and Environmental Engineering, University of Chemistry and Technology Prague, Technická, 166 28 Prague, Czech Republic

D. Barrie Johnson
School of Natural Sciences, Bangor University, Bangor LL57 2UW, UK, and Faculty of Health and Life Sciences, Coventry University, Coventry CV1 5FB, UK

Pimluck Kijjanapanich
Department of Environmental Engineering, Faculty of Engineering, Chiang Mai University, Chiang Mai 50200, Thailand

Hung V. Le
School of Civil and Environmental Engineering, UNSW, Sydney, NSW 2052, Australia

Nhat Le-Minh
School of Civil and Environmental Engineering, UNSW, Sydney, NSW 2052, Australia

Piet N. L. Lens
Department of Microbiology, National University of Ireland, Galway, University Road, H91 TK33 Galway, Ireland

Eleftheria Ntagia
Center for Microbial Ecology and Technology (CMET), Faculty of Bioscience Engineering, Ghent University, Coupure Links 653, 9000 Gent, Belgium

Vincent O'Flaherty
Department of Microbiology, National University of Ireland, Galway, University Road, H91 TK33 Galway, Ireland

Lucie Pokorná
Department of Water Technology and Environmental Engineering, University of Chemistry and Technology Prague, Technická, 166 28 Prague, Czech Republic

Antonin Prévoteau
Center for Microbial Ecology and Technology (CMET), Faculty of Bioscience Engineering, Ghent University, Coupure Links 653, 9000 Gent, Belgium

Korneel Rabaey
Center for Microbial Ecology and Technology (CMET), Faculty of Bioscience Engineering,

Ghent University, Coupure Links 653, 9000 Gent, Belgium

Ángel Robles
Department of Chemical Engineering – School of Engineering (ETSE-UV), Universitat de València, Av. de la Universitat s/n, 46100 Burjassot, Spain

Sudeshna Saikia
Department of Microbiology, National University of Ireland, Galway, University Road, H91 TK33 Galway, Ireland

Ana Laura Santos
School of Natural Sciences, Bangor University, Bangor, LL57 2UW, UK

Joaquín Serralta
Instituto de Ingeniería del Agua y Medio Ambiente, IIAMA, Universitat Politècnica de València, Camino de Vera s/n, 46022 Valencia, Spain

Ralf Steudel
Formerly at Institute of Chemistry, Technical University Berlin, D-10623 Berlin, Germany

Jean-Philippe Steyer
INRAE, Univ Montpellier, LBE, 102 Avenue des Étangs, 11100 Narbonne, France

Richard M. Stuetz
School of Civil and Environmental Engineering, UNSW, Sydney, NSW 2052, Australia

Richard Tichý
Institute of Landscape Ecology and University of South Bohemia, Faculty of Biological Sciences, Na Sádkách 7, 370 05 Ceské Budejovice, Czech Republic

Sergi Vinardell
Department of Chemical Engineering and Analytical Chemistry, University of Barcelona, Martí i Franquès 1, 08028 Barcelona, Spain

Dayang Wang
Guangdong Provincial Academy of Environmental Science, Guangzhou 510045, China

Shunxiang Wang
School of Materials Science and Engineering, Tongji University, Shanghai 201804, China

Tian Wang
Department of Chemical and Biochemical Engineering, Technical University of Denmark, 2800 Lyngby, Denmark

Min Wu
Department of Engineering, Aarhus University, Inge Lehmanns Gade 10, 8000 Aarhus C, Denmark

Lei Zheng
College of Water Sciences, Beijing Normal University, Beijing 100875, China

Shurong Zhang
College of Water Sciences, Beijing Normal University, Beijing 100875, China

Part I
Introduction

Chapter 1

Environmental technologies to treat sulfur pollution: *How to read this book?*

Piet N. L. Lens

1.1 INTRODUCTION

This book is the 2nd edition of the book '*Environmental technologies to treat sulfur pollution: principles and engineering*', that was published in 2000. This 2nd edition captures the research progress made in the last 20 years in the fascinating world of the sulfur cycle, and in particular how the sulfur cycle on the one hand generates enormous pollution problems at a high economic cost, but on the other hand also offers elegant tools to abate pollution or even recover resources.

The 2nd edition is thus an extension of the 1st edition, and both the 1st and 2nd edition have been published open access. To guide the reader through the relevant chapters of both books, Table 1.1 overviews thematically the table of content of both books, and indicates which chapters are new to the 2nd edition, which chapters have been updated and which chapters were only present in the 1st edition. Table 1.1 illustrates the vast array of topics covered when considering the principles and engineering of environmental technologies to treat sulfur pollution, and refers the reader to the relevant chapters of both the 1st and 2nd book volumes.

1.2 THE SULFUR CYCLE

Sulfur is an essential element for life on earth. Its redox conversions are of significance in the biogeochemical sulfur cycle. These sulfur conversions involve

DOI: 10.2166/9781789060966_003

Table 1.1 Overview of the topics covered, and respective chapter numbers, in the 1st and 2nd edition of the book *environmental technologies to treat sulfur pollution: principles and engineering.*

Chapter Topic	Chapter Number in this 2nd Edition/Reference to 1st Edition	Chapter Number in 1st Edition (Lens & Hulshoff Pol, 2000)
I – The Sulfur Cycle		
The chemical sulfur cycle	2	1
The geochemical sulfur cycle	Middelburg (2000)	2
The biological sulfur cycle	3	3
Agricultural aspects of the sulfur cycle	Ernst (2000)	15
II – Sulfur-related problems		
Sulfur transformations during sewage transport	4	6
Corrosion and sulfur-related bacteria	5	Little *et al.* (2000); *Chapter 21* Vincke *et al.* (2000); *Chapter 22*
Sulfur problems in activated sludge systems	Eikelboom (2000)	19
Inhibition of anaerobic digestion	O'Flaherty and Colleran (2000)	20
Methods of Odour Measurement and Assessment	10	13
III – Technologies to desulfurise resources		
Catalytic removal of sulfur from diesel oil by hydrotreating	Reinhoudt and Moulijn (2000)	4
Bioleaching of sulphide minerals	Boon (2000)	5

IV – Treatment of water polluted by sulfurous compounds		
Biological removal of sulfurous compounds from organic wastewaters	6	7
Biological removal of sulfurous compounds from inorganic wastewaters	7	8
Bioelectrochemical systems and the conversion of sulfur compounds	8	–
Process control and neural networks	9	9
V – Treatment of gases polluted by sulfurous compounds		
Survey of H_2S and SO_2 removal processes	Lagas (2000)	10
Biological treatment of H_2S and SO_X from gas streams	11	Janssen *et al.* (2000); *Chapter 11*
Biological treatment of gases polluted by volatile sulfur compounds	Herrygers *et al.* (2000)	12
VI – Use of sulfur cycle conversions in biomining and soil/sediment treatment		
Bioleaching of ores	Boon (2000)	5
Metal effects on sulfur cycle bacteria	Hao (2000)	17
Treatment of contaminated sediments and solid waste	15	14
VII – Use of sulfur cycle conversions in advanced wastewater treatment and resource recovery		
Role of sulfur radicals in advanced oxidation processes	12	–
Biodegradation of sulfonated aromatic compounds	Tan and Field (2000)	16
Biological removal of nitrogen compounds	13	18
Biomanufacturing of sulfur nanoparticles	14	–

the metabolism of several different specific groups of bacteria (i.e. sulfate-reducing bacteria, phototrophic sulfur bacteria and colourless sulfur bacteria) specialised to use sulfur compounds in their different redox states. In Part I of this book, the reader is introduced to the chemistry of sulfur compounds (Chapter 2) and the organisms involved in their biotransformations (Chapter 3). For details on the geochemistry of the sulfur cycle, the reader is referred to the 1st edition of the book (Middelburg, 2000). Details on plant-sulfur interactions and agricultural aspects of the sulfur cycle can be found in Ernst (2000).

1.3 SULFUR-RELATED PROBLEMS

Potential adverse effects of sulfurous compounds on the operation of established environmental technologies are addressed in the next part of the book, i.e. corrosion of sewers (Chapter 4) and materials (Chapter 5), activated sludge bulking (Eikelboom, 2000), odour emissions from aerobic and anaerobic treatment facilities (Chapter 10) as well as sulfide inhibition in anaerobic digesters (O'Flaherty & Colleran, 2000). The chapters included in Part II describe the mechanisms of how sulfur pollution can lead to these adverse effects and process failures. Then, remediation techniques to overcome the various adverse effects are given.

1.4 TECHNOLOGIES TO DESULFURISE RESOURCES

Besides its importance from a basic scientific point of view, the sulfur cycle also plays a significant role in environmental technology and bioremediation. The subsequent parts of the book focus on a whole set of, mainly biological, technologies developed to minimise the impact of or completely overcome pollution by sulfur compounds. In this respect, prevention of pollution is a first approach. Part III describes possibilities to avoid pollution by desulfurisation of resources, i.e. oil (Reinhoudt & Moulijn, 2000) as well as coal and minerals (Boon, 2000). These were very well covered in the 1st edition of the book, and the reader is referred to these chapters for further details.

1.5 TREATMENT OF POLLUTION BY SULFUROUS COMPOUNDS

Subsequently, environmental engineering aspects of techniques to treat wastewaters (Part IV), off gases (Part V) and solid wastes and soils (Part VI) are presented. The chapters of these sections present the chemistry, microbiology and process technology of various treatment options. Special attention is given to novel biotechnological processes. Some of these technologies allow a complete removal of sulfur from waste streams by its conversion into insoluble elemental sulfur.

Thus, sulfur can be reused, which fits well in the concept of sustainable environmental protection, which is based on remediation techniques focusing on resource recovery (See Part VII).

The treatment processes for wastewater and off-gas treatment are very well investigated. Sulfur compounds play crucial roles in the treatment of wastewater rich in organic matter (Chapter 6), and the control of sulfide as potential inhibition of biogas production has received a lot of attention from the process monitoring and control field (Chapter 9). Sulfate reduction in inorganic wastewaters (e.g. mine water), that contain high metal concentrations but do not contain organic matter, is an excellent way to treat these wastewaters and recover the metals as sulfidic precipitates (Chapter 7). The latter can also be achieved in the emerging bioelectrical systems (Chapter 8). Heavy metals can potentially be toxic to the sulfur cycle bacteria (Hao, 2000), and thus need to be carefully monitored and controlled.

For off-gases and biogas, there are a lot of chemical (Lagas, 2000) and biological (Chapter 11, Janssen *et al.*, 2000) technologies to remove hydrogen sulfide and sulfur dioxides. Different methods need to be applied for the treatment of volatile organic sulfur compounds (Herrygers *et al.*, 2000) and odours (Chapter 10).

Remediation of sulfur contaminated soils and sediments and treatment of solid wastes have received the least attention (Chapter 15). Especially, biological treatment processes for wastes from the construction sector, e.g. gypsum or construction and demolition debris, are in the early stages of development (Chapter 15). On the other hand, the use of bioleaching for releasing metals from sulfidic ores has been well studied, and is being applied at full scale (Boon, 2000).

1.6 USE OF SULFUR CYCLE CONVERSIONS IN ADVANCED WASTEWATER TREATMENT AND RESOURCE RECOVERY

Part VII of the book covers the theory and practical aspects of novel process applications of the sulfur cycle conversions focusing on advanced wastewater treatment and resource recovery. The conversions occurring in the sulfur cycle are not restricted to conversions involving solely the element sulfur. The sulfur cycle can also be utilised for the degradation of (recalcitrant) organic matter (Tan & Field, 2000), and the removal of heavy metals (Hao, 2000) or nitrogen (Chapter 13) from wastewaters, soils and sediments.

In the last two decades, special interest has grown in using sulfate radicals for organic matter degradation using advanced oxidation processes (Chapter 12). Similarly, several applications aiming towards resource recovery have emerged in the last two decades, as outlined at the end of Chapter 6, and with Chapter 14 illustrating this with the (bio)manufacturing of sulfur nanomaterials for environmental and biomedical applications.

REFERENCES

Boon M. (2000). Bioleaching of sulfide minerals. In: Environmental Technologies to Treat Sulfur Pollution: Principles and Engineering. P. N. L. Lens and L. Hulshoff Pol (eds), IWA Publishing, London, pp. 105–130.

Eikelboom D. H. (2000). Sulfur-storing bacteria and bulking of activated sludge. In: Environmental Technologies to Treat Sulfur Pollution: Principles and Engineering. P. N. L. Lens and L. Hulshoff Pol (eds), IWA Publishing, London, pp. 449–466.

Ernst W. H. O. (2000). Agricultural aspects of sulfur. In: Environmental Technologies to Treat Sulfur Pollution: Principles and Engineering. P. N. L. Lens and L. Hulshoff Pol (ed.), IWA Publishing, London, pp. 355–376.

Hao O. J. (2000). Metal effects on sulfur cycle bacteria and metal removal by sulfate reducing bacteria. In: Environmental Technologies to Treat Sulfur Pollution: Principles and Engineering. P. N. L. Lens and L. Hulshoff Pol (eds), IWA Publishing, London, pp. 393–414.

Herrygers V., Van Langenhove H. and Smet E. (2000). Biological treatment of gases polluted by volatile sulfur compounds. In: Environmental Technologies to Treat Sulfur Pollution: Principles and Engineering. P. N. L. Lens and L. Hulshoff Pol (eds), IWA Publishing, London, pp. 281–304.

Janssen A. J. H., Dijkman H. and Janssen G. (2000). Novel biological processes for the removal of H_2S and SO_2 from gas streams. In: Environmental Technologies to Treat Sulfur Pollution: Principles and Engineering. P. N. L. Lens and L. Hulshoff Pol (eds), IWA Publishing, London, pp. 265–280.

Lagas J. A. (2000). Survey of H_2S and SO_2 removal processes. In: Environmental Technologies to Treat Sulfur Pollution: Principles and Engineering. P. N. L. Lens and L. Hulshoff Pol (eds), IWA Publishing, London, pp. 237–264.

Lens P. N. L. and Hulshoff Pol L. (ed.). (2000). Environmental Technologies to Treat Sulfur Pollution: Principles and Engineering, 1st edn, IWA Publishing, London, UK, 547 p.

Little B. J., Ray R. I. and Pope R. K. (2000). Corrosion and sulfur bacteria. In: Environmental Technologies to Treat Sulfur Pollution: Principles and Engineering. P. N. L. Lens and L. Hulshoff Pol (eds), IWA Publishing, London, pp. 491–513.

Middelburg J. J. (2000). The geochemical sulfur cycle. In: Environmental Technologies to Treat Sulfur Pollution: Principles and Engineering. P. N. L. Lens and L. Hulshoff Pol (eds), IWA Publishing, London, pp. 33–46.

O'Flaherty V. and Colleran E. (2000). Sulfur problems in anaerobic digestion. In: Environmental Technologies to Treat Sulfur Pollution: Principles and Engineering. P. N. L. Lens and L. Hulshoff Pol (eds), IWA Publishing, London, pp. 467–489.

Reinhoudt H. R. and Moulijn J. A. (2000). Catalytic removal of sulfur from diesel oil by hydrotreating. In: Environmental Technologies to Treat Sulfur Pollution: Principles and Engineering. P. N. L. Lens and L. Hulshoff Pol (eds), IWA Publishing, London, pp. 87–104.

Tan N. C. G. and Field J. A. (2000). Biodegradation of sulfonated aromatic compounds. In: Environmental Technologies to Treat Sulfur Pollution: Principles and Engineering. P. N. L. Lens and L. Hulshoff Pol (eds), IWA Publishing, London, pp. 377–392.

Vincke E., Monteny J., Beeldens A., De Belie N., Taerwe L., Van Gemert D. and Verstraete W. H. (2000). Recent developments in research on biogenic sulfuric acid attack of concrete. In: Environmental Technologies to Treat Sulfur Pollution: Principles and Engineering. P. N. L. Lens and L. Hulshoff Pol (eds), IWA Publishing, London, pp. 515–541.

Part II

The Sulfur Cycle

Chapter 2

The chemical sulfur cycle

Ralf Steudel

2.1 INTRODUCTION

Sulfur is one of the most important elements for life as well as for the chemical and pharmaceutical industries. Even in extraterrestrial space, sulfur compounds are abundant albeit in low concentrations. Sulfur contributes to only 0.07 wt% of the crust of the Earth but elemental sulfur and numerous sulfur-containing minerals occur in substantial *deposits*. Important sulfidic minerals are, for example, *pyrite* FeS_2, *galena* PbS, *zinc-blende* (*sphalerite*) ZnS, *cinnabar* HgS, *chalcopyrite* $CuFeS_2$ and *chalcocite* Cu_2S. Weathering and oxidation of the sulfides has resulted in large deposits of water-insoluble or poorly soluble sulfate minerals such as *gypsum* $Ca[SO_4] \cdot 2H_2O$, *bassanite* $Ca[SO_4] \cdot 0.5H_2O$, *anhydrite* $Ca[SO_4]$ and *baryte* $Ba[SO_4]$. Gypsum is, by volume, the most abundant sulfate mineral on Earth. Ocean water contains 2.7 g L^{-1} sulfate, river waters only ca. 0.01 g L^{-1}.

Sulfur compounds are constituents of all organisms and consequently of all biomass and materials which originated from these sources such as wood, peat, coal and crude oil as well as their derivatives. Combustion of such materials releases not only sulfur dioxide (SO_2) but also traces of carbonyl sulfide (COS). The latter is assimilated by plants as part of their sulfur metabolism. Dimethyl sulfide (DMS) is released to the atmosphere in enormous quantities by phytoplankton in the oceans, and hydrogen sulfide (H_2S) as well as SO_2 and carbon dioxide (CO_2) are emitted by volcanoes. Sulfate reduction by the

DOI: 10.2166/9781789060966_0011

ubiquitous sulfur bacteria in anoxic environments such as ponds, lakes, swamps and coastal waters also produces H_2S. The human body contains 2 g S per kg, in other words 140 g S for a person of 70 kg.

Large natural underground deposits of elemental sulfur exist in the USA, Mexico and Poland, but today elemental sulfur is mainly produced by the desulfurization of crude oil, of sour natural gas and of coal. Only on a very small scale is sulfur still mined in volcanic areas such as Indonesia. Historically, Southern Italy (Sicily) was the main origin of elemental sulfur during the industrialization of Western Europe in the early 19th century. In 1900, Sicily produced 500,000 t of elemental sulfur. Another important source of sulfur for production of sulfuric acid is *pyrite* with large deposits in many countries.

The average *atomic weight* of sulfur is 32.066 representing the natural mixture of the isotopes ^{32}S (95.0 mol%), ^{33}S (0.76%), ^{34}S (4.22%) and ^{36}S (0.02%). The relative atomic weights of most elements, however, vary slightly owing to natural variations in the abundances of their isotopes. This variation is used to determine the *origin* of a particular sample (a mineral or biological material). In the case of sulfur, the variation may be ± 0.01 units, while individual samples can be determined at an accuracy of ± 0.00015 (mainly by mass spectrometry). Due to the historic developments in analytical methods, the published atomic weights of the chemical elements have changed over the years. The artificial radioactive nuclide ^{35}S is used for labeling experiments; it decomposes with a half-life of 87.2 d by β-emission to ^{35}Cl.

More than 200 years of scientific research on sulfur and its compounds has resulted in a vast body of literature which cannot easily be searched for reliable information. Moreover, this literature contains errors and contradictions since earlier workers, not having the methods available that are standard today, often made claims that have not always been subsequently confirmed. However, reliable reviews written by experts in the field are available, above all the many volumes of *Gmelin Handbook of Inorganic Chemistry* in which the chemical literature is critically and exhaustively evaluated. On sulfur and its compounds 22 volumes have appeared, dating from 1939 to 1996 and covering the literature up to 1991. Unfortunately, no further volumes have been produced. Other reliable reviews on inorganic and analytical sulfur chemistry have been published (in alphabetical order) by Devillanova (2006), Holleman-Wiberg (2017), Karchmer (1970), Müller and Krebs (1984), Nickless (1968), Schmidt and Siebert (1973), Steudel and Eckert (2003), Steudel (2003a, b, c, d and Steudel, 2020), Steudel and Chivers (2019) as well as Szekeres (1974).

2.1.1 Oxidation states and redox potentials

The complexity of sulfur chemistry originates from the many oxidation states and coordination numbers sulfur atoms can assume, as well as from the tendency of sulfur in the oxidation state zero to catenate, forming chains and rings of an astonishing variety. Sulfur atoms and ions can adopt any coordination number between 1 (e.g. CS_2) and 8 (e.g. in solid Na_2S with antifluorite structure); sulfur oxidation states range from −2 to +6 (Table 2.1).

In Table 2.1, the nine oxidation states of sulfur are illustrated by typical examples. Most of them play a role in aqueous systems in which redox reactions occur either as a result of microbiological activity or simply following the thermodynamics of the system in non-enzymatic reactions. However, chemical systems are not always composed according to the requirements of thermodynamics. High activation enthalpies may keep exergonic reactions from proceeding at ambient temperatures, resulting in a chemical composition far from equilibrium (Licht & Davis 1997).

The equilibrium composition of an aqueous system containing just sulfur and oxygen is shown in the Pourbaix diagram in Figure 2.1. Depending on the redox potential, the pH value, the temperature and the overall concentration of sulfur, the relative stability areas of sulfide HS^-, elemental sulfur (S), as well as sulfate $[SO_4]^{2-}$ and hydrogen sulfate $[HSO_4]^-$ are shown (Garrels & Naeser, 1958; Williamson & Rimstidt, 1992). The different areas of this diagram indicate which species will predominate at a given potential and pH value. As the overall sulfur concentration decreases, the smaller the existence area of elemental sulfur becomes. Sulfite, thiosulfate and other sulfur oxoanions (e.g. polythionates) never predominate, regardless of pH and potential. In other words, these species exist in water only under non-equilibrium conditions or as minority species. The thiosulfate, polythionate and disulfite ions are typical examples of anions with mixed oxidation states of sulfur (see below).

Table 2.1 The oxidation states of sulfur atoms in common compounds.

Oxidation State	Examples
−2	dihydrogen sulfide H_2S, hydrogen sulfide ion HS^-, sulfide ion S^{2-} as in FeS
−1	disulfane H_2S_2, disulfide ion $[S_2]^{2-}$ as in pyrite FeS_2
0	elemental sulfur S_n, organic polysulfanes $R{-}S_n{-}R$
+1	dichlorodisulfane Cl—S—S—Cl
+2	sulfur dichloride SCl_2, sulfoxylate ion $[SO_2]^{2-}$
+3	dithionite ion $[S_2O_4]^{2-}$
+4	sulfur dioxide SO_2, sulfite ion $[SO_3]^{2-}$
+5	dithionate ion $[S_2O_6]^{2-}$, organic sulfonates $[R{-}SO_3]^-$
+6	sulfur trioxide SO_3, sulfate ion $[SO_4]^{2-}$, peroxosulfate ion $[SO_5]^{2-}$

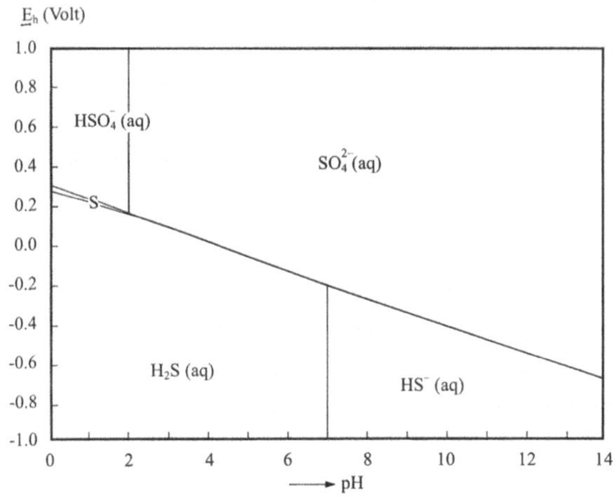

Figure 2.1 Pourbaix diagram for the binary system sulfur/oxygen in water at 25°C and 1.013 bar with the sum of the activities of all sulfur-containing ions equal to 0.1 mM.

2.1.2 Catenation of sulfur atoms

As in hydrogen sulfide H–S–H and dimethyl sulfide CH_3–S–CH_3, sulfur atoms can form two covalent bonds with other atoms or with itself to form chain-like units –S–S–S– of practically unlimited length ('catenation'). These chains may be terminated by single atoms such as H or Cl, by groups like CH_3 or SO_3H, by ions such as S^- or may 'bite their own tail' forming rings of various sizes. Corresponding examples are listed in Table 2.2. A special case are the polysulfide anions $[S_n]^{2-}$ in which the chains are terminated by negatively charged sulfur atoms which are iso-electronic with Cl atoms. Therefore, polysulfide anions $[S–S_n–S]^{2-}$ are iso-electronic with dichlorosulfanes Cl–S_n–Cl.

Table 2.2 gives those values of n which have been determined in compounds isolated in pure form (column 2) or which have been detected in mixtures by high-performance liquid chromatography (HPLC), proton nuclear magnetic resonance (^{1}H-NMR) spectroscopy or ion chromatography (column 3). From these data it is obvious that there is seemingly no limitation to the values of n. It is just that the preparation of the higher-molecular species becomes increasingly difficult since the solubility and thermal stability decrease with increasing values of n.

Polymeric sulfur (S_μ) is insoluble in all solvents (excepting liquid sulfur) and therefore is considered to consist of very long chains and/or very large rings. However, on longer extraction of S_μ by carbon disulfide at room temperature the polymer slowly dissolves (faster on heating) by depolymerization to a mixture of small homocycles such as S_6, S_7 and S_8 (Steudel *et al.*, 1984).

All low-valent compounds containing S–S bonds are light-sensitive, since homolytic cleavage occurs on irradiation with wavelengths shorter than 420 nm;

Table 2.2 Ring sizes and chain-lengths *n* in sulfur-rich compounds.

Species	Formula	Isolated as Pure Compounds	Larger Species Detected in Mixtures (Method)
Homocycles	S_n	$n=6...20$	$n=21...28$ (HPLC)
Polysulfanes	$H{-}S_n{-}H$	$n=1...8$	$n=9...35$ (^{1}H-NMR)
Polysulfides	$[S_n]^{2-}$	$n=1...10$, 12	$n=$ up to 11 (HPLC)[a]
Organic polysulfanes	$R{-}S_n{-}R$	$n=1...13$	$n=14...35$ (HPLC)
Polythionates or Sulfane disulfonates	$[O_3S{-}S_n{-}SO_3]^{2-}$	$n=1...4$	$n=5...22$ (ion chromatography)
Sulfane monosulfonates	$[S{-}S_n{-}SO_3]^{2-}$	$n=0$ (thiosulfate)	$n=1...8$ (indirect evidence only)
Polymeric sulfur	Abbreviated as S_n or $S\mu$ or S_∞	$n>10^5$	Crystal structures

[a]After derivatization.

the radicals formed trigger chain-reactions resulting in mixtures of compounds (Steudel *et al.*, 1989c). Even daylight often induces photochemical decomposition of sulfur-rich compounds; laser beams of blue and green color as sometimes used in Raman spectroscopy induce an even faster decomposition.

2.2 ELEMENTAL SULFUR AND HYDROPHOBIC SULFUR SOLS

2.2.1 Sulfur allotropes

Isolated sulfur atoms (S_1) have a very large enthalpy of formation (279 kJ mol^{-1}) and therefore cannot exist at ambient temperatures. Therefore, equations such as $H_2S + O \rightarrow H_2O + S$ are unrealistic in most cases since neither oxygen nor sulfur can exist as atoms at standard conditions. Sulfur atoms are known, however, as transient species from high-temperature vapors, from photolysis experiments and from electrical discharges in sulfur vapor. At ambient pressure and temperature elemental sulfur exists as rings of different sizes (S_n) and as polymeric chains of high molecular mass (S_μ) (Steudel & Eckert, 2003). Of all these allotropes, orthorhombic α-S_8 is most stable thermodynamically at 25°C/1.013 bar (enthalpy of formation is zero by definition). Therefore, S_8 is the main constituent of most commercial samples of elemental sulfur; it can be prepared in pure form by recrystallization of commercial sulfur (e.g. Claus sulfur) in organic solvents (Steudel & Holz, 1988). Pure α-S_8 is of greenish-yellow color and highly soluble in carbon disulfide CS_2 but much less so in chlorobenzene, chloroform and

Table 2.3 Solubility of α-S_8 in organic solvents at 25°C unless otherwise indicated (in wt%).

CS_2	34.76	$C_6H_5NH_2$	1.259	n-C_6H_{14}	0.40
1,2-$C_2H_4Br_2$	2.4 (22°C)	C_6H_{12}	1.185	n-C_7H_{16}	0.362
C_6H_6	2.093	$CHCl_3$	1.164	DMF[a]	0.191
$CH_3C_6H_5$	2.070	$C_6H_5NO_2$	0.856	$(C_2H_5)_2O$	0.181
1,4-$(CH_3)_2C_6H_4$	2.3 (22°C)	CCl_4	0.832	$(CH_3)_2CO$	0.079
1,3-$(CH_3)_2C_6H_4$	1.969	1,2-$C_2H_4Cl_2$	0.826	C_2H_5OH	0.066
C_5H_5N	1.5			CH_3OH	0.03

[a]DMF: dimethylformamide.

toluene from which it can nevertheless be recrystallized (see Table 2.3; for solubility in water, see Section 2.2.4). The other components of commercial sulfur are polymeric sulfur S_μ and traces of S_7. These originate from liquid sulfur from which all commercial sulfurs are produced. It is the small S_7 content (ca. 0.5%) that causes the 'bright-yellow' color of commercial sulfur (Steudel & Holz, 1988).

The molecular and crystal structure of α-S_8 is shown in Figure 2.2. S_8 crystallizes as orthorhombic α-S_8 at temperatures below 96°C, as monoclinic β-S_8 in the temperature region 96–120°C, and as monoclinic metastable γ-S_8 from certain organic solvents. The crystal structures of all these allotropes are quite complex

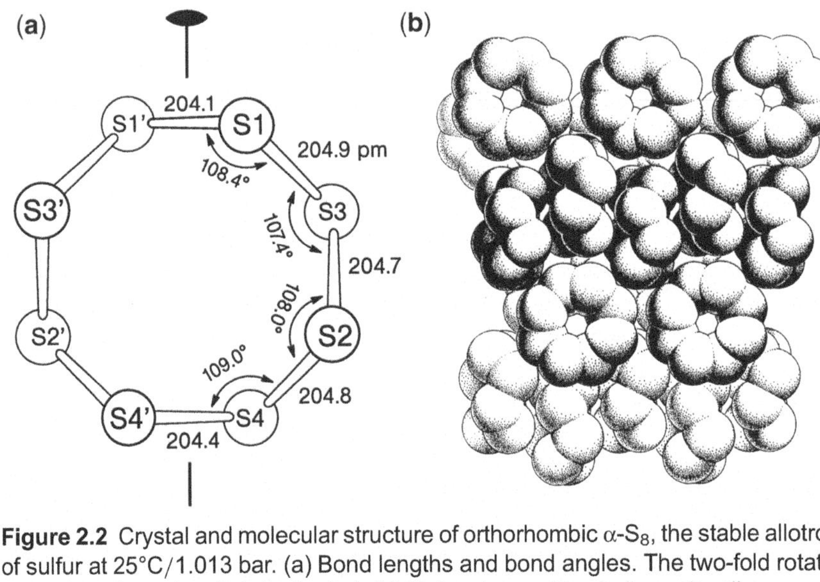

Figure 2.2 Crystal and molecular structure of orthorhombic α-S_8, the stable allotrope of sulfur at 25°C/1.013 bar. (a) Bond lengths and bond angles. The two-fold rotation axis of the S_8 molecule is indicated. (b) Molecular packing in the unit cell.

as can be seen from the structure of α-S_8 in Figure 2.2 containing 16 molecules in the unit cell (Steudel & Eckert, 2003).

By suitable chemical syntheses sulfur rings with between 6 and 20 atoms have been prepared and the following species have been obtained as pure substances (Steudel & Eckert, 2003):

$$S_6 \quad S_7 \quad S_8 \quad S_9 \quad S_{10} \quad S_{11} \quad S_{12} \quad S_{13} \quad S_{14} \quad S_{15} \quad S_{18} \quad S_{20}$$

These molecules form pale-yellow to orange-yellow solids which are soluble in carbon disulfide and, to a much lesser degree, in other organic solvents. The solubility decreases, however, with increasing ring size and with increasing symmetry of the molecule. The structures of some of these homocycles are shown in Figure 2.3. Several sulfur homocycles form more than one crystal structure, such as α-, β- and γ-S_8 (Steudel & Eckert, 2003).

Besides X-ray crystallography, Raman spectroscopy is the best method to characterize solid sulfur allotropes since their spectra are very characteristic (Eckert & Steudel, 2003). Reversed-phase high-pressure liquid chromatography (HPLC) in combination with a UV absorbance detector is used to analyze and quantify mixtures of sulfur homocycles in solutions (Strauss & Steudel, 1987).

The melting points of sulfur allotropes are found in the range 39–148°C, with S_7 having the lowest and S_{12} the highest value. α-S_8 melts at 115°C and monoclinic β-S_8 at 120°C. All thermodynamically unstable species convert rapidly to the more stable S_8 molecule on heating to temperatures close to or above the melting points. In the case of S_6, S_7, S_9, S_{11}, S_{13} and S_{15} this decomposition occurs slowly even at room temperature. The densities of the crystalline materials are in the range of 1.9–2.2 g cm^{-3} with S_6 showing the highest value. Therefore, S_8 can be turned into S_6 by application of a very high pressure via a polymeric intermediate (Crapanzano *et al.*, 2005).

All forms of pure elemental sulfur are hydrophobic and practically insoluble in water at 20°C, but neutral surfactants like Triton X-100, Tergitol 7 or sodium dodecyl sulfate added to the aqueous phase increase the solubility by several orders of magnitude, provided the surfactant concentration exceeds the critical value required for micelle formation. It is believed that the sulfur molecules are dissolved in the hydrophobic inner parts of the micelles (Steudel & Holdt, 1988).

If very small amounts of S_8 are dissolved in an excess of a polar but anhydrous solvent like methanol or acetonitrile at 20°C, an equilibrium between S_8, S_7 and S_6 is slowly established within 24 h. At a total sulfur concentration of 0.15 mg L^{-1} the three species are present in a molar ratio of approximately 1000:8:3 as determined by HPLC analysis (Tebbe *et al.*, 1982).

2.2.2 Liquid sulfur

Liquid elemental sulfur is an industrial product of outstanding importance, being produced on a huge scale (>53 million tons annually). It is used mainly for the

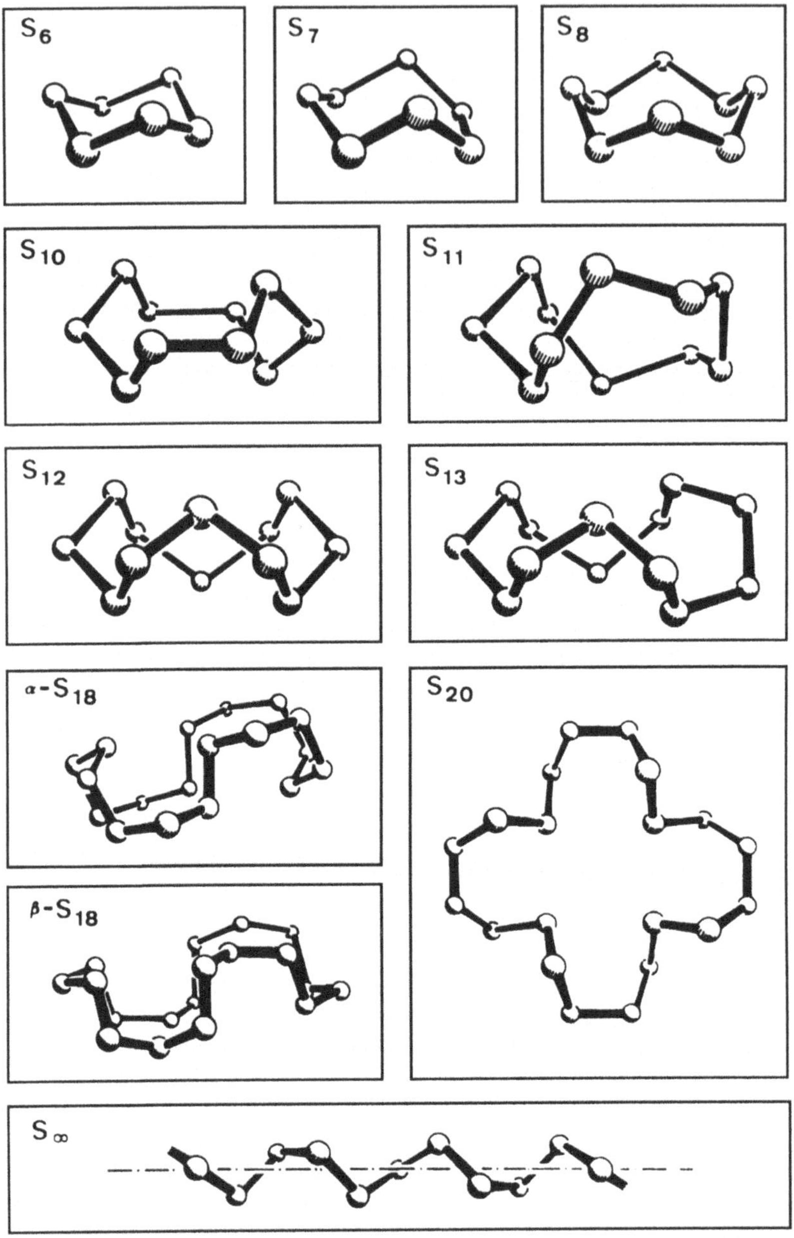

Figure 2.3 Molecular structures of sulfur rings present in various sulfur allotropes which have been prepared in a pure form; these structures have been determined by X-ray crystallography of single crystals. In addition, the structure of the helical molecules in crystalline polymeric sulfur (S_{∞}) is shown.

manufacture of sulfuric acid but after solidification also for rubber vulcanization and the production of numerous sulfur compounds. When α-S_8 is heated as a powder it transforms at 96°C into β-S_8, which melts at 120°C (Steudel *et al.*, 1984). In contrast, perfect single crystals of α-S_8 do not transform easily to β-S_8 and therefore melt at 115°C. Initially, the melt consists entirely of S_8 molecules, but a slow decomposition reaction takes place which eventually results in a complex equilibrium mixture of cyclic and chain-like molecules:

$$S_8 \rightleftharpoons \frac{8}{n} S_n \, (n = 6, 7, 9, \ldots S_\mu) \tag{2.1}$$

Depending on the purity of the system it takes up to 12 h to establish the equilibrium composition at 120°C (Steudel, 2003a). This composition has been determined after quenching to low temperatures by HPLC as well as by Raman spectroscopy; some of the HPLC data are given in Table 2.4.

The concentration of rings larger than S_{12} (termed collectively as S_x) has been calculated as difference required to reach 100%. S_μ is used as a symbol for polymeric sulfur which is insoluble in carbon disulfide. It is important to note that the chemical reactivity of S_7, for instance, is several orders of magnitude higher than that of S_8. From the temperature dependence of the concentrations the mean bond enthalpies of all non-S_8 rings compared to that of S_8 (266.6 kJ mol^{-1}) have been calculated. Small amounts of S_7, S_{12}, S_{18} and S_{20} have been prepared by recrystallization of quenched liquid sulfur (Steudel, 2003a).

The homocycles S_n ($n \neq 8$) lower the melting point of β-S_8 (120°C) to the freezing point (triple point) of equilibrated liquid sulfur of 115°C corresponding to a molar concentration of non-S_8 molecules of 5%. The density of this liquid at 115°C is only 1.80 g cm^{-3}. Even if the melt is *slowly* cooled to 20°C (as in

Table 2.4 Molecular composition of liquid sulfur after equilibration at various temperatures and quenching in liquid nitrogen (concentrations in wt%; after Steudel, 2003a).

Species	Equilibration Temperature			
	116°C	122°C	159°C	220°C
S_8	93.6	93.1	83.4	54.3
S_7	3.1	3.3	5.2	4.6
S_6	0.5	0.6	0.9	0.9
S_9	0.3	0.4	0.6	0.6
S_{10}	0.1	0.1	0.2	0.2
S_{12}	0.4	0.4	0.5	0.4
S_x	1.8	1.9	6.2	4.8
S_μ	0.2	0.2	3.0	34.2

industrial solidification processes) some of the reactive rings survive and are built into the crystal structure of S_8 as 'lattice defects' (Steudel & Holz, 1988). Therefore, most commercial sulfur samples contain traces of S_7 (up to 0.5 wt%) and varying concentrations of S_μ.

2.2.3 Gaseous sulfur

Sulfur vapor may exist either in equilibrium with solid and/or liquid sulfur (saturated vapor) or as a single phase at pressures lower than the saturation pressure (unsaturated vapor). In a vacuum, sulfur (S_8) sublimes at temperatures below its melting point; a process used for purification. The molecular composition of sulfur vapor has been determined by mass spectrometry as well as by thermodynamic and quantum-chemical calculations (Steudel *et al.*, 2003). At temperatures of up to 1000°C saturated sulfur vapors consist of all molecules S_n with n ranging from 2 to 8 which are in equilibrium with each other (Rau *et al.*, 1973):

$$S_8 \rightleftharpoons \frac{8}{n} S_n \, (n = 2 \ldots 8) \tag{2.2}$$

The lower the pressure and the higher the temperature, the smaller the average molecular size n becomes. At temperatures below 500°C the homocycles S_8, S_7 and S_6 are the main constituents of the saturated vapor. Above 800°C the S_2 molecule is the predominating species in both saturated and unsaturated sulfur vapors. The electronic structure and bonding of S_2 is analogous to that of O_2 (triplet states with two unpaired electrons).

S_2 molecules can also be prepared at moderate temperatures in solution by thermal or photochemical decomposition of suitable organic precursor molecules (Steliou, 1991; Tardif *et al.*, 1997). At ambient conditions the generated S_2 polymerizes to a mixture of S_6, S_7 and S_8 unless a trapping reagent (e.g. an unsaturated organic compound) is present.

2.2.4 Sulfur sols from elemental sulfur (Weimarn sols)

All allotropes of sulfur are practically insoluble in water at 20°C. The solubility of α-S_8 in pure water increases with temperature by more than 78 times: from 6.1 nM S_8 at 4°C to 478 nM S_8 at 80°C (1 nM = 10^{-9} M). The following thermodynamic values for the solubilization of α-S_8 in water have been calculated from analytical data determined by reaction of the saturated solution with cyanide followed by photometric determination of the formed thiocyanate ions: solubility constant $K° = 3.01 \pm 1.04 \times 10^{-8}$; Gibbs energy, enthalpy and entropy of solution: $\Delta G° = 42.93 \pm 0.73$ kJ mol^{-1}, $\Delta H° = 47.4 \pm 3.6$ kJ mol^{-1}, $\Delta S° = 15.0 \pm 11.7$ J mol^{-1} K^{-1}, respectively. The solubility of S_8 in sea water is only 61 ± 13% of the solubility in pure water regardless of the temperature (Kamyshny 2009).

Therefore, if a solution of S_8 in an organic solvent like ethanol or acetone is either diluted by much water or poured into an excess of water at 20°C with stirring, the sulfur is expected to precipitate. However, it takes a few minutes before the clear solution gets turbid and then looks like milk as shown in Figure 2.4. In this way 'sulfur sols' are prepared (Steudel, 1999; Steudel, 2003b). Since sulfur is a hydrophobic material these sols are termed hydrophobic or Weimarn sulfur sols. They are of milky-white appearance when freshly prepared and consist of tiny droplets of *liquid sulfur* dispersed in the aqueous phase ('colloidal solution' or 'emulsion'). On aging the sulfur droplets eventually crystallize and the crystals settle to the bottom of the vessel, but even after 11 days the solution will not be completely clear yet if demineralized water is used.

It is highly informative to analyze the molecular self-organization that takes place when such a sulfur sol is prepared. In an organic solvent the S_8 molecules are dispersed molecularly, i.e. as single molecules surrounded by solvent molecules. If these S_8 molecules are transferred into an aqueous phase they start forming clusters (dimers, trimers, etc.) owing to their hydrophobic nature. This cluster formation, by 'hydrophobic interaction', reduces the surface area of the hydrophobic particles exposed to the hydrophilic environment because a cluster of n S_8 molecules has a smaller outer surface than n single S_8 molecules combined. As the cluster of S_8 molecules grows it eventually forms a tiny liquid-like droplet in which other hydrophobic molecules may also dissolve, e.g. hydrophobic solvent molecules or the hydrophobic parts of proteins which might be present in the mixture (as in cultures of sulfur bacteria). Very small droplets of

Figure 2.4 Weimarn sulfur sol in water prepared by addition of an ethanolic S_8 solution to an excess of water at 20°C. From right to left, the age of the sol increases from 1 day through 2, 3, 4, 5, 7, 9 to 11 days resulting in crystallization of the sulfur droplets and settlement of the crystals to the bottom. On the far left: pure water for comparison (adapted from Steudel, 1996a).

liquid sulfur show a tendency to supercooling, i.e. they can stay liquid at 20°C for quite some time. Hydrophobic sulfur sols prepared in this way are stable for a few days but eventually the sulfur droplets crystallize (especially on agitation) and the crystals settle to the bottom of the vessel. To grow a crystal of 0.1 mg α-S_8 requires approximately 2×10^{17} S_8 molecules. The coarsening kinetics of Weimarn and related sulfur sols has been studied by Garcia & Druschel (2014).

Hydrophobic sulfur sols can be stabilized by the addition of surfactants like Triton X-100 or sodium dodecyl sulfate to the aqueous phase whereas di- or trivalent cations destabilize the sol and accelerate precipitation (Steudel, 1999; Steudel, 2003b). The reactions of elemental sulfur with *hot* water are discussed in Section 2.7.

2.3 SULFIDE AND POLYSULFIDES

2.3.1 Hydrogen sulfide and sulfide ions

H_2S (b.p. −60°C) is a colorless, extremely poisonous, noxious gas which is heavier than air. H_2S ignites spontaneously in air at 260°C and forms explosive mixtures with air at concentrations between 4 and 44 vol% H_2S. The maximum allowed concentration of H_2S at a workplace is only 10 ppmv. H_2S is most conveniently prepared from ionic metal sulfides by protonation of the anions with dilute acids under anoxic conditions.

H_2S is a constituent of certain types of natural gas ('sour gas') from which it is recovered by scrubbing. The H_2S liberated from the scrubbing liquor by heating is oxidized to elemental (liquid) sulfur by the Claus process. H_2S is also formed in huge quantities on desulfurization of crude oil by the HDS process (hydrodesulfurization) carried out in refineries. Furthermore, H_2S is a frequent component of biogas and of certain wastewaters and waste gases. Finally, H_2S is present in volcanic exhalations and is abundant in the water exiting from hydrothermal vents on ocean floors. In anaerobic aqueous environments, H_2S may be formed by sulfate reducing bacteria (Barton, 1995).

The solubility of H_2S in water at 20°C is only 0.40 g/100 g H_2O (0.12 M) at a total pressure ($H_2S + H_2O$) of 1.013 bar. As expected, the solubility decreases with increasing temperature. In aqueous solution the very weak acid H_2S is only partly deprotonated depending on the pH value:

$$H_2S + H_2O \rightleftharpoons [H_3O]^+ + HS^- \quad K_1 = 1.0 \times 10^{-7} \ (20^\circ C) \tag{2.3}$$

$$HS^- + H_2O \rightleftharpoons [H_3O]^+ + S^{2-} \quad K_2 = 10^{-14} \text{ to } 10^{-17} \ (20^\circ C) \tag{2.4}$$

The values of K_1 and K_2 depend on the salt concentration and the ionic strength. For instance, in seawater at 25°C: $K_1 = 3.24 \times 10^{-7}$ and $pK_1 = 6.51$. With rising temperatures up to 120°C, $K_1(H_2S)$ increases, followed by a decrease at higher temperatures, which is explained by a change in hydration of the anion (see below).

The extremely small second dissociation constant of H_2S is difficult to determine and depends on the ionic strength of the solution too. $K_2(H_2S)$ decreases with higher ionic strength, since ion pairs such as $[NaS]^-$ are formed between sulfide ions and suitable cations at higher concentrations. Values of between 10^{-14} and 10^{-17} have been reported in the literature. An average value of $K_2(H_2S) = 10^{-16}$ mol L^{-1} means that sulfide ions S^{2-} virtually do not exist in solutions of pH ≤ 12, and the upper value of 10^{-17} mol L^{-1} would indicate that many solubility products of insoluble metal sulfides reported in the literature may be in error (May *et al.*, 2018).

Therefore, the reactions of 'sulfide' in water at pH values up to 12 are practically the reactions of either H_2S or of the hydrogen sulfide ion HS^-. In a solution of pH = 7 with 1 mM total sulfide, the molar ratio HS^-/H_2S is 1:1 while at pH = 10 practically all sulfide is present as HS^-. At a physiological pH of 7.4 and 37°C sulfide is present as 81% HS^- and 19% H_2S. Thus, if one talks about 'sulfide solutions' one usually means solutions containing both H_2S and HS^- rather than S^{2-} ions. In water, HS^- is more reactive than H_2S.

The two acid dissociation constants of H_2S also depend on the temperature. Only K_1 will be discussed here. According to various investigations K_1 has its maximum value of 2.0×10^{-7} near 100°C while 0.50×10^{-7} has been determined at 0°C. At 200°C, K_1 is about as small as at 0°C (0.65×10^{-7}) (McCampbell Hamilton, 1991).

When aqueous sodium or potassium hydroxide is saturated with H_2S, solutions of the hydrogen sulfides NaHS and KHS are formed, respectively. From such solutions, hydrated sulfides crystallize upon addition of equivalent amounts of base followed by cooling. The salts $Na_2S \cdot 9H_2O$ and $K_2S \cdot 5H_2O$ are manufactured in this way. In air, slow autoxidation takes place with yellow discoloration due to the formation of polysulfides and thiosulfate (see below). Dehydration of these salts is accompanied by partial decomposition to NaHS and NaOH, for example. Therefore, the anhydrous sulfides are produced by non-aqueous routes. Na_2S, for instance, is made by reduction of $Na_2[SO_4]$ by low-ash coal such as anthracite at 700–1100°C (Steudel, 2020).

Hydrogen sulfide ions react with certain metal cations to form insoluble sulfides, e.g. the black iron(II) sulfide FeS:

$$Fe^{2+} + HS^- \rightleftharpoons FeS\downarrow + H^+ \tag{2.5a}$$

Since the aqueous iron ions are hydrated, a more realistic description of this multi-step reaction would be:

$$[Fe(H_2O)_6]^{2+} + HS^- \rightarrow [Fe(SH)(H_2O)_5]^+ + H_2O \tag{2.5b}$$

$$[Fe(SH)(H_2O)_5]^+ \rightarrow\rightarrow\rightarrow FeS\downarrow + H^+ + 5H_2O \tag{2.5c}$$

The latter equation (2.5c) describes a series of reactions (symbolized by the three arrows) in which di- and then polynuclear complexes are formed from which eventually FeS precipitates. In hydrochloric acid the reverse reactions take place.

Depending on their solubility products, metal sulfides are soluble in hydrochloric acid (ZnS, CdS, Al_2S_3, FeS, CoS, NiS, MnS) or insoluble in non-oxidizing acids (HgS, As_2S_3, Sb_2S_3, Bi_2S_3, Cu_2S, Ag_2S). All sulfides are soluble in oxidizing acids such as nitric acid with formation of sulfates. Deuterated hydrogen sulfide can be prepared from D_2O and Al_2S_3. Using tabulated solubility products, one should make sure that these have been calculated using the correct pK_a values of H_2S; literature values are given in Table 2.5.

The sulfides of the alkali and alkaline earth metals are soluble even in neutral water, but these solutions are subject to hydrolysis and therefore strongly alkaline, e.g.:

$$Na_2S + H_2O \rightleftharpoons NaHS + NaOH \tag{2.6}$$

These alkaline sulfide solutions react with SO_2 almost quantitatively to thiosulfate (Gmelin, 1960):

$$2NaHS + 4SO_2 + 4NaOH \rightarrow 3Na_2[S_2O_3] + 3H_2O \tag{2.7a}$$

or

$$2HS^- + 4[HSO_3]^- \rightarrow 3[S_2O_3]^{2-} + 3H_2O \tag{2.7b}$$

Even mild oxidants turn aqueous H_2S and especially hydrogen sulfide ions into elemental sulfur which precipitates from the solution. The rate of autoxidation of such solutions depends on the concentration, the pH value and the temperature (Chen & Morris, 1972; Chiu & Meerhan 1977; O'Brien & Birkner, 1977). Elemental sulfur is only formed in near neutral solutions. Alkaline sulfide solutions and solid $Na_2S \cdot xH_2O$ are oxidized by air to polysulfides, thiosulfate and eventually sulfate (x has values between 7 and 9 in commercial sodium sulfide). Therefore, commercial sodium sulfide always contains these impurities, which can only be removed by careful recrystallization from deoxygenated water in an inert atmosphere (Steudel *et al.*, 1989a). The initial reactions of the

Table 2.5 Solubility products K_L of water-insoluble sulfides and minerals at 25°C: p$K_L = -\log K_L$ (after www.aqion.de/site/17).

	K_L		K_L
MnS	10.19	*Galena* (PbS)	26.77
precipitated FeS	17.91	*Greenockite* (CdS)	29.92
Mackinawite (FeS)	18.64	*Covellite* (CuS)	36.26
Millerite (NiS)	22.03	*Chalcocite* (Cu_2S)	48.6
Wurtzite (ZnS)	23.67	*Cinnabar* (HgS)	55.08
Sphalerite (ZnS)	25.61		

autoxidation are as follows (Degrand & Lund, 1979):

$$HS^- + O_2 \rightarrow HS^\bullet + [O_2]^{\bullet -} \tag{2.8}$$

$$2HS^\bullet \rightarrow H_2S_2 \rightleftharpoons [S_2]^{2-} + 2H^+ \tag{2.9}$$

The further oxidation of disulfide ions yields thiosulfate as explained in Section 2.3.2. The two radicals formed by a 1-electron transfer in reaction (2.8) may combine to give the sulfinate anion which is highly reactive and is eventually oxidized to hydrogen sulfate:

$$HS^\bullet + [O_2]^{\bullet -} \rightarrow [HSO_2]^- \tag{2.10}$$

$$[HSO_2]^- + O_2 \rightarrow [HSO_4]^- \tag{2.11}$$

In addition, HS^- reacts with O_2 to the radical anion $[SO_2]^{\bullet -}$ (Zhu *et al.*, 1991). The disulfide anion $[S_2]^{2-}$ is the most abundant polysulfide on Earth since it is the anionic component of pyrite FeS_2. On strong heating to 1200°C, FeS_2 decomposes endothermically to FeS and S_2. On the other hand, solid FeS reacts with gaseous H_2S exothermically and exergonically to pyrite FeS_2 and H_2 ($\Delta H° = -45.1$ kJ mol^{-1} at 25°C).

The catalytic oxidation of hydrogen sulfide ions by metal cations will be discussed in Section 2.3.4. In the Earth's atmosphere H_2S is rapidly oxidized to SO_2 and eventually to H_2SO_4 by hydroxyl radicals which are produced at daytime by photodecomposition of ozone and reaction of the resulting singlet oxygen atoms (^{1}D) with H_2O (Möller, 2010):

$$O_3 + h \cdot \nu \rightarrow O(^1D) + O_2 \tag{2.12}$$

$$O(^1D) + H_2O \rightarrow 2OH^\bullet \tag{2.13}$$

$$H_2S + OH^\bullet \rightarrow HS^\bullet + H_2O \tag{2.14}$$

$$HS^\bullet + O_2 \rightarrow HSO_2^\bullet \tag{2.15}$$

$$HSO_2^\bullet + OH^\bullet \rightarrow SO_2 + H_2O \tag{2.16}$$

The oxidation products are eventually transferred into the aqueous phase (clouds and precipitation). SO_2 oxidation in the atmosphere is also achieved by photochemically produced hydroxyl radicals:

$$SO_2 + OH^{\bullet} + M \rightarrow HSO_3^{\bullet} + M \tag{2.17}$$

$$HSO_3^{\bullet} + O_2 \rightarrow HO_2^{\bullet} + SO_3 \tag{2.18}$$

$$SO_3 + 2H_2O \rightarrow H_2SO_4 \cdot H_2O \tag{2.19}$$

2.3.2 Polysulfides and polysulfanes

Alkaline sulfide solutions dissolve elemental sulfur with the formation of chain-like polysulfide anions which form equilibrium mixtures of many molecular sizes:

$$HS^- + S_8 \rightleftharpoons [H{-}S{-}S_7{-}S]^- \rightleftharpoons H^+ + [S_9]^{2-} \tag{2.20}$$

$$HS^- + [S_9]^{2-} \rightleftharpoons H^+ + 2[S_5]^{2-} \tag{2.21}$$

$$HS^- + [S_5]^{2-} \rightleftharpoons H^+ + 2[S_3]^{2-} \tag{2.22}$$

$$[S_9]^{2-} + [S_5]^{2-} \rightleftharpoons 2[S_7]^{2-} \tag{2.23}$$

etc.

These reversible reactions strongly depend on the pH and on the total sulfur concentration. Low pH values result in the formation of H_2S and precipitation of elemental sulfur, whereas very high pH values favor shorter chains over longer polysulfide anions. Polysulfide anions can therefore exist in water only at pH > 6. The maximum chain-length observed in water is 11 but there is no *a priori* reason why larger anions should not exist at least in small concentrations. A crystalline dodecasulfide with the anion $[S_{12}]^{2-}$ has recently been prepared from a non-aqueous solvent (Liebing *et al.*, 2019). If an aqueous solution of Na_2S is saturated with sulfur the *average polysulfide chain-length* is 5.5 at 25°C, and tetra-, penta- and hexasulfide anions predominate in such mixtures (Steudel & Chivers, 2019).

The molecular structures of numerous polysulfide anions have been determined by X-ray crystallography of corresponding salts and can sometimes be derived from the structures of the sulfur allotropes shown in Figure 2.3 (Steudel, 2003c). The conformation of the anions, however, may depend on the cations as the structures of the two octasulfide anions in Figure 2.5 demonstrate. The negative charges are mainly located at the terminal atoms (Steudel & Steudel, 2013).

The best method to determine single polysulfide anions in solution is HPLC analysis after derivatization. Kamyshny *et al.* (2004) published a *derivatization protocol* to turn the labile aqueous polysulfide dianions, within less than two

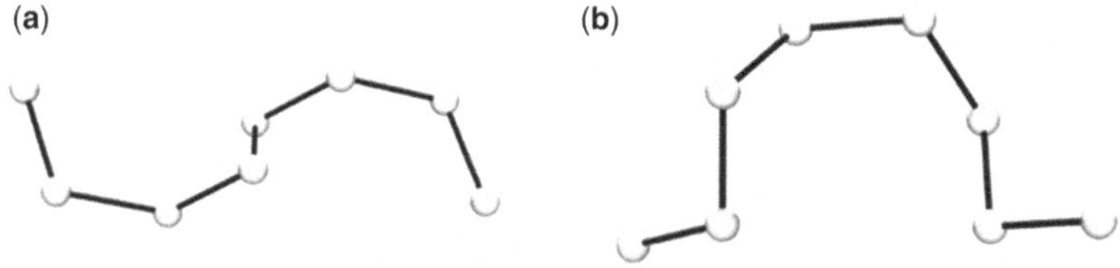

Figure 2.5 Two alternative conformations of the $[S_8]^{2-}$ ion, detected in (a) $[Ni(N\text{-}MeIm)_6][S_8]$ and (b) $[Et_3NH]_2[S_8]$ (adapted from Steudel & Chivers, 2019).

seconds, into the corresponding thermally more stable methyl derivatives using methyl triflate (trifluoromethyl sulfonate) in a *single-phase alkylation reaction* at 25°C intended to freeze the equilibrium mixture discussed above:

$$[S_n]^{2-} + 2CF_3SO_3CH_3 \rightarrow Me_2S_n + 2[CF_3SO_3]^- \quad (2.24)$$

Reverse-phase HPLC was used to separate and determine the dimethyl polysulfanes (Figure 2.6). Polysulfanes are the covalent derivatives of the ionic polysulfides.

When solid K_2S_5 was dissolved in water using a phosphate buffer the anions detected after derivatization with $CF_3SO_3CH_3$ were (in the order of decreasing concentrations): $[S_5]^{2-} >> [S_6]^{2-} > [S_4]^{2-} >> [S_7]^{2-} > [S_3]^{2-} > [S_8]^{2-} > [S_2]^{2-}$. When an aqueous solution of Na_2S_4 with natural abundance of isotopes and a solution of $Na_2{}^{34}S_4$ were mixed the isotopic exchange between the various anions was *ca.* 70% complete within 8 s.

In aqueous polysulfide solutions *saturated with elemental sulfur* (S_8) the ratio of the activities of any two polysulfide species is constant at equilibrium and invariant to pH change, i.e. the distribution of the polysulfide dianions remains constant regardless of the overall concentration.

In *unsaturated* (dilute) polysulfide solutions decreasing pH values favor larger chain-lengths until sulfur precipitation occurs at pH < 7. S_8 is thought to be ejected from either $[S_9]^{2-}$ or $[HS_9]^-$ by protonation and elimination of either $[HS]^-$ or H_2S (or from longer polysulfide chains, of course).

The extensive investigation of aqueous polysulfides at different concentrations and temperatures resulted in the thermodynamic data shown in Table 2.6, demonstrating a preference of the pentasulfide ion which has the smallest Gibbs energy of formation ΔG_f^o.

The reactions shown in equations (2.20) through (2.23) are typical nucleophilic displacements, which are quite common in sulfur chemistry. Both HS^- and S^{2-} anions are strong nucleophiles (Lewis bases) which are capable of opening S_8 rings and which tend to also cleave the S–S bonds of chain-like species. Other strong nucleophiles are the anions $[CN]^-$ and $[SO_3]^{2-}$ which react with sulfur-rich compounds to thiocyanate $[SCN]^-$ and thiosulfate $[S_2O_3]^{2-}$,

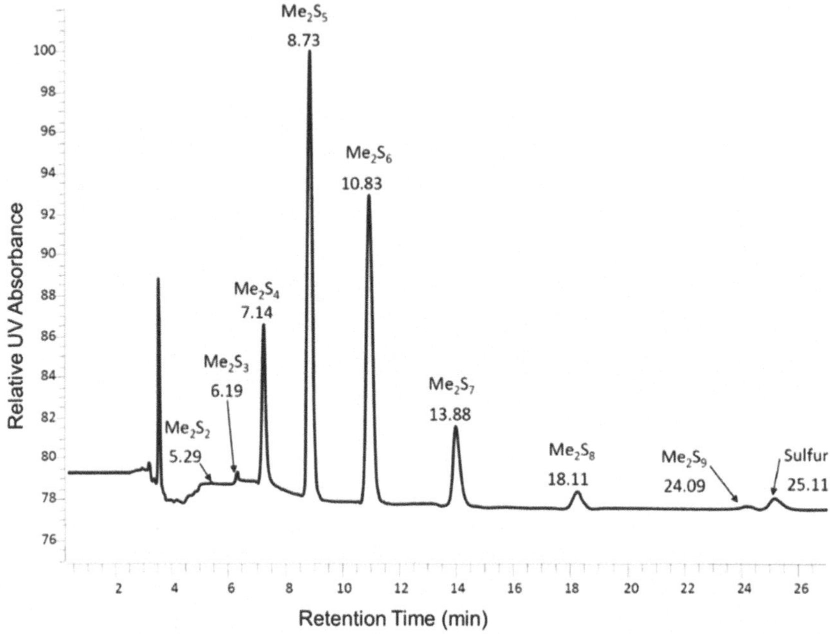

Figure 2.6 Chromatographic separation (HPLC) of a mixture of dimethyl polysulfanes (Me_2S_n) and S_8 obtained by alkylation of aqueous polysulfide anions. A UV absorbance detector was used (adapted from Steudel & Chivers, 2019).

respectively. Reactions of this type are very fast if all species involved are present in a homogeneous solution.

Aqueous polysulfide solutions are subject to rapid autoxidation when exposed to air (Kleinjahn *et al.*, 2005; Steudel *et al.*, 1986). The major products are thiosulfate and elemental sulfur, which are formed according to the equation:

$$[S_x]^{2-} + \frac{3}{2}O_2 \rightarrow [S_2O_3]^{2-} + \frac{(x-2)}{8}S_8\downarrow \quad (2.25)$$

Elemental sulfur is produced only if the average sulfur content x of the polysulfide solution is larger than 2. Since polysulfide solutions are yellow to orange whereas thiosulfate is colorless, the discoloration of the original solution and the formation of a precipitate (for $x > 2$) indicate that reaction (2.25) takes place. For this reason, thiosulfate is usually a component of aqueous and non-aqueous polysulfide solutions but also of solid polysulfides exposed to air. The thiosulfate anion can be detected by either Raman spectroscopy or ion chromatography.

Table 2.6 Gibbs energies ΔG_f^o and enthalpies ΔH_f^o of formation (in kJ mol^{-1}) as well as entropies S^o (in J mol^{-1} K^{-1}) of aqueous polysulfide dianions $[S_n]^{2-}$ with $n = 2$–8 derived from the analysis of the reaction products of solid elemental sulfur (S_8) with aqueous hydrogen sulfide according to equations (2.20) through (2.23). The value in brackets has been extrapolated (after Kamyshny *et al.*, 2007).

Parameter	$n=2$	$n=3$	$n=4$	$n=5$	$n=6$	$n=7$	$n=8$
ΔG_f^o	77.4	71.1	67.1	66.0	67.4	70.7	74.9
ΔH_f^o	13.0	6.6	9.0	9.6	13.3	16.5	23.8
S^o	(−22)	9	63	100	139	171	213

Hot aqueous polysulfide solutions contain additional species originating from the homolytic dissociation of the dianions, e.g.:

$$[S_6]^{2-} \rightleftharpoons 2[S_3]^{\bullet -} \tag{2.26}$$

Di- and trisulfide *radical anions* have been detected by electron spin resonance (ESR), UV-Vis and Resonance Raman spectroscopy. For example, $[S_3]^{\bullet -}$ is blue due to an absorption band at 610 nm. It is preferentially observed in non-aqueous polysulfide solutions, and it is the blue pigment in the silicate mineral *lapis lazuli*. Polysulfide radical anions are more reactive than the dianions, and they are probably responsible for the rapid autoxidation of polysulfide solutions in air by reacting with O_2 instantaneously to $[S_3O_2]^{\bullet -}$ as a precursor of the observed thiosulfate; the intermediate $[S_3O_2]^{\bullet -}$ has been observed by mass spectrometry (Steudel & Chivers, 2019).

The blue trisulfide radical anion has also been detected by Resonance Raman spectroscopy using a red laser beam (Steudel & Chivers, 2019). Its symmetric stretching vibration (ν_1) is observed near 540 cm^{-1} depending on the sample. In general, *Raman spectroscopy* is a powerful tool in inorganic sulfur chemistry as can be seen from Figure 2.7 which shows the five well resolved SS stretching modes of the chain-like hexasulfide dianion in crystalline Cs_2S_6 in the range 400–510 cm^{-1}. Below 300 cm^{-1} the four SSS bending modes of the anion can be seen. The torsion vibrations are expected below 100 cm^{-1}. For Raman spectra of ionic tetra- and pentasulfides, see El Jaroudi *et al.* (2000).

Mass spectrometry has also been applied to analyze aqueous polysulfide solutions. However, the dianions $[S_n]^{2-}$ ($n < 8$) are not stable in the gas phase but dissociate into radical monoanions due to Coulomb repulsion of the excess negative charges (*Coulomb explosion*); see equation (2.26). Therefore, the electrospray mass spectra of sodium trisulfide, tetrasulfide and a mixture of potassium polysulfides, dissolved in water, recorded in the negative ion mode have been found to be quite similar, showing peaks only due to radical anions $[S_n]^{\bullet -}$ with not more than three sulfur atoms (Bogdándi *et al.*, 2019).

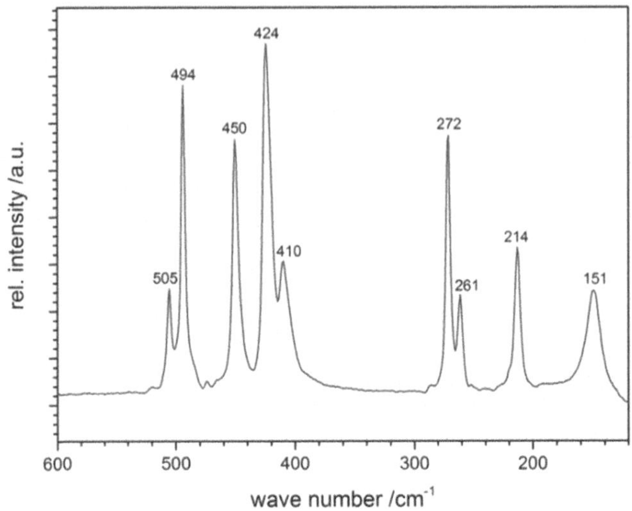

Figure 2.7 Raman spectrum of solid Cs_2S_6 showing the five SS stretching vibrations of the anion in the region 400–510 cm^{-1} and the four SSS bending modes at <300 cm^{-1} (adapted from Steudel & Chivers, 2019).

Longer-chain polysulfide ions with up to nine sulfur atoms were detected, however, by negative ion mass spectrometry after the dianions were protonated to $[HS_n]^-$ during electrospray using methanolic solutions with an ammonium acetate buffer (Gun *et al.*, 2004). In that case the overall charge of the gaseous anions is reduced by 50% and Coulomb explosion does not take place.

If a sodium polysulfide solution is poured into ice-cold concentrated hydrochloric acid a yellow oil composed of **hydrogen polysulfanes** is formed in addition to crystalline NaCl:

$$Na_2S_n + 2HCl \rightarrow 2NaCl + H_2S_n \tag{2.27}$$

The polysulfane mixture H_2S_n is almost insoluble in water; it may contain all species from H_2S up to H_2S_{35}, especially after aging, as has been demonstrated by ^{1}H-NMR spectroscopy (Hahn, 1985). The spectra and other properties of polysulfanes H–S_n–H show that these molecules are chain-like but only the structures of H_2S_2 and H_2S_3 have been determined accurately by both microwave spectroscopy and electron diffraction in the gas phase (Steudel, 2003c).

The two acidity constants of hydrogen polysulfanes are not known with certainty. An older determination (Schwarzenbach & Fischer, 1960) has been shown to be partly in error since too simple a composition of the starting polysulfide mixture Na_2S_n had been assumed (McCampbell Hamilton 1991). The pK_a values of H_2S_n, however, should be smaller by orders of magnitude than the corresponding values of H_2S. This follows from the chemical behavior of polysulfide anions in aqueous solution: it is known that these anions are completely deprotonated in alkaline solutions and hydrogen polysulfide ions $[HS_n]^-$ with $n > 1$ have never been observed in condensed phases. In the gas phase hydrogen polysulfides are very strong proton donors (Brønsted acids) (Otto & Steudel, 1999).

As pure materials, hydrogen polysulfanes H_2S_n are unstable even at 20°C and slowly decompose to H_2S and S_8, especially in the presence of alkaline substances or rough surfaces of glass or quartz (e.g. powders) which act as catalysts (Fehér, 1975):

$$H_2S_n \rightarrow H_2S\uparrow + \frac{(n-1)}{8} S_8 \tag{2.28}$$

The elemental sulfur remains dissolved in the sulfane mixture. This slightly exothermic reaction is reversible: if H_2S gas is bubbled into liquid sulfur at temperatures of >120°C the formation of hydrogen polysulfanes has been demonstrated spectroscopically. Reaction (2.28) is responsible for the fact that the solubility of H_2S in liquid sulfur increases with temperature, while usually the solubility of gases in liquids decreases with temperature.

Organic derivatives of the polysulfanes are discussed in Section 2.8.

2.3.3 Polysulfido complexes of transition metals and ion pairs

Polysulfide anions, due to their high negative charge, are excellent ligands for coordination of metal cations. Therefore, certain 'insoluble' transition metal sulfides such as Cu_2S and Ag_2S dissolve in aqueous polysulfide solutions forming chelate complexes with bi- or even tridentate polysulfido ligands (forming two or three bonds to a metal ion). Often such complexes contain heterocyclic MS_n units with ring sizes from 3 to 10 (metallacycles). In the case of copper the following anions have been isolated and structurally characterized by X-ray diffraction on single crystals: $[Cu_2S_{20}]^{4-}$, $[Cu_3S_{12}]^{3-}$, $[Cu_3S_{18}]^{3-}$, $[Cu_4S_{12}]^{2-}$, $[Cu_4S_{15}]^{2-}$ and $[Cu_6S_{17}]^{2-}$ (Figure 2.8).

Other metal cations such as Zn^{2+}, Ni^{2+}, Mn^{2+}, Hg^{2+}, and Ag^+ form mononuclear and/or polynuclear complexes too (Müller & Diemann, 1987). These anionic species are soluble in water and in polar organic solvents. Complexes with S-donor ligands (e.g. the thiolate anion of cysteine) play an important role as redox catalysts in biochemical reactions, e.g. in rubredoxins and ferredoxins (Ueyama & Nakamura, 1992).

Alkali metal ions (M^+) and polysulfide dianions form cyclic *ion pairs* of type $[MS_n]^-$ in aqueous solution at moderate and high concentrations and low or moderate temperatures in equilibrium with the free but solvated cations and anions. Such ion pairs have been observed by mass spectrometry in the gas phase and have also been computationally characterized (Steudel & Steudel, 2013). The presence of ion pairs in solution depends on the overall salt concentration. In

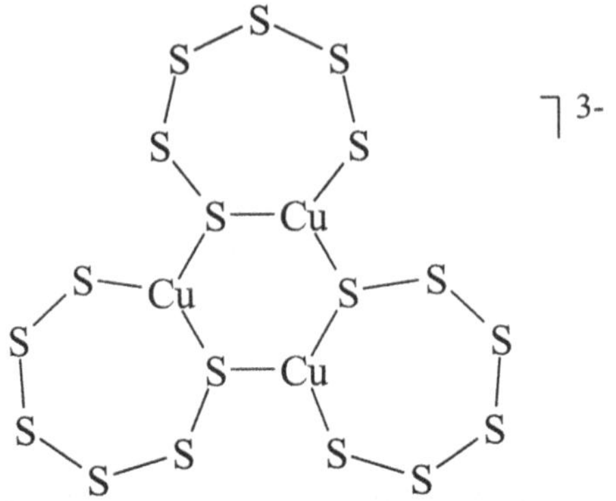

Figure 2.8 Structure of the anion $[Cu_3(S_6)_3]^{3-}$ containing tridentate hexasulfido ligands (schematic).

general, the attractive force f between two opposite ionic charges q_i at a distance d is given by Coulomb's law as $f = q_1 \times q_2/4\pi\varepsilon_0 \times \varepsilon \times d^2$ with ε_0 being the dielectric constant of vacuum and ε the relative dielectric constant of the medium (solution). It follows that ion-pair formation is favored by high ionic charges (dianions are more likely to form ion pairs than monoanions of the same size), small ionic radii, a small dielectric constant of the medium (tetrahydrofuran with $\varepsilon \approx 8$ favors ion pair formation more than water with $\varepsilon = 78$), low temperatures (ion-pair formation is exothermic) and high ion concentrations according to the law of mass action. Even KCl is not completely dissociated in aqueous solution at 25°C, as indicated by the activity coefficient γ which decreases from 0.965 at a molar concentration of 0.001 M via 0.901 at 0.01 M and 0.769 at 0.1 M to 0.606 at 1.0 M.

2.3.4 Oxidation of sulfide and polysulfide ions by metal ions

The removal of sulfide ions from aqueous solutions by oxidation to elemental sulfur is one of the most important desulfurization processes (Steudel, 1996b). This process can be achieved either chemically or microbiologically (i.e. enzymatically; see Kleinjan *et al.*, 2003). In both cases transition metal complexes are involved. The metal ion must be able to exist in at least two oxidation states between which it can shuttle back and forth. Suitable metal ions are, for instance, iron(III/II), copper(II/I), manganese(IV/III/II), and vanadium (V/IV). The complexes of these metal ions should be soluble at the pH of the sulfide solution, and the following overall reaction is expected to occur (M: any redox-active metal):

$$HS^- + 2M^{n+} \rightarrow \frac{1}{8}S_8 + H^+ + 2M^{(n-1)+} \quad (2.29)$$

The reduced metal ion is then re-oxidized by dioxygen and consequently serves as a catalyst only:

$$4M^{(n-1)+} + O_2 + 2H^+ \rightarrow 4M^{n+} + 2OH^- \quad (2.30)$$

The net reaction can therefore be represented by the equation:

$$2HS^- + O_2 \rightarrow \frac{1}{4}S_8 + 2OH^- \quad (2.31)$$

As O_2 is a paramagnetic molecule, whereas all other species in equation (2.31) are diamagnetic, this direct oxidation of HS^- by O_2 is forbidden by the *law of spin conservation* and therefore requires a paramagnetic metal ion as a catalyst as described by reactions (2.29) and (2.30).

Therefore, equation (2.31) just summarizes the chemical changes in the system but does not represent the reaction mechanism. If aqueous hydrogen sulfide ions (HS^-) are oxidized by a *one-electron oxidant* (M^{n+}), the initial product will be the radical $HS^\bullet$ which is expected to dimerize to the disulfide ion:

$$2HS^\bullet \rightarrow [S_2]^{2-} + 2H^+ \quad (2.32)$$

The disulfide ion is then oxidized either by a metal ion or by another $HS^\bullet$ radical:

$$[S_2]^{2-} + M^{n+} \rightarrow [S_2]^{\bullet -} + M^{(n-1)+} \quad (2.33)$$

$$[S_2]^{2-} + HS^\bullet \rightarrow [S_2]^{\bullet -} + HS^- \quad (2.34)$$

The disulfide radical anion $[S_2]^{\bullet -}$ dimerizes rapidly to the tetrasulfide ion $[S_4]^{2-}$ which may be further oxidized and so forth, eventually yielding a mixture of polysulfide anions of different chain-lengths. It has in fact been observed that oxidation of hydrogen sulfide ions by dioxygen in the presence of heme, or by iron(III) or ruthenium(III) complexes in the absence of dioxygen, yields polysulfides from which elemental sulfur may be formed according to the reverse reactions of the above equations (2.20)–(2.23). The electrochemical oxidation of aqueous sulfide at $pH = 9.3$ also yields polysulfide anions initially from which elemental sulfur is subsequently formed (Szynkarczuk *et al.*, 1994).

The reactions in equations (2.29) through (2.34) are the basis of aqueous desulfurization plants working by the Stretford, Sulfolin, Lo-Cat, SulFerox or Bio-SR (biological sulfur removal) processes (Steudel, 1996b).

Metal cations are also known to catalyze the oxidation of organic thiols (RSH) by dioxygen to yield disulfanes (RSSR) and water. Thiyl radicals ($RS^•$) are likely intermediates in these reactions.

2.4 SULFITES, THIOSULFATES, DITHIONITES AND DITHIONATES

2.4.1 Sulfur dioxide, sulfite and disulfite ions as well as sulfurous and sulfonic acids

Sulfur dioxide is one of the most important chemicals in industry since it is an intermediate in the large-scale production of sulfuric acid. It is mainly produced by combustion of liquid sulfur but also by the oxidative roasting process of sulfidic minerals such as FeS_2, ZnS, PbS, $CuFeS_2$ and other sulfidic minerals used for metal production. Another source of SO_2 is $Ca[SO_4]$ which is reduced by coke in the presence of SiO_2. Finally the waste products $Fe[SO_4]$ (from TiO_2 production) and spent sulfuric acid are decomposed thermally to recover the sulfur as SO_2 (Steudel, 2020).

Combustion of coal, crude oil, diesel oil, heating oil, gasoline and wood also generates SO_2 since all solid and liquid fossil fuels contain sulfur compounds. Therefore, power plants using fossil fuels must have a flue gas desulfurization unit usually based on scrubbing of the flue gas by an aqueous slurry of $Ca[OH]_2$ or $Ca[CO_3]$ producing gypsum $Ca[SO_4] \cdot 2H_2O$ or by a solution of $Na_2[SO_3]$ resulting in $Na[HSO_3]$ from which the SO_2 can be recovered by heating (Wellman-Lord process).

Sulfur dioxide is a colorless, toxic and corrosive gas with a melting and boiling point of, respectively, −75°C and −10°C. It is highly soluble in water (ca. 45 vol SO_2/vol H_2O at 15°C) with formation of what is usually called *sulfurous acid* (H_2SO_3). Since the equilibrium

$$SO_2 + H_2O \rightleftharpoons H_2SO_3 \tag{2.35}$$

is basically on the left-hand side, the concentration of undissociated H_2SO_3 is very small. In fact, the molecule H_2SO_3 has never been detected in water but exists in condensed phases in deprotonated form only:

$$SO_{2(aq)} + 2H_2O \rightleftharpoons [H_3O]^+ + [SO_3H]^- \quad pK_a = 1.86\ (25^\circ C) \tag{2.36}$$

According to Raman spectra, this solution contains SO_2 as well as the ions $[H_3O]^+$, $[SO_3H]^-$, $[HSO_3]^-$ and, at higher concentrations, disulfite ions $[S_2O_5]^{2-}$ in equilibrium.

Anions of composition $[S,H,O_3]^-$ exist as two tautomers with either an OH or SH bond (Horner & Connick, 1986; Steudel & Steudel, 2009a):

$$\underset{\text{sulfonate}}{[H\text{–}SO_3]^-} \rightleftharpoons \underset{\text{hydrogen sulfite}}{[HO\text{–}SO_2]^-} \tag{2.37}$$

The coordination geometry at the sulfur atoms in these isomers is either distorted tetrahedral (C_{3v} symmetry in $[H{-}SO_3]^-$) or distorted pyramidal (C_s symmetry of $[HO{-}SO_2]^-$). The relative concentrations of hydrogen sulfite $[HO{-}SO_2]^-$ and sulfonate ions $[H{-}SO_3]^-$ in water have been determined by ^{17}O-NMR, Raman, UV-Vis and XANES spectroscopy. At room temperature, hydrogen sulfite $[HO{-}SO_2]^-$, often abbreviated as $[SO_3H]^-$, dominates but at higher temperatures sulfonate ions are increasingly formed by tautomerization. Heating of aqueous SO_2 solutions to 200°C in an ampoule results in exothermic disproportionation to H_2SO_4 and S_8. With large cations (M) such as $[NH_4]^+$, Rb^+ and Cs^+, crystalline sulfonates $M[HSO_3]$ have been prepared while solid hydrogen sulfites $M[SO_3H]$ are unknown. On cooling of aqueous SO_2 solutions the gas-hydrate $SO_2 \cdot 6H_2O$ is formed as crystals.

The second dissociation constant of sulfurous acid, i.e. the dissociation of $[SO_3H]^-$ in water, is very small: $K_a = 0.625 \times 10^{-7}$ (p$K_a = 7.2$).

In concentrated aqueous sulfite solutions, another equilibrium is established:

$$[H{-}SO_3]^- + [HO{-}SO_2]^- \rightleftharpoons [S_2O_5]^{2-} + H_2O \quad K_c = 0.045\ (25°C) \tag{2.38}$$

Therefore, the determination of reliable pK_a values for sulfurous and sulfonic acids is difficult. The disulfite anion has the connectivity $[O_2S{-}SO_3]^{2-}$ with sulfur in the oxidation states +3 and +5, respectively. Solid sodium disulfite is produced by saturating aqueous NaOH with SO_2 followed by cooling; it decomposes at 400°C with liberation of SO_2.

Sulfite solutions are strong reducing agents which are oxidized by dioxygen, iron(III) salts or iodine to sulfate. In alkaline solution, the autoxidation is particularly fast although this reaction is formally forbidden by the law of spin conservation. In practice, however, the reaction is strongly catalyzed by even traces of metal cations such as Cu^{2+}, Mn^{2+}, Fe^{3+} and Co^{2+} some of which are present in most aqueous systems. Small concentrations of ethanol suppress the autoxidation (Connick *et al.*, 1995; Connick & Zhang, 1996).

On acidification of sulfites, SO_2 is evolved according to the reverse reaction (2.36). $Na_2[SO_3]$ is a commercial reducing agent and antioxidant, e.g. in wine.

In Europe and North America, the phenomenon of *acid rain* has basically been overcome by modern desulfurization units in power plants while in Asia it is still a considerable environmental problem. On the downside, many agricultural farms in Europe now suffer from a *sulfur deficit in soil*, which needs to be compensated by sulfate-containing fertilizers since sulfur is an essential element to plants together with N, P and K. This is despite the fact that enormous amounts of sulfur compounds are constantly emitted into the atmosphere from natural sources, mainly dimethyl sulfide Me_2S (DMS, ca. 4×10^7 t/a) from the oceans. DMS is produced by phytoplankton (floating microorganisms) and is oxidized in air to SO_2 and methane sulfonic acid $MeSO_3H$ (MSA). The oxidation of DMS in the atmosphere is complex due to several parallel reaction channels. Ultimate products are MSA, SO_2 and sulfuric acid while dimethyl sulfoxide (DMSO) and dimethyl sulfone ($DMSO_2$) are intermediates. Key reaction partners are hydroxyl radicals and O_2, but NO may also play a role. The two initial reactions are addition of an $OH^\bullet$ radical and abstraction of an H atom, respectively:

$$[CH_3\text{–}S(OH)\text{–}CH_3]^\bullet \leftarrow CH_3\text{–}S\text{–}CH_3 + OH^\bullet \rightarrow [CH_3\text{–}S\text{–}CH_2]^\bullet + H_2O \tag{2.39}$$

While the left channel leads to DMSO, $DMSO_2$ and eventually MSA, the channel on the right-hand side produces ultimately SO_2 and H_2SO_4. Since there exists also a nocturnal reaction channel starting by attack of $NO_3^\bullet$ on DMS with H abstraction and formation of HNO_3, the lifetime of DMS in the atmosphere is short (ca. 1 day); see Mardyukov and Schreiner (2018).

Therefore, the strong acid MSA ($MeSO_3H$) is one of the main constituents of acid rain besides H_2SO_4 and HNO_3. For reviews on anthropogenic SO_2 emissions to the atmosphere and the resulting acid rain, see Brandt and van Eldick (1995) and Möller (2010).

2.4.2 Thiosulfates and thiosulfuric acid

Sodium thiosulfate is produced commercially by refluxing a concentrated aqueous $Na_2[SO_3]$ solution with elemental sulfur:

$$Na_2[SO_3] + \frac{1}{8}S_8 \rightleftharpoons Na_2[S_2O_3] \tag{2.40}$$

This process is another example for a whole series of nucleophilic displacement reactions. The nucleophile $[SO_3]^{2-}$ first attacks an S_8 molecule with ring opening and initial formation of a chain-like sulfane monosulfonate dianion $[S\text{–}S_7\text{–}SO_3]^{2-}$. This ion is rapidly attacked by other sulfite ions with stepwise formation of thiosulfate ions until the sulfur chain has been turned into eight thiosulfate anions with one S–S bond each. Thus, the number of these bonds stays constant. According to photoelectron spectroscopy the oxidation states of the sulfur atoms in the thiosulfate anion are −2 and +6. Therefore, $[S_2O_3]^{2-}$ is also a nucleophile and protonation takes place at the terminal sulfur atom first (see below).

The reverse reaction (2.40) takes place when aqueous thiosulfate is decomposed by strong acids. The first reaction product is the highly reactive hydrogen thiosulfate anion which has been isolated as ammonium salt from methanolic solution at low temperatures (Steudel & Prenzel, 1989):

$$[NH_4]_2[S_2O_3] + H_2SO_4 \rightarrow [NH_4][HS_2O_3] + [NH_4][HSO_4] \tag{2.41}$$

The connectivity of the hydrogen thiosulfate ion is $[HS\text{–}SO_3]^-$ as shown by Raman spectroscopy. The free thiosulfuric acid $H_2S_2O_3$ is probably present in such solutions too, but due to its thermal instability $H_2S_2O_3$ has been isolated in a nonaqueous system at very low temperatures only (Hopfinger *et al.*, 2018). The microsolvation of $H_2S_2O_3$ and its anions by water molecules has been studied computationally (Steudel & Steudel, 2009b).

In acidic solution, the ions $[HS_2O_3]^-$ and $[S_2O_3]^{2-}$ react with exchange of sulfur atoms to give a mixture of sulfane monosulfonate ions resulting in the growth of a chain of sulfur atoms (the dianions attack the monoanions):

$$[HS_2O_3]^- + [S_2O_3]^{2-} \rightleftharpoons [HS_3O_3]^- + [SO_3]^{2-} \tag{2.42}$$

$$[HS_3O_3]^- + [S_2O_3]^{2-} \rightleftharpoons [HS_4O_3]^- + [SO_3]^{2-} \tag{2.43}$$

etc.

After the sulfur chain of the anions $[HS_nO_3]^-$ has reached a length of at least 7, 8 or 9 atoms, the cyclic molecules S_6, S_7 and S_8 can be split off. These homocycles have been detected in such mixtures by HPLC analysis after extraction with organic solvents (Steudel *et al.*, 1988):

$$[H\text{–}S_n\text{–}SO_3]^- \rightleftharpoons \textit{cyclo}\text{–}S_n + [SO_3H]^- \ (n > 5) \tag{2.44}$$

Since $[HSO_3]^-$ decomposes to H_2O and SO_2 in acidic solution and since the SO_2 formed will partly escape as a gas, reaction (2.44) proceeds to completion despite its reversibility. The precipitation of sulfur also shifts equilibrium (2.44) to the right. However, the reaction can be reversed in alkaline solution. Side reactions yield polythionates as will be shown in Section 2.5.

A highly characteristic reaction of thiosulfate anions is their facile oxidation to tetrathionate ('oxidative coupling') which – when carried out with elemental iodine (I_2) – is used in quantitative analysis of either one of the species. The following reaction mechanism provides more examples for nucleophilic displacements:

$$I_2 + [S\text{–}SO_3]^{2-} \rightarrow I^- + [I\text{–}S\text{–}SO_3]^- \tag{2.45}$$

$$[I\text{–}S\text{–}SO_3]^- + [S_2O_3]^{2-} \rightarrow I^- + [O_3S\text{–}S\text{–}S\text{–}SO_3]^{2-} \tag{2.46}$$

The progress of these reactions can be followed by the decoloring of the formerly brown iodine solution, especially if starch is used as an indicator which reacts with I_2 to a deep-blue complex. The tetrathionate anion is stable in neutral and acidic solutions. Reduction of tetrathionate by two electrons yields thiosulfate again.

Metal ions such as Fe^{3+}, Cu^{2+} or Au^{3+} oxidize thiosulfate ions by a 1-electron transfer first to the radical anion $[S_2O_3]^{\bullet-}$ which subsequently dimerizes to the tetrathionate ion. *Cytochrome c* and hexacyanoferrate(III) ions $[Fe(CN)_6]^{3-}$ are also suitable oxidants to transform thiosulfate into tetrathionate.

Reduction of thiosulfate by dithionite yields sulfide and sulfite by cleavage of all S–S bonds (Münchow & Steudel, 1994):

$$[S_2O_3]^{2-} + [S_2O_4]^{2-} + 2H_2O \rightarrow HS^- + 3[SO_3H]^- \quad (2.47)$$

In slightly acidic, concentrated solution thiosulfate reacts with SO_2 at 20°C to trithionate (Fehér, 1975):

$$3SO_2 + 2[S_2O_3]^{2-} \rightarrow 2[S_3O_6]^{2-} + \frac{1}{8}S_8 \quad (2.48)$$

This remarkable redox reaction may be used to prepare trithionate.

2.4.3 Dithionites and dithionous acid

Reduction of aqueous SO_2 by strong reducing agents like sodium amalgam, zinc dust, formic acid or electrochemically is used to produce the corresponding dithionites of which the commercial product $Na_2[S_2O_4] \cdot 2\,H_2O$ is best known:

$$2[SO_3H]^- + 2e^- \rightleftharpoons [S_2O_4]^{2-} + 2OH^- \quad (2.49)$$

In reactions where dithionite is used as a reductant the reverse reaction takes place. Aqueous dithionite rapidly reacts with molecular oxygen and therefore needs to be handled and stored in an inert atmosphere. This high reactivity is due to the partial dissociation of the dithionite ion in solution:

$$[S_2O_4]^{2-} \rightleftharpoons 2[SO_2]^{\bullet -} \quad (2.50)$$

In water this equilibrium is largely on the left-hand side, but the radical anions have nevertheless been detected by ESR spectroscopy. In polar organic solvents such as DMF, MeCN and DMSO, however, the dissociation is almost complete, even at room temperature.

Acidification of aqueous dithionite solutions results in hydrolytic decomposition which proceeds at 20°C and pH values near 6 basically according to the following equation (Drozdova *et al.*, 1998):

$$2[S_2O_4]^{2-} + H_2O \rightarrow 2[SO_3H]^- + [S_2O_3]^{2-} \quad (2.51)$$

In this disproportionation reaction the sulfur atoms change their oxidation state from +3 in dithionite to +4 in sulfite and −2 as well as +6 in thiosulfate. Consequently, salts with the anion $[HS_2O_4]^-$ as well as the free dithionous acid $H_2S_2O_4$ are unknown as pure compounds. Since thiosulfate is also unstable in acidic solution and decomposes to sulfur, SO_2 and polythionate anions, these products are also formed during acid decomposition of dithionite. The gas phase structures of $[HS_2O_4]^-$ and $H_2S_2O_4$ have, however, been predicted computationally (Drozdova *et al.*, 1998).

2.4.4 Dithionates and dithionic acid

Oxidation of aqueous sulfite by manganese(IV) oxide yields dithionate and sulfate:

$$3[SO_3H]^- + 2MnO_2 \rightarrow Mn[S_2O_6] + Mn[SO_4] + 3OH^- \quad (2.52)$$

The connectivity of dithionate anions is $[O_3S\text{–}SO_3]^{2-}$ with the sulfur atoms in the oxidation state +5. Iron(III) ions oxidize sulfite to dithionate too. This reaction occurs in flue gas scrubbers in which gypsum is produced from limestone because traces of Fe^{3+} are always present in minerals. From barium dithionate and dilute sulfuric acid aqueous dithionic acid can be obtained (Fehér, 1975):

$$Ba[S_2O_6] + H_2SO_4 \rightarrow H_2S_2O_6 + Ba[SO_4]\downarrow \quad (2.53)$$

$H_2S_2O_6$ is a medium strong two-protonic acid which decomposes above 50°C by disproportionation:

$$H_2S_2O_6 \rightarrow H_2SO_4 + SO_2 \quad (2.54)$$

The analogous decomposition occurs with solid dithionates but only above 200°C.

2.5 POLYTHIONATES AND HYDROPHILIC SULFUR SOLS

2.5.1 Polythionates and polythionic acids

Polythionates with anions of the type $[O_3S\text{–}S_n\text{–}SO_3]^{2-}$ are also known as polysulfane disulfonates. The corresponding sodium and potassium salts with $n = 1\ldots4$ have been prepared by various routes (Fehér, 1975):

$$\text{Trithionate:}\quad SCl_2 + 2[HSO_3]^- \rightarrow [S_3O_6]^{2-} + 2HCl \quad (2.55)$$

$$\text{Tetrathionate:}\quad 2[S_2O_3]^{2-} + I_2 \rightarrow [S_4O_6]^{2-} + 2I^- \quad (2.56)$$

$$\text{Pentathionate:}\quad SCl_2 + 2[HS_2O_3]^- \rightarrow [S_5O_6]^{2-} + 2HCl \quad (2.57)$$

$$\text{Hexathionate:}\quad S_2Cl_2 + 2[HS_2O_3]^- \rightarrow [S_6O_6]^{2-} + 2HCl \quad (2.58)$$

Polythionate salts are water soluble and metastable at acidic pH values. They have been detected in the waters of crater lakes by ion chromatography. The anhydrous polythionic acids (polysulfane disulfonic acids), however, are unstable and therefore unknown as pure compounds.

The most reactive sites of polythionate anions are the highly polar $S\text{–}SO_3$ bonds (the four-coordinate sulfur atoms are positively charged). Therefore, in alkaline solutions and in the presence of strong nucleophiles like sulfide or sulfite anions, polythionates are decomposed by nucleophilic displacement reactions, e.g.:

$$[S_4O_6]^{2-} + OH^- \rightarrow [HSSSO_3]^- + [SO_4]^{2-} \tag{2.59}$$

$$[S_4O_6]^{2-} + HS^- \rightarrow [HSSSO_3]^- + [S_2O_3]^{2-} \tag{2.60}$$

The nucleophile attacks the sulfonate S atom and the initially formed sulfane monosulfonate ions $[HS_nSO_3]^-$ decompose as outlined in Section 2.4.2. The strong reducing agent dithionite turns polythionates into thiosulfate and sulfane monosulfonates (Münchow & Steudel, 1994):

$$[S_nO_6]^{2-} + [S_2O_4]^{2-} + 2H_2O \rightarrow [HSS_{n-3}O_3]^- + [HS_2O_3]^- + 2[SO_3H]^- \tag{2.61}$$

In other words, polythionates can only exist at acidic pH values and in the absence of strong nucleophiles and reducing agents. Strong oxidizing agents such as hydrogen peroxide or elemental chlorine convert polythionates into sulfate.

Solid polythionates have been characterized by their strong absorption bands in the infrared spectrum as well as by crystal structure determinations. In solution these ions can be separated analytically by ion-pair chromatography and all species with chain-lengths up to 22 sulfur atoms have been identified (Steudel & Holdt, 1986; Steudel *et al.*, 1987).

2.5.2 Hydrophilic sulfur sols (Raffo and Selmi sols)

Chemically prepared hydrophilic sulfur sols consist of elemental sulfur together with hydrophilic compounds such as polythionates or similar constituents which turn the hydrophobic sulfur particles into a hydrophilic material by adsorption through hydrophobic interaction. Depending on the method of preparation, two types of (purely inorganic) sulfur sols are defined: Raffo (or Lamer) sols are prepared by decomposition of concentrated sodium thiosulfate solutions with concentrated sulfuric acid (Steudel *et al.*, 1988). Selmi sols, on the other hand, are obtained by reaction of aqueous sulfite (or SO_2) with sulfide ions (or H_2S) (Steudel et al, 1989b).

When a **Raffo sol** is prepared from aqueous thiosulfate and concentrated H_2SO_4 the reactions occurring are similar to those outlined in Section 2.4.2. In addition, condensation and/or oxidation reactions of sulfane monosulfonate anions to polythionates take place:

$$2[H{-}S_n{-}SO_3]^- \rightarrow H_2S + [O_3S{-}S_{2n-1}{-}SO_3]^{2-} \tag{2.62}$$

$$2[H{-}S_n{-}SO_3]^- + [O] \rightarrow H_2O + [O_3S{-}S_{2n}{-}SO_3]^{2-} \tag{2.63}$$

The concentrated sulfuric acid or other sulfur oxoanions can serve as oxidants [O]. By salting-out with sodium chloride, a yellow precipitate is obtained which looks like sulfur but is mainly composed of long-chain polythionates together with some elemental sulfur in the form of the homocycles S_6 to S_{14}. The total sulfur content is ca. 85wt% but only 17% of this sulfur is present as elemental sulfur. The remaining 15wt% is composed of oxygen, hydrogen and sodium. Therefore, this material should not be termed 'elemental sulfur' or S°, especially since its reactivity is quite different from that of α-S_8. The precipitated particles of the sol can be re-dissolved in water forming a colloidal solution (peptization) and this sol can be precipitated once again by addition of NaCl. It is believed that the long-chain polythionate anions (in the form of sodium salts) form micelle-like particles in water and the elemental sulfur rings are dissolved in their core. The sulfonate groups are likely to cover the surface of the micelles making the particles hydrophilic despite their high sulfur content (Figure 2.9).

The particle size is in the range of 0.1–0.5 μm and the sulfur content of the aqueous sol can be very high. By partial evaporation of the solvent, sulfur concentrations of up to 600 g L^{-1} have been obtained. Evaporation of Raffo sols

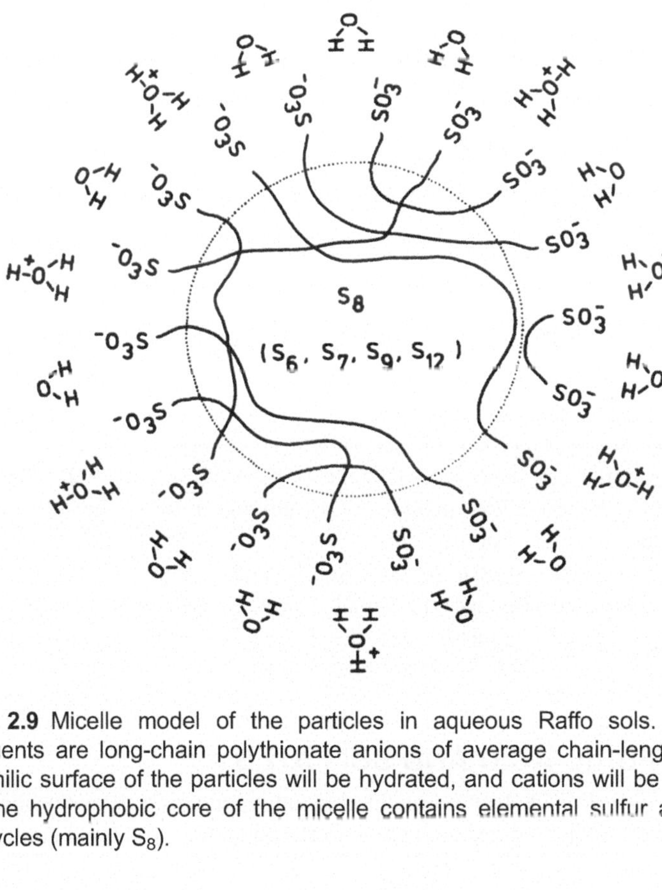

Figure 2.9 Micelle model of the particles in aqueous Raffo sols. The main constituents are long-chain polythionate anions of average chain-length 29. The hydrophilic surface of the particles will be hydrated, and cations will be present as well. The hydrophobic core of the micelle contains elemental sulfur as different homocycles (mainly S_8).

to dryness yields a yellow solid product which looks like elemental sulfur but is soluble in water. Ion-pair chromatography and reversed-phase HPLC are suitable methods to analyze sulfur sols (Steudel *et al.*, 1988).

Selmi sols, on the other hand, are prepared from aqueous H_2S and SO_2 solutions at acidic pH values (Steudel *et al.*, 1989b). Their composition and properties are analogous to those of Raffo sols but the average chain-length of the polythionate anions is smaller. The reactions resulting in these products are very complex. Raffo and Selmi sols are thermodynamically unstable and slowly decompose at room temperature, the Selmi sol more rapidly than the Raffo sol. This aging process involves the conversion of longer-chain polythionates to shorter ones with simultaneous formation of S_8. In other words, the micelles are destroyed and elemental sulfur precipitates within several days:

$$[O_3S\text{–}S_n\text{–}SO_3]^{2-} \rightarrow \textit{cyclo}\text{–}S_8\downarrow + [O_3S\text{–}S_{n-8}\text{–}SO_3]^{2-} \quad (2.64)$$

From the model in Figure 2.9 it follows that the particles of hydrophilic sulfur sols are negatively charged. This equal charge results in repulsion between neighboring micelles which stabilizes the sol. However, when multivalent cations are added, the repulsion is diminished since the electric field is shielded by the cations and a kind of insoluble 'salt' is formed by Coulomb attraction of neighboring sol particles to one and the same cation. Therefore, even small concentrations of two- or three-valent cations initiate precipitation (Steudel, 1999). All reagents and pH values that destroy polythionates (see Section 2.5.1) also destroy Raffo and Selmi sols.

Hydrophilic sulfur sols are also produced by enzymatic oxidation of sulfide ions or other reduced sulfur compounds by a variety of sulfur bacteria (Brune, 1989; Takakuwa, 1992). However, the physical properties and chemical composition of this biologically produced sulfur (often abbreviated as S°) are different from the hydrophilic sulfur sols discussed above (Dahl, 1999; Janssen, 1996; Prange *et al.*, 1999, 2002, 2004; Steudel *et al.*, 1987). Bacterial sulfur globules produced by phototrophic bacteria consist of sulfur-rich components and do not always contain substantial amounts of S_8 (Prange *et al.* 1999, 2002, 2004). Therefore, they should not be termed 'elemental sulfur' although most of the sulfur is in the oxidation state zero. Hydrophilic sulfur particles produced by mixed cultures of *Thiobacilli* have been shown to be composed of a core of elemental sulfur covered by a layer of polymers (presumably proteins) which are equally charged resulting in Coulomb repulsion which stabilizes the globules (Janssen *et al.*, 1999; Kleinjan *et al.*, 2003). Evaporation to dryness yields α-S_8.

2.6 SULFURIC ACID AND SULFATES

Anhydrous sulfuric acid (100%) is a colorless oily liquid (m.p. 10.4°C) which boils at 290–317°C with partial decomposition to H_2O and SO_3. Solid and liquid H_2SO_4

are strongly associated by hydrogen bonds which cause the high melting temperature and the high viscosity.

H_2SO_4 is produced industrially on an enormous scale by catalytic oxidation of SO_2 with air followed by dissolution of SO_3 in H_2SO_4 to give polysulfuric acids $H_2S_nO_{3n+1}$ ('oleum') which are subsequently diluted by water to sulfuric acid of desired concentration. The commercial 'concentrated' acid contains 4% water. This liquid as well as the 100% acid are oxidants, especially when hot. They are also dehydrating agents since the hydrate $H_2SO_4 \cdot H_2O$ is formed in a strongly exothermic reaction through hydrogen bonds. This hydrate of m.p. 8.5°C is a salt with the structure of an oxonium hydrogen sulfate: $[H_3O][HSO_4]$. Therefore, concentrated sulfuric acid conducts electricity as does the 100% acid through the following auto-dissociation equilibrium (Steudel, 2020):

$$2H_2SO_4 \rightleftharpoons [H_3SO_4]^+ + [HSO_4]^- \tag{2.65}$$

In water, sulfuric acid is a very strong acid and its first pK_a value is not known with certainty but has been estimated in the range −4 to −8. The pK_a(2) value of $[HSO_4]^-$ is +1.92. Therefore, in dilute solution the acid is completely ionized in the first step but only partly in the second step. At pH = 1 the concentration ratio of $[SO_4]^{2-}$ to $[HSO_4]^-$ is ca. 0.1. The sulfate anion is rather inert chemically and needs to be activated for biological reduction at ambient temperatures: The phosphosulfate ions APS and PAPS shown in Figure 2.10 are activated forms containing the sulfate anion attached to a phosphate unit as a mixed anhydride.

The energy gained on the exothermic hydrolysis of the P–O–S bridge promotes reactions of the liberated sulfate ion which therefore is termed as 'activated'.

Sulfate aerosols in the upper atmosphere influence the climate on Earth. These sols consist of the hydrated salts $[H_3O][HSO_4]$ and $[NH_4][HSO_4]$ and originate mainly from the volcanic SO_2 exhaustions which reach the stratosphere. For example, in June of 2011 the layer volcano Nabro in Eritrea ejected approximately 1.3×10^6 tons of SO_2 into the atmosphere some of which reached altitudes of 9–14 km; in general, magmatic gases released from volcanoes today contain water vapor and carbon dioxide as the main components, with smaller contributions of SO_2, H_2S, HCl, HF, CO, H_2 and N_2, but also traces of volatile metal chlorides and SiF_4 (Möller, 2010).

SO_2 is oxidized in the atmosphere by hydroxyl radicals to $HSO_3^{\bullet}$ radicals which dimerize to $H_2S_2O_6$ and this acid disproportionates to SO_2 and H_2SO_4 (see Section 2.3.1). Ammonia is a well-known pollutant in the atmosphere too, although in low concentration. The sulfate aerosols reflect sunlight resulting in a drop in temperature at the ground which may last for 2 or 3 years before sedimentation and rainfall cleaned the atmosphere.

Water-insoluble sulfates occur in nature as oxidation products of metal sulfides which were formed in ancient times during periods when the atmosphere of Earth

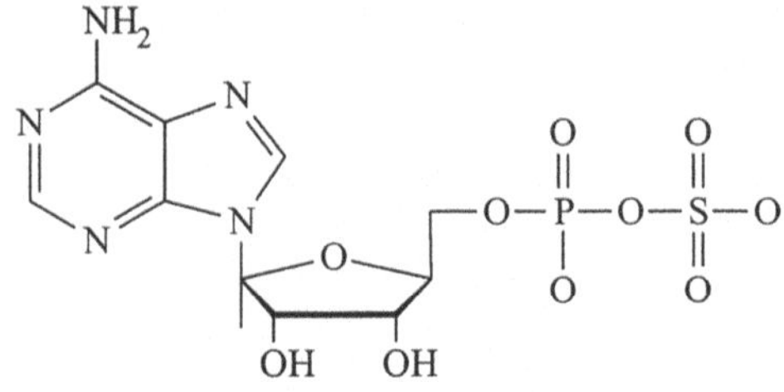

Adenosine Phosphosulfate (APS)

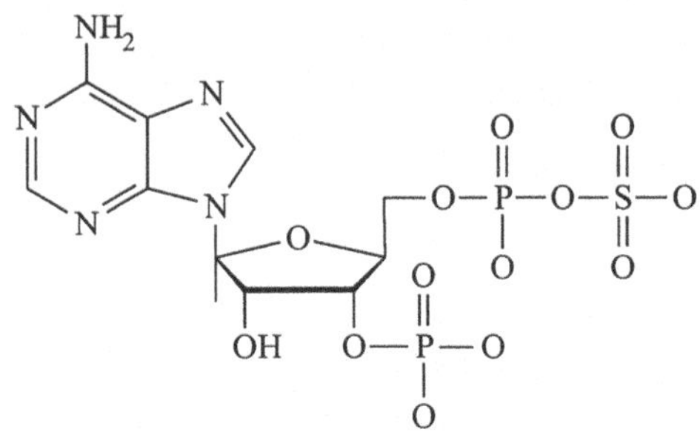

3′-Phosphoadenosine Phosphosulfate (PAPS)

Figure 2.10 Structures of the phosphosulfate anions APS and PAPS.

Table 2.7 Solubility products K_L of water-insoluble sulfate minerals at 25°C: $pK_L = -\log K_L$ (after www.aqion.de/site/17).

Name	Formula	pK_L
Anhydrite	$Ca[SO_4]$	4.36
Gypsum	$Ca[SO_4] \cdot 2H_2O$	4.56
Celestite	$Sr[SO_4]$	6.63
Anglesite	$Pb[SO_4]$	7.79
Barite	$Ba[SO_4]$	9.97

was mainly oxygen-poor or even reducing. Examples of natural sulfates and their solubility products are given in Table 2.7.

2.7 DISPROPORTIONATION OF ELEMENTAL SULFUR IN WATER

As shown in Figure 2.1 elemental sulfur is thermodynamically unstable in water at pH values exceeding 7. However, the corresponding reactions are slow at 25°C and in the absence of a catalyst. At higher temperatures, however, several equilibrium reactions are established. The primary reaction in basic solutions is the following disproportionation (Licht & Davis 1997):

$$\frac{1}{2}S_8 + 4OH^- \rightleftharpoons [S_2O_3]^{2-} + 2HS^- + H_2O \quad (2.66)$$

At pH >11.5 this reversible reaction occurs already at 20°C. At pH = 7.6, a temperature of 80°C is necessary but at 100–120°C the equilibrium is established rapidly (Giggenbach, 1974). Secondary reactions are the formation of polysulfide ions:

$$\frac{n}{8}S_8 + HS^- + OH^- \rightleftharpoons [S_{n+1}]^{2-} + H_2O \quad (2.67)$$

followed by disproportionation of the polysulfides, e.g.:

$$[S_5]^{2-} + 3OH^- \rightleftharpoons [S_2O_3]^{2-} + 3HS^- \quad (2.68)$$

and by hydrolysis of thiosulfate:

$$[S_2O_3]^{2-} + OH^- \rightleftharpoons [SO_4]^{2-} + HS^- \quad (2.69)$$

The latter two reactions require temperatures above 160°C unless enzymatically catalyzed. All these reactions result in a drop of pH value (consumption of OH^-) and the whole reaction sequence comes to a standstill unless a buffer is used. The final products are sulfide and sulfate, in accordance with Figure 2.1.

In acidic and neutral solutions, the disproportionation of sulfur proceeds according to the equation:

$$\frac{1}{2}S_8 + 4H_2O \rightleftharpoons 3H_2S + [HSO_4]^- + H^+ \quad (2.70)$$

This endergonic equilibrium ($\Delta G° > 0$) is rapidly established above 150°C. After sulfuric acid has been formed, the following secondary equilibrium will be established too at this temperature (comproportionation):

$$\frac{1}{8}S_8 + 2H_2SO_4 \rightleftharpoons 3SO_2 + 2H_2O \quad (2.71)$$

2.8 ORGANIC DERIVATIVES OF THE TYPE R–S_n–R (ORGANOPOLYSULFANES)

2.8.1 Synthetic polysulfanes

Bis(organo)oligo- or polysulfanes R_2S_n are often termed *polysulfides* although they are not ionic materials such as sodium polysulfides, for instance. The recommended nomenclature for R–S_n–R is *polysulfanes* since they are derivatives of the unsubstituted sulfanes H–S_n–H. These in turn are analogues of the alkanes C_nH_{2n+2}, the silanes Si_nH_{2n+2} and the phosphanes P_nH_{2n+1}. Organic polysulfanes are much more stable thermally than the parent species H_2S_n. Consequently, numerous derivatives with different R groups have been prepared. These compounds are either chain-like or cyclic and species with up to 35 sulfur atoms have been detected by HPLC (Table 2.2) but more sulfur-rich members may exist also. The molecular structures of derivatives with up to 11 sulfur atoms have been elucidated by X-ray crystallography and their reactions have been studied extensively (for a comprehensive review see: Steudel, 2002).

2.8.2 Naturally occurring polysulfanes

Although it has been known for a long time that organic mono- and disulfanes like methionine and cystine occur in organisms, the widespread natural occurrence of tri- and higher polysulfanes with organic substituents has been demonstrated only relatively recently (Steudel, 2002). The reason for this may be that mono- and disulfanes are more stable chemically and thermally than higher-rank polysulfanes as a result of their much lower homolytic dissociation enthalpy $D°$ of the central sulfur-sulfur bonds, as shown by the following values for dimethyl oligosulfanes (Benson, 1978):

$$\text{Me–S–S–Me} \quad 310\,\text{kJmol}^{-1}$$

$$\text{Me–S–S–S–Me} \quad 226\,\text{kJmol}^{-1}$$

$$\text{Me–S–S–S–S–Me} \quad 141\,\text{kJmol}^{-1}$$

Table 2.8 Naturally occurring chain-like bis-organyl polysulfanes $R^1–S_n–R^2$ ($n > 2$) with different substituents $R^{1,2}$.

R^1	R^2	*n*	Source
methyl	methyl	4	*Lentinus edodes*
methyl	methyl	3	(Shiitake mushroom)
methyl	2-butyl	3	oil made from
2-butyl	2-butyl	3,4	*Ferula asafoetida* (Afghanistan)
allyl	allyl	3–6	garlic oil
alanyl	alanyl	3,4	wool hydrolysate (acidic)
3-oxoundecyl	3-oxoundecyl	3,4	*Dictyopteris plagiogramma* (Hawaiian algae)
methyl	complex structure	3	Calichemicin (*Micromonospora echinospora*)

Table 2.9 Cyclic polysulfanes with three or more neighboring sulfur atoms isolated from organisms.

Compound		Source
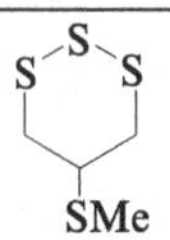	5-Methylthio-1,2,3-trithiane	*Chara globularia* (green algae)
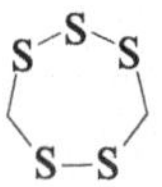	Lenthionine	*Chondria californica* (red algae) *Parkia speciosa* (Mimosaceae species) *Lentinus edodes* (Shiitake mushroom)
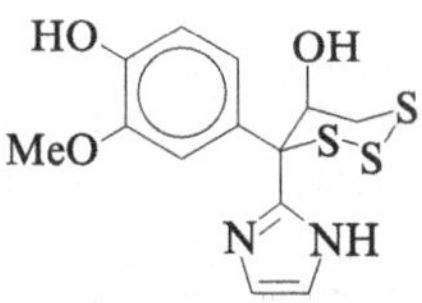		*Aplidium* sp. D (ascidian)
	Hexathiepane	*Lentinus edodes* (Shiitake mushroom)
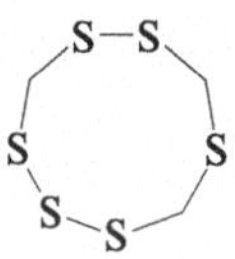 Hexathionane	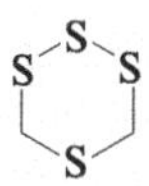 1,2,3,5-Tetrathiane	
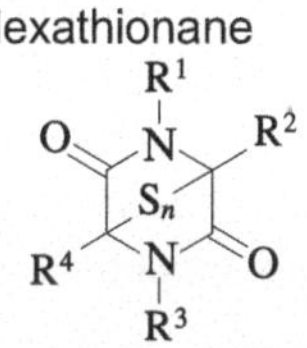	3,6-Epipolythiopiperazine-2,5-dione	
$n = 3$; $R^1 = R^3 = Me$; $R^2 = CH_2OH$; $R^4 = CH_2Ph$: Sporidesmine E $n = 4$; $R^1 = R^3 = Me$; $R^2 = CH_2OH$; $R^4 = CH_2Ph$: Sporidesmine G		Fungus *Pithomyces chartarum* *Hyalodendron* sp. (fungus); *Penicillium turbatum* *Pithomyces chartarum*
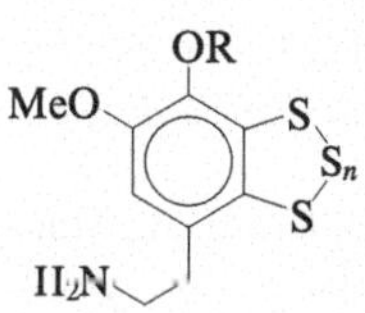	$n = 1$; R = H Lissoclinotoxin A $n = 3$; R = Me Varacin	*Lissoclinum perforatum* (ascidian) *Lissoclinum vareau* (ascidian)

Like other compounds with S–S bonds the organic polysulfanes are easily attacked by nucleophiles such as sulfide, sulfite and cyanide anions. For example, hydrogen sulfide cleaves cystine with formation of the corresponding thiolate (cysteinate) and 'persulfide' anions:

$$\text{R–S–S–R} + \text{HS}^- \rightleftharpoons [\text{R–S}]^- + [\text{R–S–S}]^- + \text{H}^+ \tag{2.72}$$

Sulfide and persulfide ions are important signaling molecules in biology (Filipovic *et al.*, 2018).

In the presence of O_2 the persulfide undergoes autoxidation to the corresponding cysteine S-sulfonate which is also known as the Bunte salt of cysteine (Steudel & Albertsen, 1992):

$$[\text{R–S–S}]^- + \frac{3}{2}\text{O}_2 \rightarrow [\text{R–S–SO}_3]^- \tag{2.73}$$

In Tables 2.8 and 2.9 some polysulfanes containing three or more neighboring (or cumulated) sulfur atoms are listed which have been isolated from various natural sources like algae, mushrooms, higher plants or animals. These compounds most probably have some function beneficial to the particular organism but in most cases this function is unknown. It has been speculated that deterrence of predators is the main function. For example, *varacin* has strong bacteriostatic properties. The beneficial properties of the organosulfur compounds contained in onions, garlic and other *Allium species* to human health are well known (Block, 1992). One needs to be aware, however, of the possibility that thermally unstable ingredients of plants and animals may decompose during workup resulting in secondary products published in the literature.

Unusual peptides have been isolated from genetically engineered bacteria which contain trisulfane units between cysteine residues instead of disulfane units. For example, the human growth hormone with one or both disulfane units replaced by trisulfane groups was biosynthesized by *E. coli* after genetic engineering (Canova-Davis *et al.*, 1996). Synthetic biology provides numerous examples of chemically modified natural substances.

REFERENCES

Barton L. E. (ed.) (1995) Sulfate Reducing Bacteria, Plenum, New York.

Benson S. W. (1978). Thermochemistry and Kinetics of Sulfur-Containing Molecules and Radicals. *Chemical Reviews*, **78**, 23–35.

Block E. (1992). Organosulfur Chemistry of the Species Allium and their Importance for the Organic Chemistry of Sulfur. *Angewandte Chemie International Edition*, **31**, 1135–1178.

Bogdándi V., Ida T., Sutton T. R., Bianco C., Ditrói T., Koster G., Henthorn H. A., Minnion M., Toscano J. P., van der Vliet A., Pluth M. D., Feelisch M., Fukuto J. M., Akaike T. and Nagy P. (2019). Speciation of reactive sulfur species and their reactions with alkylating agents: do we have any clue about what is present inside the cell? *British Journal of Pharmacology*, **176**, 646–670.

Brandt C. and van Eldick R. (1995). Transition Metal-Catalyzed Oxidation of Sulfur(IV) Oxides. Atmospheric-Relevant Processes and Mechanisms. *Chemical Reviews*, **95**, 119–190.

Brune D. C. (1989). Sulfur oxidation by phototrophic bacteria. *Biochimica et Biophysica Acta*, **975**, 189–221.

Canova-Davis E., Baldonado I. P., Chloupek R. C., Ling V. T., Gehant R., Olson K. and Gillece-Castro B. L. (1996). Confirmation by Mass Spectrometry of a Trisulfide Variant in Methionyl Human Growth Hormone Biosynthesized in *Echerichia coli*. *Analytical Chemistry*, **68**, 4044–4051.

Chen K. Y. and Morris J. C. (1972). Kinetics of Oxidation of Aqueous Sulfide by O_2. *Environm. Science and Technology*, **6**, 529–537.

Chiu G. and Meerhan E. J. (1977). Monodisperse sulfur sols from the air oxidation of hydrogen sulfide. *Journal of Colloid and Interface Science*, **62**, 1–7.

Connick R. E. and Zhang Y.-X. (1996). Kinetics and Mechanism of the Oxidation of $[HSO_3]^-$ by O_2. Part 2: The Manganese(II) Catalyzed Reaction. *Inorganic Chemistry*, **35**, 4613–4621.

Connick R. E., Zhang Y.-X., Lee S., Adamic R. and Chieng P. (1995). Kinetics and Mechanism of the Oxidation of $[HSO_3]^-$ by O_2. 1. The Uncatalyzed Reaction. *Inorganic Chemistry*, **34**, 4543–4553.

Crapanzano L., Crichton W. A., Monaco G., Bellissent R. and Mezouar M. (2005). Alternating sequence of ring and chain structures in sulfur at high pressure and temperature. *Nature Materials*, **41**, 550–532.

Dahl C. (1999). Deposition and Oxidation of Polymeric Sulfur in Prokaryotes. In: Biochemical Principles and Mechanisms of Biosynthesis and Biodegradation of Polymers A. Steinbüchel, (ed.), Wiley-VCH, Weinheim, p. 27.

Degrand C. and Lund H. (1979). Electrochemical Synthesis of Sulfinic Acids. *Acta Chemica Scandinavica Part B*, **33**, 512–514.

Devillanova F. A. (ed.) (2006). Handbook of Chalcogen Chemistry. Royal Society of Chemistry, London.

Drozdova Y., Steudel R., Hertwig R. H., Koch W. and Steiger T. (1998). Structures and Energies of Various Isomers of Dithionous Acid, $H_2S_2O_4$, and of its Anion $[HS_2O_4]^-$. *Journal of Physical Chemistry* Part A, **102**, 990–996 and references cited therein.

Eckert B. and Steudel R. (2003). Molecular Spectra of Sulfur Molecules and Solid Sulfur Allotropes, *Topics in Current Chemistry*, **231**, 31–98.

El Jaroudi O., Picquenard E., Demortier A., Lelieur J.-P. and Corset J. (2000). Polysulfide Anions II: Structure and Vibrational Spectra of the S_4^{2-} and S_5^{2-} Anions. Influence of the Cations on Bond Length, Valence and Torsion Angle. *Inorganic Chemistry*, **39**, 2593–2603.

Fehér F. (1975). Schwefel, Selen, Tellur. In: Handbuch der Präparativen Anorganischen Chemie, 3rd ed., vol. **1** G. Brauer (ed.), Enke, Stuttgart, pp. 356, 362, 395/6 and 398.

Filipovic M. R., Zivanovic J., Alvarez B. and Banerjee R. (2018). Chemical Biology of H_2S Signaling through Persulfidation. *Chemical Reviews*, **118**, 1253–1337.

Garcia A. A., Jr. and Druschel G. K. (2014). Elemental sulfur coarsening kinetics. *Geochemical Transactions*, **15**, 11.

Garrels R. M. and Naeser C. R. (1958). Equilibrium distribution of dissolved sulfur species in water at 25°C and 1 atm total pressure. *Geochimica. Cosmochimica Acta*, **15**, 113–130.

Giggenbach W. F. (1974). Equilibria Involving Polysulfide Ions in Aqueous Sulfide Solutions up to 240°C. *Inorganic Chemistry*, **13**, 1724–1729.

Gmelin (1960). Handbuch der anorganischen Chemie, Schwefel, Teil B, Lieferung 2, VCH, Weinheim, pp. 532–538 and 895–896.

Gun J., Modestov A. D., Kamshny A., Jr., Ryzkov D., Gitis V., Goifman A., Lev O., Hultsch V., Grischek T. and Worch E. (2004). Electrospray Ionization Mass Spectrometric Analysis of Aqueous Polysulfide Solutions. *Microchimica Acta*, **146**, 229–237.

Hahn J. (1985). Zusammensetzung von Rohsulfan, Nachweis der Sulfane H_2S_9 bis H_2S_{35}. *Zeitschrift für Naturforschung*. Part B, **40**, 263–272.

Holleman-Wiberg (2017). Lehrbuch der Anorganischen Chemie, 103rd ed., de Gruyter, Berlin (Germany).

Hopfinger M., Zischka F., Seifert M. and Kornath A. J. (2018). Preparation and Characterization of Pure Thiosulfuric Acid. *Zeitschrift fuer Anorganische und Allgemeine Chemie*, **644**, 12–13.

Horner D. A. and Connick R. E. (1986). Equilibrium Quotient for the Isomerization of Bisulfite Ion from $[HSO_3]^-$ to $[SO_3H]^-$. *Inorganic Chemistry*, **25**, 2414–2417.

Janssen A. J. H. (1996). Formation and colloidal behavior of elemental sulfur produced from the biological oxidation of hydrogen sulfide. Doctoral Dissertation, Landbouwuniversiteit Wageningen (NL).

Janssen A. J. H., Lettinga G. and de Keizer A. (1999). Removal of hydrogen sulfide from wastewater and waste gases by biological conversion to elemental sulfur. *Colloids Surfaces A: Physicochemical and Engineering Aspects*, **151**, 389–397.

Kamyshny A. (2009). Solubility of cyclooctasulfur in pure water and sea water at different temperatures. *Geochimica et Cosmochimica Acta*, **73**, 6022–6028.

Kamyshny A., Jr., Goifman A., Gun J., Rizkov D. and Lev O. (2004). Equilibrium Distribution of Polysulfide Ions in Aqueous Solutions at 25°C: A New Approach for the Study of Polysulfides' Equilibria. *Environmental Science and Technology*, **38**, 6633–6644.

Kamyshny A., Jr., Gun J., Rizkov D., Voitsekovski T. and Lev O. (2007). Equilibrium Distribution of Polysulfide Ions in Aqueous Solutions at Different Temperatures by Rapid Single-Phase Derivatization. *Environmental Science and Technology*, **41**, 2395–2400.

Karchmer J. H. (ed.) (1970). The Analytical Chemistry of Sulfur and its Compounds, Parts I and II, Wiley, New York.

Kleinjan W. J., de Keizer A. and Janssen A. J. H. (2003). Biologically Produced Sulfur. *Topics in Current Chemistry*, **230**, 167–188.

Kleinjan W. E., de Keizer A. and Janssen A. J. H. (2005). Kinetics of the chemical oxidation of polysulfide anions in aqueous solution. *Water Research*, **39**, 4093–4100.

Licht S. and Davis J. (1997). Disproportionation of aqueous sulfur and sulfide: Kinetics of polysulfide decomposition. *Journal of Physical Chemistry Part B*, **101**, 2540–2545.

Liebing P., Kühling M., Swanson C., Feneberg M., Hilfert L., Goldhahn R., Chivers T. and Edelmann F. T. (2019). Catenated and spirocyclic polychalcogenides from potassium carbonate and elemental chalcogens. *Chemical Communication*, (DOI: 10.1039/c9cc08347b).

Mardyukov A. and Schreiner P. R. (2018). Atmospherically Relevant Radicals Derived from the Oxidation of Dimethyl Sulfide, *Accounts of Chemical Research*, **51**, 475–483.

May P. M., Batka D., Hefter G., Königsberger E. and Rowland D. (2018). Goodbye to S^{2-} in aqueous solution. *Chemical Communication*, **54**, 1980–1983.

McCampbell Hamilton E. (1991). The Chemistry of Low Valent Sulfur Compounds in the Sulfur-Water System, Ph.D. dissertation, University of Minnesota (USA).

Möller D. (2010). Chemistry of the Climate System, de Gruyter, Berlin/New York, pp. 132–136 and 538–555.

Müller A. and Diemann E. (1987). Polysulfide Complexes of Metals. *Advances in Inorganic Chemistry*, **31**, 89–122.

Krebs B. and Müller A. (eds) (1984). Sulfur – Its Significance for Chemistry, for the Geo-, Bio- and Cosmosphere and Technology, Elsevier, Amsterdam.

Münchow V. and Steudel R. (1994). The Decomposition of Aqueous Dithionite and its Reactions with Polythionates $[S_nO_6]^{2-}$ ($n = 3–5$) Studied by Ion-Pair Chromatography. *Zeitschrift fuer Anorganische und Allgemeine Chemie*, **620**, 121–126.

Nickless G. (ed.) (1968). Inorganic Sulphur Chemistry. Elsevier, Amsterdam.

O'Brien D. J. and Birkner F. B. (1977). Kinetics of Oxygenation of Reduced Sulfur Species in Aqueous Solution. *Environm. Science and Technology*, **11**, 1114–1120.

Otto A. H. and Steudel R. (1999). The Gas Phase Acidities of the Sulfanes H_2S_n ($n = 1–4$). *European Journal of Inorganic Chemistry*, pp. 2057–2061.

Prange A., Arzberger I., Engemann C., Modrow H., Schumann O., Trüper H. G., Steudel R., Dahl C. and Hormes J. (1999). In situ analysis of sulfur in the sulfur globules of phototrophic sulfur bacteria by X-ray absorption near edge spectroscopy. *Biochimica et Biophysica Acta*, **1428**, 446–454.

Prange A., Chauvistre R., Modrow H., Hormes J., Trüper H. G. and Dahl C. (2002). Quantitative speciation of sulfur in bacterial sulfur globules: X-ray absorption spectroscopy reveals at least three different species of sulfur, *Microbiology (Reading, England)*, **148**, 267–276.

Prange A., Engelhardt H., Trüper H. G. and Dahl C. (2004). The role of the sulfur globule proteins of *Allochromatium vinosum*: mutagenesis of the sulfur globule protein genes and expression studies by real-time RT-PCR. *Archives of Microbiology*, **182**, 165–174.

Rau H., Kutty T. R. N. and Guedes de Carvalho J. R. F. (1973). Thermodynamics of sulphur vapour. *Journal of Chemical Thermodynamics*, **5**, 833–844.

Schmidt M. and Siebert W. (1973). Sulphur. In: Comprehensive Inorganic Chemistry, A. F. Trotman-Dickenson (eds), Vol. **2**, Pergamon, Oxford, pp. 795–933.

Schwarzenbach G. and Fischer A. (1960). Die Acidität der Sulfane und die Zusammensetzung wässriger Polysulfidlösungen. *Helvetica Chimica Acta*, **43**, 1365–1391.

Steliou K. (1991). Diatomic Sulfur. *Accounts of Chemical Research*, **24**, 341–350.

Steudel R. (1996a). Das gelbe Element und seine erstaunliche Vielseitigkeit, *Chemie in unserer Zeit*, **30**, 226–234.

Steudel R. (1996b). Mechanism for the Formation of Elemental Sulfur from Aqueous Sulfide in Chemical and Microbiological Desulfurization Processes. *Industrial & Engineering Chemistry Research*, **35**, 1417–1423 and references cited therein.

Steudel R. (1999). The Chemistry of Aqueous Sulfur Sols – Models for Bacterial Sulfur Globules? In: Biochemical Principles and Mechanisms of Biosynthesis and Biodegradation of Polymers, A. Steinbüchel (ed.), Wiley-VCH, Weinheim, pp. 17–25.

Steudel R. (2002). The Chemistry of Organic Polysulfanes R_2S_n. *Chemical Reviews*, **102**, 3905–3946.

Steudel R. (2003a). Liquid Sulfur. *Topics in Current Chemistry*, **230**, 81–116.

Steudel R. (2003b). Aqueous Sulfur Sols. *Topics in Current Chemistry*, **230**, 153–166.

Steudel R. (2003c). Inorganic Polysulfanes H_2S_n with $n > 1$. *Topics in Current Chemistry*, **231**, 99–126.

Steudel R. (2003d). Inorganic Polysulfides and Radical Anions. *Topics in Current Chemistry*, **231**, 127–152.

Steudel R. (2020). Chemistry of the Non-Metals, 2nd ed., de Gruyter, Berlin/Boston.

Steudel R. and Albertsen A. (1992). Determination of cysteine-S-sulfonate by ion-pair chromatography and its formation by autoxidation of cysteine persulfide. *Journal of Chromatography*, **606**, 260–263.

Steudel R. and Chivers T. (2019). The role of polysulfide dianions and radical anions in the chemical, physical and biological sciences, including sulfur-based batteries. *Chemical Society Reviews*, **48**, 3279–3319.

Steudel R. and Eckert B. (2003). Solid Sulfur Allotropes. *Topics in Current Chemistry*, **230**, 1–80.

Steudel R. and Holdt G. (1986). Ion-pair chromatographic separation of polythionates $[S_nO_6]^{2-}$ with up to thirteen sulfur atoms. *Journal of Chromatography*, **361**, 379–384.

Steudel R. and Holdt G. (1988). Solubilization of Elemental Sulfur in Water by Cationic and Anionic Surfactants. *Angewandte Chemie International Edition*, **27**, 1358–1359.

Steudel R. and Holz B. (1988). Detection of Reactive Sulfur Molecules (S_6, S_7, S_9, S_μ) in Commercial Sulfur, in Sulfur Minerals, and in Sulfur Melts Slowly cooled to 20°C. *Zeitschrift fuer Naturforschung* Part B, **43**, 581–589.

Steudel R. and Prenzel A. (1989). Raman Spectroscopic Discovery of the Hydrogenthiosulfate Anion $[HS\text{-}SO_3]^-$ in Solid $[NH_4][HS_2O_3]$. *Zeitschrift fuer Naturforschung* Part B, **44**, 1499–1502.

Steudel R. and Steudel Y. (2009a). Sulfur Dioxide and Water: Structures and Energies of the Hydrated Species $SO_2 \cdot nH_2O$, $[HSO_3]^- \cdot nH_2O$, $[SO_3H]^- \cdot nH_2O$ and $H_2SO_3 \cdot nH_2O$ ($n = 0$–8). *European Journal of Inorganic Chemistry*, 1393–1405.

Steudel R. and Steudel Y. (2009b). Microsolvation of Thiosulfuric Acid ($H_2S_2O_3$) and of its Tautomeric Anions $[HSSO_3]^-$ and $[SSO_2(OH)]^-$ Studied by B3LYP-PCM and G3X (MP2) Calculations. *Journal of Physical Chemistry A*, **113**, 9920–9933 and 11096.

Steudel R. and Steudel Y. (2013). Polysulfide Chemistry in Sodium–Sulfur Batteries and Related Systems: A Computational Study by G3X(MP2) and PCM Calculations. *Chemistry A European Journal*, **19**, 3162–3176.

Steudel R., Paßlack-Stephan S. and Holdt G. (1984). Thermal Polymerization and Depolymerization Reactions of 10 Sulfur Allotropes (including polymeric sulfur), Studied by HPLC and DSC, *Zeitschrift fuer Anorganische und Allgemeine Chemie*, **517**, 7–42.

Steudel R., Holdt G. and Nagorka R. (1986). On the Autoxidation of AqueousSodium Polysulfide. *Zeitschrift für Naturforschung Part B*, **41**, 1519–1522 and references cited therein.

Steudel R., Holdt G., Göbel T. and Hazeu W. (1987). Chromatographic Separation of Higher Polythionates $[S_nO_6]^-$ ($n = 3$–22) and Their Detection in Cultures of *Thiobacillus ferrooxidans*; Molecular Composition of Bacterial Sulfur Secretions. *Angewandte Chemie International Edition*, **26**, 151–153.

Steudel R., Göbel T. and Holdt G. (1988). The Molecular Composition of Hydrophilic Sulfur Sols Prepared by Acid Decomposition of Thiosulfate. *Zeitschrift für Naturforschung Part B*, **44**, 203–218.

Steudel R., Holdt G. and Göbel T. (1989a). Ion-pair chromatographic separation of inorganic sulfur anions including polysulfide. *Journal of Chromatography*, **475**, 442–446.

Steudel R., Göbel T. and Holdt G. (1989b). The Molecular Nature of the Hydrophilic Sulfur Prepared from Aqueous Sulfide and Sulfite (Selmi Sulfur Sol). *Zeitschrift für Naturforschung Part B*, **44**, 526–530.

Steudel R., Albertsen J. and Zink K. (1989c). Photoinduced ESR Signals in Orthorhombic cyclo-Octasulfur (α-S_8) and in Quenched Liquid Sulfur at Low Temperatures. *Berichte der Bunsengesellschaft fuer Physikalische Chemie*, **93**, 502–509.

Steudel R., Steudel Y. and Wong M. W. (2003). Speciation and Thermodynamics of Sulfur Vapor. *Topics in Current Chemistry*, **230**, 117–134.

Strauss R. and Steudel R. (1987). Rapid Chromatographic Separation and Determination of Homocyclic Sulfur Molecules S_n ($n = 6$–28) by HPLC. *Fresenius' Zeitschrift fur Analytische Chemie*, **326**, 543–546.

Szekeres L. (1974). The Analytical Chemistry of Sulfur Acids. *Talanta*, **21**, 1–44.

Szynkarczuk J., Komorowski P. G. and Donini J. C. (1994). Redox Reactions of Hydrosulfide Ions on the Platinum Electrode – I. The Presence of Intermediate Polysulfide Ions and Sulfur Layers. *Electrochimica Acta*, **39**, 2285–2289.

Takakuwa S. (1992). Biochemical Aspects of Microbial Oxidation of Inorganic Sulfur Compounds. In: Organic Sulfur Chemistry: Biochemical Aspects, S. Oae and T. Okuyama (eds), CRC Press, Tokyo, p. 1.

Tardif S. L., Rys A. Z., Abrams C. B., Abu-Yousef I. A., Lesté-Lassere P. B. F., Schultz E. K. V. and Harpp D. N. (1997). Recent Chemistry of the Chalcogen Diatomics. *Tetrahedron*, **36**, 12225–12236.

Tebbe F. N., Wasserman E., Peet W. G., Vatvars A. and Hayman A. C. (1982). Composition of elemental sulfur in solution: equilibrium of S_6, S_7 and S_8 at ambient temperatures. *Journal of the American Chemical Society*, **104**, 4971–4972.

Ueyama N. and Nakamura A. (1992). Sulfur Ligand in Metalloproteins and Metalloenzymes. In: Organic Sulfur Chemistry: Biochemical Aspects, S. Oae and T. Okuyama (eds), CRC Press, Boca Raton, p. 45.

Williamson M. A. and Rimstidt J. D. (1992). Correlation between structure and thermodynamic properties of aqueous sulfur species. *Geochimica et Cosmochimica Acta*, **56**, 3867–3880.

Zhu J., Petit K., Colson A. O., DeBolt S. and Sevilla M. D. (1991). Reactions of $HS^{\bullet}$ and $S^{\bullet -}$ with Molecular Oxygen, H_2S, HS^-, and S^{2-}: Formation of SO^-, $[HSSH]^{\bullet -}$, $[HSS]^{\bullet 2-}$ and $HSS^{\bullet}$. *Journal of Physical Chemistry*, **95**, 3676–3681.

Chapter 3

A biochemical view on the biological sulfur cycle

Christiane Dahl

3.1 INTRODUCTION

Sulfur is the tenth most abundant element on Earth (Steudel & Chivers, 2019) and is cycled dynamically between the geosphere and biosphere. Sulfate or sulfide in water and soil and sulfur dioxide in the atmosphere constitute the majority of sulfur in nature (Middelburg, 2000) with the oceans as a major reservoir. Here, sulfur is present in very large quantities as dissolved sulfate and also as sedimentary minerals like gypsum (Sievert *et al.*, 2007). Among sulfur-containing minerals, pyrite (FeS_2) is especially ubiquitous (Johnson *et al.*, 2012; Muyzer & Stams, 2008).

Sulfur exhibits high reactivity in reduced forms and occurs in several stable oxidation states, ranging from −2 (as in sulfide or reduced organic sulfur) to +6 in sulfate. The most stable form of sulfur in the presence of oxygen is sulfate, while the reduced inorganic forms of sulfur with oxidation states of −2 and zero (as in elemental sulfur) are quite common in anoxic environments. Smaller but significant roles are played by polysulfide, polythionates, thiosulfate, sulfoxides as well as elemental sulfur. Sulfur compounds of mixed valence states like thiosulfate or tetrathionate occur transiently. Organic sulfur compounds like the volatile dimethylsulfide are also important on a global scale. The latter transports sulfur from the oceans to terrestrial regions, and it also affects atmospheric chemistry and the climate system (Charlson *et al.*, 1987; Kettle & Andreae, 2000).

DOI: 10.2166/9781789060966_0055

Sulfur is an essential nutrient for life and abundant in all organisms, where it is present in the amino acids methionine and cysteine, in (poly-)peptides, enzyme cofactors, redox-active co-substrates, such as glutathione, as well as in redox-active enzyme cofactors (iron-sulfur clusters), sulfolipids or carbohydrates.

In contrast, the biological roles of inorganic sulfur compounds are comparatively restricted: either they serve as sources for sulfur assimilation and incorporation into the above mentioned organic compounds, or they are employed as donors or acceptors of electrons for dissimilatory energy-conserving electron transport. The latter pathways establish an electrochemical membrane potential which in turn can be used for adenosine triphosphate (ATP) synthesis, nicotinamide adenine dinucleotide (NAD) reduction or other energy-consuming purposes. Dissimilatory sulfur-based energy conservation goes along with mass transformations and is restricted almost exclusively to prokaryotes, while assimilatory sulfur metabolism is not only very common in prokaryotes, but also occurs in plants, algae and fungi. The turnover of assimilatory pathways is comparatively low and the sulfur compounds are incorporated into organic compounds, mainly – but not exclusively – at the sulfide level.

This review summarizes the current knowledge about the various parts of the biological sulfur cycle. I will focus on new developments arising from studies performed during the 20 years that have passed since Thomas Brüser, Piet Lens and Hans G. Trüper published their chapter about the biological sulfur cycle in the first edition of this book (Brüser *et al.*, 2000). For further details about particular aspects, the reader is referred to more comprehensive reviews on dissimilatory reduction of sulfur compounds (Grein *et al.*, 2013; Muyzer & Stams, 2008; Rabus *et al.*, 2015; Venceslau *et al.*, 2014) and dissimilatory sulfur oxidation (Dahl, 2015, 2017; Frigaard & Dahl, 2009; Kappler, 2008, 2011; Tanabe *et al.*, 2019). More detailed overviews of the biogeochemical sulfur cycle in marine sediments have been published by, e.g. Canfield (2001), Jørgensen and Kasten (2006), Wasmund *et al.* (2017) and Jørgensen *et al.* (2019). Organosulfur compound and assimilatory sulfur metabolism will not be dealt with here and the reader is referred to information provided by others for this part of the sulfur cycle (Cook *et al.*, 2008; Kappler & Schäfer, 2014; Kappler *et al.*, 2019; Leustek *et al.*, 2000; Schäfer *et al.*, 2010).

3.2 IMPORTANT INORGANIC SULFUR COMPOUNDS OF THE BIOLOGICAL SULFUR CYCLE

Sulfide, polysulfides, thiosulfate, polythionates, elemental sulfur, bisulfite and sulfate are the most common inorganic sulfur compounds that occur in natural environments. Sulfur compounds of intermediate oxidation states may serve either as electron acceptors or as electron donors in redox processes. In contrast, sulfide and sulfate cannot be further reduced or oxidized, respectively. Sulfide and sulfate are, therefore, the final products of most sulfur compound reduction or oxidation pathways, respectively. The oxidation states of relevant compounds are listed in Table 3.1.

It is important to recall that free inorganic sulfur compounds are not only produced by biological processes and that abiotic processes and the biological sulfur cycle are interconnected. The abiotic formation of inorganic sulfur substrates can strongly influence the microbial species composition in natural habitats. At this point the various known origins of biologically important inorganic sulfur compounds are briefly summarized.

Hydrogen **sulfide**, the most reduced form of sulfur, is a poisonous, corrosive and foul-smelling gas (Jiang *et al.*, 2016). It is ubiquitous in natural aqueous environments and is often accompanied by **polysulfides.** Hydrogen sulfide occurs naturally in large amounts in underground deposits of "sour" natural gas. In addition, it is found in minerals in soils and rocks or in sulfide-enriched springs and is produced as the end-point of microbial sulfate/sulfur reduction. It is highly soluble in water, where it can dissociate into hydrogen sulfide (HS^-). Except at the highest pH values, the concentration of sulfide (S^{2-}) ions is negligible (Philips & Philips, 2000). Chemical oxidation of hydrogen sulfide ions results in disulfide ions which may be further oxidized to higher inorganic polysulfides and eventually to elemental sulfur (Dahl *et al.*, 2002; Steudel, 1996a). Sulfide is also released by desulfuration of organic compounds that contain reduced sulfur (e.g. proteins). **Sulfide** is formed as the main product of sulfate respiration by bacteria that dominate under anaerobic conditions when organic matter and sulfate are not limiting. Sulfur compounds other than sulfate (e.g. thiosulfate and elemental sulfur) can also serve as electron acceptors for respirations that lead to sulfide (Rabus *et al.*, 2015).

Elemental sulfur naturally occurs in a huge variety of environments, such as volcanic areas including sulfidic springs, deep-sea hydrothermal vents, and hydrocarbon seeps, salt marshes or marine sediments (Cosmidis *et al.*, 2019; Jørgensen & Nelson, 2004; Kamyshny & Ferdelman, 2010; Zopfi *et al.*, 2004). Elemental sulfur can be formed through abiogenic processes when sulfide is

Table 3.1 Inorganic sulfur compounds of biological relevance.

Compound	Chemical Formula	Sulfur Oxidation State
Sulfide	H_2S, HS^-	−2
Polysulfides	$^-S\text{-}S_n\text{-}S^-$	−1 (terminal S)/0 (inner S)
Thiosulfate	$S_2O_3^{2-}$	−1 (sulfane S)/+5 (sulfone S)
Polythionates	$^-{}_3OS\text{-}S_n\text{-}SO_3^-$	0 (inner S)/+5 (sulfone S)
Elemental sulfur	S_n rings or chains	0
Sulfite	HSO_3^-, SO_3^{2-}	+4
Sulfate	SO_4^{2-}	+6

oxidized by molecular oxygen, possibly catalyzed by oxidized metals (Luther *et al.*, 2011). At standard conditions, homocyclic, orthorhombic crystalline α-sulfur (α -S_8) (cyclo-octasulfur) is the thermodynamically most stable form (Steudel, 1996a, b). Commercially-available sulfur consists mainly of S_8 rings and polymeric sulfur. Polymeric sulfur consists of very long helically wound chains of almost all sizes (Steudel, 2000). The bright yellow color of elemental sulfur is due to traces of S_7 rings. Besides S_8, biological samples may also contain S_6, S_7 and S_{12} rings, while bigger rings of up to S_{20} are accessible through chemical synthesis (Steudel, 2000). Regardless of the molecular size, all sulfur allotropes are hydrophobic, are not wetted by water and have a very low solubility in water. In the environment, the presence of elemental sulfur is often associated with microbial oxidation of reduced sulfur compounds (Kleinjan *et al.*, 2003). Elemental sulfur can be formed by sulfide-oxidizing bacteria (Frigaard & Dahl, 2009; Wasmund *et al.*, 2017). In some species, elemental sulfur is the final oxidation product and can be further metabolized by other bacteria. In others, for example in many anoxygenic phototrophic bacteria, intra- or extracellular sulfur globules are formed as transient intermediates during sulfur compound oxidation to sulfate (Dahl, 2017; Frigaard & Dahl, 2009).

Thiosulfate is mainly formed as a product of sulfide oxidation, but can also be formed by the spontaneous non-biological reaction of sulfite with sulfur or polythionates. A further origin of thiosulfate is the biological leaching process where thiosulfate is formed by oxidation of sulfidic minerals like FeS_2, MoS_2 or WS_2 (Vera *et al.*, 2013). Release of thiosulfate as a product of taurine fermentation has also been observed (Cook *et al.*, 2008). Thiosulfate may furthermore arise as an intermediate or end product of dimethyl sulfide oxidation (de Zwart *et al.*, 1997; Koch & Dahl, 2018). Thiosulfate is not only used as an effective electron acceptor in anaerobic respiratory processes (Burns & DiChristina, 2009), but also widely used as an electron donor by sulfur-oxidizing bacteria, because it is chemically stable at neutral to alkaline conditions and non-toxic.

Tetrathionate also can be a natural substrate of sulfur-oxidizing bacteria. It originates from incomplete oxidation processes and can be produced from thiosulfate by a number of bacteria in one oxidation step (Denkmann *et al.*, 2012). Tetrathionate can furthermore serve as a terminal electron acceptor, a reaction with special relevance for *Salmonella enterica* serotype Typhimurium in the human gut (Winter *et al.*, 2010).

Sulfite is highly reactive. Its nucleophility and strong reducing capacity contribute to its toxicity and antimicrobial action culminating in its widespread use as food preservative and disinfectant (Kappler & Dahl, 2001). In nature it occurs as a consequence of geological and industrial processes and of the anaerobic mineralization of organic matter by dissimilatory sulfate reduction. Various prokaryotes unable to respire sulfate are able to grow at the expense of sulfite reduction by employing a respiratory and presumably also detoxifying process (Simon & Kroneck, 2013). On the other hand, many bacteria also grow by oxidizing sulfite and some of them seem to be specialized in sulfite oxidation, e.g. *Sulfitobacter* species (Sorokin, 1995). Sulfite is an intermediate in most sulfur compound redox pathways and in some cases is released into the medium to some extent (Neutzling *et al.*, 1985). Such a "leakage" could make sulfite available to other bacteria.

Sulfate is formed under aerobic conditions from most sulfur compounds which are sensitive to oxidation processes. Under oxic as well as under anoxic conditions sulfate is produced by chemotrophic (aerobic and anaerobic) and phototrophic (anaerobic) sulfur compound oxidizing bacteria. Huge amounts of sulfate are dissolved in the oceans (~28 mM) or in minerals like gypsum. Most freshwaters contain about 50 times less sulfate. Soil water concentrations vary depending on climate conditions and mineral composition.

3.3 THE BIOLOGICAL SULFUR CYCLE

Biochemical oxidation and reduction of sulfur compounds constitute the biological sulfur cycle (Figure 3.1) that is not only ongoing today but also had a profound impact on Earth evolution (Colman *et al.*, 2019). Sulfur cycling is tightly interwoven with other important element cycles (carbon, nitrogen, iron and manganese) and has therefore profound implications for ecosystem level processes.

Organisms participate in the sulfur cycle in two fundamentally different ways: the **assimilation** of sulfur compounds serves for the biosynthesis of sulfur-containing cell constituents, while during **dissimilation** sulfur compounds serve as electron donors or acceptors for energy-conserving processes. The vast part of sulfate assimilation is contributed by phototrophic organisms, i.e. plants, algae and cyanobacteria. On a global scale, land vegetation and oceans contribute approximately equal parts to this process (Giordano *et al.*, 2008). Sulfate uptake, reduction and incorporation of sulfide into biomolecules are all highly regulated steps and induced only in the absence of external reduced sulfur compounds.

Under anoxic conditions in the absence of nitrate, sulfate and elemental sulfur are used as electron acceptors of anaerobic respiratory processes. In fact, the dissimilatory reduction of sulfate is the primary driver of biogeochemical sulfur cycling (Rabus *et al.*, 2015; Wasmund *et al.*, 2017). In addition, some sulfur compound respirers can use sulfite, thiosulfate, organic sulfoxides, inorganic polysulfides and/or organic disulfanes as terminal electron acceptors (Hedderich *et al.*, 1999; Simon & Kroneck, 2013). All these processes lead to the release of large amounts of sulfide, which may be chemically oxidized to thiosulfate and polythionates at the oxic/anoxic interface.

In turn, hydrogen sulfide serves as an electron donor for a huge array of chemo- and photolithotrophic bacteria and archaea such as *Acidithiobacillus* or *Acidianus* species (Dahl, 2017; Frigaard & Dahl, 2009; Kletzin *et al.*, 2004; Mangold *et al.*, 2011; Wang *et al.*, 2019). Reduced sulfur compound oxidation can occur aerobically as well as anaerobically, with either oxygen, nitrate or manganese

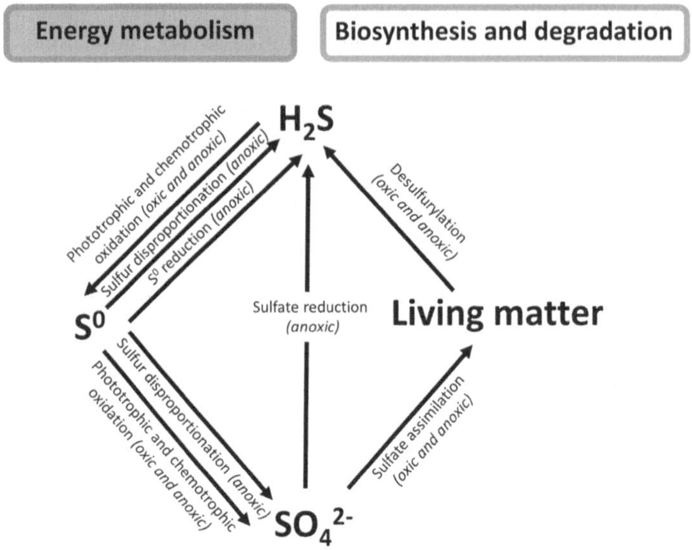

Figure 3.1 Simple view of the biological sulfur cycle. Sulfur, an essential element for life is taken up by sulfate by prokaryotes, eukaryotic microorganisms, plants and in reduced form by animals. Decomposition of dead organic matter releases the sulfur in reduced form, as sulfide. Sulfate-reducing prokaryotes (e.g. *Desulfovibrio* spp.) use sulfate as respiratory electron acceptor and produce large amounts of sulfide which can in turn be oxidized by anaerobic phototrophic sulfur oxidizers (e.g. *Allochromatium* or *Chlorobium* spp.) or by chemolithotrophs (e.g. *Acidithiobacillus* or *Beggiatoa* spp.) under oxic or anoxic conditions. More specialized groups can reduce (e.g. *Desulfuromonas* spp.) or disproportionate (e.g. *Desulfovibrio sulfodismutans*) elemental sulfur.

[Mn(IV)] acting as electron acceptors, or in anoxygenic, anaerobic photosynthesis (Dahl *et al.*, 2008; Henkel *et al.*, 2019). The microbial oxidation of sulfide establishes a complex network of pathways in the sulfur cycle, leading to intermediate sulfur species, and partly back to sulfate. The intermediates include elemental sulfur, polysulfides, thiosulfate, and sulfite, which are all substrates for further microbial oxidation, reduction or disproportionation. New microbiological discoveries, such as long-distance electron transfer through sulfide oxidizing cable bacteria (Jørgensen *et al.*, 2019), add to the complexity.

It should be emphasized here, that the notion that the two major branches of the sulfur cycle, i.e. sulfide oxidation and sulfate reduction, are necessarily performed by distinct sets of organisms has recently been severely shaken, when the Deltaproteobacterium *Desulfurivibrio alkaliphilus* was shown to grow by sulfide oxidation with nitrate (Thorup *et al.*, 2017). The organism has the genetic equipment of a sulfate reducer but still it is unable to grow by sulfate reduction.

Rather it oxidizes sulfide to sulfur which is then oxidized by a reversal of the sulfate reduction pathway or disproportionated. Notably, *D. alkaliphilus* is closely related to cable bacteria and may be suited as a blueprint for understanding the physiology of these organisms, for which pure cultures do not yet exist.

The disproportionation of inorganic sulfur compounds is another important process with considerable impact on the biogeochemical sulfur cycle. It is a microbiologically catalyzed chemolithotrophic process in which sulfur compounds of intermediary redox state, like elemental sulfur, thiosulfate, and sulfite, serve as both electron donor and acceptor resulting in the formation of hydrogen sulfide and sulfate as end products (Finster, 2008). Strong evidence has been provided that disproportionating bacteria participate in and enhance the rate of pyrite formation and that they are partly responsible for the isotopic signatures of sulfidic minerals in recent and old sediments. Furthermore, the activity of elemental sulfur disproportionation can be traced back in time as long as 3.5 billion (10^9) years and would thus be one of the oldest biological processes on Earth (Finster, 2008).

One of the most intriguing recent discoveries in sulfur biogeochemistry is the ubiquity of cryptic sulfur cycling (Canfield *et al.*, 2010). Cryptic sulfur cycling means rapid recycling of sulfur species, i.e. simultaneous consumption and production of sulfate such that the process does not leave an imprint on the sulfur budget of the ambient environment. Such a cycle can still have massive impact on other intertwined element cycles. In fact, there is growing evidence that sulfur cycling also occurs outside of classical sulfur-compound dominated environments like marine sediments or geologically active zones in environments that were traditionally viewed as largely sulfate-free, fermentative and dominated by methanogenesis (Wasmund *et al.*, 2017). For example, cryptic sulfur cycling is inferred to occur in sediments from Aarhus Bay (Baltic Sea) (Holmkvist *et al.*, 2011) and Black Sea sediments (Mikucki *et al.*, 2009) that were considered to be devoid of active sulfate reduction. Cryptic sulfur cycling has furthermore been described for subglacial lakes where sulfate-sulfur is apparently reduced and re-oxidized back via coupling to reductive iron cycling (Mikucki *et al.*, 2009), for sulfide-poor groundwater of the deep terrestrial subsurface (Bell *et al.*, 2020) as well as for marine oxygen minimum zones (Canfield *et al.*, 2010; Johnston *et al.*, 2014) where extensive sulfur cycles are presumably tied to reductive nitrogen cycling.

3.4 DISSIMILATORY REDUCTION OF OXIDIZED SULFUR COMPOUNDS

3.4.1 Dissimilatory reduction of sulfate

Sulfate-reducing prokaryotes are widespread in anoxic habitats where they use sulfate as a terminal electron acceptor mainly for the degradation of organic compounds. The process results in the production of sulfide. The role of sulfate reduction should not be underestimated as it accounts for more than 50% of organic carbon mineralization in marine sediments (Jørgensen, 1982). Sulfate reducers can grow on a large range of substrates including hydrogen, organic compounds like ethanol, malate, short and long-chain fatty acids, aromatic compounds, sugars, amino acids, alkanes, alkenes and one-carbon compounds such as carbon monoxide or methanethiol (Muyzer & Stams, 2008; Rabus *et al.*, 2015). In addition, the anaerobic oxidation of methane can be coupled to sulfate reduction (Boetius *et al.*, 2000; Nauhaus *et al.*, 2007; Orphan *et al.*, 2001). This process is performed by consortia of sulfate-reducing bacteria that thrive on intermediates produced by methanotrophic archaea performing reverse methanogenesis (Bhattarai *et al.*, 2019; Milucka *et al.*, 2013).

The sulfate-reducing prokaryotes are a very heterogeneous group. Many ecologically relevant sulfate reducers belong to the class Deltaproteobacteria (e.g. the genera *Desulfovibrio*, *Desulfobulbus* or *Desulfonema*) with members of the family Desulfobacteraceae often being especially active and abundant. Sulfate reducers are furthermore found among the Gram-positives and the classes *Thermodesulfobacteria* and *Nitrospira* and also include archaea of the genus *Archaeoglobus* (Rabus *et al.*, 2015). In addition, recent molecular ecology surveys based on metagenomics revealed further unexpected diversity of sulfate reducers and localized the trait in as many as 13 newly identified groups including 8 candidate phyla (Anantharaman *et al.*, 2018).

All sulfate reducers described so far use the same pathway for sulfate respiration (Figure 3.2). All involved enzymes reside in the cytoplasm into which sulfate is transported in an active process (Kreke & Cypionka, 1995).

The sulfate anion is very stable and consequently its reduction to bisulfite (HSO_3^-) has a very low redox potential ($E^{0\prime} = -516$ mV (Thauer *et al.*, 1977). Therefore, sulfate is first converted to a mixed anhydride which can be more easily reduced. This activation of sulfate with ATP to form adenosine 5′-phosphosulfate (APS) and pyrophosphate is catalyzed by ATP sulfurylase (Sat, sulfate adenylyltransferase). The reaction is an endergonic process ($\Delta G^{0\prime} = +46$ kJ mol^{-1}) and requires the hydrolysis of the second reaction product pyrophosphate ($\Delta G^{0\prime} = -33$ kJ mol^{-1}) and the subsequent reduction of APS ($\Delta G^{0\prime} = -60$ kJ mol^{-1}) for high efficiency. ATP sulfurylases from sulfate reducers are usually homo-oligomeric enzymes (Gavel *et al.*, 1998; Parey *et al.*, 2013).

The APS/HSO_3^- pair has a redox potential $E^{0\prime}$ of −60 mV and this two-electron reduction is catalyzed by APS reductase, AprBA. AprA is a flavoprotein binding an flavin adenine dinucleotide (FAD) cofactor and AprB binds two [4Fe-4S] clusters. The physiological electron donor for AprBA is the membrane-bound QmoABC complex (Pires *et al.*, 2003), thus coupling the oxidation of menaquinol to the reduction of APS and allowing chemiosmotic energy conservation (Duarte *et al.*, 2016; Ramos *et al.*, 2012).

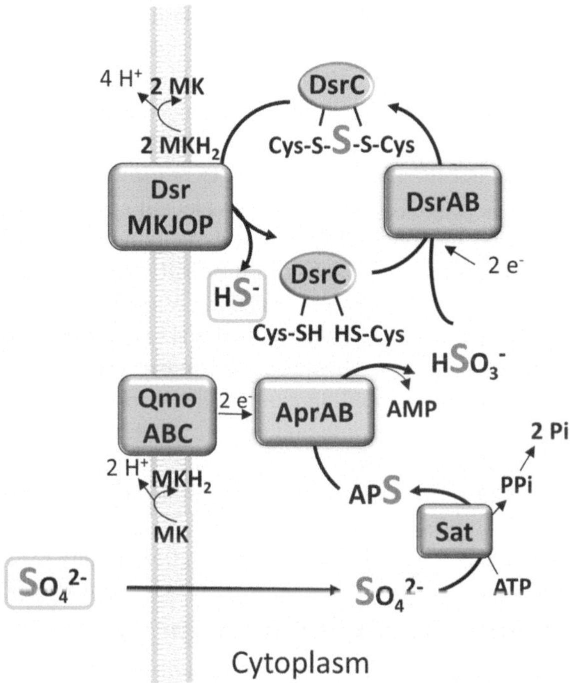

Figure 3.2 Pathway of dissimilatory sulfate reduction. The pathway involves import of sulfate, activation of sulfate to adenosine-5′-phosphosulfate (APS) at the expense of ATP catalyzed by ATP sulfurylase (sulfate adenylyl transferase, Sat), followed by reduction of APS to sulfite and release of AMP. This reaction is catalyzed by APS reductase (AprBA) and electrons are provided from menaquinol (MKH_2) via the QmoABC complex. Sulfite is reduced to a DsrC trisulfide by sulfite reductase (DsrAB). The origin of the electrons for this reduction is unclear for most sulfate reducers. In the last step the DsrC trisulfide is reduced to sulfide and the DsrC protein is regenerated. Here, the electrons most probably also stem from menaquinol and the reaction is catalyzed by the DsrMKJOP complex.

The reduction of sulfite is carried out by dissimilatory sulfite reductase. This enzyme is composed of two related subunits DsrA and DsrB in a $\alpha_2\beta_2$ arrangement. Both subunits carry an iron-tetrahydrophorphyrin (siroheme) that is coupled to a [4Fe-4S] cluster through its cysteine heme axial ligand. The siroheme-[4Fe-4S] sites in the DsrB subunits are catalytically active, while they appear to have a merely structural role in DsrA (Oliveira *et al.*, 2008; Santos *et al.*, 2015; Schiffer *et al.*, 2008). Both DsrA and DsrB also include a ferredoxin domain that binds a second [4Fe-4S] cluster that is most likely responsible for electron transfer between the physiological electron donor and the catalytic site. During sulfite reduction, the protein DsrC works as the physiological partner of

DsrAB. A DsrC trisulfide, in which a sulfur atom is bridging two conserved cysteine residues, is released as the product of the DsrAB-catalyzed reaction (Santos *et al.*, 2015). As a consequence, DsrAB catalyzed (bi)sulfite reduction requires only two electrons, with the other two originating from reduced DsrC. The origin of these two electrons is still a major obstacle (Rabus *et al.*, 2015; Santos *et al.*, 2015). A number, albeit by far not the majority, of sulfate reducers encode the iron-sulfur flavoprotein DsrL, an NAD(P)H-reactive enzyme with the capacity to provide electrons from reduced nicotinamide adenine dinucleotides for DsrAB-catalyzed reduction of sulfite (Löffler *et al.*, 2020). In the final step of dissimilatory sulfate reduction, the transmembrane complex DsrMK(JOP) probably serves in electron transfer and reductive release of sulfide from the DsrC protein trisulfide (Grein *et al.*, 2013; Santos *et al.*, 2015). The electrons needed most likely originate from the menaquinone pool, thus enabling coupling of ΔpH generation across the membrane directly to sulfite reduction and energy conservation.

In addition to the reduction of sulfate, most sulfate reducers can use many other electron acceptors like other sulfur compounds (thiosulfate, sulfite and sulfur), nitrate and nitrite, fumarate or dimethylsulfoxide for growth and can also grow through fermentative reactions (Muyzer & Stams, 2008; Rabus *et al.*, 2015).

3.4.2 Dissimilatory reduction of sulfur cycle intermediates

Sulfur compounds of intermediate redox state, such as sulfite, elemental sulfur, polysulfide, or polythionates, serve as respiratory electron acceptors for energy conservation in an ever increasing range of microorganisms (Wasmund *et al.*, 2017). On the other hand, the same compounds may be oxidized as electron donors or serve as substrates for disproportionation (see below). We should be aware that sulfur compounds of intermediate redox state usually do not accumulate to high concentrations due to rapid turnover by microorganisms and also by chemical reactions (Jørgensen, 1990b; Wasmund *et al.*, 2017; Zopfi *et al.*, 2004). Sulfite for example is especially prone to reactions with elemental sulfur yielding thiosulfate.

3.4.2.1 Dissimilatory reduction of sulfite

As evident from section 3.4.1, sulfite is an intermediate in dissimilatory sulfate reduction. It therefore does not appear astonishing that many sulfate reducing prokaryotes can also use externally supplied sulfite. In addition, various organisms unable to respire sulfate can still grow at the expense of sulfite reduction by employing a respiratory process. Espilonproteobacteria like several *Campylobacter* species, *Sulfurospirillum deleyianum* and *Wolinella succinogenes* (Kern *et al.*, 2011) belong to this group that also comprises Firmicutes (e.g. *Desulfitobacterium*) and Gammaproteobacteria like species of the genus *Shewanella* (Shirodkar *et al.*, 2011; Simon & Kroneck, 2013). These bacteria catalyze sulfite respiration with periplasmic multiheme cytochrome *c* instead of the cytoplasmically located siroheme-containing DsrAB sulfite reductases (Kern *et al.*, 2011; Shirodkar *et al.*, 2011).

Sulfite respiration has also been described for *Salmonella typhimurium* and in this case a siroheme-containing cytoplasmic enzyme named AsrABC is used as the catalyst (Huang & Barrett, 1990, 1991). AsrA and AsrC are ferredoxin-like subunits while AsrB may be a flavoprotein. Binding of siroheme to AsrC was also proposed and it is possible that the holoenzyme catalyzes reduction of sulfite with nicotinamide adenine dinucleotide hydride (NADH) in the cytoplasm.

Another type of sulfite reductase (Fsr, coenzyme F_{420}-dependent sulfite reductase) is present in *Methanocaldococcus jannaschii* and other hydrogenotrophic methanogenic archaea. This enzyme again contains siroheme, iron-sulfur clusters and an FAD cofactor that is reduced by coenzyme F_{420}. Sulfite reduction as catalyzed by this enzyme is a detoxification process and additionally allows the use of sulfite as a sulfur source for biosynthetic purposes (Johnson & Mukhopadhyay, 2005).

3.4.2.2 Dissimilatory reduction of thiosulfate

The use of thiosulfate and sulfite as electron acceptors is more widespread than dissimilatory sulfate reduction (Simon & Kroneck, 2013). In fact, most sulfate-reducing prokaryotes cannot only use sulfite but also thiosulfate as an alternative electron acceptor (Rabus *et al.*, 2015). In sediments, thiosulfate reduction is primarily carried out by sulfate-reducing bacteria and it has been reported that they prefer thiosulfate over sulfate as an electron donor (Jørgensen, 1990b).

The ability to respire thiosulfate is conferred by the enzyme thiosulfate reductase which catalyzes the reaction $S_2O_3^{2-} + 2\ e^- + 2\ H^+ \rightarrow HS^- + HSO_3^-$. Thiosulfate reductase activity is not restricted to sulfate-reducing bacteria but can be found in other types of environmentally abundant organisms such as *Shewanella* species (Burns & DiChristina, 2009). Thiosulfate reduction is also prevalent in the human gut, where mitochondria in the colonic mucosa detoxify sulfide produced by intestinal bacteria to thiosulfate (Goubern *et al.*, 2007; Libiad *et al.*, 2014). The thiosulfate is used as a respiratory substrate by many colonic bacteria such as species of the genera *Salmonella*, *Proteus*, *Citrobacter* and *Edwardsiella* (Barrett & Clark, 1987).

In *Salmonella typhimurium*, thiosulfate respiration is catalyzed by the membrane-bound PhsABC complex. The thiosulfate reduction site is located in the periplasm (Stoffels *et al.*, 2012). The *S. typhimurium* enzyme consists of three subunits, PhsABC. The PhsA subunit is a peripheral membrane protein oriented towards the periplasm containing an active site bis(molybdopterin guanine dinucleotide) molybdenum cofactor. PhsC is an integral membrane protein containing two hemes *b*. PhsB is predicted to possess four iron-sulfur centers that transfer electrons between PhsC and PhsA. The naphtoquinone menaquinone is the sole electron donor to thiosulfate reductase. The reduction of thiosulfate by menaquinol is highly endergonic and consequently, thiosulfate reductase was found to depend on the proton motive force across the cytoplasmic membrane (Stoffels *et al.*, 2012).

3.4.2.3 Dissimilatory reduction of tetrathionate

Tetrathionate has long been known to be used by some bacteria as an electron acceptor under anaerobic conditions (Barrett & Clark, 1987). In a single step requiring input of two electrons, tetrathionate is reduced to two molecules of thiosulfate ($E^{0\prime} = +198$ mV) (Kurth *et al.*, 2015). Tetrathionate reduction is prevalent in the gut, where tetrathionate is used as a respiratory substrate by a number of species (Oltman *et al.*, 1979; Winter *et al.*, 2010). The high relevance of tetrathionate as *in vivo* electron acceptor for bacterial pathogenesis is emphasized by the finding that *Salmonella typhimurium* induces host-driven formation of tetrathionate from thiosulfate by reactive oxygen species produced during inflammation. This tetrathionate is then used as a terminal electron acceptor providing the pathogen with a growth advantage over the commensal microbiota lacking this capacity (Winter *et al.*, 2010). The food-borne intestinal pathogen *Campylobacter jejuni* may follow a similar strategy (Kurth *et al.*, 2016a; Liu *et al.*, 2013). Tetrathionate reduction has also been detected in *Wolinella succinogenes*, a close relative of *C. jejuni* (Kurth *et al.*, 2017). Tetrathionate respiration has also been reported for environmentally abundant organisms such as *Shewanella* species.

So far, three distinct types of tetrathionate reductases have been described, the iron-sulfur molybdoenzyme TtrABC found in *S. typhimurium* (Hensel *et al.*, 1999), the octaheme Otr enzyme from *Shewanella oneidensis* (Mowat *et al.*, 2004), and the diheme cytochrome *c* TsdA from *C. jejuni* (Kurth *et al.*, 2016a; Liu *et al.*, 2013). It should be noted that Otr was later structurally characterized and shown to reduce nitrite and hydroxylamine to ammonia (Atkinson *et al.*, 2007).

3.4.2.4 Dissimilatory reduction of sulfur and polysulfides

Microbial reduction of elemental sulfur to hydrogen sulfide has been known for many years and is widespread among bacteria and archaea. Many of these organisms are hyperthermophilic (Hedderich *et al.*, 1999). In addition to some sulfate-reducers, numerous bacteria incapable of sulfate reduction can use sulfur as a respiratory electron donor, e.g. species of the genera *Desulfurella, Desulfuromonas, Geobacter, Shewanella* or *Thermotoga* (Moser & Nealson, 1996). Due to its low solubility in water, elemental sulfur itself is not well suited as a substrate of bacterial respiration. Elemental sulfur is readily converted to polysulfide in aqueous solutions of sulfide, a product of sulfur respiration, and polysulfides are intermediates of sulfur respiration in most of the known sulfur reducers. In fact, the actual electron acceptor of sulfur respiration in the best studied model organism, the Epsilonproteobacterium *W. succinogenes*, is polysulfide (Hedderich *et al.*, 1999; Klimmek *et al.*, 2004). Polysulfide reduction is catalyzed by PsrABC (Krafft *et al.*, 1995), a membrane-bound iron-sulfur molybdoprotein resembling PhsABC thiosulfate reductase and TtrABC tetrathionate reductase. In *W. succinogenes,* polysulfide reduction is facilitated by a periplasmic sulfur-binding protein (Sud) (Klimmek *et al.*, 1998).

The situation in the hyperthermophilic archaeon *Pyrococcus furiosus* appears to be complex. The originally described cytoplasmic coenzyme A-dependent nicotinamide dinucleotide phosphate hydride (NADPH) sulfur oxidoreductase later turned out not to be essential for sulfur reduction (Bridger *et al.*, 2011). Probably, two sulfide dehydrogenase isoenzymes provide a compensatory NADPH-dependent sulfur reduction system (Nsr). Apart from these, a membrane-bound oxidoreductase complex is indispensable for growth on sulfur (Bridger *et al.*, 2011), that is proposed to catalyze ferredoxin-dependent reduction of NADP and thereby provide reducing equivalents for sulfur reduction. The substrate for Nsr has been proposed to be colloidal sulfur, i.e. short S^0 chains and S_8.

It should be emphasized that recently the application of metagenomics to marine sediments has revealed potential for sulfur and possibly also thiosulfate reduction in taxa completely undescribed and not previously connected with the sulfur cycle, such as the archaeal phylum Thorarchaeota (Lazar *et al.*, 2017; Seitz *et al.*, 2016).

3.5 DISSIMILATORY OXIDATION OF REDUCED SULFUR COMPOUNDS

Concerning their physiology, two groups of lithoautotrophic sulfur-oxidizing bacteria can be differentiated: the first are phototrophic prokaryotes that use sulfur compounds as electron donors for CO_2 fixation in the light (Dahl, 2017; Frigaard & Dahl, 2009). The second are chemotrophic sulfur-oxidizing prokaryotes that use the energy derived from the oxidation of sulfur compounds with either oxygen, nitrate or Mn(IV) oxide as respiratory electron acceptors to fix carbon dioxide (Dahl *et al.*, 2008; Henkel *et al.*, 2019). It should be noted that sulfur compounds are also an important energy source for symbiotic associations of chemoautotrophic sulfur bacteria with marine organisms including unicellular protists, meduzoans, bivalves and nematodes (Abouna *et al.*, 2015; Frenkiel *et al.*, 1996; Himmel *et al.*, 2009; Ott *et al.*, 2004). Although first discovered at hydrothermal vents, symbiosis with sulfur oxidizers is by far not limited to these highly specialized environments (Dubilier *et al.*, 2008; Petersen *et al.*, 2016). In addition, a considerable number of heterotrophs are able to use sulfur compounds as a supplemental electron source during organoheterotrophic growth (Koch & Dahl, 2018).

Sulfur-oxidizing prokaryotes are metabolically and phylogenetically extremely diverse and occur in a wide range of habitats covering the complete pH range from acidic to extremely alkaline environments, as well as a huge temperature range from hot geothermal aquatic over temperate to low-temperature marine Arctic sediments (Dahl *et al.*, 2008). Inorganic sulfur compound-metabolizing archaea, typically of the order *Sulfolobales* within the *Crenarchaeota*, are largely found in volcanic environments due to the abundance of the substrate(s) in these environments. Among the Bacteria, the ability to oxidize reduced sulfur compounds occurs in many different phyla including the Proteobacteria (e.g. the genera *Beggiatoa*, *Thiobacillus*, *Thiomicrospira*, *Thioploca*, *Thiothrix*, *Thiovulum*, *Arcobacter*, *Sulfurovum* or *Sulfurimonas*), Chlorobi, Aquificae, and Chloroflexi (Dahl, 2017; Wasmund *et al.*, 2017). Over recent years, metagenome studies have indicated even greater diversity via the detection of genes related to sulfur oxidation not only in Proteobacteria that had before not been associated with sulfur oxidation (Lenk *et al.*, 2011, 2012; Pham *et al.*, 2008; Thomas *et al.*, 2014), but also in new phylogenetic lineages, e.g. the candidate phylum Rokubacteria (Anantharaman *et al.*, 2016, 2018). Furthermore, evidence is accumulating that sulfur-oxidizing communities in a given habitat are not necessarily dominated by classical photo- or chemolithoautotrophic sulfur-oxidizing specialists. Rather, sulfur-oxidizing capabilities were assigned to organisms grouped into families known for metabolic versatility, e.g. the alphaproteobacterial Rhodobacteraceae, Rhodocyclaceae or Rhizobicaceae (Bell *et al.*, 2020; Lenk *et al.*, 2012).

A universal mechanism for the oxidation of reduced sulfur compounds does not exist. Rather, various modules for sulfur compound oxidation can occur alone or in different combinations enabling complete oxidation of the reduced sulfur substrate to sulfate and even providing redundancy of pathways.

Many sulfur-oxidizing prokaryotes thrive in environments that are characterized by steep and opposing gradients of the sulfur compounds they use as electron donors and their respiratory electron acceptor, commonly oxygen. Different strategies have been developed to overcome the spatial separation of oxygen and sulfide including internal storage of nitrate as an alternative electron acceptor and elemental sulfur, an intermediate of sulfide and thiosulfate oxidation. This strategy is for example applied by large sulfur bacteria of the family Beggiatoaceae and in several species, e.g. *Thioploca*, supported by the ability to move towards the limiting substrate by gliding motility.

The most intriguing strategy of adaptation to gradients of electron donors and acceptors is probably implemented in the so-called cable bacteria. Cable bacteria are filamentous multicellular organisms of the deltaproteobacterial family Desulfobulbaceae (Nielsen *et al.*, 2010; Pfeffer *et al.*, 2012) found worldwide in marine and freshwater sediments as well as in groundwater (Burdorf *et al.*, 2017; Müller *et al.*, 2016, 2019; Risgaard-Petersen *et al.*, 2015). To date, the Desulfobulbaceae have been known to contain sulfate-reducing or sulfur-disporortionating species but not canonical sulfide oxidizers. Cable bacteria appear in redox gradients and are electrically conductive: they transfer electrons from sulfide oxidation at one end over centimeter distances to oxygen or nitrate reduction at the other end (Bjerg *et al.*, 2018; Marzocchi *et al.*, 2014).

The mechanism for the necessary long-distance electron transfer that couples oxidative and reductive processes at the far ends of the filaments has also not yet been finally clarified, but it appears to involve periplasmic cytochromes and yet-unidentified conductive periplasmic fibers (Meysman *et al.*, 2019). Recent genome sequencing revealed that cable bacteria likely oxidize sulfide by a reversal of the canonical sulfate reduction pathway (Kjeldsen *et al.*, 2019). In that respect, they resemble their close relative *Desulfurivibrio alkaliphilus* which grows by nitrate-dependent sulfide oxidation and sulfur disproportionation (Thorup *et al.*, 2017). A two-step process is implied where sulfide is initially oxidized to sulfur and then either disproportionated or oxidized to sulfate (Kjeldsen *et al.*, 2019). A recent study on groundwater cable bacteria revealed that cable bacteria not only use sulfide oxidation coupled to oxygen or nitrate reduction for energy conservation but that sulfur disproportionation might constitute the energy metabolism for cells in large parts of the cable bacterial filaments (Müller *et al.*, 2019).

3.5.1 Oxidation of thiosulfate

Thiosulfate oxidation is conducted by a very large number of photo- and chemotrophic bacteria. Two pathways, both of which proceed outside of the cytoplasm (if present, in the periplasm), can be differentiated. In the first, two thiosulfate anions are oxidized to tetrathionate. In the second, multiple steps lead to the end product sulfate. In a number of bacteria both pathways coexist (Hensen *et al.*, 2006; Koch & Dahl, 2018).

3.5.1.1 Oxidation of thiosulfate to tetrathionate

The very simple oxidation of two molecules of thiosulfate to tetrathionate according to the equation $2\ S_2O_3^{2-} \rightarrow S_4O_6^{2-} + 2\ e^-$ is widespread among prokaryotes. It either can occur as an intermediate step in the oxidation of reduced sulfur compounds to sulfate (Dam *et al.*, 2007; Ghosh & Dam, 2009; Lu & Kelly, 1988; Wang *et al.*, 2019; Wentzien *et al.*, 1994) or lead to the formation of tetrathionate as a dead end product, as for example described for phototrophic bacteria (Hensen *et al.*, 2006) or the Alphaproteobacterium *Hyphomicrobium denitrificans* (Koch & Dahl, 2018). Two different enzymes catalyzing the reaction have been described in some detail (Figure 3.3).

The first is the membrane-bound *doxDA*-encoded thiosulfate:quinone oxidoreductase (TQO) which was first characterized from the thermoacidophilic archaeon *Acidianus ambivalens* (Müller *et al.*, 2004). The enzyme transfers electrons into the respiratory chain via quinones. Genes encoding TQO have been identified in several bacterial genera, e.g. *Acidithiobacillus* (Wang *et al.*, 2019). Nothing is known about the catalytic mechanism of the enzyme.

The second enzyme catalyzing tetrathionate formation from thiosulfate is TsdA, a diheme cytochrome *c* first characterized from *A. vinosum* (Brito *et al.*, 2015;

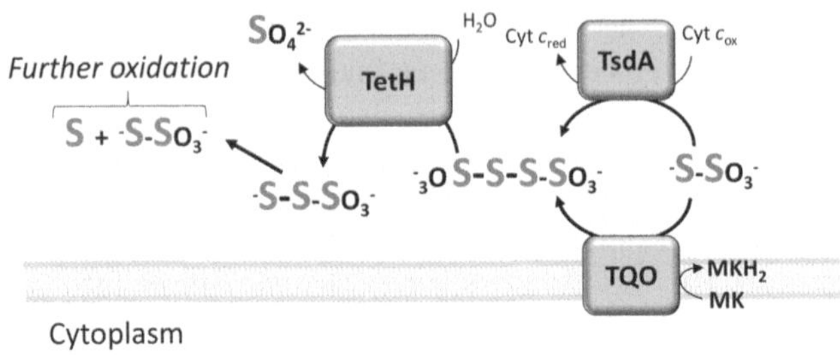

Figure 3.3 Tetrathionate formation and hydrolysis in sulfur oxidizing prokaryotes. TsdA, thiosulfate dehydrogenase; TQO, thiosulfate:quinone oxidoreductase; TetH, tetrathionate hydrolase. Decomposition of disulfanemonosulfonic acid is proposed to occur spontaneously.

Denkmann *et al.*, 2012). The crystal structure of the enzyme revealed His/Cys iron coordination for Heme 1, the active site of the enzyme (Brito *et al.*, 2015). The reaction mechanism of TsdAs includes a thiosulfate covalently bound to the active site cysteine (Grabarczyk *et al.*, 2015b; Kurth *et al.*, 2016b). The diheme cytochrome *c* TsdB, which may be fused to TsdA, has been identified as the electron accepting unit (Kurth *et al.*, 2016b). Whilst the enzyme from the sulfur oxidizer *A. vinosum* is strongly biased towards catalyzing thiosulfate oxidation (Brito *et al.*, 2015), TsdA from *Campylobacter jejuni* acts primarily as a tetrathionate reductase and enables the organism to use tetrathionate as an alternative electron acceptor for anaerobic respiration (Liu *et al.*, 2013), see section 3.4.2.3.

Further thiosulfate dehydrogenases are much less characterized. A homotetramer of 45 kDa catalyzing tetrathionate formation from thiosulfate was described for *Acidithiobacillus ferrooxidans* (Janiczek *et al.*, 2007), but the gene (AFE_0042) identified later encodes a periplasmic 25 kDa protein (Kikumoto *et al.*, 2013). The enzyme does not transfer electrons to ubiquinone or horse heart cytochrome *c*. A closely related protein is encoded in the vicinity (AFE_0050) in the same organism. Conspicuously, both genes reside in a cluster of genes together with *doxDA* encoding TQO and genes for rhodanese-like sulfurtransferases (Wang *et al.*, 2019). Related genes also occur outside of the Gammaproteobacteria, in sulfur-oxidizing Beta- (e.g. *Sulfuriferula* and *Thiomonas*) and Alphaproteobacteria (e.g. *Acidiphilium*). Clearly more work is necessary to clarify the mechanism and relevance of these enzymes.

The pyrroloquinoline quinone (PQQ)-containing periplasmic enzyme thiol dehydrotransferase from the Betaproteobacterium *Advenella kashmirensis* has also been reported to play a role in thiosulfate-to-tetrathionate conversion (Pyne *et al.*, 2018). This enzyme has dual activity catalyzing PQQ-dependent thiosulfate dehydrogenation as well as PQQ-independent thiol transfer (see sections 3.5.1.3 and 3.5.2 and Figure 3.4b).

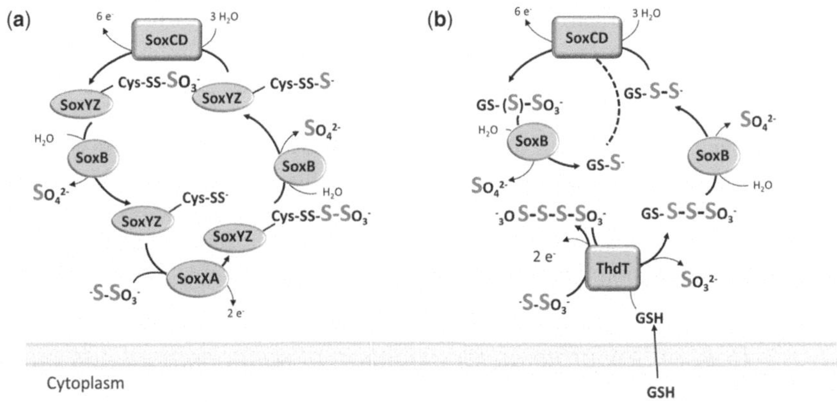

Figure 3.4 The Sox system operating in the periplasm of sulfur-oxidizing bacteria. (a). The Sox system including SoxCD completely oxidizes thiosulfate to sulfate. (b) In *Adevenalla kashimirensis* the enzyme ThdT has a dual function as thiosulfate dehydrogenase forming tetrathionate and as thiol dehydrotransferase mediating reaction of tetrathionate and glutathione and release of sulfite. Subsequent oxidation of the glutathione disulfane adduct is proposed to proceed via iterative action of SoxB and SoxCD (Pyne *et al.*, 2018).

3.5.1.2 Oxidation of thiosulfate to sulfate: the Sox system

The complete oxidation of thiosulfate to sulfate is always initiated, and in many cases also completely performed, in the bacterial periplasm and involves the well-studied thiosulfate-oxidizing Sox multienzyme system. Among the many organisms pursuing this pathway, some form sulfur deposits as intermediates either extracellularly or as intracellular sulfur globules.

In organisms that do not form sulfur deposits, thiosulfate oxidation requires four different proteins: SoxB, SoxXA, SoxYZ and SoxCD (Friedrich *et al.*, 2001) (Figure 3.4a). The heterodimeric SoxYZ protein acts as the central player and serves as a carrier of pathway intermediates (Sauvé *et al.*, 2007). These intermediates are not simply bound to a cysteine residue located near the carboxy-terminus of the SoxY subunit, the true carrier species is a SoxYZ-S-sulfane adduct (Grabarczyk & Berks, 2017). The *c*-type cytochrome SoxXA catalyzes the oxidative formation of a disulfide linkage between the sulfane sulfur of thiosulfate and the persulfurated active site cysteine residue of SoxY. Then, the sulfone group is hydrolytically released as sulfate. This reaction is catalyzed by SoxB (Sauvé *et al.*, 2009; Grabarczyk *et al.*, 2015a) and leaves the original sulfane sulfur of thiosulfate bound to SoxY. The last steps are oxidation of the SoxY-bound sulfane sulfur to a sulfone by the hemomolybdoprotein SoxCD and again hydrolytic release of sulfate (Zander *et al.*, 2010).

In organisms forming sulfur as an intermediate, SoxCD is usually not present. This truncated system is incapable of oxidizing SoxY-bound sulfane sulfur which is instead transferred to zero-valent sulfur stored extracellularly or intracellularly in the periplasm. This process occurs via an unresolved mechanism, in some species possibly involving the rhodanese-like protein SoxL (Welte *et al.*, 2009). It may be important to note in this regard, that polysulfurated SoxY($S_{3\text{-}4}$)Z species occur as intermediates of thiosulfate oxidation catalyzed by a reconstituted Sox system *in vitro* (Grabarczyk & Berks, 2017). Such polysulfurated species could serve as direct donors for sulfur deposits. For complete oxidation to sulfate, truncated Sox systems must be combined with cytoplasmic sulfur oxidation systems (Tanabe *et al.*, 2019), see section 3.5.7 and Figure 3.5.

3.5.1.3 Role of Sox proteins for oxidation of sulfur compounds other than thiosulfate

In 2001, Friedrich and coworkers put forward the idea that Sox proteins could well be involved in the oxidation of sulfur compounds other than thiosulfate and showed that sulfite, sulfur and hydrogen sulfide served as *in vitro* substrates for the reconstituted Sox system of *Paracoccus pantotrophus* (Friedrich *et al.*, 2001; Rother *et al.*, 2001). At the same time, it was reported for *Rhodovulum sulfidophilum* that the Sox machinery is not only essential for thiosulfate oxidation, but also indispensable for the oxidation of sulfide *in vivo* (Appia-Ayme *et al.*, 2001). In the purple sulfur bacterium *Allochromatium vinosum*, the presence of SoxYZ is needed in parallel to the cytoplasmic enzymes for effective oxidation of sulfite (Dahl *et al.*, 2013). An important role for SoxYZ independent of other Sox proteins is furthermore strongly indicated by the presence of *soxYZ* genes in a number of phototrophic sulfur bacteria that are unable to metabolize thiosulfate (Dahl *et al.*, 2013).

In more recent analyses of *Sulfurimonas* species, members of the class Campylobacteria among the Epsilonproteobacteria, revealed two separate *sox* clusters, $soxXY_1Z_1AB$ and $soxCDY_2Z_2$ (Grote *et al.*, 2012; Sievert *et al.*, 2008). Sulfur-oxidizing, nitrate-reducing *Sulfurimonas* spp. are widespread in sediments, hydrothermal vent fields, aquifers and subsurface environments such as oil reservoirs and their role in the natural sulfur cycle should not be underestimated. The two different copies of *soxYZ* show a high level of divergence, the gene sets appear to be differentially regulated and the proteins are therefore proposed to be involved in oxidation of different sulfur compounds (Götz *et al.*, 2019; Meier *et al.*, 2017; Pjevac *et al.*, 2018). Remarkably, a *Sulfurimonas* strain incapable of thiosulfate oxidation was described that contains only the $soxCDY_2Z_2H$ cluster and oxidizes elemental sulfur to sulfate and thiosulfate as end products (Lahme *et al.*, 2019). In this strain, the lack of SoxB might be overcome by the product of the *soxH* gene, a putative metallohydrolase that may release sulfate from SoxY (Lahme *et al.*, 2019).

The $GSSSSO_3^-$ adduct formed during tetrathionate degradation in *Advenella* (see sections 3.5.1.1 and 3.5.2) has been proposed as a further substrate for Sox proteins. The adduct appears as a suitable substrate for SoxB that may

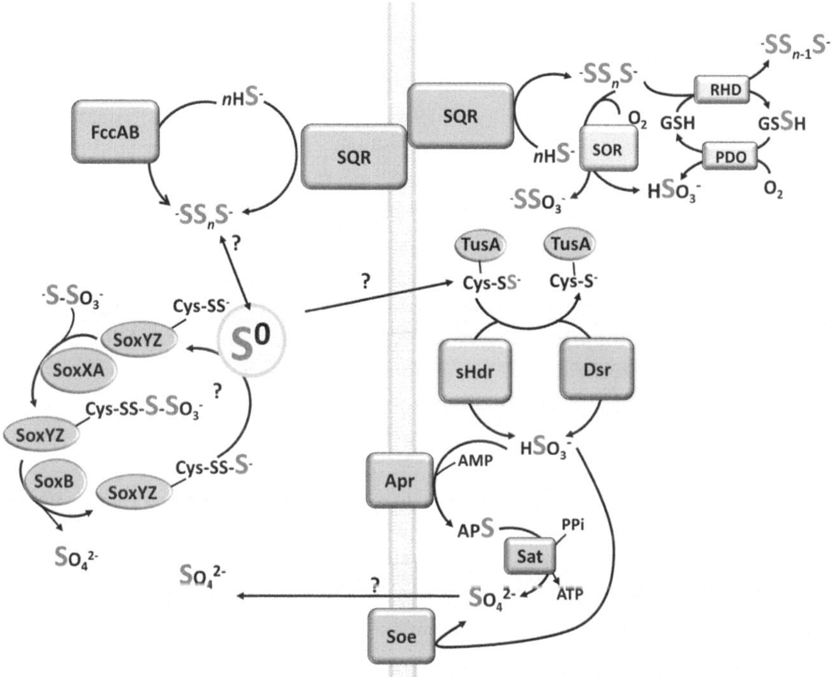

Figure 3.5 Sulfur oxidation pathways in sulfur oxidizers forming zero-valent sulfur as an intermediate and sulfide oxidation/detoxification in the cytoplasm. For clarity reactions are not given with exact stoichiometries. Sulfide oxidation in the periplasm is catalyzed by SQR (sulfide:quinone oxidoreductase) and/or flavocytochrome *c* (FccAB). The sulfur in the resulting polysulfides is accumulated zero-valent sulfur, which is also the product of thiosulfate oxidation mediated by a Sox system lacking SoxCD. Sulfur is transferred into the cytoplasm where it is delivered to the oxidizing system (either Dsr or sHdr) via a sulfur relay system (here summarized as TusA). Sulfide oxidation via SQR can also occur in the cytoplasm as a supplementary or as a detoxification reaction. Rhodanese and sulfur/persulfide dioxygenase (PDO) catalyze subsequent steps. Electrons resulting from the SQR-catalyzed reaction can be fed into the quinone pool. Sulfite is formed by direct reaction with molecular oxygen preventing energy conservation. GSH (GSSH): reduced glutathione (persulfide). Sulfur oxygenase reductase (SOR) also resides in the cytoplasm and catalyzes a dioxygen-dependent disproportionation of elemental sulfur and/or poylsulfides to hydrogen sulfite, thiosulfate, and hydrogen sulfide. Here, SOR is depicted without reductase activity (Rühl *et al.*, 2017).

hydrolytically release the terminal sulfone group as sulfate. SoxCD would then oxidize the first remaining sulfane sulfur, again followed by hydrolytic release of sulfur and one more cycle resulting in oxidation and release of the last remaining sulfane sulfur and regeneration of reduced glutathione (Pyne *et al.*, 2018).

3.5.2 Tetrathionate oxidation

Many sulfur oxidizers can oxidize tetrathionate to sulfate and/or tetrathionate occurs as an intermediate during the oxidation of thiosulfate to sulfate (Dam *et al.*, 2007; Ghosh & Dam, 2009; Wang *et al.*, 2019). The name "S_4I" (S_4 intermediate) has been coined for the latter pathway which occurs in many beta- and gammaproteobacterial chemolithotrophs, e.g. species of the genus *Acidithiobacillus* (Bugaytsova & Lindström, 2004; Dam *et al.*, 2007; Kikumoto *et al.*, 2013; Pyne *et al.*, 2017, 2018; Rameez *et al.*, 2019; Visser *et al.*, 1996). Two different enzymes initiating tetrathionate degradation have so far been described.

The first is the enzyme tetrathionate hydrolase, TetH that has been isolated from various acidophilic *Acidithiobacillus* species, *Acidiphilium acidophilum* and the archaeon *Acidianus ambivalens*. Tetrathionate hydrolases are monomeric and/or homodimeric, membrane-associated or soluble enzymes with pH optima in the acidic range acting outside of the cytoplasm (Bugaytsova & Lindström, 2004; de Jong *et al.*, 1997; Mangold *et al.*, 2011; Protze *et al.*, 2011). The enzymes exhibit similarity to proteins of the PQQ dehydrogenase family, however they neither appear to contain PQQ nor do they need it for activity (Kanao *et al.*, 2010; Protze *et al.*, 2011). The initial products of TetH-catalyzed tetrathionate hydrolysis are sulfate and disulfane monosulfonic acid ($^-S\text{-}S\text{-}SO_3^-$) (Figure 3.3). The latter is highly reactive and decomposes spontaneously to sulfur and thiosulfate (Steudel *et al.*, 1987). Alternatively, reactions of the disulfane monosulfonic acid with itself were proposed that result in successive chain elongation, finally leading to the production of elemental sulfur and sulfite. Indeed, the comparatively well-studied model organism *A. ferrooxidans* forms sulfur deposits on tetrathionate (Hazeu *et al.*, 1988; Steudel *et al.*, 1987). Sulfur as well as sulfite could, in principle, serve as substrates for Sox proteins allowing their further oxidation (Friedrich *et al.*, 2001), however *sox* genes are not present in all organisms pursuing the S_4 intermediate pathway (Wang *et al.*, 2019). Many questions therefore remain to be resolved.

The second enzyme with a described capacity for initiating tetrathionate degradation is the periplasmic enzyme thiol dehydrotransferase from *Advenella kashmirensis* (Pyne *et al.*, 2018) (Figure 3.4b). This enzyme cannot only catalyze tetrathionate formation from thiosulfate (see section 3.5.1.1), but also promotes the reaction of tetrathionate with reduced glutathione (GSH) yielding a glutathione sulfodisulfane adduct and sulfite according to the equation $GSH + S_4O_6^{2-} \rightarrow GSSSSO_3^- + SO_3^{2-}$. Subsequent oxidation of the glutathione derivative may then proceed via iterative action of SoxB and SoxCD (Pyne *et al.*, 2018), see sections 3.5.1.1 and 3.5.1.2 and Figure 3.4b.

3.5.3 Oxidation of sulfide and polysulfides

Initial oxidation of sulfide is typically catalyzed by sulfide quinone oxidoreductases (SQRs) and flavocytochrome *c* sulfide dehydrogenases (Frigaard & Dahl, 2009)

(Figure 3.5). Both enzymes belong to the disulfide reductase protein family, both depend on FAD as a cofactor for activity and both produce polysulfide as the initial product. Accumulation of polysulfides leads to sulfur precipitation outside the cell or the sulfur is stored internally as sulfur globules (Brito *et al.*, 2009; Cherney *et al.*, 2010; Frigaard & Dahl, 2009; Griesbeck *et al.*, 2000, 2002; Marcia *et al.*, 2009; Schütz *et al.*, 1999).

Virtually nothing is known about the oxidation of polysulfides and their transformation into sulfur deposits. Theoretically, this could be a purely chemical process as longer polysulfides are in equilibrium with elemental sulfur (Steudel *et al.*, 1990).

3.5.3.1 Sulfide:quinone oxidoreductase

SQRs do not only occur in prokaryotes but also in mitochondria of lower and higher eukaryotes. In the latter as well as in many heterotrophic prokaryotes, SQRs do not primarily serve in energy conservation by feeding electrons into electron transport chains on the level of quinones, but instead their main purpose is sulfide detoxification. Along this line, mitochondrial SQR interacts with a rhodanese that is proposed to transfer the zero-valent sulfane sulfur formed during the reaction to glutathione (GSH) to produce glutathione persulfide (GSSH) (Libiad *et al.*, 2014). Persulfide dioxygenase (PDO) then oxidizes the sulfane sulfur in GSSH to sulfite (Kabil & Banerjee, 2012). In heterotrophic bacteria, genes for related SQRs are usually genetically linked with PDO genes (Guimaraes *et al.*, 2011; Liu *et al.*, 2014a; Shen *et al.*, 2015; Xia *et al.*, 2017). PDO is a cytoplasmic soluble enzyme and accordingly SQR interacting with it has been localized on the cytoplasmic side of the membrane (Gao *et al.*, 2017) (Figure 3.5).

In dissimilatory sulfur-oxidizing prokaryotes, SQR has been proposed to be located at the outer surface of the cytoplasmic membrane (Christel *et al.*, 2016; Mangold *et al.*, 2011; Quatrini *et al.*, 2009), the same localization as has been experimentally verified for the enzyme from *Rhodobacter capsulatus* (Schütz *et al.*, 1999). In accordance, polysulfides, the primary products of sulfide oxidation in purple and green sulfur bacteria instantly accumulate outside of the cells, in the medium (Marnocha *et al.*, 2016; Prange *et al.*, 2004).

3.5.3.2 Flavocytochrome c *and multitude of sulfide-oxidizing systems*

In a variety of sulfide oxidizing species, flavocytochrome *c* is present as a soluble protein in the periplasm or as a membrane-bound enzyme (Frigaard & Dahl, 2009). It consists of a flavoprotein (FccB) and a hemoprotein (FccA) and has sulfide:cytochrome *c* oxidoreductase activity *in vitro*.

Many prokaryotes have a variety of sulfur-oxidizing modules, for example *Marichromatium purpuratum* encodes three different SQRs as well as FccBA (Dahl, 2017). *A. vinosum* mutants deficient in either flavocytochrome *c* (Reinartz *et al.*, 1998), *sox* genes (Hensen *et al.*, 2006) or both (unpublished) still exhibit sulfide oxidation rates comparable to the wildtype, indicating that SQR plays the major role. It is possible that FccAB is advantageous under certain growth conditions and that each of the multiple sulfide-oxidizing modules conveys adaptation to different environmental sulfide concentrations.

3.5.4 Oxidation of external sulfur

Elemental sulfur is virtually insoluble in water and how exactly it is made available as a substrate still requires more research. Most information is available for *Acidithiobacilli* that attach to sulfur by glycocalx-like extracellular polymeric substances, specifically lipopolysaccharides (Bryant *et al.*, 1984; Espejo & Romero, 1987; Gehrke *et al.*, 1998). Biofilms formed on the surface of elemental sulfur by cells of the thermoacidophilic archaeal species *Acidianus* also comprise an extracellular matrix that contains carbohydrates, proteins and lipids. The lipids are suggested to mediate initial attachment of the cells to sulfur while sugars like glucose and mannose are thought to be important for further colonization of sulfur and a mature biofilm architecture (Zhang *et al.*, 2019). For the purple sulfur bacterium *A. vinosum*, intimate physical cell-sulfur contact was also established as a prerequisite for sulfur uptake (Franz *et al.*, 2007). After attachment of the cells, sulfur needs to be mobilized. In *A. ferrooxidans* and *Acidiphilum* spp. this step involves thiols of specific outer-membrane proteins and transport into the cytoplasm as persulfide sulfane sulfur (Rohwerder & Sand, 2003). Glutathione plays a catalytic role in this step.

3.5.5 Biogenic sulfur globules

A large portion of sulfur oxidizers form sulfur deposits as intermediates, both extracellularly and intracellularly (Dahl, 2020; Marnocha *et al.*, 2016, 2019). Formation of extracellular sulfur globules is typical for green sulfur bacteria and even name-giving for sulfur oxidizers of the gammaproteobacterial family Ectothiorhodospiraceae while intracellular sulfur globules typically occur in magnetotactic sulfur oxidizers, in purple sulfur bacteria of the family Chromatiaceae and in sulfur oxidizing bacterial endosymbionts (Dahl, 2020). Extracellular biogenic sulfur globules have properties clearly distinguishing them from abiotic elemental sulfur, such that the green sulfur bacterium *C. tepidum* can grow on self-produced sulfur globules but other forms of elemental sulfur do not support growth. In this case, the biogenic sulfur globules are enveloped by a layer of proteins and polysaccharides that enables cell-sulfur interactions and probably keeps the sulfur in a bioavailable form (Marnocha *et al.*, 2019).

Usually the diameter of sulfur globules is in the range of 1–3 μm. The sulfur can comprise 20 to 34% of the cell dry mass (Dahl, 2020; Nelson & Castenholz, 1981). The general target compartment for sulfur storage is the bacterial periplasm. Intracellular sulfur globules are encased by a protein envelope consisting of one or more structural proteins resembling cytoskeletal keratins or plant cell wall proteins. The chemical nature of the sulfur in the globules has been the subject of intensive controversy (Dahl, 2020; Dahl & Prange, 2006; George *et al.*, 2002, 2008). In most cases, the sulfur appears to be present as S_8 rings in a nano-crystalline form possibly mixed with linear polysulfides (Dahl, 2020).

3.5.6 Sox-independent, cytoplasmic oxidation of sulfane sulfur to sulfite

Except for sulfide detoxification, all pathways described so far take place outside of the cytoplasm. As outlined above in section 3.5.1.2, the periplasmic enzyme SoxCD catalyzes the oxidation of carrier-bound sulfane sulfur to the level of a sulfone group and feeds electrons into respiratory or light-driven electron transport at the level of *c*-type cytochromes. If SoxCD is not present, the oxidation of zero-valent sulfane sulfur to sulfite is achieved in the cytoplasm by multienzyme systems allowing more efficient energy conservation.

How the sulfur is transferred into the cytoplasm for further oxidation is currently still an enigma. For organisms forming periplasmic sulfur deposits, low molecular weight organic persulfides such as glutathione amide persulfide have been proposed as carrier molecules (Dahl, 2017; Frigaard & Dahl, 2009). It is not known how these proposed carrier molecules may be generated nor have any transporters for such molecules been characterized. In *Hyphomicrobium denitrificans*, a gene encoding a predicted sulfur compound transporter resides in the same cluster as *soxYZ* and other *sox* genes and in the immediate vicinity of the genes for the cytoplasmic sulfane sulfur-oxidizing system. This predicted transporter has been suggested to pick up sulfur from periplasmic SoxYZ and to shuttle it over the cytoplasmic membrane onto cytoplasmic sulfur-binding proteins (Koch & Dahl, 2018). Experimental evidence supporting this proposal has not been provided so far.

Once transported into the cytoplasm, rhodanese-, DsrE- and TusA-like sulfurtransferases constitute a persulfide shuttling system that delivers sulfur very specifically to the target enzymes catalyzing its oxidation to sulfite (Dahl, 2015; Liu *et al.*, 2014b; Tanabe *et al.*, 2019; Venceslau *et al.*, 2014). Two fundamentally different pathways have been described to achieve this purpose (Figure 3.5). Both pathways have in common the possibility that at least some of the electrons released can be shuttled onto NAD^+. This would considerably reduce the need for energy-demanding reverse electron flow in lithoautotrophic sulfur oxidizers.

3.5.6.1 rDsr pathway

The rDsr pathway with reverse acting dissimilatory sulfite reductase rDsrAB as its key enzyme has been characterized in quite some detail. A typical *dsr* gene cluster can consist of up to 15 different genes and may be linked with the genes for the sulfurtransferases providing DsrAB with oxidizable sulfur. In the model organism *A. vinosum*, a rhodanese-like protein and a DsrE-like protein were identified as entry points for sulfur delivery to the TusA protein. TusA moves the sulfur onto DsrEFH from where it is transferred to the final sulfur-accepting protein DsrC. It is assumed but not yet experimentally proven that the membrane-bound DsrMKJOP electron-transporting complex oxidizes persulfurated DsrC, thus generating DsrC trisulfide which then serves as the substrate for reverse acting dissimilatory sulfite reductase. The electrons released by the oxidation of the sulfur bridged in the DsrC trisulfide to sulfite are probably transferred onto NAD^+. This reaction is catalyzed by the iron-sulfur flavoprotein DsrL (Löffler *et al.*, 2020).

3.5.6.2 sHdr pathway

Many sulfur-oxidizing prokaryotes such as species of the bacterial genera *Acidithiobacillus* and *Aquifex* or archaeal sulfur oxidizers like *Sulfolobus* or *Acidianus* do not have the genetic capacity for rDsr-driven generation of sulfite. Instead, these organisms run a new pathway named the sulfur-oxidizing heterodisulfide reductase-like (sHdr) pathway due to the involvement of polypeptides resembling the HdrA, B and C subunits of heterodisulfide reductase from methanogenic archaea (Cao *et al.*, 2018; Koch & Dahl, 2018). The rDsr and sHdr pathways occur virtually exclusively. Very few organisms bear the genetic potential for both oxidation routes (Berben *et al.*, 2019; Koch & Dahl, 2018).

The core genes *shdrC1B1AhyphdrC2B2* that probably encode a hexameric $HdrA_2C_1C_2B_1B_2$ complex (Ernst *et al.*, 2020), are usually accompanied by genes for rhodanese, DsrE- and TusA-like proteins as well as by a gene for a lipoate-binding protein (LbpA). Just like the sHdr proteins themselves, the latter has been proven to be indispensable for sulfur oxidation in the model organism *H. denitrificans* (Cao *et al.*, 2018). The exact function of lipoate within the catalytic cycle of the Hdr-LbpA-based sulfur oxidation system is currently unclear, but it may either serve as a substrate binding entity during sulfur oxidation or exclusively function as a redox switch. In the second case, the sulfur substrate would be oxidized on one of the sulfur-binding proteins encoded in *shdr* gene clusters. In any event, all suggestions of the sHdr mechanism that have so far been put forward (Cao *et al.*, 2018; Ernst *et al.*, 2020; Tanabe *et al.*, 2019) include the reduction of lipoate to dihydrolipoate at the end of the reaction cycle. Its reoxidation and concomitant reduction of NAD^+ would be catalyzed by dihydroliponamide dehydrogenase.

3.5.6.3 Formation of sulfite via reactions involving molecular oxygen

Two different enzymes have been described in sulfur-oxidizing prokaryotes that catalyze reactions of reduced sulfane sulfur with molecular oxygen. Such reactions clearly prevent energy conservation because they can neither be coupled with electron transfer nor with substrate-level phosphorylation. Still, evidence has been collected that these enzymes may be important at least in acidophiles (Wang *et al.*, 2019).

3.5.6.3.1 Sulfur dioxygenase

The role of sulfur dioxygenases (SDOs) in chemolithotrophic sulfur oxidizers has been best studied for *Acidithiobacillus* species. The actual substrate for the SDO-catalyzed reaction is sulfane sulfur bound to glutathione, i.e. glutathione persulfide (GSSH, and also GSS_nH with $n > 1$) (Rohwerder & Sand, 2003). *Acidithiobacillus* species contain one or two copies of *sdo* genes. The encoded proteins appear to reside in the cytoplasm (Wu *et al.*, 2017). The *in vivo* role of the various SDO proteins was studied in mutant strains of *A. ferrooxidans* and *A. caldus* yielding somewhat inconclusive results (Wang *et al.*, 2014; Wu *et al.*, 2017), but showing that the absence of SDOs neither impaired elemental sulfur oxidation activities nor prevented growth on externally provided elemental sulfur (Wu *et al.*, 2017). It has been put forward that SDOs in *Acidithiobacillus* spp. are mainly involved in sulfane sulfur oxidation in the cytoplasm. Nothing is known about how the sulfur is distributed between the energy-conserving sHdr system and the SDOs that clearly prevent energy conservation.

Enzymes with sulfide dioxygenase or persulfide dioxygenase (PDO) activity also occur in mitochondria of animals and plants as well as in a number or heterotrophic bacteria (Kabil & Banerjee, 2012; Sattler *et al.*, 2015; Xia *et al.*, 2017). These proteins work together with cytoplasmically oriented sulfide. quinone oxidoreductases and play an important role in the detoxification of intracellular sulfide (Guimaraes *et al.*, 2011; Liu *et al.*, 2014a).

3.5.6.3.2 Sulfur oxygenase reductase

Sulfur oxygenase reductase (SOR) is another enzyme catalyzing the reaction of elemental sulfur with molecular oxygen (Figure 3.5). This enzyme, first reported in several acidophilic and thermophilic archaea, catalyzes the disproportionation of cytoplasmic elemental sulfur and/or polysulfides to thiosulfate, sulfite, and sulfide (Kletzin, 1989; Kletzin *et al.*, 2004). The enzyme from *Thioalkalivibrio paradoxus* does not have reductase activity (Rühl *et al.*, 2017). The *sor* gene is also present in some, though not in all *Acidithiobacillus* species and it has been put forward that that the enzyme is supplementary, but not necessary, for cytoplasmic elemental sulfur oxidation (Chen *et al.*, 2012; Christel *et al.*, 2016; You *et al.*, 2011).

3.5.7 Oxidation of sulfite

Sulfite-oxidizing enzymes (SOEs) are molybdenum-containing enzymes that exist in almost all forms of life where they carry out important functions in protecting cells and organisms against sulfite-induced damage (Kappler, 2008, 2011). In lithotrophic sulfur oxidizers they are furthermore important for transfer of the two electrons released during oxidation of sulfite to sulfate into respiratory or photosynthetic electron transport.

3.5.7.1 Oxidation of sulfite outside of the cytoplasm

Most bacterial SOEs reside in the bacterial periplasm and react with *c*-type cytochrome electron acceptors (Kappler & Enemark, 2014). This allows feeding respiratory electron transport at the level of cytochrome *c* oxidase. As already pointed out by others (Kappler & Enemark, 2014), the energy yields from this reaction are low, and usually sulfite oxidation alone will not support cell growth.

SorAB sulfite dehydrogenase from *Starkeya novella* was the first SOE characterized from a bacterial host (Kappler *et al.*, 2000). It consists of the catalytic molybdoprotein SorA and a small c-type cytochrome, SorB. Over the past two decades, at least six other bacterial SOEs have been purified and only one of these might contain a heme redox center as an integral part of the enzyme. Thus, most of the bacterial SOEs contain only molybdenum redox centers (Kappler & Enemark, 2014).

Even sulfur-oxidizers containing sulfite-generating enzyme systems in the cytoplasm (rDsr or sHdr) may have to rely on further oxidation of sulfite outside of the cytoplasm because they lack cytoplasmic sulfite-oxidizing enzyme systems. In these cases, a gene encoding a TauE-like potential sulfite exporter may be encoded in the vicinity of the *shdr* genes (Koch & Dahl, 2018).

3.5.7.2 Oxidation of sulfite in the cytoplasm

Direct, one step oxidation of sulfite to sulfate can also occur in the bacterial cytoplasm (Figure 3.5) as has first been shown for the purple sulfur bacterium *A. vinosum*. In this organism, the membrane-bound, polysulfide reductase-like iron-sulfur molybdoprotein SoeABC appears to be the major sulfite-oxidizing agent (Dahl *et al.*, 2013). Genes encoding proteins related to SoeABC not only occur in purple and green phototrophic sulfur bacteria but also in a large number of chemotrophs, indicating general importance for this enzyme (Dahl *et al.*, 2013; Wang *et al.*, 2019).

It is firmly established that an array of anoxygenic phototrophic as well as of chemotrophic sulfur bacteria oxidize sulfite in the cytoplasm using an indirect pathway via adenosine-5′-phosphosulfate catalyzed by APS reductase (AprBA) and ATP sulfurylase (Sat) (Figure 3.5). This pathway constitutes the reversal of sulfate activation to APS and APS reduction to sulfite occurring in dissimilatory sulfate-reducers. The electrons released during the formation of APS from sulfite and adenosine monophosphate (AMP) can be fed into electron transfer chains via two different modules, depending on the organism. The first is the transmembrane protein AprM (Dahl *et al.*, 2013; Meyer & Kuever, 2007) and the second is the QmoABC complex that serves the reverse function, i.e. electron transport from quinol to APS in sulfate reducers (Grein *et al.*, 2013; Ramos *et al.*, 2012).

3.6 SULFUR DISPROPORTIONATION

The disproportionation of sulfur compounds in energy metabolism was discovered in sulfate-reducing bacteria, when *Desulfovibrio desulfodismutans* was found to metabolize thiosulfate and sulfite according to SO_3^{2-} or $S_2O_3^{2-}$ to HS^- plus SO_4^{2-} (Bak & Pfennig, 1987). In fact, thiosulfate and sulfite disproportionation is common among sulfate reducers, albeit not always coupled to growth (Krämer & Cypionka, 1989). While thiosulfate and sulfite disproportionation are exergonic reactions, the disproportionation of elemental sulfur is endergonic under standard conditions. Nevertheless, sulfur disproportionation according to:

$$4\,S^0 + 4\,H_2O \rightarrow SO_4{}^{2-} + 3\,HS^- + 5\,H^+ \;\; \Delta G^{0\prime} = 10.2\,\text{kJ}\,\text{mol}^{-1}$$

can support growth in the environment when the produced hydrogen sulfide is scavenged by iron and manganese oxides (Finster, 2008; Thamdrup *et al.*, 1993). Growth coupled to elemental sulfur disproportionation is found predominantly in the family Desulfobulbaceae and often goes along with autotrophy (Finster *et al.*, 2013). Both thiosulfate and elemental sulfur disproportionation are very relevant processes in marine and freshwater sediments and may belong to the earliest metabolic processes on earth (Jørgensen, 1990a, b; Fike *et al.*, 2008; Thamdrup *et al.*, 1993)

Studies of the biochemistry of sulfite and thiosulfate disproportionation of a few isolates indicate reversal of the sulfate reduction pathway during disproportionation. It has been proposed for two *Desulfovibrio* strains that ATP is formed by substrate level phosphorylation via ATP sulfurylase (Krämer & Cypionka, 1989). In addition, a reversed electron transport step is probably necessary for the reduction of thiosulfate to sulfide and sulfite fueled by electrons stemming from the oxidation of sulfite to APS (Finster, 2008; Krämer & Cypionka, 1989). In a second study, somewhat contradictory results were obtained that do not support an initial reductive cleavage of thiosulfate (Cypionka *et al.*, 1998). Once produced, sulfite is further oxidized along the APS reductase pathway and possibly also via direct sulfite:oxidoreductase catalyzed oxidation (Frederiksen & Finster, 2004). Clearly, more work is necessary to fully understand the biochemistry underlying the microbial disproportionation of sulfur compounds.

ACKNOWLEDGEMENTS

Funding by the Deutsche Forschungsgemeinschaft (grants Da 351/6-2 and Da 351/8-1) is gratefully acknowledged.

REFERENCES

Abouna S., Gonzalez-Rizzo S., Grimonprez A. and Gros O. (2015). First description of sulphur-oxidizing bacterial symbiosis in a Cnidarian (Medusozoa) living in sulphidic shallow-water environments. *Plos One*, **10**, e0127625.

Anantharaman K., Brown C. T., Hug L. A., Sharon I., Castelle C. J., Probst A. J., Thomas B. C., Singh A., Wilkins M. J., Karaoz U., Brodie E. L., Williams K. H., Hubbard S. S. and Banfield J. F. (2016). Thousands of microbial genomes shed light on interconnected biogeochemical processes in an aquifer system. *Nature Communications*, **7**, 13219.

Anantharaman K., Hausmann B., Jungbluth S. P., Kantor R. S., Lavy A., Warren L. A., Rappé M. S., Pester M., Loy A., Thomas B. C. and Banfield J. F. (2018). Expanded diversity of microbial groups that shape the dissimilatory sulfur cycle. *The ISME Journal*, **12**, 1715–1728.

Appia-Ayme C., Little P. J., Matsumoto Y., Leech A. P. and Berks B. C. (2001). Cytochrome complex essential for photosynthetic oxidation of both thiosulfate and sulfide in *Rhodovulum sulfidophilum*. *Journal of Bacteriology*, **183**, 6107–6118.

Atkinson S. J., Mowat C. G., Reid G. A. and Chapman S. K. (2007). An octaheme *c*-type cytochrome from *Shewanella oneidensis* can reduce nitrite and hydroxylamine. *FEBS Letters*, **581**, 3805–3808.

Bak F. and Pfennig N. (1987). Chemolithotrophic growth of *Desulfovibrio sulfodismutans* sp. nov. by disproportionation of inorganic sulfur compounds. *Archives of Microbiology*, **147**, 184–189.

Barrett E. L. and Clark M. A. (1987). Tetrathionate reduction and production of hydrogen sulfide from thiosulfate. *Microbiological Reviews*, **51**, 192–205.

Bell E., Lamminmäki T., Alneberg J., Andersson A. F., Qian C., Xiong W., Hettich R. L., Frutschi M. and Bernier-Latmani R. (2020). Active sulfur cycling in the terrestrial deep subsurface. *The ISME Journal*, **14**, 1260–1272.

Berben T., Overmars L., Sorokin D. Y. and Muyzer G. (2019). Diversity and distribution of sulfur oxidation-related genes in *Thioalkalivibrio*, a genus of chemolithoautotrophic and haloalkaliphilic sulfur-oxidizing bacteria. *Frontiers in Microbiology*, **10**, 160.

Bhattarai S., Cassarini C. and Lens P. N. L. (2019). Physiology and distribution of archaeal methanotrophs that couple anaerobic oxidation of methane with sulfate reduction. *Microbiology and Molecular Biology Reviews*, **83**, e00074–00018.

Bjerg J. T., Boschker H. T. S., Larsen S., Berry D., Schmid M., Millo D., Tataru P., Meysman F. J. R., Wagner M., Nielsen L. P. and Schramm A. (2018). Long-distance electron transport in individual, living cable bacteria. *Proceedings of the National Academy of Sciences of the USA*, **115**, 5786–5791.

Boetius A., Ravenschlag K., Schubert C. J., Rickert D., Widdel F., Gieseke A., Amann R., Jorgensen B. B., Witte U. and Pfannkuche O. (2000). A marine microbial consortium apparently mediating anaerobic oxidation of methane. *Nature*, **407**, 623–626.

Bridger S. L., Clarkson S. M., Stirrett K., DeBarry M. B., Lipscomb G. L., Schut G. J., Westpheling J., Scott R. A. and Adams M. W. (2011). Deletion strains reveal metabolic roles for key elemental sulfur-responsive proteins in *Pyrococcus furiosus*. *Journal of Bacteriology*, **193**, 6498–6504.

Brito J. A., Sousa F. L., Stelter M., Bandeiras T. M., Vonrhein C., Teixeira M., Pereira M. M. and Archer M. (2009). Structural and functional insights into sulfide:quinone oxidoreductase. *Biochemistry*, **48**, 5613–5622.

Brito J. A., Denkmann K., Pereira I. A. C., Archer M. and Dahl C. (2015). Thiosulfate dehydrogenase (TsdA) from *Allochromatium vinosum*: Structural and functional insights into thiosulfate oxidation. *Journal of Biological Chemistry*, **290**, 9222–9238.

Brüser T., Lens P. and Trüper H. G. (2000). The biological sulfur cycle. In: Environmental Technologies to Treat Sulfur Pollution, P. Lens and L. H. Pol (eds), IWA Publishing, London, pp. 47–86.

Bryant R. D., Costerton J. W. and Laishley E. J. (1984). The role of *Thiobacillus albertis* glycocalyx in the adhesion of cells to elemental sulfur. *Canadian Journal of Microbiology*, **30**, 81–90.

Bugaytsova Z. and Lindström E. B. (2004). Localization, purification and properties of a tetrathionate hydrolase from *Acidithiobacillus caldus*. *European Journal of Biochemistry*, **271**, 272–280.

Burdorf L. D. W., Tramper A., Seitaj D., Meire L., Hidalgo-Martinez S., Zetsche E. M., Boschker H. T. S. and Meysman F. J. R. (2017). Long-distance electron transport occurs globally in marine sediments. *Biogeosciences*, **14**, 683–701.

Burns J. L. and DiChristina T. J. (2009). Anaerobic respiration of elemental sulfur and thiosulfate by *Shewanella oneidensis* MR-1 requires *psrA*, a homolog of the *phsA* gene of *Salmonella enterica* serovar *typhimurium* LT2. *Applied and Environmental Microbiology*, **75**, 5209–5217.

Canfield D. E. (2001). Biogeochemistry of sulfur isotopes. *Reviews in Mineralogy & Geochemistry*, **43**, 607–636.

Canfield D. E., Stewart F. J., Thamdrup B., De Brabandere L., Dalsgaard T., DeLong E. F., Revsbech N. P. and Ulloa O. (2010). A cryptic sulfur cycle in oxygen-minimum zone waters off the chilean coast. *Science*, **330**, 1375–1378.

Cao X., Koch T., Steffens L., Finkensieper J., Zigann R., Cronan J. E. and Dahl C. (2018). Lipoate-binding proteins and specific lipoate-protein ligases in microbial sulfur oxidation reveal an atpyical role for an old cofactor. *eLife*, **7**, e37439.

Charlson R., Lovelock J., Andreae M. and Warren S. (1987). Oceanic phytoplankton, atmospheric sulphur, cloud albedo and climate. *Nature*, **326**, 655–661.

Chen L., Ren Y., Lin J., Liu X., Pang X. and Lin J. (2012). *Acidithiobacillus caldus* sulfur oxidation model based on transcriptome analysis between the wild type and sulfur oxygenase reductase defective mutant. *PLoS One*, **7**, e39470.

Cherney M. M., Zhang Y., Solomonson M., Weiner J. H. and James M. N. (2010). Crystal structure of sulfide:quinone oxidoreductase from *Acidithiobacillus ferrooxidans*: insights into sulfidotrophic respiration and detoxification. *Journal of Molecular Biology*, **398**, 292–305.

Christel S., Fridlund J., Buetti-Dinh A., Buck M., Watkin E. L. and Dopson M. (2016). RNA transcript sequencing reveals inorganic sulfur compound oxidation pathways in the acidophile *Acidithiobacillus ferrivorans*. *FEMS Microbiology Letters*, **363**, fnw057.

Colman D. R., Lindsay M. R., Amenabar M. J. and Boyd E. S. (2019). The intersection of geology, geochemistry, and microbiology in continental hydrothermal systems. *Astrobiology*, **19**, 1505–1522.

Cook A. M., Smits T. H. M. and Denger K. (2008). Sulfonates and organotrophic sulfite metabolism. In: Microbial Sulfur Metabolism, C. Dahl and C. G. Friedrich (eds), Springer-Verlag, Berlin, Heidelberg, pp. 170–183.

Cosmidis J., Nims C. W., Diercks D. and Templeton A. S. (2019). Formation and stabilization of elemental sulfur through organomineralization. *Geochimica et Cosmochimica Acta*, **247**, 59–82.

Cypionka H., Smock A. M. and Böttcher M. E. (1998). A combined pathway of sulfur compound disproportionation in *Desulfovibrio desulfuricans*. *FEMS Microbiology Letters*, **166**, 181–186.

Dahl C. (2015). Cytoplasmic sulfur trafficking in sulfur-oxidizing prokaryotes. *IUBMB Life*, **67**, 268–274.

Dahl C. (2017). Sulfur metabolism in phototrophic bacteria. In: Modern Topics in the Phototrophic Prokaryotes: Metabolism, Bioenergetics and Omics, P. C. Hallenbeck (ed.), Springer International Publishing, Cham, pp. 27–66.

Dahl C. (2020). Bacterial intracellular sulfur globules. In: Bacterial Organelles and Organelle-like Inclusions, D. Jendrossek (ed.), Springer, Heidelberg Dordrecht London New York, in press.

Dahl C. and Prange A. (2006). Bacterial sulfur globules: occurrence, structure and metabolism. In: Inclusions in Prokaryotes, J. M. Shively (ed.), Springer, Berlin Heidelberg, pp. 21–51.

Dahl C., Prange A. and Steudel R. (2002). Natural polymeric sulfur compounds. In: Miscellaneous Biopolymers and Biodegradation of Synthetic Polymers, A. Steinbüchel (ed.), Wiley-VCH, Weinheim, pp. 35–62.

Dahl C., Friedrich C. G. and Kletzin A. (2008). Sulfur oxidation in prokaryotes. In: Encyclopedia of Life Sciences (ELS), John Wiley & Sons, Ltd., Chichester, http://www.els.net/ [DOI: 10.1002/9780470015902.a9780470021155].

Dahl C., Franz B., Hensen D., Kesselheim A. and Zigann R. (2013). Sulfite oxidation in the purple sulfur bacterium *Allochromatium vinosum*: Identification of SoeABC as a major player and relevance of SoxYZ in the process. *Microbiology (Reading, England)*, **159**, 2626–2638.

Dam B., Mandal S., Ghosh W., Das Gupta S. K. and Roy P. (2007). The S4-intermediate pathway for the oxidation of thiosulfate by the chemolithoautotroph *Tetrathiobacter kashmirensis* and inhibition of tetrathionate oxidation by sulfite. *Research Microbiological*, **158**, 330–338.

de Jong G. A. H., Hazeu W., Bos P. and Kuenen J. G. (1997). Isolation of the tetrathionate hydrolase from *Thiobacillus acidophilus*. *European Journal of Biochemistry*, **243**, 678–683.

Denkmann K., Grein F., Zigann R., Siemen A., Bergmann J., van Helmont S., Nicolai A., Pereira I. A. C. and Dahl C. (2012). Thiosulfate dehydrogenase: a wide-spread unusual acidophilic *c*-type cytochrome. *Environmental Microbiology*, **14**, 2673–2688.

de Zwart J. M. M., Sluis J. M. R. and Kuenen J. G. (1997). Competition for dimethyl sulfide and hydrogen sulfide by *Methylophaga sulfidovorans* and *Thiobacillus thioparus* T5 in continuous cultures. *Applied and Environmental Microbiology*, **63**, 3318–3322.

Duarte A. G., Santos A. A. and Pereira I. A. (2016). Electron transfer between the QmoABC membrane complex and adenosine 5′-phosphosulfate reductase. *Biochimica et Biophysica Acta*, **1857**, 380–386.

Dubilier N., Bergin C. and Lott C. (2008). Symbiotic diversity in marine animals: the art of harnessing chemosynthesis. *Nature Reviews Microbiology*, **6**, 725–740.

Ernst C., Kayashta K., Koch T., Venceslau S. S., Pereira I. A. C., Demmer U., Ermler U. and Dahl C. (2020). Structural and spectroscopic characterization of a HdrA-like subunit from *Hyphomicrobium denitrificans*. *FEBS Journal*, doi 10.1111/febs.15505.

Espejo R. T. and Romero P. (1987). Growth of *Thiobacillus ferrooxidans* on elemental sulfur. *Applied and Environmental Microbiology*, **53**, 1907–1912.

Fike D. A., Gammon C. L., Ziebis W. and Orphan V. J. (2008). Micron-scale mapping of sulfur cycling across the oxycline of a cyanobacterial mat: a paired nanoSIMS and CARD-FISH approach. *The ISME Journal*, **2**, 749–759.

Finster K. (2008). Microbiological disproportionation of inorganic sulfur compounds. *Journal of Sulfur Chemistry*, **29**, 281–292.

Finster K. W., Kjeldsen K. U., Kube M., Reinhardt R., Mussmann M., Amann R. and Schreiber L. (2013). Complete genome sequence of *Desulfocapsa sulfexigens*, a marine deltaproteobacterium specialized in disproportionating inorganic sulfur compounds. *Standards in Genomic Sciences*, **8**, 58–68.

Franz B., Lichtenberg H., Hormes J., Modrow H., Dahl C. and Prange A. (2007). Utilization of solid "elemental" sulfur by the phototrophic purple sulfur bacterium *Allochromatium vinosum*: A sulfur K-edge XANES spectroscopy study. *Microbiology*, **153**, 1268–1274.

Frederiksen T. M. and Finster K. (2004). The transformation of inorganic sulfur compounds and the assimilation of organic and inorganic carbon by the sulfur disproportionating bacterium *Desulfocapsa sulfoexigens*. *Antonie van Leeuwenhoek*, **85**, 141–149.

Frenkiel L., Gros O. and Mouëza M. (1996). Gill ultrastructure in *Lucina pectinata* (Bivalvia: Lucinidae) with reference to hemoglobin in bivalves with symbiotic sulphur-oxidizing bacteria. *Marine Biology*, **125**, 511–524.

Friedrich C. G., Rother D., Bardischewsky F., Quentmeier A. and Fischer J. (2001). Oxidation of reduced inorganic sulfur compounds by bacteria: emergence of a common mechanism? *Applied and Environmental Microbiology*, **67**, 2873–2882.

Frigaard N. U. and Dahl C. (2009). Sulfur metabolism in phototrophic sulfur bacteria. *Advances in Microbial Physiology*, **54**, 103–200.

Gao R., Liu H. and Xun L. (2017). Cytoplasmic localization of sulfide:quinone oxidoreductase and persulfide dioxygenase of *Cupriavidus pinatubonensis* JMP134. *Applied and Environmental Microbiology*, **83**, e01820–01817.

Gavel O. Y., Bursakov S. A., Calvete J. J., George G. N., Moura J. J. G. and Moura I. (1998). ATP sulfurylases from sulfate-reducing bacteria of the genus *Desulfovibrio*. A novel metalloprotein containing cobalt and zinc. *Biochemistry*, **37**, 16225–16232.

Gehrke T., Telegdi J., Thierry D. and Sand W. (1998). Importance of extracellular polymeric substances from *Thiobacillus ferrooxidans* for bioleaching. *Applied and Environmental Microbiology*, **64**, 2743–2747.

George G. N., Pickering I. J., Yu E. Y. and Prince R. C. (2002). X-ray absorption spectroscopy of bacterial sulfur globules. *Microbiology*, **148**, 2267–2268.

George G. N., Gnida M., Bazylinski D. A., Prince R. C. and Pickering I. J. (2008). X-ray absorption spectroscopy as a probe of microbial sulfur biochemistry: the nature of bacterial sulfur globules revisited. *Journal of Bacteriology*, **190**, 6376–6383.

Ghosh W. and Dam B. (2009). Biochemistry and molecular biology of lithotrophic sulfur oxidation by taxonomically and ecologically diverse bacteria and archaea. *FEMS Microbiology Reviews*, **33**, 999–1043.

Giordano M., Norici A., Ratti S. and Raven J. A. (2008). Role of sulfur for algae: acquisition, metabolism, ecology and evolution. In: Sulfur Metabolism in Phototrophic Organisms, R. Hell, C. Dahl, D. B. Knaff and T. Leustek (eds), Springer, Dordrecht, pp. 397–415.

Götz F., Pjevac P., Markert S., McNichol J., Becher D., Schweder T., Mussmann M. and Sievert S. M. (2019). Transcriptomic and proteomic insight into the mechanism of cyclooctasulfur- versus thiosulfate-oxidation by the chemolithoautotroph *Sulfurimonas denitrificans*. *Environ. Microbiol*, **21**, 244–258.

Goubern M., Andriamihaja M., Nubel T., Blachier F. and Bouillaud F. (2007). Sulfide, the first inorganic substrate for human cells. *FASEB Journal*, **21**, 1699–1706.

Grabarczyk D. B. and Berks B. C. (2017). Intermediates in the Sox sulfur oxidation pathway are bound to a sulfane conjugate of the carrier protein SoxYZ. *PLoS One*, **12**, e0173395.

Grabarczyk D. B., Chappell P. E., Johnson S., Stelzl L. S., Lea S. M. and Berks B. C. (2015a). Structural basis for specificity and promiscuity in a carrier protein/enzyme system from the sulfur cycle. *Proceedings of the National Academy of Sciences of the USA*, **112**, E7166–E7175.

Grabarczyk D. B., Chappell P. E., Eisel B., Johnson S., Lea S. M. and Berks B. C. (2015b). Mechanism of thiosulfate oxidation in the SoxA family of cysteine-ligated cytochromes. *Journal of Biological Chemistry*, **290**, 9209–9221.

Grein F., Ramos A. R., Venceslau S. S. and Pereira I. A. C. (2013). Unifying concepts in anaerobic respiration: Insights from dissimilatory sulfur metabolism. *Biochimica et Biophysica Acta*, **1827**, 145–160.

Griesbeck C., Hauska G. and Schütz M. (2000). Biological sulfide oxidation: sulfide:quinone reductase (SQR), the primary reaction. In: Recent Research Developments in Microbiology, S. G. Pandalai (ed.), Research Signpost, Trivandrum, pp. 179–204.

Griesbeck C., Schütz M., Schödl T., Bathe S., Nausch L., Mederer N., Vielreicher M. and Hauska G. (2002). Mechanism of sulfide-quinone oxidoreductase investigated using site-directed mutagenesis and sulfur analysis. *Biochemistry*, **41**, 11552–11565.

Grote J., Schott T., Bruckner C. G., Glöckner F. O., Jost G., Teeling H., Labrenz M. and Jürgens K. (2012). Genome and physiology of a model Epsilonproteobacterium responsible for sulfide detoxification in marine oxygen depletion zones. *Proceedings of the National Academy of Sciences of the USA*, **109**, 506–510.

Guimaraes B. G., Barbosa R. L., Soprano A. S., Campos B. M., de Souza T. A., Tonoli C. C., Leme A. F., Murakami M. T. and Benedetti C. E. (2011). Plant pathogenic bacteria utilize biofilm growth-associated repressor (BigR), a novel winged-helix redox switch, to control hydrogen sulfide detoxification under hypoxia. *Journal of Biological Chemistry*, **286**, 26148–26157.

Hazeu W., Batenburg-van der Vegte W. H., Bos P., van der Pas R. K. and Kuenen J. G. (1988). The production and utilization of intermediary elemental sulfur during the oxidation of reduced sulfur compounds by *Thiobacillus ferrooxidans*. *Archives of Microbiology*, **150**, 574–579.

Hedderich R., Klimmek O., Kröger A., Dirmeier R., Keller M. and Stetter K. O. (1999). Anaerobic respiration with elemental sulfur and with disulfides. *FEMS Microbiology Reviews*, **22**, 353–381.

Henkel J. V., Dellwig O., Pollehne F., Herlemann D. P. R., Leipe T. and Schulz-Vogt H. N. (2019). A bacterial isolate from the Black Sea oxidizes sulfide with manganese(IV) oxide. *Proceedings of the National Academy of Sciences of the USA*, **116**, 12153–12155.

Hensel M., Hinsley A. P., Nikolaus T., Sawers G. and Berks B. C. (1999). The genetic basis of tetrathionate respiration in *Salmonella typhimurium*. *Molecular Microbiology*, **32**, 275–287.

Hensen D., Sperling D., Trüper H. G., Brune D. C. and Dahl C. (2006). Thiosulphate oxidation in the phototrophic sulphur bacterium *Allochromatium vinosum*. *Molecular Microbiology*, **62**, 794–810.

Himmel D., Maurin L. C., Gros O. and Mansot J. L. (2009). Raman microspectrometry sulfur detection and characterization in the marine ectosymbiotic nematode *Eubostrichus dianae* (Desmodoridae, Stilbonematidae). *Biology of the Cell*, **101**, 43–54.

Holmkvist L., Ferdelman T. G. and Jørgensen B. B. (2011). A cryptic sulfur cycle driven by iron in the methane zone of marine sediment (Aarhus Bay, Denmark). *Geochimica et Cosmochimica Acta*, **75**, 3581–3599.

Huang C. J. and Barrett E. L. (1990). Identification and cloning of genes involved in anaerobic sulfite reduction by *Salmonella typhimurium*. *Journal of Bacteriology*, **172**, 4100–4102.

Huang C. J. and Barrett E. L. (1991). Sequence analysis and expression of the *Salmonella typhimurium asr* operon encoding production of hydrogen sulfide from sulfite. *Journal of Bacteriology*, **173**, 1544–1553.

Janiczek O., Zemanova J. and Mandl M. (2007). Purification and some properties of thiosulfate dehydrogenase from *Acidithiobacillus ferrooxidans*. *Preparative Biochemistry & Biotechnology*, **37**, 101–111.

Jiang J., Chan A., Ali S., Saha A., Haushalter K. J., Lam W. L., Glasheen M., Parker J., Brenner M., Mahon S. B., Patel H. H., Ambasudhan R., Lipton S. A., Pilz R. B. and Boss G. R. (2016). Hydrogen sulfide - Mechanisms of toxicity and development of an antidote. *Scientific Reports*, **6**, 20831.

Johnson E. F. and Mukhopadhyay B. (2005). A new type of sulfite reductase, a novel coenzyme F-420-dependent enzyme, from the methanarchaeon *Methanocaldococcus jannaschii*. *Journal of Biological Chemistry*, **280**, 38776–38786.

Johnson D. B., Kanao T. and Hedrich S. (2012). Redox transformations of iron at extremely low pH: fundamental and applied aspects. *Frontiers in Microbiology*, **3**, 96.

Johnston D. T., Gill B. C., Masterson A., Beirne E., Casciotti K. L., Knapp A. N. and Berelson W. (2014). Placing an upper limit on cryptic marine sulphur cycling. *Nature*, **513**, 530–533.

Jørgensen B. B. (1982). Mineralization of organic matter in the sea bed – the role of sulphate reduction. *Nature*, **296**, 643–645.

Jørgensen B. B. (1990a). A thiosulfate shunt in the sulfur cycle of marine sediments. *Science*, **249**, 152–154.

Jørgensen B. B. (1990b). The sulfur cycle of freshwater sediments: Role of thiosulfate. *Limnology and Oceanography*, **35**, 1329–1342.

Jørgensen B. B. and Kasten S. (2006). Sulfur cycling and methane oxidation. In: Marine Geochemistry, H. D. Schulz and M. Zabel (eds), Springer, Berlin, pp. 271–309.

Jørgensen B. B. and Nelson D. C. (2004). Sulfide oxidation in marine sediments: geochemistry meets microbiology. In: Sulfur Biochemistry – Past and Present, J. P. Amend, K. J. Edwards and T. W. Lyons (eds), Geological Society of America, Special Paper 379, Boulder, Colorado, pp. 63–81.

Jørgensen B. B., Findlay A. J. and Pellerin A. (2019). The biogeochemical sulfur cycle of marine sediments. *Frontiers in Microbiology*, **10**, 849.

Kabil O. and Banerjee R. (2012). Characterization of patient mutations in human persulfide dioxygenase (ETHE1) involved in H_2S catabolism. *Journal of Biological Chemistry*, **287**, 44561–44567.

Kamyshny A. and Ferdelman T. G. (2010). Dynamics of zero-valent sulfur species including polysulfides at seep sites on intertidal sand flats (Wadden Sea, North Sea). *Marine Chemistry*, **121**, 17–26.

Kanao T., Matsumoto C., Shiraga K., Yoshida K., Takada J. and Kamimura K. (2010). Recombinant tetrathionate hydrolase from *Acidithiobacillus ferrooxidans* requires exposure to acidic conditions for proper folding. *FEMS Microbiology Letters*, **309**, 43–47.

Kappler U. (2008). Bacterial sulfite-oxidizing enzymes – enzymes for chemolithotrophy only? In: Microbial Sulfur Metabolism, C. Dahl and C. G. Friedrich (eds), Springer, Berlin Heidelberg, pp. 151–169.

Kappler U. (2011). Bacterial sulfite-oxidizing enzymes. *Biochimica et Biophysica Acta*, **1807**, 1–10.

Kappler U. and Dahl C. (2001). Enzymology and molecular biology of prokaryotic sulfite oxidation (minireview). *FEMS Microbiology Letters*, **203**, 1–9.

Kappler U. and Enemark J. H. (2014). Sulfite-oxidizing enzymes. *Journal of Biological Inorganic Chemistry*, **20**, 253–264.

Kappler U. and Schäfer H. (2014). Transformations of dimethylsulfide. *Metal Ions in Life Sciences*, **14**, 279–313.

Kappler U., Bennett B., Rethmeier J., Schwarz G., Deutzmann R., McEwan A. G. and Dahl C. (2000). Sulfite:cytochrome *c* oxidoreductase from *Thiobacillus novellus* – Purification, characterization, and molecular biology of a heterodimeric member of the sulfite oxidase family. *Journal of Biological Chemistry*, **275**, 13202–13212.

Kappler U., Nasreen M. and McEwan A. (2019). New insights into the molecular physiology of sulfoxide reduction in bacteria. *Advances in Microbial Physiology*, **75**, 1–51.

Kern M., Klotz M. G. and Simon J. (2011). The *Wolinella succinogenes mcc* gene cluster encodes an unconventional respiratory sulphite reduction system. *Molecular Microbiology*, **82**, 1515–1530.

Kettle A. J. and Andreae M. O. (2000). Flux of dimethylsulfide from the oceans: a comparison of updated datasets and flux models. *Journal of Geophysical Research*, **105**, 26793–26808.

Kikumoto M., Nogami S., Kanao T., Takada J. and Kamimura K. (2013). Tetrathionate-forming thiosulfate dehydrogenase from the acidophilic, chemolithoautotrophic bacterium *Acidithiobacillus ferrooxidans*. *Applied and Environmental Microbiology*, **79**, 113–120.

Kjeldsen K. U., Schreiber L., Thorup C. A., Boesen T., Bjerg J. T., Yang T., Dueholm M. S., Larsen S., Risgaard-Petersen N., Nierychlo M., Schmid M., Boggild A., van de Vossenberg J., Geelhoed J. S., Meysman F. J. R., Wagner M., Nielsen P. H., Nielsen L. P. and Schramm A. (2019). On the evolution and physiology of cable bacteria. *Proceedings of the National Academy of Sciences of the USA*, **116**, 19116–19125.

Kleinjan W. E., de Keizer A. and Janssen A. J. H. (2003). Biologically produced sulfur. In: Elemental Sulfur and Sulfur-rich Compounds I, R. Steudel (ed.), Springer, Berlin, pp. 167–187.

Kletzin A. (1989). Coupled enzymatic production of sulfite, thiosulfate, and hydrogen sulfide from sulfur: purification and properties of a sulfur oxygenase reductase from the facultatively anaerobic archaebacterium *Desulfurolobus ambivalens*. *Journal of Bacteriology*, **171**, 1638–1643.

Kletzin A., Urich T., Müller F., Bandeiras T. M. and Gomes C. M. (2004). Dissimilatory oxidation and reduction of elemental sulfur in thermophilic archaea. *Journal of Bioenergetics and Biomembranes*, **36**, 77–91.

Klimmek O., Kreis V., Klein C., Simon J., Wittershagen A. and Kröger A. (1998). The function of the periplasmic Sud protein in polysulfide respiration of *Wolinella succinogenes*. *European Journal of Biochemistry*, **253**, 263–269.

Klimmek O., Dietrich W., Dancea F., Lin Y. J., Pfeiffer S., Löhr F., Rüterjans H., Gross R., Simon J. and Kröger A. (2004). Sulfur respiration. In: Respiration in Bacteria and Archaea, D. Zannoni (ed.), Springer, Dordrecht, pp. 217–232.

Koch T. and Dahl C. (2018). A novel bacterial sulfur oxidation pathway provides a new link between the cycles of organic and inorganic sulfur compounds. *The ISME Journal*, **12**, 2479–2491.

Krafft T., Gross R. and Kröger A. (1995). The function of *Wolinella succinogenes psr* genes in electron transport with polysulphide as the terminal electron acceptor. *European Journal of Biochemistry*, **230**, 601–606.

Krämer M. and Cypionka H. (1989). Sulfate formation via ATP sulfurylase in thiosulfate- and sulfite-disproportionating bacteria. *Archives of Microbiology*, **151**, 232–237.

Kreke B. and Cypionka H. (1995). Energetics of sulfate transport in *Desulfomicrobium baculatum*. *Archives of Microbiology*, **163**, 307–309.

Kurth J. M., Dahl C. and Butt J. N. (2015). Catalytic protein film electrochemistry provides a direct measure of the tetrathionate/thiosulfate reduction potential. *Journal of the American Chemical Society*, **137**, 13232–13235.

Kurth J. M., Butt J. N., Kelly D. J. and Dahl C. (2016a). Influence of heme envionment on the catalytic properties of tetrathionate reductase (TsdA) from *Campylobacter jejuni*. *Bioscience Reports*, **36**, e00422.

Kurth J. M., Brito J. A., Reuter J., Flegler A., Koch T., Franke T., Klein E. M., Rowe S. F., Butt J. N., Denkmann K., Pereira I. A. C., Archer M. and Dahl C. (2016b). Electron accepting units of the diheme cytochrome *c* TsdA, a bifunctional thiosulfate dehydrogenase/tetrathionate reductase. *Journal of Biological Chemistry*, **291**, 24804–24818.

Kurth J. M., Schuster A., Seel W., Herresthal S., Simon J. and Dahl C. (2017). TsdC, a unique lipoprotein from *Wolinella succinogenes* that enhances tetrathionate reductase activity of TsdA. *FEMS Microbiology Letters*, **364**, fnx003.

Lahme S., Callbeck C. M., Eland L. E., Wipat A., Enning D., Head I. M. and Hubert C. R. J. (2020). Comparison of sulfide-oxidizing *Sulfurimonas* strains reveals a new mode of thiosulfate formation in subsurface environments. *Environmental Microbiology*, **22**, 1784–1800.

Lazar C. S., Baker B. J., Seitz K. W. and Teske A. P. (2017). Genomic reconstruction of multiple lineages of uncultured benthic archaea suggests distinct biogeochemical roles and ecological niches. *The ISME Journal*, **11**, 1058.

Lenk S., Arnds J., Zerjatke K., Musat N., Amann R. and Mussmann M. (2011). Novel groups of Gammaproteobacteria catalyse sulfur oxidation and carbon fixation in a coastal, intertidal sediment. *Environmental Microbiology*, **13**, 758–774.

Lenk S., Moraru C., Hahnke S., Arnds J., Richter M., Kube M., Reinhardt R., Brinkhoff T., Harder J., Amann R. and Mussmann M. (2012). *Roseobacter* clade bacteria are abundant in coastal sediments and encode a novel combination of sulfur oxidation genes. *The ISME Journal*, **6**, 2178–2187.

Leustek T., Martin M. N., Bick J. A. and Davies J. P. (2000). Pathways and regulation of sulfur metabolism revealed through molecular and genetic studies. *Annual Review of Plant Physiology and Plant Molecular Biology*, **51**, 141–165.

Libiad M., Yadav P. K., Vitvitsky V., Martinov M. and Banerjee R. (2014). Organization of the human mitochondrial hydrogen sulfide oxidation pathway. *Journal of Biological Chemistry*, **289**, 30901–30910.

Liu Y. W., Denkmann K., Kosciow K., Dahl C. and Kelly D. J. (2013). Tetrathionate stimulated growth of *Campylobacter jejuni* identifies TsdA as a new type of bi-functional tetrathionate reductase that is widely distributed in bacteria. *Molecular Microbiology*, **88**, 173–188.

Liu H., Xin Y. and Xun L. (2014a). Distribution, diversity, and activities of sulfur dioxygenases in heterotrophic bacteria. *Applied and Environmental Microbiology*, **80**, 1799–1806.

Liu L. J., Stockdreher Y., Koch T., Sun S. T., Fan Z., Josten M., Sahl H. G., Wang Q., Luo Y. M., Liu S. J., Dahl C. and Jiang C. Y. (2014b). Thiosulfate transfer mediated by DsrE/TusA homologs from acidothermophilic sulfur-oxidizing archaeon *Metallosphaera cuprina*. *Journal of Biological Chemistry*, **289**, 26949–26959.

Löffler M., Feldhues J., Venceslau S. S., Kammler L., Grein F., Pereira I. A. C. and Dahl C. (2020). DsrL mediates electron transfer between NADH and rDsrAB in *Allochromatium vinosum*. *Environmental Microbiology*, **22**, 783–795.

Lu W. P. and Kelly D. P. (1988). Respiration-driven proton translocation in *Thiobacillus versutus* and the role of the periplasmic thiosulphate-oxidizing enzyme system. *Archives of Microbiology*, **149**, 297–302.

Luther G. W., 3rd, Findlay A. J., Macdonald D. J., Owings S. M., Hanson T. E., Beinart R. A. and Girguis P. R. (2011). Thermodynamics and kinetics of sulfide oxidation by oxygen: a look at inorganically controlled reactions and biologically mediated processes in the environment. *Frontiers in Microbiology*, **2**, 62.

Mangold S., Valdés J., Holmes D. S. and Dopson M. (2011). Sulfur metabolism in the extreme acidophile *Acidithiobacillus caldus*. *Frontiers in Microbiology*, **2**, 17.

Marcia M., Ermler U., Peng G. H. and Michel H. (2009). The structure of *Aquifex aeolicus* sulfide: quinone oxidoreductase, a basis to understand sulfide detoxification and respiration. *Proceedings of the National Academy of Sciences of the USA*, **106**, 9625–9630.

Marnocha C. L., Levy A. T., Powell D. H., Hanson T. E. and Chan C. S. (2016). Mechanisms of extracellular S(0) globule production and degradation in *Chlorobaculum tepidum* via dynamic cell-globule interactions. *Microbiology (Reading, England)*, **162**, 1125–1134.

Marnocha C. L., Sabanayagam C. R., Modla S., Powell D. H., Henri P. A., Steele A. S., Hanson T. E., Webb S. M. and Chan C. S. (2019). Insights into the mineralogy and surface chemistry of extracellular biogenic S^0 globules produced by *Chlorobaculum tepidum*. *Frontiers in Microbiology*, **10**, 271.

Marzocchi U., Trojan D., Larsen S., Meyer R. L., Revsbech N. P., Schramm A., Nielsen L. P. and Risgaard-Petersen N. (2014). Electric coupling between distant nitrate reduction and sulfide oxidation in marine sediment. *The ISME Journal*, **8**, 1682–1690.

Meier D. V., Pjevac P., Bach W., Hourdez S., Girguis P. R., Vidoudez C., Amann R. and Meyerdierks A. (2017). Niche partitioning of diverse sulfur-oxidizing bacteria at hydrothermal vents. *The ISME Journal*, **11**, 1545–1558.

Meyer B. and Kuever J. (2007). Phylogeny of the alpha and beta subunits of the dissimilatory adenosine-5′-phosphosulfate (APS) reductase from sulfate-reducing prokaryotes – origin and evolution of the dissimilatory sulfate-reduction pathway. *Microbiology*, **153**, 2026–2044.

Meysman F. J. R., Cornelissen R., Trashin S., Bonne R., Martinez S. H., van der Veen J., Blom C. J., Karman C., Hou J. L., Eachambadi R. T., Geelhoed J. S., Wael K., Beaumont H. J. E., Cleuren B., Valcke R., van der Zant H. S. J., Boschker H. T. S. and Manca J. V. (2019). A highly conductive fibre network enables centimetre-scale electron transport in multicellular cable bacteria. *Nature Communications*, **10**, 4120.

Middelburg J. J. (2000). The geochemical sulfur cycle. In: Environmental Technologies to Treat Sulfur Pollution, P. Lens and W. Hulshoff Pol (eds), IWA Publishing, London, pp. 33–46.

Mikucki J. A., Pearson A., Johnston D. T., Turchyn A. V., Farquhar J., Schrag D. P., Anbar A. D., Priscu J. C. and Lee P. A. (2009). A contemporary microbially maintained subglacial ferrous "ocean". *Science*, **324**, 397–400.

Milucka J., Widdel F. and Shima S. (2013). Immunological detection of enzymes for sulfate reduction in anaerobic methane-oxidizing consortia. *Environmental Microbiology*, **15**, 1561–1571.

Moser D. P. and Nealson K. H. (1996). Growth of the facultative anaerobe *Shewanella putrefaciens* by elemental sulfur reduction. *Applied and Environmental Microbiology*, **62**, 2100–2105.

Mowat C. G., Rothery E., Miles C. S., McIver L., Doherty M. K., Drewette K., Taylor P., Walkinshaw M. D., Chapman S. K. and Reid G. A. (2004). Octaheme tetrathionate reductase is a respiratory enzyme with novel heme ligation. *Nature Structural & Molecular Biology*, **11**, 1023–1024.

Müller F. H., Bandeiras T. M., Urich T., Teixeira M. and Kletzin A. (2004). Coupling of the pathway of sulphur oxidation to dioxygen reduction: characterization of a novel membrane-bound thiosulphate:quinone oxidoreductase. *Molecular Microbiology*, **53**, 1147–1160.

Müller H., Bosch J., Griebler C., Damgaard L. R., Nielsen L. P., Lueders T. and Meckenstock R. U. (2016). Long-distance electron transfer by cable bacteria in aquifer sediments. *The ISME Journal*, **10**, 2010–2019.

Müller H., Marozava S., Probst A. J. and Meckenstock R. U. (2019). Groundwater cable bacteria conserve energy by sulfur disproportionation. *The ISME Journal*, **14**, 623–634.

Muyzer G. and Stams A. J. (2008). The ecology and biotechnology of sulphate-reducing bacteria. *Nature Reviews Microbiology*, **6**, 441–454.

Nauhaus K., Albrecht M., Elvert M., Boetius A. and Widdel F. (2007). *In vitro* cell growth of marine archaeal–bacterial consortia during anaerobic oxidation of methane with sulphate. *Environmental Microbiology*, **9**, 187–196.

Nelson D. C. and Castenholz R. W. (1981). Use of reduced sulfur compounds by *Beggiatoa* sp. *Journal of Bacteriology*, **147**, 140–154.

Neutzling O., Pfleiderer C. and Trüper H. G. (1985). Dissimilatory sulphur metabolism in phototrophic "non-sulphur" bacteria. *Journal of General Microbiology*, **131**, 791–798.

Nielsen L. P., Risgaard-Petersen N., Fossing H., Christensen P. B. and Sayama M. (2010). Electric currents couple spatially separated biogeochemical processes in marine sediment. *Nature*, **463**, 1071–1074.

Oliveira T. F., Vonrhein C., Matia P. M., Venceslau S. S., Pereira I. A. C. and Archer M. (2008). The crystal structure of *Desulfovibrio vulgaris* dissimilatory sulfite reductase bound to DsrC provides novel insights into the mechanism of sulfate respiration. *Journal of Biological Chemistry*, **283**, 34141–34149.

Oltman L. F., Claasen V. P., Kastelein P., Reijinders W. N. M. and Stouthammer A. H. (1979). The influence of tungstate on the formation and activities of four reductases of Proteus mirabilis: identification of two new molybdoenzymes, chlorate reductase and tetrathionate reductase. *FEBS Letters*, **106**, 43–46.

Orphan V. J., Hinrichs K. U., Ussler W., 3rd, Paull C. K., Taylor L. T., Sylva S. P., Hayes J. M. and Delong E. F. (2001). Comparative analysis of methane-oxidizing archaea and sulfate-reducing bacteria in anoxic marine sediments. *Applied and Environmental Microbiology*, **67**, 1922–1934.

Ott J., Bright M. and Bulgheresi S. (2004). Marine microbial thiotrophic ectosymbioses. In: Oceanography and Marine Biology – An Annual Review, R. N. Gibson, R. J. A. Atkinson and J. D. M. Gordon (eds), CRC Press, Boca Raton, pp. 95–118.

Parey K., Fritz G., Ermler U. and Kroneck P. M. (2013). Conserving energy with sulfate around 100°C. Structure and mechanism of key metal enzymes in hyperthermophilic *Archaeoglobus fulgidus. Metallomics: Integrated Biometal Science*, **5**, 302–317.

Petersen J. M., Kemper A., Gruber-Vodicka H., Cardini U., van der Geest M., Kleiner M., Bulgheresi S., Mussmann M., Herbold C., Seah B. K., Antony C. P., Liu D., Belitz A. and Weber M. (2016). Chemosynthetic symbionts of marine invertebrate animals are capable of nitrogen fixation. *Nature Microbiology*, **2**, 16195.

Pfeffer C., Larsen S., Song J., Dong M., Besenbacher F., Meyer R. L., Kjeldsen K. U., Schreiber L., Gorby Y. A., El-Naggar M. Y., Leung K. M., Schramm A., Risgaard-Petersen N. and Nielsen L. P. (2012). Filamentous bacteria transport electrons over centimetre distances. *Nature*, **491**, 218–221.

Pham V. H., Yong J. J., Park S. J., Yoon D. N., Chung W. H. and Rhee S. K. (2008). Molecular analysis of the diversity of the sulfide:quinone reductase (*sqr*) gene in sediment environments. *Microbiology*, **154**, 3112–3121.

Philips D. and Philips S. L. (2000). High temperature dissociation constants of HS^- and the standard thermodynamic values for S_2^-. *Journal of Chemical and Engineering Data*, **45**, 981–987.

Pires R. H., Lourenco A. I., Morais F., Teixeira M., Xavier A. V., Saraiva L. M. and Pereira I. A. C. (2003). A novel membrane-bound respiratory complex from *Desulfovibrio desulfuricans* ATCC 27774. *Biochimica et Biophysica Acta*, **1605**, 67–82.

Pjevac P., Meier D. V., Markert S., Hentschker C., Schweder T., Becher D., Gruber-Vodicka H. R., Richter M., Bach W., Amann R. and Meyerdierks A. (2018). Metaproteogenomic profiling of microbial communities colonizing actively venting hydrothermal chimneys. *Frontiers in Microbiology*, **9**, 680.

Prange A., Engelhardt H., Trüper H. G. and Dahl C. (2004). The role of the sulfur globule proteins of *Allochromatium vinosum*: mutagenesis of the sulfur globule protein genes and expression studies by real-time RT PCR. *Archives of Microbiology*, **182**, 165–174.

Protze J., Müller F., Lauber K., Nass B., Mentele R., Lottspeich F. and Kletzin A. (2011). An extracellular tetrathionate hydrolase from the thermoacidophilic archaeon *Acidianus ambivalens* with an activity optimum at pH 1. *Frontiers in Microbiology*, **2**, 68.

Pyne P., Alam M. and Ghosh W. (2017). A novel *soxO* gene, encoding a glutathione disulfide reductase, is essential for tetrathionate oxidation in *Advenella kashmirensis*. *Microbiology Research*, **205**, 1–7.

Pyne P., Alam M., Rameez M. J., Mandal S., Sar A., Mondal N., Debnath U., Mathew B., Misra A. K., Mandal A. K. and Ghosh W. (2018). Homologs from sulfur oxidation (Sox) and methanol dehydrogenation (Xox) enzyme systems collaborate to give rise to a novel pathway of chemolithotrophic tetrathionate oxidation. *Molecular Microbiology*, **109**, 169–191.

Quatrini R., Appia-Ayme C., Denis Y., Jedlicki E., Holmes D. S. and Bonnefoy V. (2009). Extending the models for iron and sulfur oxidation in the extreme acidophile *Acidithiobacillus ferrooxidans*. *BMC Genomics*, **10**, 394.

Rabus R., Venceslau S. S., Wohlbrand L., Voordouw G., Wall J. D. and Pereira I. A. (2015). A post-genomic view of the ecophysiology, catabolism and biotechnological relevance of sulphate-reducing prokaryotes. *Advances in Microbial Physiology*, **66**, 55–321.

Rameez M. J., Pyne P., Mandal S., Chatterjee S., Alam M., Bhattacharya S., Mondal N., Sarkar J. and Ghosh W. (2019). Two pathways for thiosulfate oxidation in the alphaproteobacterial chemolithotroph *Paracoccus thiocyanatus* SST. *Microbiology Research*, **230**, 126345.

Ramos A. R., Keller K. L., Wall J. D. and Pereira I. A. C. (2012). The membrane QmoABC complex interacts directly with the dissimilatory adenosine 5′-phosphosulfate reductase in sulfate reducing bacteria. *Frontiers in Microbiology*, **3**, 137.

Reinartz M., Tschäpe J., Brüser T., Trüper H. G. and Dahl C. (1998). Sulfide oxidation in the phototrophic sulfur bacterium *Chromatium vinosum*. *Archives of Microbiology*, **170**, 59–68.

Risgaard-Petersen N., Kristiansen M., Frederiksen R. B., Dittmer A. L., Bjerg J. T., Trojan D., Schreiber L., Damgaard L. R., Schramm A. and Nielsen L. P. (2015). Cable bacteria in freshwater sediments. *Applied and Environmental Microbiology*, **81**, 6003–6011.

Rohwerder T. and Sand W. (2003). The sulfane sulfur of persulfides is the actual substrate of the sulfur-oxidizing enzymes from *Acidithiobacillus* and *Acidiphilium* spp. *Microbiology*, **149**, 1699–1709.

Rother D., Heinrich H. J., Quentmeier A., Bardischewsky F. and Friedrich C. G. (2001). Novel genes of the *sox* gene cluster, mutagenesis of the flavoprotein SoxF, and evidence for a general sulfur-oxidizing system in *Paracoccus pantotrophus* GB17. *Journal of Bacteriology*, **183**, 4499–4508.

Rühl P., Poll U., Braun J., Klingl A. and Kletzin A. (2017). A sulfur oxygenase from the haloalkaliphilic bacterium *Thioalkalivibrio paradoxus* with atypically low reductase activity. *Journal of Bacteriology*, **199**, e00675–00616.

Santos A. A., Venceslau S. S., Grein F., Leavitt W. D., Dahl C., Johnston D. T. and Pereira I. A. (2015). A protein trisulfide couples dissimilatory sulfate reduction to energy conservation. *Science*, **350**, 1541–1545.

Sattler S. A., Wang X., Lewis K. M., DeHan P. J., Park C. M., Xin Y., Liu H., Xian M., Xun L. and Kang C. (2015). Characterizations of two bacterial persulfide dioxygenases of the metallo-beta-lactamase superfamily. *Journal of Biological Chemistry*, **290**, 18914–18923.

Sauvé V., Bruno S., Berks B. C. and Hemmings A. M. (2007). The SoxYZ complex carries sulfur cycle intermediates on a peptide swinging arm. *Journal of Biological Chemistry*, **282**, 23194–23204.

Sauvé V., Roversi P., Leath K. J., Garman E. F., Antrobus R., Lea S. M. and Berks B. C. (2009). Mechanism for the hydrolysis of a sulfur-sulfur bond based on the crystal structure of the thiosulfohydrolase SoxB. *Journal of Biological Chemistry*, **284**, 21707–21718.

Schäfer H., Myronova N. and Boden R. (2010). Microbial degradation of dimethylsulphide and related C_1-sulphur compounds: organisms and pathways controlling fluxes of sulphur in the biosphere. *Journal of Experimental Botany*, **61**, 315–334.

Schiffer A., Parey K., Warkentin E., Diederichs K., Huber H., Stetter K. O., Kroneck P. M. H. and Ermler U. (2008). Structure of the dissimilatory sulfite reductase from the hyperthermophilic archaeon *Archaeoglobus fulgidus*. *Journal of Molecular Biology*, **379**, 1063–1074.

Schütz M., Maldener I., Griesbeck C. and Hauska G. (1999). Sulfide-quinone reductase from *Rhodobacter capsulatus*: requirement for growth, periplasmic localization, and extension of gene sequence analysis. *Journal of Bacteriology*, **181**, 6516–6523.

Seitz K. W., Lazar C. S., Hinrichs K. U., Teske A. P. and Baker B. J. (2016). Genomic reconstruction of a novel, deeply branched sediment archaeal phylum with pathways for acetogenesis and sulfur reduction. *The ISME Journal*, **10**, 1696–1705.

Shen J., Keithly M. E., Armstrong R. N., Higgins K. A., Edmonds K. A. and Giedroc D. P. (2015). *Staphylococcus aureus* CstB Is a novel multidomain persulfide dioxygenase-sulfurtransferase involved in hydrogen sulfide detoxification. *Biochemistry*, **54**, 4542–4554.

Shirodkar S., Reed S., Romine M. and Saffarini D. (2011). The octahaem SirA catalyses dissimilatory sulfite reduction in *Shewanella oneidensis* MR-1. *Environmental Microbiology*, **13**, 108–115.

Sievert S. M., Kiene R. P. and Schulz-Vogt H. N. (2007). The sulfur cycle. *Oceanography*, **20**, 117–123.

Sievert S. M., Scott K. A., Klotz M. G., Chain P. S. G., Hauser L. J., Hemp J., Hugler M., Land M., Lapidus A., Larimer F. W., Lucas S., Malfatti S. A., Meyer F., Paulsen I. T., Ren Q. and Simon J. (2008). Genome of the epsilonproteobacterial chemolithoautotroph *Sulfurimonas denitrificans*. *Applied and Environmental Microbiology*, **74**, 1145–1156.

Simon J. and Kroneck P. M. (2013). Microbial sulfite respiration. *Advances in Microbial Physiology*, **62**, 45–117.

Sorokin D. Y. (1995). *Sulfitobacter pontiacus* gen. nov., sp. nov. – A new heterotrophic bacterium from the Black Sea, specialized on sulfite oxidation. *Microbiology*, **64**, 295–305.

Steudel R. (1996a). Mechanism for the formation of elemental sulfur from aqueous sulfide in chemical and microbiological desulfurization processes. *Industrial & Engineering Chemistry Research*, **35**, 1417–1423.

Steudel R. (1996b). Das gelbe Element und seine erstaunliche Vielseitigkeit. *Chemie Unserer Zeit*, **30**, 226–234.

Steudel R. (2000). The chemical sulfur cycle. In: Environmental Technologies to Treat Sulfur Pollution, P. Lens and W. Hulshoff Pol (eds), IWA Publishing, London, pp. 1–31.

Steudel R. and Chivers T. (2019). The role of polysulfide dianions and radical anions in the chemical, physical and biological sciences, including sulfur-based batteries. *Chemical Society Reviews*, **48**, 3279–3319.

Steudel R., Hold G., Göbel T. and Hazeu W. (1987). Chromatographic separation of higher polythionates $S_nO_6^{2-}$ (n = 3…22) and their detection in cultures of *Thiobacillus ferrooxidans*; Molecular composition of bacterial sulfur secretions. *Angewandte Chemie International Edition*, **26**, 151–153.

Steudel R., Holdt G., Visscher P. T. and van Gemerden H. (1990). Search for polythionates in cultures of *Chromatium vinosum* after sulfide incubation. *Archives of Microbiology*, **155**, 432–437.

Stoffels L., Krehenbrink M., Berks B. C. and Unden G. (2012). Thiosulfate reduction in *Salmonella enterica* is driven by the proton motive force. *Journal of Bacteriology*, **194**, 475–485.

Tanabe T. S., Leimkühler S. and Dahl C. (2019). The functional diversity of the prokaryotic sulfur carrier protein TusA. *Advances in Microbial Physiology*, **75**, 233–277.

Thamdrup B., Finster K., Hansen J. W. and Bak F. (1993). Bacterial disproportionation of elemental sulfur coupled to chemical reduction of iron or manganese. *Applied and Environmental Microbiology*, **59**, 101–108.

Thauer R. K., Jungermann K. and Decker K. (1977). Energy conservation in chemotrophic anaerobic bacteria. *Bacteriological Reviews*, **41**, 100–180.

Thomas F., Giblin A. E., Cardon Z. G. and Sievert S. M. (2014). Rhizosphere heterogeneity shapes abundance and activity of sulfur-oxidizing bacteria in vegetated salt marsh sediments. *Frontiers in Microbiology*, **5**, 309.

Thorup C., Schramm A., Findlay A. J., Finster K. W. and Schreiber L. (2017). Disguised as a sulfate reducer: growth of the Deltaproteobacterium *Desulfurivibrio alkaliphilus* by sulfide oxidation with nitrate. *MBio*, **8**, e00671–00617.

Venceslau S. S., Stockdreher Y., Dahl C. and Pereira I. A. C. (2014). The "bacterial heterodisulfide" DsrC is a key protein in dissimilatory sulfur metabolism. *Biochimica et Biophysica Acta*, **1837**, 1148–1164.

Vera M., Schippers A. and Sand W. (2013). Progress in bioleaching: fundamentals and mechanisms of bacterial metal sulfide oxidation–part A. *Applied and Environmental Microbiology*, **97**, 7529–7541.

Visser J. M., de Jong G. A., Robertson L. A. and Kuenen J. G. (1996). Purification and characterization of a periplasmic thiosulfate dehydrogenase from the obligately autotrophic *Thiobacillus* sp. W5. *Archives of Microbiology*, **166**, 372–378.

Wang H., Liu S., Liu X., Li X., Wen Q. and Lin J. (2014). Identification and characterization of an ETHE1-like sulfur dioxygenase in extremely acidophilic *Acidithiobacillus* spp. *Applied Microbiology and Biotechnology*, **98**, 7511–7522.

Wang R., Lin J.-Q., Liu X.-M., Pang X., Zhang C.-J., Yang C.-L., Gao X.-Y., Lin C.-M., Li Y.-Q., Li Y., Lin J.-Q. and Chen L.-X. (2019). Sulfur oxidation in the acidophilic autotrophic *Acidithiobacillus* spp. *Frontiers in Microbiology*, **9**, 3290.

Wasmund K., Mussmann M. and Loy A. (2017). The life sulfuric: microbial ecology of sulfur cycling in marine sediments. *Environmental Microbiology Reports*, **9**, 323–344.

Welte C., Hafner S., Krätzer C., Quentmeier A. T., Friedrich C. G. and Dahl C. (2009). Interaction between Sox proteins of two physiologically distinct bacteria and a new protein involved in thiosulfate oxidation. *FEBS Letters*, **583**, 1281–1286.

Wentzien S., Sand W., Albertsen A. and Steudel R. (1994). Thiosulfate and tetrathionate degradation as well as biofilm generation by *Thiobacillus intermedius* and *Thiobacillus versutus* studied by microcalorimetry, HPLC and ion-pair chromatography. *Archives of Microbiology*, **161**, 116–125.

Winter S. E., Thiennimitr P., Winter M. G., Butler B. P., Huseby D. L., Crawford R. W., Russell J. M., Bevins C. L., Adams L. G., Tsolis R. M., Roth J. R. and Baumler A. J. (2010). Gut inflammation provides a respiratory electron acceptor for *Salmonella*. *Nature*, **467**, 426–429.

Wu W., Pang X., Lin J., Liu X., Wang R., Lin J. and Chen L. (2017). Discovery of a new subgroup of sulfur dioxygenases and characterization of sulfur dioxygenases in the sulfur metabolic network of *Acidithiobacillus caldus*. *PLoS One*, **12**, e0183668.

Xia Y., Lu C., Hou N., Xin Y., Liu J., Liu H. and Xun L. (2017). Sulfide production and oxidation by heterotrophic bacteria under aerobic conditions. *The ISME Journal*, **11**, 2754–2766.

You X. Y., Guo X., Zheng H. J., Zhang M. J., Liu L. J., Zhu Y. Q., Zhu B., Wang S. Y., Zhao G. P., Poetsch A., Jiang C. Y. and Liu S. J. (2011). Unraveling the *Acidithiobacillus caldus* complete genome and its central metabolisms for carbon assimilation. *Journal of Genetics and Genomics*, **38**, 243–252.

Zander U., Faust A., Klink B. U., de Sanctis D., Panjikar S., Quentmeier A., Bardischewsky F., Friedrich C. G. and Scheidig A. J. (2010). Structural basis for the oxidation of protein-bound sulfur by the sulfur cycle molybdohemo-enzyme sulfane dehydrogenase SoxCD. *Journal of Biological Chemistry*, **286** 8349–8360.

Zhang R., Neu T. R., Li Q., Blanchard V., Zhang Y., Schippers A. and Sand W. (2019). Insight into interactions of thermoacidophilic archaea with elemental sulfur: Biofilm dynamics and EPS analysis. *Frontiers in Microbiology*, **10**, 896.

Zopfi J., Ferdelman T. G. and Fossing H. (2004). Distribution and fate of sulfur intermediates – sulfite, tetrathionate, thiosulfate and elemental sulfur – marine sediments. In: Sulfur Biogeochemistry – Past and Present, J. P. Amend, K. J. Edwards and T. W. Lyons (eds), Geological Society of America, Special Paper 379, Boulder, Colorado, pp. 97–116.

Part III

Sulfur-Related Problems

Chapter 4

Sulfur transformations in sewer networks: effects, prediction and mitigation of impacts

Thorkild Hvitved-Jacobsen

4.1 INTRODUCTION

Sulfur transformations in sewer networks are predominantly of importance because of the formation and occurrence of hydrogen sulfide. Hydrogen sulfide related to wastewater in sewers is a worldwide problem as it leads to concrete and metal corrosion, odor nuisance and human health impacts. The solution to these problems has its starting point in an understanding of the sewer network as a reactor for physical, chemical and biological sulfur processes that govern the formation of hydrogen sulfide and its control. In this respect, it becomes important to predict the interacting sulfur processes in terms of both simple empirical equations and more advanced models that are based on a conceptual understanding of the sulfur processes, the latter is of particular importance when applied to the complex constructed and functioning sewers of today.

Problems related to sewer gases have been known for centuries. However, no more than about 100 years ago, based on sound scientific and technical knowledge, it was recognized that corrosion and odor problems in wastewater collection systems are closely related to anaerobic conditions and a corresponding formation of hydrogen sulfide. In brief, sulfide formation takes place by the activity of anaerobic sulfate-reducing bacteria that are mainly found in sewer biofilms from where sulfide penetrates the wastewater phase (see also Chapter 5). It is the subsequent emission of hydrogen sulfide into the sewer atmosphere that

DOI: 10.2166/9781789060966_0099

results in corrosion of concrete and metal constructions in the wastewater collection system, and furthermore ventilation results in odor nuisance in the urban atmosphere. Although it is rather simple to discover this cause-effect relationship, the complex sewer networks require that the entire sulfur and carbon cycles be considered and investigated to predict and efficiently mitigate the adverse effects of hydrogen sulfide.

The design characteristics and operation mode of a sewer network determine to a great extent the redox conditions in the wastewater phase. Aerobic and anaerobic conditions are particularly of interest, whereas the anoxic state in principle only exists if nitrate – or oxidized inorganic nitrogen substances – is artificially added to the wastewater. If dissolved oxygen (DO) or nitrate is not present in the wastewater phase, strictly anaerobic conditions occur. The full flowing pressure sewers where dissolved oxygen is fast depleted are therefore in general exposed to anaerobic conditions and sulfide problems. In contrast, the air-water oxygen transfer, i.e. reaeration, is in a gravity sewer typically sufficient to maintain aerobic conditions and depress the occurrence of hydrogen sulfide in the water phase. Under such conditions, aerobic degradation of easily biodegradable organic matter is an important process and sulfur transformations are typically not an essential issue, although sulfide formation in the anaerobic parts of sewer biofilms may still proceed.

In brief, under anaerobic conditions sulfate is the dominating external electron acceptor for the microorganisms in the sewer biofilms resulting in formation of hydrogen sulfide. A subsequent release of hydrogen sulfide to the wastewater, sewer gas phase and the urban atmosphere may cause serious impacts. It is particularly under such conditions that insight into sulfur transformations is important and assessment and mitigation of the adverse impacts of sulfide become central.

4.2 SEWER NETWORK CHARACTERISTICS AND RELATED POTENTIAL FOR SULFUR TRANSFORMATIONS

A sewer is not just a collector and transport system for wastewater. It is a process reactor for wastewater constituents including microorganisms. The quality of wastewater, the constructive details and the operational characteristics of the sewer determine jointly the conditions for sulfide formation with molecular hydrogen sulfide, H_2S, in the sewer air phase and the urban atmosphere as the main harmful sulfur constituent (ASCE & WEF, 2007; USEPA, 1985).

4.2.1 Microbial and chemical process characteristics of sewer networks

Sulfide formation in sewers is a microbial, anaerobic, heterotrophic process. When dissolved oxygen and nitrate in the wastewater phase are depleted, the slow growing sulfate-reducing bacteria that primarily are found in the sewer biofilms and deposits use sulfate as the electron acceptor. Biodegradable organic matter (organic carbon) in the wastewater is the energy and carbon source for these heterotrophic bacteria (see also Chapter 3). Without taking into account organic matter for growth of the microorganisms, fermentation and methane formation, the hydrogen sulfide producing process is schematically:

$$SO_4^{2-} + \text{organic carbon} \rightarrow HCO_3^-(\text{carbon dioxide}) + H_2S$$

The potential for a build-up of fixed microbial structures as biofilms in the sewer supports its role as a process reactor. The sewer is in this respect a diverse system and microbial and chemical processes proceed in the following five compartments of the sewer where furthermore transfer of substances across interfaces between the different compartments takes place (Figure 4.1):

- The water phase with soluble substances and suspended organic and inorganic particles, including microorganisms.
- The sewer biofilm, i.e. slime. These terms are typically used when the microorganisms are attached to the submerged solid parts of the sewer but microorganisms attached to surfaces above the water table also form biofilms.
- The sewer sediments, i.e. deposits.

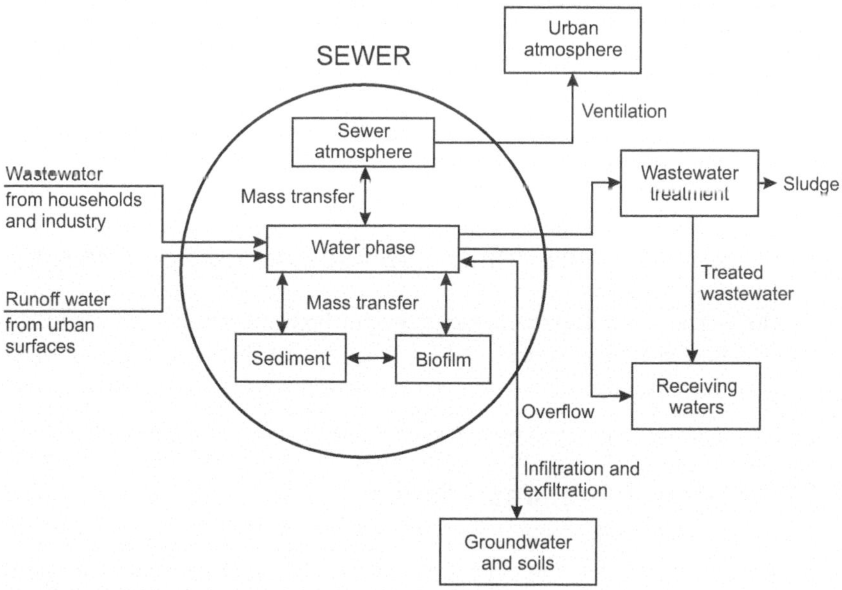

Figure 4.1 Overview of water, solids and gas flows in a sewer and their interactions with the surroundings (source with permission: Hvitved-Jacobsen *et al.*, 2013).

- The sewer walls exposed to the sewer atmosphere with the adhered microorganisms, i.e. biofilm.
- The sewer atmosphere.

The constructive and process related characteristics determine why sewers and sulfur constituents operate in combination and thereby underline why a sewer is a chemical and biological process reactor. In addition to the pipe system, it is important to notice that ancillary sewer constructions like inlet and outlet structures, drop structures, sewer falls, and junctions typically result in specific conditions for the chemical, biological and physical processes (Beceiro *et al.*, 2017).

When focusing on the collection system and omitting the influence of wastewater, the sewer network problems related to sulfide formation are typically caused by:

- Sewer constructions with low aeration potential that support in-sewer anaerobic conditions – or more correctly a low redox potential – causing formation of hydrogen sulfide and odorous volatile organic compounds (VOCs).
- Sewer constructions, e.g. sewer drops, with enhanced turbulent conditions and thereby – in addition to aeration – also risk for emission of molecular hydrogen sulfide and VOCs to the sewer gas phase.
- Openings and leaks in the collection system from where sewer gases can be emitted to an adjacent urban area.

In brief and when focusing on the process conditions in sewers, the following four aspects are central for occurrence of sulfide problems:

- Wastewater, in particular its quality as substrate for microorganisms, the pH value and the redox conditions.
- Sewer network, its construction and functioning as a reactor for microbial, chemical and physical processes.
- Microbial processes, their kinetic and stoichiometric characteristics related to sulfur constituents.
- Movement and transport characteristics of both wastewater and sewer gas, including interfacial processes for exchange of substances between water, air and solids.

These four aspects are to a greater extent explained below in Sections 4.2.2–4.2.5.

4.2.2 Wastewater characteristics

The wastewater in a sewer with its contents of organic and inorganic substances determines via particular microbial activity the type and the extent of sulfur transformation. In addition to organic matter as substrate for the heterotrophic microorganisms, it is particularly the occurrence or absence of dissolved oxygen and oxidized inorganic nitrogen constituents that govern the redox conditions. In this connection, it is important to note that there is a close relation between the type and functioning of a sewer and the potential for sulfur constituents to undergo transformation. Furthermore, the pH value of the wastewater affects the kinetics for a number of sulfur reactions and the relative amount of volatile, molecular sulfide, H_2S.

4.2.3 Sewer networks

The sewer networks that are relevant in this case are:

- Sanitary sewers (also named separate sewers), which are designed and constructed for collection and transport of wastewater from households, commercial buildings or industries.
- Combined sewers that furthermore receive runoff water from urban surfaces during rain periods.

The storm sewers that convey only runoff water from urban surfaces are basically not of relevance in this context.

The constructive and operational characteristics of a sewer determine the types of processes that may dominate sulfur transformations. In this respect there are two main types of sewers, namely gravity sewers and pressure sewers:

- A gravity sewer is designed with a sloping bottom and the flow occurs by gravitation. The water surface in a gravity sewer is thereby exposed to the sewer atmosphere. Air-water oxygen transfer, i.e. reaeration of the wastewater, and emission of volatile substances from the wastewater phase may take place. If reaeration of the wastewater phase dominates, sulfur transformations are subject to aerobic conditions.
- In a pressure sewer, the driving force for the flow of wastewater is pumping and it is therefore also named a pumping sewer or a force main. The sewer is characterized by full flowing movement of the wastewater, which prevents reaeration from taking place. Dissolved oxygen is therefore typically fast depleted resulting in anaerobic sulfur transformations.

4.2.4 Microbial and chemical processes

The fact that sulfur in sewer systems occurs in oxidation states from −2 (hydrogen sulfide, H_2S and HS^-) to +6 (sulfate, SO_4^{2-}) indicates that varying aerobic and anaerobic conditions, i.e. the redox conditions, via microbial reactions may result in rather drastic consequences for sulfur constituents. It is therefore important to recall which sulfur components may primarily exist in a sewer and where in the sewer they are typically found. From a process point of view the inorganic sulfur constituents dominate, whereas organic bound sulfur, in for example proteins and amino acids, plays a minor role (Table 4.1). Although Table 4.1 is a simple

Table 4.1 Overview of process characteristics for inorganic sulfur constituents related to their occurrence in sewer systems (Figure 4.1).

Sulfur Components	Oxidation State of S	The Sewer Compartment where the Components Primarily Occur	Major Process Characteristic of Components
Hydrogen sulfide, H_2S and HS^-	−2	Sewer biofilms, suspended water phase (H_2S and HS^-) and sewer atmosphere (H_2S)	Produced in anaerobic sewer biofilms and oxidized under aerobic conditions
Metal sulfides, e.g. FeS and ZnS	−2	Suspended water phase, sediments and biofilm	Precipitated naturally or by adding metal salts
Elemental sulfur, S^0	0	Sewer walls exposed to the sewer atmosphere	Intermediate product related to concrete corrosion
Thiosulfate, $S_2O_3^{2-}$	+5 and −1*	Sewer walls exposed to the sewer atmosphere	Intermediate product mainly related to concrete corrosion
Sulfate, SO_4^{2-}	+6	Suspended water phase and sewer biofilm	Available via the incoming wastewater and reduced under anaerobic conditions
Sulfuric acid, H_2SO_4 and HSO_4^-	+6	Sewer walls exposed to the sewer atmosphere	Produced under aerobic conditions and reactive in concrete corrosion

*The S-S bond is different from the S-O bond in thiosulfate resulting in the highest oxidation state for the central S atom (see also Chapter 2). The oxidation states of the two S atoms are with reference to IUPAC (2014) in the literature also denoted as +4 and 0, respectively.

interpretation of the real situation, it gives a first estimate of which sulfur constituents may dominate the different compartments of a sewer. At a more detailed level, quite different types of bacteria are able to take part in sulfur transformations (Karunakaran *et al.*, 2016).

Sulfur transformations are (with reference to Table 4.1) primarily controlled by redox reactions, i.e. processes where electrons at molecular scale are transferred from one atom to another. Transfer of electrons results in oxidation (electron donor) and reduction (electron acceptor) of the constituents involved. Sulfur components in the different oxidation states as shown in Table 4.1 are in this respect central, either as reactants or products. The redox conditions depend on the availability of an electron acceptor, i.e. the substance that receives electrons in the redox reaction. Table 4.2 outlines the main process conditions related to a sewer type where the electron acceptor typically is active.

Table 4.2 Overview of central sulfur associated redox processes in sewers related to an electron acceptor.

Redox Process and Electron Acceptor	Electron Donor Process	Important type of Transformation	Typical Sewer Conditions for the Redox Process
Aerobic: + oxygen	Sulfide oxidation, $H_2S \rightarrow S^0$, $S_2O_3^{2-}$ or SO_4^{2-}	Chemoautotrophic and chemical sulfide oxidation	Gravity sewer with fairly large flow and moderate temperature (1) at corroding concrete walls (2) in the water phase (3) in the wetted biofilm
Anoxic*: − oxygen + nitrate or nitrite	Sulfide oxidation**, $H_2S \rightarrow S^0$	Chemoautotrophic sulfide oxidation	Sewer where nitrate typically is added for sulfide control
Anaerobic − oxygen − nitrate and nitrite + sulfate	Biodegradable organic matter → CO_2 and low molecular organics	Heterotrophic sulfate respiration ($SO_4^{2-} \rightarrow H_2S$) in mainly biofilms	(1) pressure sewer (2) full flowing gravity sewer (3) gravity sewer with low slope and high temperature

*The term anoxic is used here to describe process conditions where dissolved oxygen is depleted and inorganic oxidized nitrogen species, typically nitrate or nitrite, are electron acceptors.

**Elemental sulfur, S^0, can be further oxidized to sulfate, SO_4^{2-}, but is apparently rather stable (Yang *et al.*, 2005).

Table 4.2 focuses on redox processes that are directly related to sulfur transformations. However, several other processes in sewers indirectly affect transformation of sulfur and its occurrence in the different compartments of the sewer. As an example, hydrolysis and fermentation under anaerobic conditions in a pressure sewer produce easily biodegradable organic matter that is used as the substrate for the heterotrophic sulfate respiration producing hydrogen sulfide in anaerobic sewer biofilms.

4.2.5 Transport characteristics

The water and gas flows in a sewer determine the locations for where the problematic substances are found. Furthermore, the interfacial exchange of substances between the solid biofilm, the water phase and the sewer atmosphere is of central importance for how and where the different constituents react. As an example, transport processes determine how reaeration in a gravity sewer maintains aerobic conditions in the wastewater and the upper part of the biofilm. Furthermore, the interfacial transport also controls how hydrogen sulfide formation in the sewer biofilm is transferred to the water phase and subsequently to the atmosphere. The ultimate occurrence of molecular hydrogen sulfide, H_2S, in the sewer atmosphere is the key problem as it is from the gas phase that concrete corrosion and malodors arise.

4.2.6 Formulation of the sulfur cycle in sewer networks

The overall result of what is described for sulfur transformations in sewers in Sections 4.2.1–4.2.5 is schematically expressed in Figure 4.2. Regarding sulfur transformations, it is important to consider interacting processes with other components, in particular the electron acceptors and the carbon cycle, as an integral part of the sulfur cycle. Although rather complex, Figure 4.2 is still a simple description of the sulfur cycle including associated physical, chemical and biological processes.

The basic understanding of sulfur transformations in sewers has its starting point in research projects carried out during the late 1940s (Parker, 1945a, b, 1951; Pomeroy & Bowlus, 1946). Results from investigations during the period 1950 through 1990 gradually improved this insight. Several research projects during the past 30 years have focused on understanding of the process kinetics and quantification of the processes and phenomena shown in Figure 4.2. The outcome is a more detailed understanding of the sulfur cycle including interactions with other constituents. The knowledge gained has been crucial for a conceptual formulation of the sewer processes in model terms (Section 4.5.4). Table 4.3 shows selected research papers that have been central for quantification of the processes shown in Figure 4.2.

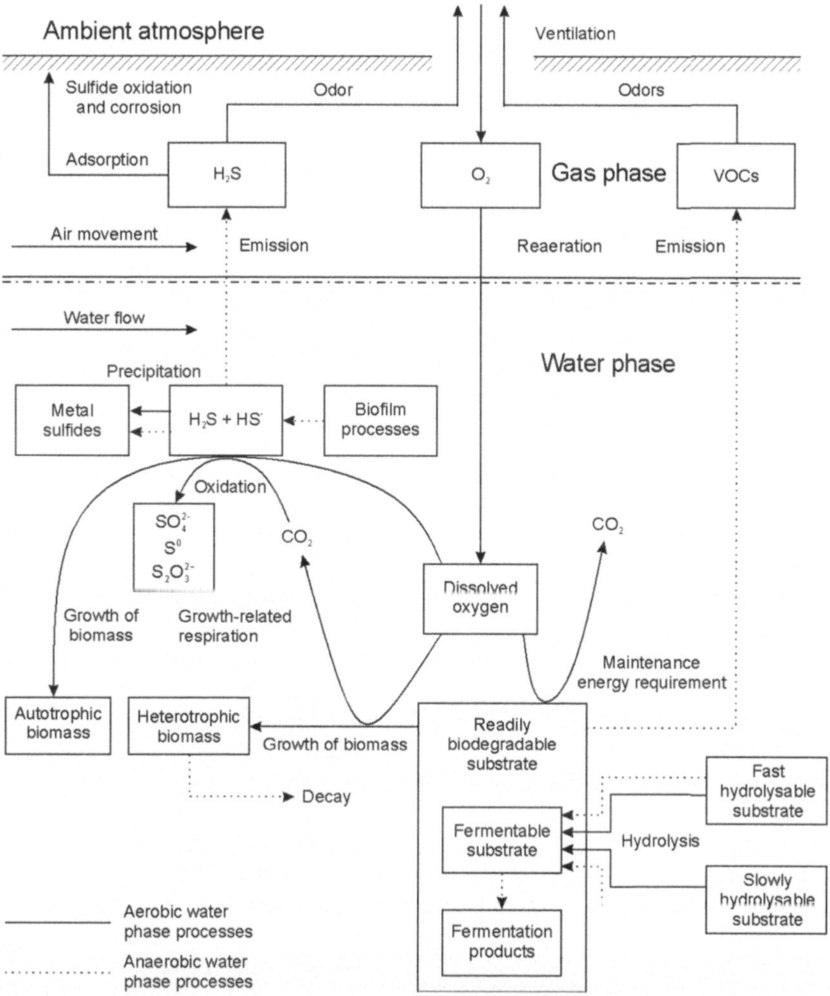

Figure 4.2 Conceptual formulated overview of aerobic, anaerobic and transport processes for carbon and sulfur in a sewer. The anoxic transformations and detailed interactions with metals and concrete are for simplicity reasons not shown (source with permission: Hvitved-Jacobsen *et al.*, 2013).

4.3 EFFECTS OF HYDROGEN SULFIDE IN SEWERS

The nature of the three main sulfur related problems of sewers, namely corrosion, odor and human health impacts, are briefly outlined below. These problems are initiated by the formation of sulfide and thereby particularly dependent on the redox conditions and the sewer type as outlined in Section 4.2 and depicted in Table 4.2:

Table 4.3 Selected research papers related to the formulation of the sewer processes shown in Figure 4.2.

Sewer Process and Modeling Approach	Selected Research Papers
Overall process description and modeling	Hvitved-Jacobsen *et al.* (1998) Hvitved-Jacobsen *et al.* (2002) Vollertsen *et al.* (2011) Hvitved-Jacobsen *et al.* (2013) Vollertsen *et al.* (2015)
Air-water mass transfer and gas phase transport	Yongsiri *et al.* (2004) Yongsiri *et al.* (2005) Madsen *et al.* (2006)
Sulfur transformations in biofilms and water phase	Nielsen and Hvitved-Jacobsen (1988) Nielsen *et al.* (1998) Nielsen *et al.* (2004)
Sulfide and iron interactions	Nielsen *et al.* (2005a) Nielsen *et al.* (2008) Kiilerich *et al.* (2018a) Kiilerich *et al.* (2018b)
Sulfur related corrosion processes	Jensen *et al.* (2009a) Jensen *et al.* (2009b) Jensen *et al.* (2011) Grengg *et al.* (2015) Yuan *et al.* (2015)
Aerobic-anoxic-anaerobic transient reactions	Tanaka *et al.* (2000a) Tanaka *et al.* (2000b) Yang *et al.* (2004) Yang *et al.* (2005)
Stoichiometric and kinetic model parameters	Vollertsen and Hvitved-Jacobsen (1999) Nielsen *et al.* (2005b)

- Concrete corrosion has its origin in the anaerobic sulfate respiration process in the sewer biofilms but is not active until H_2S is transferred to the water phase, emitted from there and subsequently under aerobic conditions oxidized by chemoautotrophic microorganisms to sulfuric acid at the sewer walls. Furthermore, hydrogen sulfide may corrode metals in, for example, ancillary sewer constructions.
- Odor is likewise not a problem until H_2S is emitted from the water phase and typically by ventilation released to the urban atmosphere. Volatile organic and inorganic substances that are primarily produced under anaerobic conditions in sewers may also give rise to odor problems. This aspect is given minor attention in this chapter which focuses on sulfur.

- Human health problems related to hydrogen sulfide in the gas phase ultimately have life threatening effects. H_2S is primarily a risk for sewer workers and occurs at high concentrations where the characteristic smell of H_2S is lost. Therefore, precautions are required to protect the workers, e.g. by continuous monitoring of H_2S and use of safety equipment.

In addition to these three main types of effects it should be mentioned that the following impacts are closely related to the presence of anaerobic conditions in wastewater of sewers:

- Inflow into treatment plants of anaerobic wastewater with sulfide.
- Combined sewer overflows with anaerobic wastewater into receiving waters.

It is important to note that as long sulfide remains in the water phase, these three main types of sulfur related problems do not occur. The pH dependent equilibrium in the water phase between the molecular and volatile form of sulfide, $H_2S(aq)$, and the ionic form (HS^-) is therefore crucial. Alkaline conditions in the wastewater may therefore correspondingly reduce hydrogen sulfide problems. Furthermore, it is interesting to note that concrete corrosion that reduces H_2S in the sewer atmosphere correspondingly decreases the potential for odor and human health problems. Although control of hydrogen sulfide is important for alleviation of the three types of problems, it is from an applied point of view relevant to discuss the negative impacts in more detail and thereby to assess to what extent mitigation is needed. It is, however, not the objective of this chapter to report specific regulatory requirements, standards or recommendations published by different authorities. The subsequent assessment can be discussed but it is in general relevant to consider, in particular from a practitioner's point of view:

(1) Corrosion: Corrosion is a material-related phenomenon, particularly degradation of concrete pipes and structures, mortar in brick constructions and installations with metals in ancillary structures. The cost of rehabilitating a part of a sewer system versus the cost of implementing and operating mitigation can in principle be balanced. In the case of severe corrosion, rehabilitation may be demanded after 2–4 years of operation, while moderate corrosion may allow acceptable function and service level of the sewer section for decades (Figure 4.3). However, in several parts of the world – particularly in warm countries – concrete corrosion is considered a serious obstacle for management of wastewater collection systems and therefore a phenomenon of importance to be assessed in detail (Grengg *et al.*, 2015; Jensen *et al.*, 2009a; Jiang *et al.*, 2016; Vollertsen *et al.*, 2008; Wells & Melchers, 2015).
(2) Odor: The level of an odor problem in the surroundings is a public concern that in the planning stage of a sewer system is typically difficult to locate and predict in detail. However, in a planning and rehabilitation process, odor problems – in contrast to corrosion – must in principle be

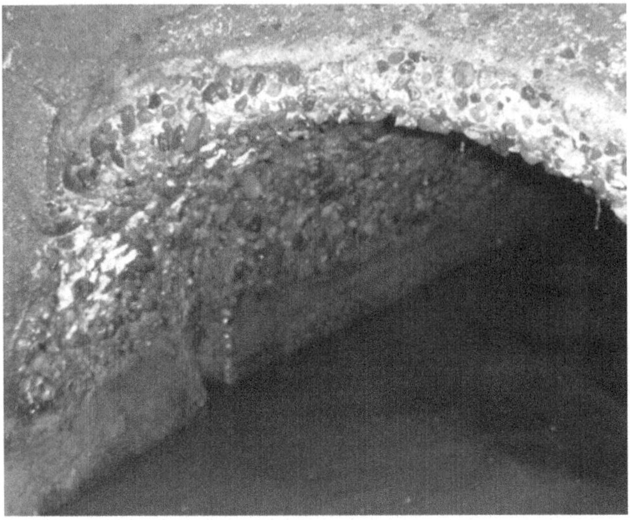

Figure 4.3 Corroded sewer pipe where the cement in the surface layer is degraded leaving stones and gravel from the concrete aggregate.

considered a zero-tolerance phenomenon to avoid depressed city areas and odor complaints from the public. From a pragmatic point of view the type of city affected is important, e.g. defined in terms of a residential, commercial, institutional or industrial area. Furthermore, it is crucial to take into account that numerous volatile organic carbon constituents (VOCs) – in general anaerobically produced by fermentation – may contribute to odor problems in addition to hydrogen sulfide (Stuetz & Frechen, 2001; WERF, 2009; Chapter 10).

(3) Human health impacts: H_2S is a gas with potential health effects at rather low concentration levels in terms of headache symptoms, respiratory injury and irritation of eyes, nose and throat. These effects depend on human sensitivity and time of exposure. At high concentrations, i.e. higher than about 300 ppm, hydrogen sulfide becomes almost immediately life threatening. Regarding the working environment, it is therefore very important to implement both regulations and safety measures.

Sulfide build-up in wastewater of a sewer is often used as a simple indicator for all types of hydrogen sulfide problems (Table 4.4). It is obvious that the complex effects of both process and structural related factors make such an assessment of sulfide impacts imprecise. It is therefore crucial to first identify the nature of the main influencing factors for sulfide problems. Secondly, it becomes important to exploit this knowledge to develop and apply assessment tools in terms of both simple empirical equations and complex formulated sewer process models.

Table 4.4 Total sulfide concentration in wastewater of a sewer expressed as a simple indicator for potential problems related to corrosion, malodors and health.

Total Sulfide Concentration (gS m^{-3})	Problems Typically Reported
<0.5	Minor
0.5–2	Medium
>2	Considerable

4.4 FACTORS AFFECTING SULFIDE RELATED PROBLEMS IN SEWERS

Anaerobic conditions, i.e. the absence of dissolved oxygen (DO) and oxidized nitrogen compounds like nitrate and nitrite, are required for sulfate reduction (Table 4.2). The most important factors determining sulfate reduction and thereby sulfide production will be outlined in the following subsections. These factors are crucial for prediction of potential impacts and they provide corresponding knowledge on mitigation. The different factors govern the extent of the sulfur processes and they provide details on the importance of external conditions like wastewater redox potential, DO, temperature, flow and biodegradable organic matter content.

Sulfide is primarily produced in submerged biofilms fixed at pipe surfaces and sewer sediment surfaces in pressure sewers, where dissolved oxygen is generally easily depleted. In gravity sewers it may therefore typically occur under the following conditions that often jointly exist:

- In full-flowing sewers.
- Under slow-flowing conditions, i.e. when the flow rate is typically <0.3 m s^{-1}.
- In large diameter pipe sewers with a high water table and insufficient reaeration.
- At relatively high water temperatures, i.e. typically >15–20°C.

The permanent occurrence of sulfide in bulk water of gravity sewers is not common in Northern Europe, owing to a moderate wastewater temperature even during summer periods and a general high capacity of the gravity sewers compared with actual loads of wastewater. However, changing aerobic and anaerobic conditions may exist, e.g. caused by the daily variations in flow, wastewater quality and temperature (Gudjonsson *et al.*, 2001). In warmer climates, e.g. Central and Southern Europe, the southern United States, Africa and Australia, sulfide problems under gravity sewer conditions are common (ASCE, 1989).

It is not possible to elaborate simple and general valid criteria that show the extent of sulfide related problems although certain parameters may give some indication. According to USEPA (1974), sulfide is usually not present in the bulk water phase of a gravity sewer at DO concentrations >1 gO_2 m^{-3}. Furthermore, experience based on information from the literature seems to indicate that if the DO concentration exceeds 0.2–0.5 gO_2 m^{-3}, sulfide may occur in the water phase but problems are not typical. When sulfide problems are identified, the sulfide concentration in the bulk water is, as shown in Table 4.4, a very simple indicator for the corresponding problems (in networks with a free water surface). Sulfide concentrations of 0.5, 3 and 10 gS m^{-3} are often considered as low, moderate and high, respectively, in terms of problems that are typically reported. For long pressure mains or high temperatures of wastewater in gravity sewers, concentrations significantly higher than 10 gS m^{-3} have been reported (Pomeroy & Parkhurst, 1977; Thistlethwayte, 1972).

The occurrence of sulfide and corresponding problems in a sewer network relate to a complex balance between several wastewater quality parameters. Furthermore, the extent of the problem depends on the design of the sewer network and the local climate. The DO concentration level in the bulk water phase, the pH value, the relevant microbial process rates and interfacial exchange rates for oxygen and sulfide outline, in general, the nature of the central factors. It is crucial to "organize" such factors in a context that can be applied when predicting and mitigating the sulfide problem. Below, eight main quality and system related factors that influence the occurrence of sulfide are briefly discussed as a basis for the development of prediction tools for sulfide problems and development of control methods (Sections 4.5 and 4.6).

4.4.1 Presence of sulfate

Sulfate is typically found in all types of municipal wastewaters in concentrations higher than 5–15 gS m^{-3}. Such concentration levels do not limit sulfide formation in relatively thin biofilms due to full penetration (Nielsen & Hvitved-Jacobsen, 1988). In sewer sediments, however, where sulfate may penetrate deeper sediment layers, the potential for sulfate reduction may increase with increasing sulfate concentration in the bulk water phase. Although sulfate is the most common electron acceptor for microorganisms producing sulfide in sewer networks, other oxidized sulfur compounds, for example thiosulfate and sulfite, may act as sulfur sources for such bacteria (Nielsen, 1991). In case of intensive sulfide formation it is possible that sulfate has been depleted below the actual concentration for non-limited growth.

4.4.2 Temperature

Although sulfide problems in sewer networks are particularly common in climate regions with high temperatures, they may also occur under temperate climate conditions in pressure mains during winter at wastewater temperatures around 5–12°C (Hvitved-Jacobsen *et al.*, 1995; Nielsen *et al.*, 1998). At such relatively low temperatures, the sulfide production rate is generally low, and the anaerobic residence time should typically exceed 0.5–2 hours before sulfide production becomes significant.

The temperature dependency of the microbial sulfate reduction rate for single bacteria is high, corresponding to a temperature coefficient of about 1.13, i.e. a change in the rate occurs with a factor $Q_{10} = 3.4$ per 10°C of temperature increase. Because the diffusion rate of substrate into biofilms or sediments is typically limiting the sulfide formation rate, the value of the temperature coefficient is reduced to about 1.03 (Nielsen *et al.*, 1998). The development of sulfate reducing microorganisms that are adapted to low and to high temperatures, respectively, may result in an apparent long-term reduced temperature dependency between winter and summer.

4.4.3 Dissolved oxygen

The DO mass balance is crucial for the net formation of sulfide in the biofilm and for the occurrence of sulfide in the wastewater phase. In principle, anaerobic conditions occur in full-flowing gravity sewers and rising mains where DO is absent, cf. further details in the introduction to Section 4.4. In situations where aerobic wastewater, typically with moderate DO concentration levels, flows into pressure sewers, the DO concentration is typically quickly depleted, often after 5–30 minutes, depending on the DO concentration, temperature and the aerobic respiration rate of the wastewater. Although sulfate reduction requires strict anaerobic conditions, it is important to note that sulfide can be transported to and also temporarily exist at those parts of the sewer where the DO concentration is not zero.

4.4.4 pH

The pH value of wastewater in sewer networks is in several ways important for sulfur transformations, the impacts of sulfide and its control. It is first of all important that sulfate-reducing bacteria are active between pH 6 and 9. In practice, however, a significant inhibition of these bacteria in sewer networks will require a pH value of more than 10. Furthermore, the relative distribution between the two sulfide components, H_2S and HS^-, with an oxidation level of −2, depends on the pH value. This aspect is crucial because an increase in the pH value will reduce the harmful effects of sulfide in the volatile form, H_2S. The pH value is furthermore important for efficient precipitation of sulfide with iron salts and thereby central for a common way of controlling sulfide, cf. Section 4.6.2 (Nielsen *et al.*, 2008). It is furthermore important that both chemical and biological oxidation kinetics of sulfide are pH-dependent. The pH value is therefore also of importance for another predominant control method for sulfide (Nielsen *et al.*, 2003, 2004, 2005a, b, 2006).

4.4.5 Area-to-volume ratio of sewer pipes

Sulfide is produced in the biofilm at the submerged parts of the sewer – and in the sewer sediment if it is permanently present. The corresponding water phase concentration of sulfide is caused by diffusion of sulfide from the biofilm and sediment surfaces (Liu *et al.*, 2015). The observed water phase concentration is therefore related to the biofilm area-to-water volume (A/V) ratio. In a pressure main and a full-flowing gravity sewer, the governing parameters correspond to the surface area and the volume of the pipe. In a partly flowing gravity sewer pipe, A and V are the wetted pipe surface area and the bulk water volume, respectively. At constant residence times, small diameter pressure sewer pipes, having an A/V ratio larger than bigger diameter pipes, typically result in higher sulfide concentrations in the water phase.

4.4.6 Quality and quantity of biodegradable organic matter

As heterotrophic microorganisms, the sulfate-reducing biomass requires biodegradable organic matter as the substrate for growth and as an electron donor (Hvitved-Jacobsen *et al.*, 1995). Wastewater from, for example, food industries with relatively high concentrations of readily biodegradable organics are suitable substrates for the sulfate-reducing bacteria and the sulfate reduction rate resulting in formation of sulfide is therefore typically higher than for wastewater from households (Raunkjaer *et al.*, 1994). Several specific organics, for example formate, lactate and ethanol, are good substrates for sulfate-reducing bacteria (Nielsen & Hvitved-Jacobsen, 1988).

4.4.7 Anaerobic residence time in the sewer network

The anaerobic residence time of the wastewater during transport in a sewer is a central factor that determines the conditions for microbial sulfate reduction and thereby the concentration of sulfide in the water phase. The residence time is determined by the amount of wastewater inflow compared with the volume (length and diameter) of the pipe. The level of sulfide formation in a given sewer network is therefore subject to the diurnal variation of the inflowing wastewater and to the precipitation pattern in a combined sewer catchment (Figure 4.4).

4.4.8 Flow velocity

The sulfide production in a sewer depends on the active thickness of an anaerobic biofilm. If the flow velocity in a pressure main frequently exceeds 0.8–1 m s^{-1}, the corresponding biofilm is relatively smooth and thin, typically 100–300 μm (USEPA, 1974). However, high flow velocities also reduce the diffusion layer at the biofilm-water interface and thereby the resistance against transport of substrates and products (Kiilerich *et al.*, 2018a). In total, a high flow velocity will normally – considering constant residence time – reduce the potential for sulfide formation in pressure mains. The flow rate conditions in gravity sewers also affect the air-water exchange processes, for example the emission of hydrogen sulfide and reaeration.

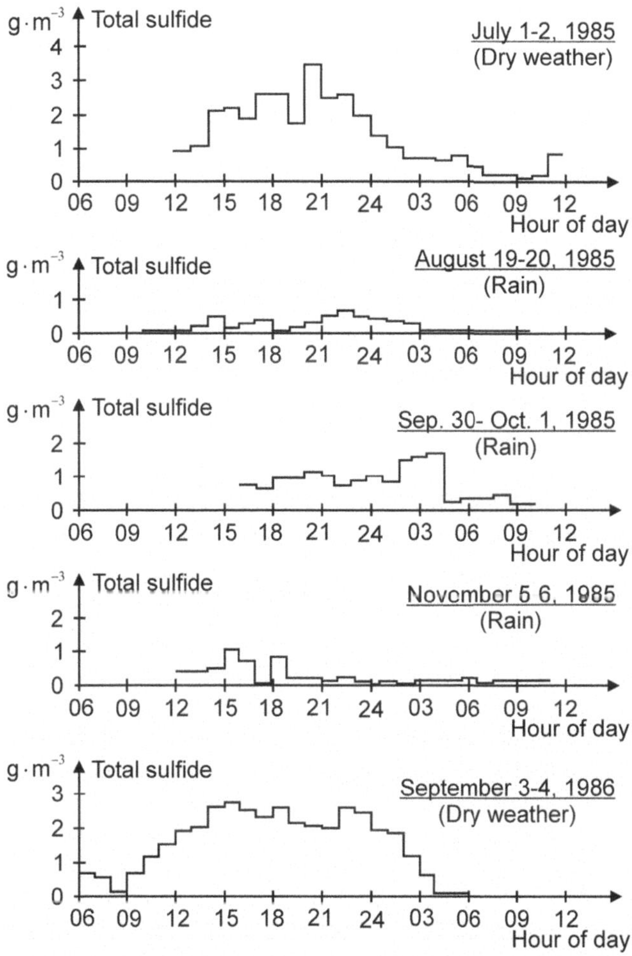

Figure 4.4 Diurnal variation of total sulfide in the wastewater outlet from a 4 km pressure main. The sewer operates at a combined sewer catchment in North Jutland (Denmark). The typical diurnal variation of the residence time during a dry-weather period is between 6 and 14 hours. During a rain event, the residence time in the sewer can be as low as about 1 hour (source with permission: Hvitved-Jacobsen *et al.*, 2013).

4.5 PREDICTION OF SULFIDE RELATED ADVERSE EFFECTS IN SEWERS

Although sewer deterioration is historically known, the evidence of sulfide related problems is a rather late discovery. The first indication of concrete corrosion caused by sulfide was probably reported for an outfall sewer in Los Angeles (USA) in 1895, where it was observed that the mortar used in the joints of the brickwork swelled during its disintegration (Olmsted & Hamlin, 1900). The knowledge we have today on sulfur transformations in sewers, and particularly the quantification of the problems related to hydrogen sulfide, has in terms of what is described in Section 4.4 gradually developed since that time and in particular during the period after World War II.

The following description aims at giving a brief overview of different types and levels of prediction tools for hydrogen sulfide and its adverse effects in sewers. Further details can be found in the literature mentioned. An overview of and comparison between different types of equations and models is found in Hvitved-Jacobsen *et al.* (2013). A central point for grouping prediction tools is the potential for reaeration of the wastewater in the sewer. As a result, the rather simple equations and models for pressure sewers and gravity sewers are, therefore, differently designed.

4.5.1 Empirical equations for sulfide formation in pressure sewers and full flowing gravity sewers

A number of simple empirical equations have been developed for the prediction of the sulfide production rate, r_a, of the biofilm in pressure pipes and full flowing gravity sewers under conditions where DO and nitrate are absent (Table 4.5). It is a fundamental fact that the generation of sulfide in terms of a biofilm surface rate, r_a, is dependent on the concentration and the quality of organic matter (biochemical oxygen demand [BOD], total chemical oxygen demand [COD] or soluble COD) and the temperature. Based on the biofilm surface rate, the concentration of sulfide in the wastewater is under typical sewer conditions determined by the area/volume ratio of the sewer pipe and the anaerobic residence time in the sewer.

4.5.2 Simple formulated "risk models" for sulfide build-up in gravity sewers

In contrast to pressure sewers and full flowing gravity sewers, the prediction of sulfide formation and related adverse effects in gravity sewers becomes more complex, in particular because reaeration of the wastewater flow is central.

Figure 4.5 is an example of how risk for sulfide build-up in a gravity sewer has been predicted (ASCE, 1989; USEPA, 1974). The formulation of the problem is expressed in very simple terms and has its clear limitations. However, it indicates that parameters like wastewater flow and pipe slope of the sewer play a central role for both the development of the biofilm and reaeration (Section 4.4.8).

The so-called "Z-formula" is another example of a risk model formulated by Pomeroy (1970) for sulfide build-up in gravity sewers. Although the model is simple, it reflects in addition to the temperature that rather few parameters give a fundamental understanding of which factors and processes limit sulfide build-up in

Table 4.5 Empirical equations for prediction of the biofilm surface formation rate of sulfide, r_a ($gS\ m^{-2}\ h^{-1}$), in pressure sewers and full flowing gravity sewers.

Biofilm Surface Rate, r_a	**Reference**	**Nomenclature**
$M\ BOD_5\ 1.07^{T-20}$	Pomeroy (1959) USEPA (1974) Pomeroy and Parkhurst (1977)	Empirical rate constant: $M = 10^{-3}$ ($m\ h^{-1}$) BOD_5 ($gO_2\ m^{-3}$) temperature: T (°C)
$0.5\ 10^{-3}\ u\ BOD_5^{0.8}\ S_{SO4}^{0.4}\ 1.14^{T-20}$	Thistlethwayte (1972)*	Flow velocity: u ($m\ s^{-1}$) sulfate concentration: S_{SO4} ($g_{SO4}\ m^{-3}$)
$0.23\ 10^{-3}\ COD\ 1.07^{T-20}$	Boon and Lister (1975)	COD ($gO_2\ m^{-3}$)
$a\ (COD_S - 50)^{0.5}\ 1.03^{T-20}$	Hvitved-Jacobsen *et al.* (1988) Nielsen *et al.* (1998)	Empirical rate constant: a** ($gS\ gO_2^{-0.5}\ m^{-0.5}\ h^{-1}$) soluble COD: COD_s ($gO_2\ m^{-3}$)

*Developed for conditions with limiting sulfate concentration and flow.

**$a = 0.001$–0.002 domestic wastewater without sewage from food industries; $a = 0.003$–0.006 wastewater from mixed domestic and industrial sources; $a = 0.007$–0.010 wastewater with biodegradable organic matter from mainly foodstuff industries.

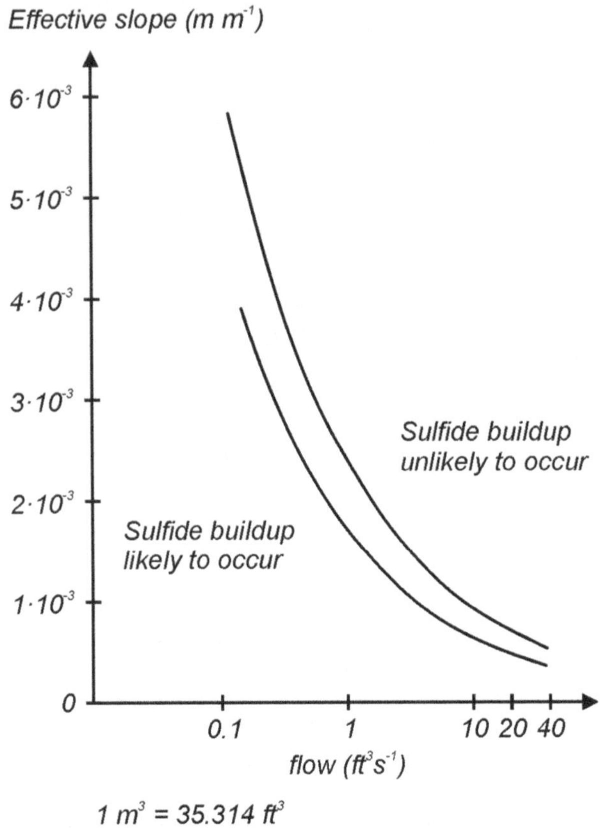

Figure 4.5 Flow-to-slope relationships expressed as guideline to sulfide forecasting and recommendation for design of gravity sewers. The effective slope of pipe is defined for the average conditions at the 1-hr flow distance of the pipe upstream of point of reckoning (source with permission: Hvitved-Jacobsen *et al.*, 2013).

gravity sewers: the reaeration of the water phase followed by a competition for DO by sulfide being produced in the biofilm and the availability of biodegradable organic matter in the water phase. The formula is not applicable for very small flows.

$$Z = \text{EBOD}(s^{0.5}Q^{0.33})^{-1}\frac{P}{b} \quad (4.1)$$

where:

$\text{EBOD} = \text{BOD}_5\ 1.07^{T-20}$
s = slope of pipe (m per 1000 m)
Q = pipe flow (ft^3 s^{-1}); 1 m^3 = 35.314 ft^3
P = wetted pipe-wall perimeter (m)
b = pipe width at the water surface (m)

Table 4.6 Sulfide problems in gravity sewers expressed by the Z-formula (source: data from Pomeroy, 1970).

Numerical Value of the Z-formula	Estimated Magnitude of the Sulfide Problem
$Z < 5000$	Sulfide build-up is not likely to occur
$5000 < Z < 10{,}000$	Some risk of sulfide build-up
$Z > 10{,}000$	Sulfide build-up is likely to occur

Pomeroy (1970) assessed the Z-formula based on an array of 50 tests and found the results shown in Table 4.6 generally valid. This interpretation was subsequently recommended for gravity sewer pipe design.

Although the Z-formula is simple, the low requirement for data makes it useful as a tool for screening (ASCE, 1989; ASCE & WPCF, 1982; Pomeroy & Parkhurst, 1977).

4.5.3 Empirical equations for sulfide formation in gravity sewers

The sulfur cycle in partly filled gravity sewer pipes where air-water exchange processes become important is, as mentioned, more complex than in pressure sewers. The pH dependent emission of hydrogen sulfide into the sewer atmosphere, reaeration, oxidation of sulfide and precipitation are four major processes that affect the concentration of sulfide in the water phase, see Figure 4.2 (Pomeroy & Parkhurst, 1977; Tchobanoglous, 1981).

Based on intensive studies, Pomeroy and Parkhurst (1977) presented the first reliable empirical model that was developed according to these facts and later reformulated by Kienow *et al.* (1982) and Matos (1992):

$$
\begin{aligned}
r_a &= \left(\frac{A}{V}\right)^{-1}\left(\frac{dC_{S(-II)}}{dt}\right) = M'\mathrm{BOD}_5 1.07^{T-20} - N(s \cdot u)^{3/8} d_m^{-1} R C_{S(-II)} \\
&= M'\mathrm{BOD}_5 1.07^{T-20} - N(s \cdot u)^{3/8}\left(\frac{P}{b}\right)^{-1} C_{S(-II)}
\end{aligned}
\tag{4.2}
$$

where:

r_a = net formation rate of sulfide in the biofilm (gS m^{-2} h^{-1})
A/V = area-to-volume ratio, i.e. wetted pipe surface area/wet pipe volume (m^{-1})
$C_{S(-II)}$ = sulfide concentration (gS m^{-3})
t = time (h)
M' = empirical rate constant
BOD_5 = biochemical oxygen demand (gO_2 m^{-3})
T = temperature (°C)
N = empirical sulfide loss coefficient

s = slope (m m^{-1})
u = flow velocity (m s^{-1})
d_m = hydraulic mean depth of the water phase, i.e. the cross-sectional area of the water volume divided by the water surface width (m)
R = hydraulic radius, i.e. the cross-sectional area of the water volume divided by the wetted perimeter (m)
P = wetted pipe-wall perimeter (m)
b = pipe width at the water surface (m)

The first term in Equation 4.2 is basically equal to the equation by Pomeroy (1959), see Table 4.5. The second term in Equation 4.2 corresponds to the sinks for sulfide in the water phase that according to Figure 4.2 are primarily caused by oxidation in the water phase, emission into the sewer atmosphere and also, to some extent, precipitation. Equation 4.2 is discussed in more detail in Hvitved-Jacobsen *et al.* (2013).

Although the equations shown in Sections 4.5.1–4.5.3 are empirical by nature, it is important to stress that they include fundamental knowledge on the phenomena that determine the occurrence of sulfide and its adverse effects. It is also important to mention that a simple formulated equation for prediction of sulfide induced concrete corrosion has been developed (USEPA, 1974).

4.5.4 Analytical and conceptual formulated sewer process models

It is beyond the objective of this chapter to present a detailed conceptually formulated "sewer process model". Such a model that is formulated in mathematical terms includes a great number of different flows, pathways and processes for the central constituents, i.e. dissolved oxygen, organic carbon and substances with sulfur and nitrogen, which influence sulfur transformations in sewers. A process model comprises water, solids and gas phase transformations and furthermore interfacial exchange processes between the different compartments of a sewer (Section 4.2).

A "model" is in this context a mathematically formulated tool, which for its implementation requires a programming language and hardware in terms of a computer. However, the core aspect is to establish the "framework" that should be modeled in terms of a formulation that describes which transformations are of importance and to what extent they occur in terms of kinetics and stoichiometry. In a situation where hydrogen sulfide, concrete corrosion, odor and human health aspects are central, the framework is basically described in Figure 4.2. The core aspect is to designate the different pathways mathematically in a way that simulates a real situation, in principle any real situation.

Since the 1990s, two different aspects have facilitated the development and implementation of such a conceptual formulated framework in terms of the integrated chemical, biological and physical processes as the central means:

(1) Numerous research projects have given insight into the dynamics of sewer processes (Table 4.3). This understanding concerns (i) a kinetic description, (ii) a stoichiometric formulation of the transformations, and iii) a corresponding establishment of the mass balances for flows and transformations of the relevant constituents including reliable values of the process parameters. In addition to the references shown in Table 4.3, further details are given in Hvitved-Jacobsen *et al.* (2013).

(2) The increase in the speed and power of personal computers and development of corresponding programming languages. This means that the outcome of complex processes in large sewer networks can be calculated in small time and space scales.

The first version of a conceptual formulation for modeling of sewer processes, WATS (Wastewater Aerobic/anaerobic Transformations in Sewers), was developed during the late 1990s (Hvitved-Jacobsen *et al.*, 1998, 1999). Since then, further development of WATS has progressively taken place (Vollertsen *et al.*, 2005, 2008, 2011, 2015). The focal aspect of WATS is a mathematical description of the mass balances including the kinetics for the relevant processes and their dependency on central parameters (see Figure 4.2 and references shown in Table 4.3). In addition to WATS, other analytical formulations for sewer process modeling have been developed (Donckels *et al.*, 2014; Sharma *et al.*, 2008).

When applying WATS with process formulations and values for parameters, a main advantage is that it simulates the outcome of all integrated sewer processes. Furthermore, the concept is designed for gradual improvement when new findings are available. When this concept in model terms is applied for real sewer networks with actual flows and parameter values, it simulates any relevant process performance, e.g. concrete corrosion at selected locations, odor, pH and occurrence of sulfide in the water and gas phase.

4.5.5 Computational and probabilistic models for sewer deterioration and service life

During the past 10 years, computational models for assessment of sewer deterioration have been developed. It is characteristic for such models to use different types of algorithms, for example multiple linear regression (MLR) and artificial neural network (ANN), in order to establish a relationship between independent variables and a criterion variable. These computational model types have, until now, particularly focused on prediction of corrosion and related aspects, e.g. service life of sewers (Jiang *et al.*, 2016; Li *et al.*, 2019).

Furthermore, probabilistic models, for example based on stochastic finite element methods, have been developed to predict failure of sewer pipes exposed to corrosion and the Gaussian processes regression (GPR) model was found to be valuable for extension of a limited dataset to estimate the concrete corrosion rate in sewers (Alani & Faramarzi, 2015; Liu *et al.*, 2017).

4.5.6 Final comments for prediction of sulfide related impacts on sewers

It is important to consider which type of sewer process prediction tool is the most convenient in a specific case. Aspects such as the purpose of an analysis and the extent and details of available relevant data should always be taken into account. The model types described in this section are quite different. The overall results of their prediction in terms of details are also differently expressed and they typically predict both parameters and impacts with varying accuracy (Matias *et al.*, 2018).

4.6 METHODS FOR CONTROL OF SULFIDE PROBLEMS IN SEWERS

As pointed out in Section 4.3, in-sewer sulfur related processes constitute potential severe impacts, for example reduced service lifetime of a sewer network caused by concrete corrosion, and obnoxious smell in the surroundings. Anaerobic processes and the occurrence of hydrogen sulfide and VOCs are, in this respect, crucial (Figure 4.2). Destruction of a sewer may cause risk of blockages, infiltration of groundwater and exfiltration of sewage, and ventilation of sewer gas may result in malodors of public concern. The economic and engineering aspects of a malfunctioning sewer and its rehabilitation are in general considerable. It is characteristic that the adverse effects caused by the occurrence of hydrogen sulfide and VOCs are closely related to the air phase in the sewer or in its adjacent environment, and it is important to note that the location where a problem is identified is not necessarily where the generation of the problematic substances takes place. General knowledge on sulfur processes is required to assess mitigation of the adverse effects, but it is likewise important that since any sewer system has its own characteristics it should be carefully analyzed both in terms of generation and transport of the problematic substances. By this means it is possible to select which technology for mitigation is most convenient and where and how equipment for this purpose should be installed. The assessment of control methods for sulfide has, in both general and specific terms, extensively been focused on in scientific articles, manuals and textbooks (ASCE, 1989; ASCE & WEF, 2007; Boon, 1995; Gutierrez *et al.*, 2008; Hvitved-Jacobsen *et al.*, 2013; Thistlethwayte, 1972; USEPA, 1974, 1985; WERF, 2007; Zhang *et al.*, 2008).

It is useful to discuss and compare the different types of mitigation methods for hydrogen sulfide based on the sulfur cycle in sewers (Figure 4.2). By doing so, it is possible to identify where in a sewer and how a control method for sulfide is feasible. Figure 4.6 shows, based on this approach, where sulfide control is appropriate, remembering that a problem does not occur until hydrogen sulfide appears in the air phase. Figure 4.6 indicates thereby three different levels of sulfur pathways where a mitigation method has a potential "starting point" by

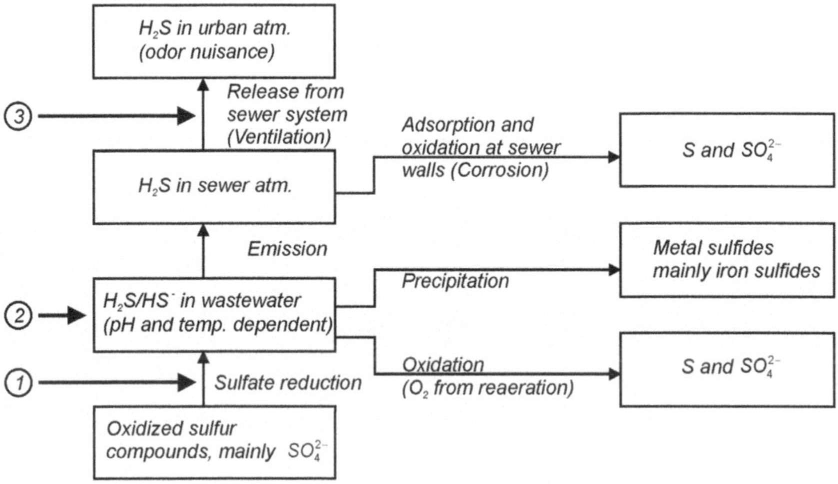

Figure 4.6 The sulfur cycle in a sewer indicating three points of the cycle where a mitigation method can change the formation, the occurrence or the flow of hydrogen sulfide. The mitigation methods corresponding to points No. 1–3 are described in Sections 4.6.1–4.6.3, respectively (source with permission: Hvitved-Jacobsen *et al.*, 2013).

closing, reducing or changing the generation or the transfer of hydrogen sulfide. These three different ways of tackling the hydrogen sulfide problem are discussed in Sections 4.6.1 to 4.6.3 below.

In addition to the use of control methods, proper design, construction and maintenance of sewer networks are important for reducing the impact of sulfide formation and other noxious volatile substances. In this respect, several handbooks produced by agencies, municipalities and consultants provide valuable guidance. Examples are the manuals from WEF (1999), ASCE and WEF (2007) and WERF (2007).

The following description of control methods for sulfide will only outline selected mitigation methods in brief. In principle there are many such methods but in the following focus is on those that are often used in practice. Further details of these and other methods can be found in the literature mentioned in the introduction to Section 4.6 and in Tables 4.7 to 4.9 below. It is finally important to stress that each sewer system has its own characteristics that may affect which mitigation method is most convenient.

4.6.1 Suppression or inhibition of sulfide formation

The mitigation methods related to point No. 1 in Figure 4.6 aim at tackling the sulfide related problems at the root by either reducing or inhibiting the activity of

Table 4.7 Control methods for suppression or inhibition of sulfide formation.

Method	Action	Reference
Suppression of the sulfate-reducing biomass activity	1. pH increase 2. Removal of biofilm 3. Use of biocides	Lin *et al.* (2017) Hvitved-Jacobsen *et al.* (2013)
Prevention of the anaerobic biological activity by changing the electron acceptor	4. Aerobic conditions by injection of oxygen (air or pure oxygen) 5. Anoxic conditions by nitrate dosing	Thistlethwayte (1972) Boon (1995) Yang *et al.* (2005) Gutierrez *et al.* (2008) Jiang *et al.* (2009)

Table 4.8 Control methods for reduction of the sulfide concentration in the water phase.

Method	Action	Reference
Reduction of sulfide by oxidation	1. Injection of oxygen 2. Dosing of nitrate	WEF (2004)
Chemical oxidation of sulfide	3. Addition of chemical oxidants, e.g. chlorine compounds, hydrogen peroxide, ozone and permanganate	ASCE and WEF (2007)
Precipitation of sulfide	4. Addition of ferrous iron (iron(II)) and ferric iron (iron(III)) salts	WERF (2007) Hvitved-Jacobsen *et al.* (2013)

Table 4.9 Control methods for reduction of foul sewer gases.

Method	Action	Reference
Treatment of sewer gas	1. Dilution, e.g. via a chimney 2. Chemical or biological treatment	Stuetz and Frechen (2001) WEF (2004) Zhang *et al.* (2008) WERF (2009) Shammay *et al.* (2016)
Increase of pH	3. Addition of an alkaline substance	Hvitved-Jacobsen *et al.* (2013) Talaiekhozani *et al.* (2016)

the sulfate reducing biomass. Except for the use of biocides, the mitigation methods outlined in Table 4.7 are frequently used in practice:

The main characteristics of the methods outlined in Table 4.7 are briefly described below.

4.6.1.1 pH increase

Referring to Section 4.4.4, a pH above 10 in the wastewater will suppress the activity of the sulfate-reducing biomass rather efficiently (Lin *et al.*, 2017). A so-called "shock dosing" of an alkaline chemical to about pH 12 for about 1 hour typically affects the anaerobic microorganisms by sloughing of the biofilm. The effect of such a shock dosing is only temporary until a new biofilm has been re-established, typically after a few days to about 2–3 weeks.

4.6.1.2 Mechanical removal of biofilm

The removal of sewer biofilms and deposits will reduce sulfide formation. Mechanical removal can be done by flushing or by use of a ball called a "cleaning pig". This technology should, like shock dosing with alkaline chemicals (see Section 4.6.1.1), be repeated regularly.

4.6.1.3 Injection of oxygen or nitrate dosing

The main characteristic of oxygen or nitrate as methods for sulfide control is their activity as electron acceptors by increasing the redox potential. Anaerobic processes, sulfate reduction, hydrolysis and fermentation (production of VOCs), requiring a redox potential <-200 mV, are thereby suppressed. An increased biological activity in the sewer is, however, under aerobic or anoxic conditions generally the result.

There are basically two ways to apply an electron acceptor: either to prevent anaerobic conditions in the sewer or to oxidize reduced substances (sulfide and VOCs) that have been produced in upstream sewer sections. The first way can be achieved by adding the electron acceptor upstream of a section where anaerobic conditions otherwise may occur. The second and generally less efficient means of control is by adding oxygen or nitrate downstream of an anaerobic section of the sewer for oxidation of reduced substances.

When oxygen is injected in the water phase of a sewer, the activity of the sulfate-reducing biomass is suppressed in the upper part of the biofilm and a DO concentration in the water phase above 0.5 $gO_2\ m^{-3}$ is typically sufficient to prevent occurrence of sulfide (Boon, 1995). Depending on the temperature, the solubility of oxygen at 1 atmosphere pressure is 8–11 $gO_2\ m^{-3}$, whereas the solubility of pure oxygen is 45–50 $gO_2\ m^{-3}$. As the DO consumption rate of wastewater is typically 20–40 $gO_2\ m^{-3}\ h^{-1}$, it is clear that compared with pure oxygen, the use of atmospheric air requires more injection points for oxygen. As the high solubility of pure oxygen refers to 1 atmosphere partial pressure of oxygen in the overlaying atmosphere, in practice the use of pure oxygen is limited to pressure sewers.

When DO is depleted, addition of nitrate generates anoxic conditions at a redox potential that – as is the case for DO – will depress the activity of the sulfate-reducing biomass in the biofilm and oxidize already produced sulfide (Abdul-Talib *et al.*, 2002; Jiang *et al.*, 2009; Yang *et al.*, 2005). The nitrate uptake rate (NUR) of wastewater is in equivalent units (electron equivalents) lower than the oxygen uptake rate (OUR). Nitrate is relatively expensive but the smaller amount of nitrate required compared with oxygen and the use of rather simple equipment for its dosing will reduce the difference in costs. Furthermore, a lower transformation rate in the sewer of the most biodegradable fractions of organic matter may be of advantage for a subsequent biological phosphorus removal step at the wastewater treatment plant.

4.6.2 Reduction of the sulfide concentration in the water phase

Methods to lower the sulfide concentration (corresponding to point No. 2 in Figure 4.6) typically operate in the water phase and often under anaerobic conditions, i.e. under conditions where sulfide is still produced. The methods (outlined in Table 4.8) are either biological or chemical in nature and aim at removing sulfide from the water phase and thereby controlling its emission. Other malodorous VOCs are generally not controlled and their formation and release to the atmosphere may continue unchanged.

4.6.2.1 Addition of electron acceptors

The addition of electron acceptors, i.e. method No. 1 and 2 in Table 4.8, is in principle identical to the corresponding methods outlined in Section 4.6.1, but used for reduction of the sulfide concentration in the water phase. These methods basically act only by biological oxidation.

4.6.2.2 Iron salt addition

Method No. 4 of Table 4.8 is the most frequently used for sulfide precipitation. Sulfide control by addition of an iron(II) salt, typically as sulfate or chloride, has been known and applied for decades (Pomeroy & Bowlus, 1946). Equation 4.3 shows the stoichiometry of the reaction by forming a precipitate of insoluble iron (II) sulfide whereby the emission of gaseous H_2S is prevented:

$$Fe^{2+} + HS^- \rightarrow FeS + H^+ \quad (4.3)$$

As only HS^- – and not H_2S – forms a precipitate with ferrous iron, the pH should in practice exceed a value of 8, e.g. by addition of $Ca(OH)_2$ as an alkaline substance, before the method is sufficiently efficient (Nielsen *et al.*, 2008).

In practice, however, a mixture of both iron(II) as a soluble Fe^{2+} salt and iron(III) as Fe^{3+} are applied and the following reaction, that proceeds both chemically and biologically by chemoautotrophic microorganisms, is important in addition to Equation 4.3:

$$2Fe^{3+} + HS^{-} \rightarrow 2Fe^{2+} + S^{0} + H^{+} \tag{4.4}$$

The two processes, Equations 4.3 and 4.4, describe theoretically the combined use of ferrous and ferric iron salts. The reactions are, in practice, complex and dependent on pH, concentrations of iron and sulfide, redox potential and the microbial composition of the wastewater (Kiilerich *et al.*, 2018b; Nielsen *et al.*, 2005a, 2008). It has furthermore been observed that the use of iron salts may reduce the diffusion of the sewer biofilm (Kiilerich *et al.*, 2018a). In spite of the complex nature of the reaction kinetics, in practice the use of iron salts for sulfide control is rather simple, efficient and relatively cheap. The method is therefore applied worldwide. A combination of iron(II) and iron(III) salt is generally preferred when applied in upstream sections where anaerobic conditions exist, whereas only iron(II) salt is recommended for control at downstream locations. This is due to the fact that when adding both types of iron salts at upstream locations, the residence time at the anaerobic sewer section is typically sufficient to compensate for the relatively low reaction rate of the autotrophic microbial conversion process shown in Equation 4.4.

It should be recalled that precipitation of iron salts in general occurs under anaerobic conditions and that volatile noxious VOCs may still be produced. Furthermore, and as a side effect, the use of iron salts in sewers enhances chemical phosphorus removal during subsequent aerobic wastewater treatment (Gutierrez *et al.*, 2010).

4.6.3 Reduction or dilution of sewer gases

Referring to point No. 3 in Figure 4.6, treatment or dilution of foul sewer gases is by nature an "end of pipe solution" that does not aim at reducing the formation of hydrogen sulfide or other malodorous volatile substances in the sewer network. Table 4.9 shows methods that tend to reduce the impact of hydrogen sulfide by focusing on its potential occurrence in the gas phase, and thereby in particular in the urban atmosphere.

A great number of technologies exist for chemical, physico-chemical and biological treatment of foul air (see method No. 2 in Table 4.9). The are many systems for process and reactor design and these depend in principle on the manufacturer of the equipment. In terms of processes, these methods can be subdivided as follows:

- Absorption of a gas in a liquid stream (wet or dry scrubbing)
- Catalytic combustion (incineration)

- Activated carbon adsorption
- Chemical oxidation
- Biological oxidation (biofilters, bioscrubbers and biotrickling filters)

Increasing the pH (method No. 3 in Table 4.9) operates due to the fact that only the molecular form of sulfide, H_2S, can be emitted to the air phase and that consequently the ionic form, HS^-, remains in the water phase. Compared with method No. 1 in Table 4.7, an increase in pH is required to a lesser extent when aiming at minimizing foul sewer gas emissions.

REFERENCES

Abdul-Talib S., Hvitved-Jacobsen T., Vollertsen J. and Ujang Z. (2002). Anoxic transformations of wastewater organic matter in sewers – process kinetics, model concept and wastewater treatment potential. *Water Science & Technology*, **45**(3), 53–60.

Alani A. M. and Faramarzi A. (2015). Predicting the probability of failure of cementitious sewer pipes using stochastic finite element method. *International Journal of Environmental Research and Public Health*, **12**(6), 6641–6656.

ASCE (1989). Sulfide in Wastewater Collection and Treatment Systems. ASCE (American Society of Civil Engineers) manuals and reports on engineering practice No. 69, pp. 324.

ASCE & WEF (2007). Gravity Sanitary Sewer Design and Construction, 2nd edn, ASCE (American Society of Civil Engineers) Manuals and Reports on Engineering Practice No. 60, WEF (Water Environment Federation) Manual of Practice No. FD-5, pp. 422.

ASCE & WPCF (1982). Gravity sanitary sewer design and construction, ASCE (American Society of Civil Engineers) manuals and reports on engineering practice No. 60 and WPCF (Water Pollution Control Federation) manual of practice no. FD-5, p. 275.

Beceiro P., Almeida M. D. C. and Matos J. (2017). Numerical modeling of air-water flows in sewer drops. *Water Science & Technology*, **76**(3–4), 642–652.

Boon A. G. (1995). Septicity in sewers: causes, consequences and containment. *Water Science & Technology*, **31**(7), 237–253.

Boon A. G. and Lister A. R. (1975). Formation of sulfide in rising main sewers and its prevention by injection of oxygen. *Progress in Water Technology*, **7**, 289–300.

Donckels B. M., Kroll S., Van Dorpe M. and Weemaes M. (2014). Global sensitivity analysis of an in-sewer process model for the study of sulfide-induced corrosion of concrete. *Water Science & Technology*, **69**(3), 647–655.

Grengg C., Mittermayr F., Baldermann A., Böttcher M. E., Leis A., Koraimann G., Grunert P. and Dietzel M. (2015). Microbiologically induced concrete corrosion: A case study from a combined sewer network. *Cement and Concrete Research*, **77**, 16–25.

Gudjonsson G., Vollertsen J. and Hvitved-Jacobsen T. (2001). Dissolved oxygen in gravity sewers – measurement and simulation. Proceedings from the 2nd International Conference on Interactions between Sewers, Treatment Plants and Receiving Waters in Urban Areas (INTERURBA II), February 19–22, 2001, Lisbon, Portugal, pp. 35–43.

Gutierrez O., Mohanakrishnan J., Sharma K. R., Meyer R. L., Keller J. and Yuan Z. (2008). Evaluation of oxygen injection as a means of controlling sulfide production in a sewer system. *Water Research*, **42**(17), 4549–4561.

Gutierrez O., Park D., Sharma K. R. and Yuan Z. (2010). Iron salts dosage for sulfide control in sewers induces chemical phosphorus removal during wastewater treatment. *Water Research*, **44**(11), 3467–3475.

Hvitved-Jacobsen T., Jütte B., Nielsen P.H. and Jensen N. Aa (1988). Hydrogen sulphide control in municipal sewers. In: Pretreatment in Chemical Water and Wastewater Treatment, H.H. Hahn and R. Klute (eds), Springer-Verlag, 239–247.

Hvitved-Jacobsen T., Raunkjaer K. and Nielsen P. H. (1995). Volatile fatty acids and sulfide in pressure mains. *Water Science & Technology*, **31**(7), 169–179.

Hvitved-Jacobsen T., Vollertsen J. and Nielsen P. H. (1998). A process and model concept for microbial wastewater transformations in gravity sewers. *Water Science & Technology*, **37**(1), 233–241.

Hvitved-Jacobsen T., Vollertsen J. and Tanaka N. (1999). Wastewater quality changes during transport in sewers – an integrated aerobic and anaerobic concept for carbon and sulfur microbial transformations. *Water Science & Technology*, **39**(2), 242–249.

Hvitved-Jacobsen T., Vollertsen J. and Matos J. S. (2002). The sewer as a bioreactor – a dry weather approach. *Water Science & Technology*, **45**(3), 11–24.

Hvitved-Jacobsen T., Vollertsen J. and Nielsen A. H. (2013). Sewer Processes: Microbial and Chemical Process Engineering of Sewer Networks, 2nd edn, CRC Press, Boca Raton, FL, USA, pp. 379.

IUPAC (2014). Compendium of Chemical Terminology, Gold Book version 2.3.3, International Union of Pure and Applied Chemistry, pp. 1622.

Jensen H. S., Nielsen A. H., Hvitved-Jacobsen T. and Vollertsen J. (2009a), Modeling of hydrogen sulfide oxidation in concrete corrosion products from sewer pipes. *Water Environment Research*, **81**(4), 365–373.

Jensen H. S., Nielsen A. H., Lens P. N. L., Hvitved-Jacobsen T. and Vollertsen J. (2009b), Hydrogen sulfide removal from corroded concrete: Comparison between surface removal rates and biomass activity. *Environmental Technology*, **30**(12), 1291–1296.

Jensen H. S., Lens P. N. L., Nielsen J. L., Bester K., Nielsen A. H., Hvitved-Jacobsen T. and Vollertsen J. (2011). Growth kinetics of hydrogen sulfide oxidizing bacteria in corroded concrete from Sewers. *Journal of Hazardous Materials*, **189**, 685–691.

Jiang G., Sharma K. R., Guisasola A., Keller J. and Yuan Z. (2009). Sulfur transformation in rising main sewers receiving nitrate dosage. *Water Research*, **43**(17), 4430–4440.

Jiang G., Keller J., Bond P. and Yuan Z. (2016). Predicting concrete corrosion of sewers using artificial neural network. *Water Research*, **92**, 52–60.

Karunakaran E., Vernon D., Biggs C., Saul A., Crawford D. and Jensen H. (2016). Enumeration of sulfate-reducing bacteria for assessing potential for hydrogen sulphide production in urban drainage systems. *Water Science & Technology*, **73**(12), 3087–3094.

Kiilerich B., Wagner M., Nielsen A. H. and Vollertsen J. (2018a), Apparent diffusion coefficients in sewer force main biofilms treated with iron salts. *Environmental Science: Water Research and Technology*, **4**(10), 1501–1510.

Kiilerich B., Nielsen A. H. and Vollertsen J. (2018b), Kinetics of sulfide precipitation with ferrous and ferric iron in wastewater. *Water Science & Technology*, **78**(5), 1071–1081.

Kienow K. E., Pomeroy R. D. and Kienow K. K. (1982). Prediction of sulfide buildup in sanitary sewers, J. Environmental Engineering Division. *Proceedings of the American Society of Civil Engineers (ASCE)*, **108**(EE5), 941–956.

Li X., Khademi F., Liu Y., Akbari M., Wang C., Bond P. L., Keller J. and Jiang G. (2019). Evaluation of data-driven models for predicting the service life of concrete sewer pipes subjected to corrosion. *Journal of Environmental Management*, **234**, 431–439.

Lin H. W., Lu Y., Ganigué R., Sharma K. R., Rabaey K., Yuan Z. and Pikaar I. (2017). Simultaneous use of caustic and oxygen for efficient sulfide control in sewers. *Science of the Total Environment*, **601–602**, 776–783.

Liu Y., Ni B., Ganigué R., Werner U., Sharma K. and Yuan Z. (2015). Sulfide and methane production in sewer sediments. *Water Research*, **70**, 350–359.

Liu Y., Song Y., Keller J., Bond P. and Jiang G. (2017). Prediction of concrete corrosion in sewers with hybrid Gaussian processes regression model. RSC (Royal Society of Chemistry) Advances, **49**(7), 30894–30903.

Madsen H. I., Hvitved-Jacobsen T. and Vollertsen J. (2006). Gas phase transport in gravity sewers – methodology for determination of horizontal gas transport and ventilation. *Water Environment Research*, **78**(11), 2203–2209.

Matias N., Matos R., Ferreira F., Vollertsen J. and Matos J. (2018). Release of hydrogen sulfide under intermittent flow conditions – the potential of simulation models. *Water Science and Technology*, **77**(3), 777–787.

Matos J. S. (1992). Aerobiose e septicidade im sistemas de drenagem de águas residuais, Ph.D. thesis, IST, Lisbon Portugal.

Nielsen P. H. (1991). Sulfur sources for hydrogen sulfide production in biofilm from sewer systems. *Water Science & Technology*, **23**, 1265–1274.

Nielsen P. H. and Hvitved-Jacobsen T. (1988). Effect of sulfate and organic matter on the hydrogen sulfide formation in biofilms of filled sanitary sewers. *Journal Water Pollution Control Federation*, **60**, 627–634.

Nielsen P. H., Raunkjaer K. and Hvitved-Jacobsen T. (1998). Sulfide production and wastewater quality in pressure mains. *Water Science & Technology*, **37**(1), 97–104.

Nielsen A. H., Vollertsen J. and Hvitved-Jacobsen T. (2003). Determination of kinetics and stoichiometry of chemical sulfide oxidation in wastewater of sewer networks. *Environmental Science and Technology*, **37**(17), 3853–3858.

Nielsen A. H., Vollertsen J. and Hvitved-Jacobsen T. (2004). Chemical sulfide oxidation of wastewater – effects of pH and temperature. *Water Science and Technology*, **50**(4), 185–192.

Nielsen A. H., Lens P., Vollertsen J. and Hvitved-Jacobsen T. (2005a), Sulfide-iron interactions in domestic wastewater from a gravity sewer. *Water Research*, **39**, 2747–2755.

Nielsen A. H., Hvitved-Jacobsen T. and Vollertsen J. (2005b), Kinetics and stoichiometry of sulfide oxidation by sewer biofilms. *Water Research*, **39**, 4119–4125.

Nielsen A. H., Vollertsen J. and Hvitved-Jacobsen T. (2006). Kinetics and stoichiometry of aerobic sulfide oxidation in wastewater from sewers – effects of pH and temperature. *Water Environment Research*, **78**(3), 275–283.

Nielsen A. H., Hvitved-Jacobsen T. and Vollertsen J. (2008). Effects of pH and iron concentrations on sulfide precipitation in wastewater collection systems. *Water Environment Research*, **80**(4), 380–384.

Olmsted F. H. and Hamlin H. (1900). Converting portions of the Los Angeles outfall sewer into a septic tank. *Engineering News*, **44**(19), 317–318.

Parker C. D. (1945a). The corrosion of concrete 1. The isolation of a species of bacterium associated with the corrosion of concrete exposed to atmospheres containing hydrogen sulphides. *Australian Journal of Experimental Biology and Medical Science*, **23**, 81–90.

Parker C. D. (1945b). The corrosion of concrete 2. The function of *Thiobacillus concretivorus* (nov. spec.) in the corrosion of concrete exposed to atmospheres containing hydrogen sulphide. *Australian Journal of Experimental Biology and Medical Science*, **23**, 91–98.

Parker C. D. (1951). Mechanics of corrosion of concrete sewers by hydrogen sulfide. *Sewage Industrial Wastes*, **23**, 1477–1485.

Pomeroy R. (1959). Generation and control of sulfide in filled pipes. *Sewage and Industrial Wastes*, **31**(9), 1082–1095.

Pomeroy R. D. (1970). Sanitary sewer design for hydrogen sulfide control. *Public Works*, **101**(10), 93–96 and 130.

Pomeroy R. and Bowlus F. (1946). Progress Report on Sulfide Control Research. *Sewage Works Journal*, **13**, 597–640.

Pomeroy R. D. and Parkhurst J. D. (1977). The forecasting of sulfide buildup rates in sewers. *Progress in Water Technology*, **9**(3), 621–628.

Raunkjaer K., Hvitved-Jacobsen T. and Nielsen P. H. (1994). Measurement of pools of protein, carbohydrate and lipid in domestic wastewater. *Water Research*, **28**(2), 251–262.

Shammay A., Sivret E. C., Le-Minh N., Fernandez R. L., Evanson I. and Stuetz R. M. (2016). Review of odour abatement in sewer networks. *Journal of Environmental Chemical Engineering*, **4**(4), 3866–3881.

Sharma K. R., Yuan Z., de Haas D., Hamilton G., Corrie S. and Keller J. (2008). Dynamics and dynamic modelling of H_2S production in sewer systems. *Water Research*, **42**(11–12), 2527–2538.

Stuetz R. and Frechen F.-B. (eds), (2001). Odours in Wastewater Treatment – Measurement, Modelling and Control, IWA Publishing, London, UK, pp. 437.

Talaiekhozani A., Bagheri M., Goli A. and Talaei K. M. (2016). An overview of principles of odor production, emission and control methods in wastewater collection and treatment systems. *Journal of Environmental Management*, **170**, 186–206.

Tanaka N., Hvitved-Jacobsen T. and Horie T. (2000a). Transformations of carbon and sulfur wastewater components under aerobic-anaerobic transient conditions in sewer systems. *Water Environment Research*, **72**(6), 651–664.

Tanaka N., Hvitved-Jacobsen T., Ochi T. and Sato N. (2000b). Aerobic-anaerobic microbial wastewater transformations and reaeration in an air-injected pressure sewer. *Water Environment Research*, **72**(6), 665–674.

Tchobanoglous G. (ed.) (1981). Wastewater engineering: Collection and pumping of wastewater, Metcalf and Eddy, Inc., McGraw Hill Book Co., New York.

Thistlethwayte D. K. B. (ed) (1972). The Control of Sulphides in Sewerage Systems, Butterworths, Sidney, Australia, pp. 173.

USEPA (1974). Process Design Manual for Sulfide Control in Sanitary Sewerage Systems. USEPA (US Environmental Protection Agency), 625/1-74-005, Technology Transfer, Washington, DC.

USEPA (1985). Odor and corrosion control in sanitary sewerage systems and treatment plants. USEPA 625/1-85/018, Washington, DC.

Vollertsen J. and Hvitved-Jacobsen T. (1999). Stoichiometric and kinetic model parameters for microbial transformations of suspended solids in combined sewer systems. *Water Research*, **33**(14), 3127–3141.

Vollertsen J., Hvitved-Jacobsen T. and Nielsen A. H. (2005). Stochastic modeling of chemical oxygen demand transformations in gravity sewers. *Water Environment Research*, **77**(4), 331–339.

Vollertsen J., Nielsen A. H., Jensen H. S., Wium-Andersen T. and Hvitved-Jacobsen T. (2008). Corrosion of concrete sewers – the kinetics of hydrogen sulfide oxidation. *Science of the Total Environment*, **394**, 162–170.

Vollertsen J., Nielsen L., Blicher T. D., Hvitved-Jacobsn T. and Nielsen A. H. (2011). A sewer process model as planning and management tool – hydrogen sulfide simulation at catchment scale. *Water Science & Technology*, **64**(2), 348–354.

Vollertsen J., Revilla N., Hvitved-Jacobsen T. and Nielsen A. H. (2015). Modeling sulfides, pH and hydrogen sulfide gas in the sewers of San Francisco. *Water Environment Research*, **87**(11), 1980–1989.

WEF (1999). Wastewater Collections Systems Management, WEF (Water Environment Federation) Manual of Practice, MOP No. 7, Fifth Edition, pp. 244.

WEF (2004). Control of odors and emissions from wastewater treatment plants. WEF (Water Environment Federation) Manual of Practice No. 25, Alexandria, VA, USA, pp 537.

Wells T. and Melchers R. E. (2015). Modelling concrete deterioration in sewers using theory and field observations. *Cement and Concrete Research*, **77**, 82–96.

WERF (2007). Minimization of Odors and Corrosion in Collection Systems – Phase 1, WERF (Water Environment Research Foundation) Co-published by IWA (International Water Association) Publishing, Final Report.

WERF (2009). Collection System Ventilation Research Report, WERF (Water Environment Research Foundation), Report No. 04-CTS-1A.

Yang W., Vollertsen J. and Hvitved-Jacobsen T. (2004). Anoxic control of odour and corrosion from sewer networks. *Water Science & Technology*, **50**(4), 341–349.

Yang W., Vollertsen J. and Hvitved-Jacobsen T. (2005). Anoxic sulfide oxidation in wastewater of sewer networks. *Water Science & Technology*, **52**(3), 191–199.

Yongsiri C., Vollertsen J. and Hvitved-Jacobsen T. (2004). Effect of temperature on air-water transfer of hydrogen sulfide. *Journal of Environmental Engineering*, **130**, 104–109.

Yongsiri C., Vollertsen J. and Hvitved-Jacobsen T. (2005). Influence of wastewater constituents on hydrogen sulfide emission in sewer networks. *Journal of Environmental Engineering*, **131**, 1676–1683.

Yuan H., Dangla P., Chatellier P. and Chaussadent T. (2015). Degradation modeling of concrete submitted to biogenic acid attack. *Cement and Concrete Research*, **70**, 29–38.

Zhang L., De Schryver P., De Gusseme B., De Muynck W., Boon N. and Verstraete W. (2008). Chemical and biological technologies for hydrogen sulfide emission control in sewer systems: A review. *Water Research*, **42**, 1–12.

Chapter 5

Corrosion and sulfur-related bacteria

Min Wu, Tian Wang and Shunxiang Wang

5.1 INTRODUCTION

Concrete and carbon steel are among the most used materials in many industries, such as infrastructure, transportation and manufacturing. Surfaces of both materials, when exposed to non-sterile environments, can be the sites of biological activities of various microorganisms. Many of the activities are deleterious. From the long-term perspective, both concrete and metal can deteriorate and may ultimately lose their original functionalities. The deterioration of materials caused by microorganisms is known as microbiologically influenced (induced) corrosion (MIC) (Enning & Garrelfs, 2014; Jia *et al.*, 2019; Little & Lee, 2007; Usher *et al.*, 2014; Wu *et al.*, 2020).

MIC of concrete has attracted considerable attention since it was first reported more than a century ago (Olmstead & Hamlin, 1900). MIC can significantly reduce the lifetime of concrete structures (Jensen, 2009; O'Connell *et al.*, 2010). The economic losses caused by MIC of concrete are tremendous, e.g. the rehabilitation costs in the US were estimated to be \$390 billion (10^9) over about 20 years (Gutiérrez-Padilla *et al.*, 2010). Besides the economic impact, health and environmental concerns caused by MIC are also important aspects to be considered (US EPA, 1985). Despite the tremendous efforts that have been made, the severity of the MIC problems in recent decades has still had a tendency to increase (Wu *et al.*, 2020). Sewer networks, among others, are the most affected

DOI: 10.2166/9781789060966_0133

concrete structures and have attracted huge attention (Grengg *et al.*, 2018; Wu *et al.*, 2020), partly due to the significance to society and the large-scale builds nowadays. The emphasis of the current chapter is on sewer structures when discussing MIC of concrete.

Similar to concrete, carbon steel is almost indispensable for many sectors. Susceptibility to corrosion is a major drawback of carbon steel. The total costs caused by corrosion of carbon steel and other metals can be enormous, and have been estimated to be 2–3% of the gross domestic product (GDP) in developed countries (Enning & Garrelfs, 2014) and even up to 3–5% in developing countries (Hou *et al.*, 2017). Although the estimation of the costs associated with MIC varies, it is generally considered that MIC accounts for a significant fraction of the total costs (Jia *et al.*, 2019; Usher *et al.*, 2014). It is noted that besides carbon steel, microorganisms can cause the corrosion of many other metals as well, e.g. aluminum alloy, magnesium, copper, silver, zinc and even stainless steel (Javaherdashti, 2017; Zhou *et al.*, 2018). Nevertheless, carbon steel is used in much larger quantities than other metallic materials (Enning & Garrelfs, 2014). Part of this chapter will be devoted to MIC of carbon steel.

Many types of microorganisms can trigger MIC, including bacteria, fungi, archaea and diatoms (Little & Lee, 2007; Sun, 2015; Usher *et al.*, 2014). Among them, bacteria have been the subject of the majority of MIC research. For MIC of concrete that occurs especially in sewers, sulfur-related bacteria, e.g. sulfate-reducing bacteria (SRB) and sulfur-oxidizing bacteria (SOB), are among the most relevant (Sun, 2015; US EPA, 1974; Wu *et al.*, 2020). For MIC of carbon steel, SRB are the most well-known and studied bacteria; while several other types, e.g. SOB, iron oxidizing bacteria (IOB), nitrate reducing bacteria (NRB) and acid producing bacteria (APB), have also attracted extensive research interest (Jia *et al.*, 2019; Little & Lee, 2007; Usher *et al.*, 2014). Sulfur-related bacteria can be involved in a number of processes, generating sulfur and sulfur compounds (Islander *et al.*, 1991; Little & Lee, 2007). The consequences include material corrosion and different types of cracking (Enning & Garrelfs, 2014; Jia *et al.*, 2019; Little & Lee, 2007; US EPA, 1974; Wu *et al.*, 2020). This chapter will focus on discussing several aspects of corrosion of concrete in sewer structures and carbon steel caused by sulfur-related bacteria.

5.2 MECHANISMS

Corrosion of concrete and carbon steel caused by sulfur-related bacteria, especially SRB and SOB, is rather complicated. A series of events and processes are involved, including biological, chemical, physiochemical and electrochemical processes (Enning & Garrelfs, 2014; Jia *et al.*, 2019; Little & Lee, 2007; US EPA, 1974; Usher *et al.*, 2014). It is worth mentioning that not all the processes/events are fully understood yet. This section reviews some important aspects related to the fundamental mechanisms involved in the corrosion processes.

5.2.1 Corrosion of concrete

The initiation and propagation of MIC of concrete exposed to sewer environments (Figure 5.1a) may be divided into several steps/stages. The commonly accepted main events, e.g. as schematically shown in Figure 5.1b, are summarized and discussed below.

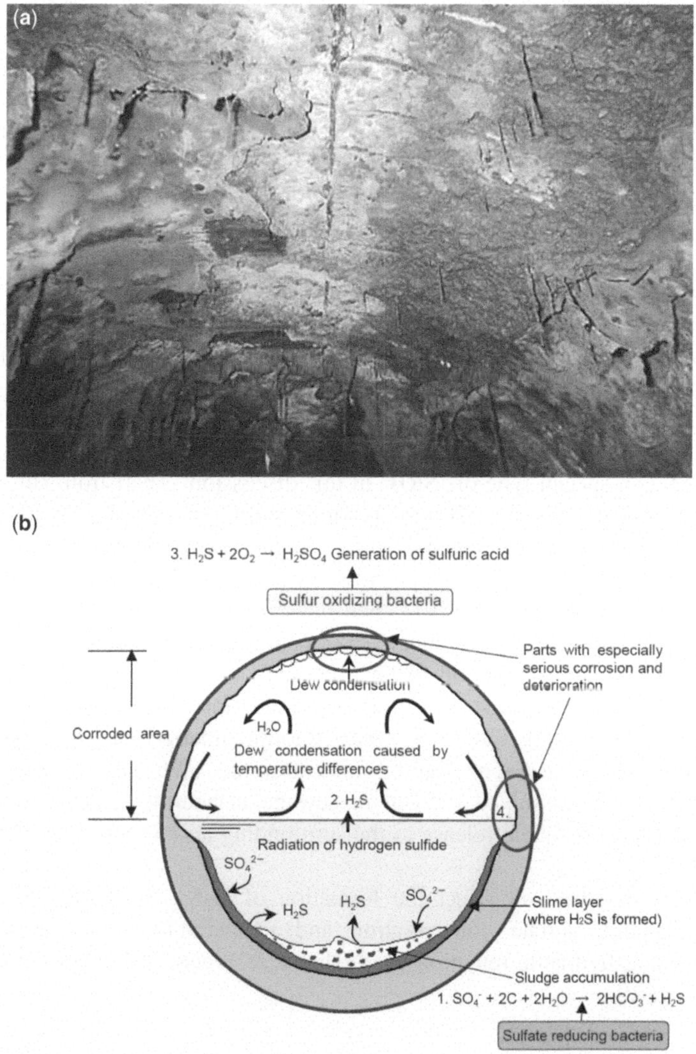

Figure 5.1 MIC of concrete: (a) crown corrosion in a sewage system (Vincke *et al.*, 2000); (b) schematic illustration of the main events involved in the MIC processes, adapted from Japan Sewage Works Agency (2007).

5.2.1.1 Formation of aqueous hydrogen sulfide

Formation of aqueous hydrogen sulfide $H_2S_{(aq)}$ may be considered as the starting point for MIC of concrete in sewer structures. Wastewater flowing in sewers contains different sulfur compounds, such as sulfate and sulfur-containing organic compounds (Sun, 2015). Under anaerobic conditions, sulfate can be reduced to sulfide by SRB residing in, for example, the biofilms and the sediments under waterline (the slime layer) (Japan Sewage Works Agency, 2007; US EPA, 1974), refer to Figure 5.1. SRB rely on the available organic substances as electron donors and the sources include organic matter already present in sewage or the fermentation products of carbohydrates, proteins and lipids (Sharma *et al.*, 2008). On the other side of the reaction, sulfate will be a terminal electron acceptor. The overall reaction can be expressed as Equation 5.1 (Japan Sewage Works Agency, 2007):

$$SO_4{}^{2-} + 2C\,(\text{Organic matter}) + 2H_2O \xrightarrow{\text{SRB}} 2HCO_3{}^{-} + H_2S_{(aq)} \tag{5.1}$$

With respect to the bacteria, *Desulfotomaculum* and *Desulfovibrio* are the two mostly reported SRB genera concerning MIC (Hao *et al.*, 1996; Satoh *et al.*, 2009). Other microbial species that can cause MIC include, for example, *proteobacteria*, *firmicutes* and *nitrospira* (Wu *et al.*, 2020). In addition, certain archaea exhibit a sulfate reducing metabolism and may contribute to the MIC as well (Santo Domingo *et al.*, 2011). Therefore, sulfate reducing archaea are considered as a special type of 'SRB' in this discussion. Apart from sulfate, it is noted that some species of SRB may use other sulfur compounds, e.g. thiosulfate, sulfite, tetrathionate and elemental sulfur, in the reduction reactions (Kleinjan *et al.*, 2003). Interested readers may find more information related to the SRB species in Jensen (2009), Sun (2015) and Peyre Lavigne *et al.* (2016). The slime layer plays an important role in concrete corrosion in sewer structures.

It is the place where the majority of the sulfate reduction takes place (US EPA, 1985; Wu *et al.*, 2018a). The slime layer may be divided into three zones (Figure 5.2): an aerobic zone starting from the sewage/biofilm interface; an anaerobic zone in the middle and an inert zone near the concrete surface (Wu *et al.*, 2018a).

The content of dissolved oxygen in sewage is a critical factor for sulfide buildup and its subsequent release to the surrounding air (Ayoub *et al.*, 2004; US EPA, 1991).

Other factors that can affect the formation of $H_2S_{(aq)}$ include, biochemical oxygen demand, sulfate concentration and temperature of the wastewater, turbulence and hydraulic retention time (House & Weiss, 2014; US EPA, 1974; Wu *et al.*, 2020).

5.2.1.2 Radiation and buildup of hydrogen sulfide

After its formation in wastewater, the $H_2S_{(aq)}$ concentration can increase and part of it may be released to the gas phase. The release of $H_2S_{(aq)}$ is heavily influenced by

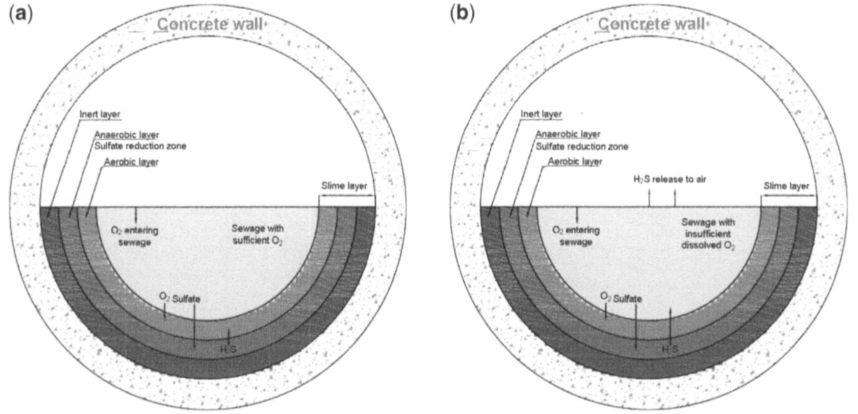

Figure 5.2 Schematic illustration of the slime layer: (a) with sufficient dissolved oxygen and (b) with insufficient dissolved oxygen, adapted based from Wu *et al.* (2018a). Note: the scale of the slime layer (in the order of 0.3–1.0 mm) is exaggerated to show detail.

certain factors, e.g. pH of the wastewater, the equilibrium conditions between the gas and the liquid phase, temperature and turbulence of the flow (House & Weiss, 2014). The transition of $H_2S_{(aq)}$ from the liquid phase to the gas phase $H_2S_{(g)}$ is governed by Henry's law (under closed conditions):

$$S_g = kP_g \tag{5.2}$$

where S_g is the concentration of the $H_2S_{(aq)}$; P_g is the partial pressure of $H_2S_{(g)}$ over the solution and k is the applicable Henry's law constant. Among others, temperature and the dissolved solute $H_2S_{(aq)}$ are the most significant factors influencing the concentrations of the two phases (US EPA, 1985). Figure 5.3 shows that as temperature increases, the gaseous $H_2S_{(g)}$ concentration also increases.

Gaseous $H_2S_{(g)}$ will start to build up in the sewer headspace after release from the solution. In general, a higher gaseous $H_2S_{(g)}$ concentration can result in a higher level of corrosion. However, it should also be mentioned that the exact relationship between the concentration of $H_2S_{(g)}$ and the MIC rate is far from being well defined (Apgar *et al.*, 2008; Vollertsen *et al.*, 2008; Wells *et al.*, 2009). The buildup of gaseous $H_2S_{(g)}$ is partly dependent on the ventilation conditions within the sewer network. Turbulence in the wastewater collection system is another important influencing factor, as the release of $H_2S_{(g)}$ is exacerbated by increasing the surface area of the water-air interface (House & Weiss, 2014; US EPA, 1974).

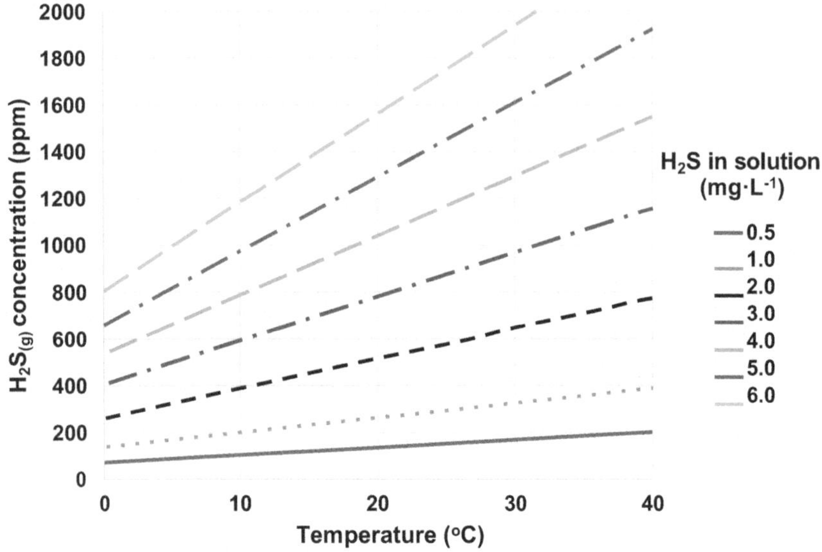

Figure 5.3 In a closed system, the concentration of gaseous $H_2S_{(g)}$ is a function of temperature and the $H_2S_{(aq)}$ concentration in the solution, redrawing based on data in (US EPA, 1985).

5.2.1.3 Generation of sulfuric acid

A number of abiotic (chemical) and biotic (biological) reactions will take place after gaseous $H_2S_{(g)}$ enters into the moisture films present on concrete surfaces. The reactions can lead to the generation of sulfuric acid and ultimately corrosion of the concrete. A three-step model has been proposed for the corrosion process and SOB play a critical role, see e.g. Islander *et al.* (1991) and Grengg *et al.* (2018).

The initial pH of newly cast concrete, i.e. in the range of 12.5–13.5 (Monteiro & Mehta, 1993), is too high for the colonization and growth of SOB. Only when the pH is lowered to a value ~9, will the microbial communities (e.g. SOB) develop (Mori *et al.*, 1992; Roberts *et al.*, 2002). Several theories have been proposed for the pH reduction, with the majority assuming abiotic reactions. The most assumed ones are carbonation of the concrete surfaces (Roberts *et al.*, 2002) and reactions of $H_2S_{(g)}$ with concrete surfaces (Joseph *et al.*, 2012). However, Yamanaka *et al.* (2002) claimed that bacteria can also directly oxidize $H_2S_{(g)}$ to form sulfuric acid, i.e. biotic reactions may start in the very beginning. In addition, fungi may have a contributing role to the initial pH reduction by producing various acids, e.g. acetic acid, oxalic acid and glucuronic acid (Gu *et al.*, 1998; Mori *et al.*, 1992). Nevertheless, the reduction of the concrete surface pH in this initial stage (or Stage 1) has been mainly attributed to abiotic reactions, where the biological activities are considered limited (Islander *et al.*, 1991; Sun, 2015).

The decrease of the pH of the concrete surfaces to ~9 in general marks the starting of the so-called Stage 2. A series of SOB can colonize the concrete surfaces and oxidize $H_2S_{(g)}$ in the presence of sufficient moisture, nutrients and oxygen (Okabe *et al.*, 2007; Vincke *et al.*, 2001). It may start with the colonization of a group of bacteria known as neutrophilic sulfur oxidizing bacteria (NSOB), e.g. *Thiobacillus thioparus*. It is important to note that depending on the local (pH) conditions, $H_2S_{(g)}$ can be oxidized into products of various sulfur states, e.g. S^0, $S_2O_3^{2-}$, $S_nO_6^{2-}$ and SO_4^{2-} (House & Weiss, 2014; Jensen, 2009). The acids produced in the processes react with the concrete surfaces and therefore, the surface pH is further lowered. As the pH gradually decreases, a number of other NSOB can colonize the concrete surfaces and become active, e.g. *Halothiobacillus neapolitanus*, *Thiomonas* spp. and *Starkeya novella* (George *et al.*, 2012; Valdebenito-Rolack *et al.*, 2011). The consequence is that the pH of the concrete surfaces will be further decreasing.

The activity of NSOB is gradually inhibited when the pH decreases to 4–5. Then acidophilic sulfur oxidizing bacteria (ASOB) can start to colonize the concrete surfaces, indicating the starting of the most destructive stage (Stage 3). One reason for this is that ASOB can directly oxidize H_2S to H_2SO_4 which will cause severe damage to concrete (Section 5.2.1.4). In addition, ASOB can also oxidize other sulfur compounds, e.g. thiosulfate ($S_2O_3^{2-}$) and elemental sulfur (S0) deposited on the sewer walls (Islander *et al.*, 1991; Jensen, 2009). Due to the formation of the strong acid H_2SO_4, the activity of ASOB further reduces the pH down to as low as 1–2. Widely reported ASOB include *Acidithiobacillus thiooxidans*, *Acidithiobacillus ferrooxidans* and *Acidiphilium* spp. (Cayford *et al.*, 2012; Nica *et al.*, 2000; Sun, 2015).

5.2.1.4 Deterioration of concrete materials

Portland cements are the most commonly used cements to produce concrete. Due to their alkaline nature, the hardened concrete is susceptible to degradation when exposed to acidic conditions (Monteiro & Mehta, 1993). The H_2SO_4 generated by SOB (Section 5.2.1.3) will react with cement hydration products. The reactions can start with the dissolution of calcium hydroxide and carbonate compounds. Decalcification of less soluble hydration products, e.g. calcium silicate hydrate, is often reported as an accompanying process (House & Weiss, 2014). Formation of gypsum is one result of the reactions (Monteny *et al.*, 2000; Wells *et al.*, 2009), i.e. Equations 5.3 to 5.5. The formed gypsum may further react with the aluminate phases and ettringite is the product (Monteny *et al.*, 2000), i.e. Equations 5.6. One important consequence related to the formation of these products can be the volume expansion which may ultimately lead to the destruction of the concrete.

$$H_2SO_4 + CaCO_3 \rightarrow CaSO_4 + H_2CO_3 \tag{5.3}$$

$$H_2SO_4 + Ca(OH)_2 \rightarrow CaSO_4 + 2H_2O \tag{5.4}$$

$$H_2SO_4 + CaO \cdot SiO_2 \cdot nH_2O \rightarrow CaSO_4 + Si(OH)_4 + (n-1)H_2O \tag{5.5}$$

$$3CaSO_4 + 3CaO \cdot Al_2O_3 \cdot 6H_2O + 26H_2O \rightarrow 3CaO \cdot Al_2O_3 \cdot 3CaSO_4 \cdot 32H_2O \tag{5.6}$$

Figure 5.4 shows a schematic illustration of the characteristics of concrete subjected to MIC. Since its formation at low pH is preferred, gypsum is usually present at the surface of the corrosion layer (Mori *et al.*, 1992). In general, ettringite can only exist in deeper sections of concrete where the pH is high enough for it to form and the gypsum can migrate into these regions (Marchand *et al.*, 2003).

Apart from sulfuric acid, other biogenic acids have also been considered relevant for the deterioration of concrete, e.g. acetic acid, propionic acid and carbonic acids (Tazawa *et al.*, 1994). As mentioned in Section 5.2.1.3, fungi are able to create acids as well. The effects of those acids (and the mechanisms accounting for the generation of them) on concrete are not well understood yet.

5.2.2 Corrosion of carbon steel

Numerous mechanisms have been proposed to explain MIC of carbon steel, see e.g. Little and Lee (2007), Gu (2012), Enning and Garrelfs (2014) and Usher *et al.* (2014). It is not the intention to cover all the possible mechanisms in this section, but to discuss the most pertinent ones for sulfur-related bacteria/microorganisms. For the most studied bacterial group, SRB, the representative mechanisms include the classical 'cathodic depolarization' theory and emerging theories such as 'chemical microbiologically influenced corrosion' and 'electrical

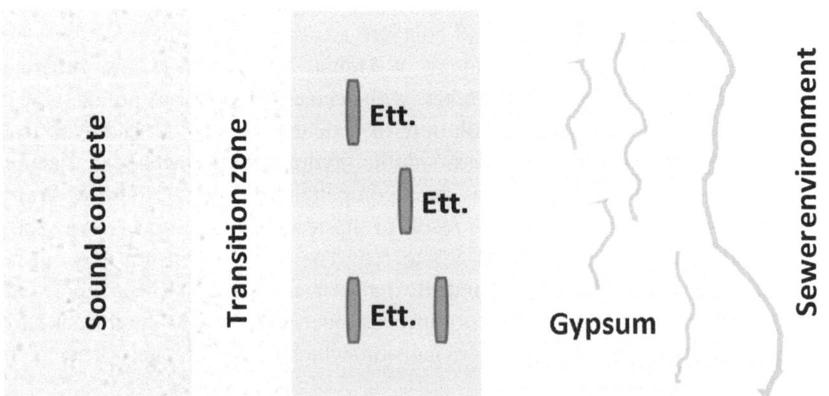

Figure 5.4 Main deterioration compositions at different layers of the concrete subjected to MIC, adapted based on Wu *et al.* (2020). (Ett.: ettringite).

microbiologically influenced corrosion' (Enning & Garrelfs, 2014). With respect to SOB, the corrosion of carbon steel generally can be attributed to the sulfuric acid produced by these microorganisms (Little *et al.*, 2000; Usher *et al.*, 2014). Most of the bacterial strains discussed in Section 5.2.1 are relevant for the corrosion of carbon steel as well, see e.g. Friedrich *et al.* (2001), Little and Lee (2007), Enning (2012) and (Jia *et al.*, 2019).

5.2.2.1 Cathodic depolarization

The cathodic depolarization theory was first proposed by von Wolzogen Kuehr and van der Vlugt (1934) to account for the iron pipe failures caused by SRB in the sulfate-rich soils of North Holland. It involves several steps and the suggested reactions include:

$$\text{Anodic reaction:}\quad 4Fe \rightarrow 4Fe^{2+} + 8e^- \tag{5.7}$$

$$\text{Water dissociation:}\quad 8H_2O \rightarrow 8H^+ + 8OH^- \tag{5.8}$$

$$\text{Cathodic reaction:}\quad 8H^+ + 8e^- \rightarrow 8H_{(ads)} \tag{5.9}$$

$$\text{Sulfate reduction:}\quad {SO_4}^{2-} + 8H_{(ads)} \xrightarrow{SRB} S^{2-} + 4H_2O \tag{5.10}$$

$$\text{Corrosion product:}\quad Fe^{2+} + S^{2-} \rightarrow FeS \tag{5.11}$$

$$\text{Corrosion product:}\quad 3Fe^{2+} + 6OH^- \rightarrow 3Fe(OH)_2 \tag{5.12}$$

$$\text{Overall reaction:}\quad 4Fe + {SO_4}^{2-} + 4H_2O \xrightarrow{SRB} 3Fe(OH)_2 + FeS + 2OH^- \tag{5.13}$$

The core of the theory is that SRB use atomic hydrogen (H(ads)) accumulated at the cathode as the sole electron donor for sulfate reduction (Enning & Garrelfs, 2014). The cathode is polarized when the electron generation rate at the anode is not matched by the electron consumption rate at the cathode. Removal of hydrogen from the cathode allows the cathodic reaction (Equation 5.9) to proceed, i.e. the cathode is depolarized. Because of this depolarization, the anodic reaction (Equation 5.7), i.e. the dissolution of iron, is enhanced, and consequently more corrosion products FeS and $Fe(OH)_2$ can be formed.

The theory was favored by many authors, but it was also the subject of much controversy, see e.g. Enning and Garrelfs (2014). A series of electrochemical investigations have been conducted to validate the relevance of the theory (Booth & Tiller, 1960; Costello, 1974; Daniels *et al.*, 1987; Dinh *et al.*, 2004). However, it turned out that the observed acceleration of cathodic reactions in the presence of SRB cultures could be explained by the reaction between sulfide and iron rather than by the consumption of the hydrogen at the cathode (Costello, 1974; Enning & Garrelfs, 2014). In the presence of hydrogen-consuming SRB, experiments with iron as the only electron donor showed that SRB were capable of using cathodic hydrogen as a substrate but this did not affect iron corrosion to any significant extent (Dinh *et al.*, 2004; Mori *et al.*, 2010). Such results put much of the electrochemical evidence for the cathodic depolarization theory under dispute (Enning & Garrelfs, 2014). Little and Lee (2007) argued that it is doubtful that the reactions would proceed following the steps described in Equations 5.7 to 5.12, though the overall reaction may be correct. In addition, the existence of a layer of atomic hydrogen on the metal surfaces is also questionable (Little & Lee, 2007). In some studies, e.g. by Gu *et al.* (2019), it was considered that gaseous H_2 can be formed via chemical or electrochemical reactions from atomic hydrogen (H(ads)). Then SRB can metabolize gaseous H_2 where a hydrogenase enzyme plays an important role (Gu *et al.*, 2019; Little & Lee, 2007). Nevertheless, the theory based on cathodic hydrogen consumption remains an important one in the context of MIC of metal surfaces and steel.

5.2.2.2 Chemical microbiologically influenced corrosion (CMIC)

The CMIC theory considers that the corrosion is linked to the biogenic sulfide generation, a process similar to that described in Section 5.2.1.1 (Equation 5.1) where organic matter is used by SRB as an electron donor. The generated sulfide when dissolved will then react with metallic iron, leading to the formation of the corrosion product iron sulfide (e.g. Equation 5.14). The stoichiometry of the overall reaction has been proposed as shown in Equation 5.15 (Enning & Garrelfs, 2014).

$$H_2S + Fe \rightarrow FeS + H_2 \tag{5.14}$$

$$2\,SO_4{}^{2-} + 3C\,(\text{Organic matter}) + 2Fe + H^+ + H_2O \xrightarrow{\text{SRB}} 3HCO_3{}^- + 2FeS \tag{5.15}$$

The effects of sulfide generated from sulfate reduction are undoubtedly the focus of most studies. However, it should be mentioned that other microbiological processes, e.g. the dissimilatory reduction of thiosulfate or sulfite, can also produce significant amounts of corrosive sulfide (Magot *et al.*, 1997). The generated biogenic sulfide may initially enhance the anodic part of the reaction, i.e. iron oxidization, by chemisorption and direct reaction with metallic iron (Enning & Garrelfs, 2014). While at a later stage when the metallic surfaces are covered with inorganic corrosion products such as FeS, the cathodic reactions may be more important drivers for the metal oxidization (Newman *et al.*, 1991). Relevant explanations accounting for the cathodic stimulation can be found in Costello (1974) and Cord-Ruwisch (2000).

Another type of damage caused by biogenic hydrogen sulfide is known as sulfide stress cracking. Hydrogen sulfide can decelerate (or poison) the combining of hydrogen atoms into molecular hydrogen at the metal surface and, as a result, the diffusion of a higher fraction of hydrogen atoms into the metal matrix may occur (Elboujdaini, 2011). Combining of absorbed atomic hydrogen molecules into hydrogen gas (H_2) within the metal can cause embrittlement of the metal (Radkevych & Pokhmurs'kyi, 2001). Cracking of metals under the effects of mechanical stress induced by the sulfide attack is referred to as sulfide stress cracking.

5.2.2.3 Electrical microbiologically influenced corrosion (EMIC)

The central concept of the EMIC theory is that SRB can directly use elemental iron as the only electron donor (Dinh *et al.*, 2004; Enning *et al.*, 2012; Enning & Garrelfs, 2014; Gu *et al.*, 2019; Jia *et al.*, 2019; Venzlaff *et al.*, 2013). This differs from the cathodic depolarization theory where cathodic hydrogen is considered as an essential intermediate in the reactions (Section 5.2.2.1). Compared with CMIC (Section 5.2.2.2), supply of electrons from the organic matter is not a condition for EMIC to occur. Figure 5.5 shows a schematic illustration of CMIC and EMIC.

A common feature noted for EMIC is that the corrosion rates can be significantly accelerated (Enning *et al.*, 2012; Usher *et al.*, 2014), while the exact mechanisms responsible for the corrosion rate increase have not yet been confirmed (Usher *et al.*, 2014). It was suggested by Mori *et al.* (2010) that an enzyme, most likely a hydrogenase, may catalyze the oxidation of metallic iron. Other assumptions include that the consumption of electrons by EMIC microorganisms may drive the cathodic reaction (Venzlaff *et al.*, 2013).

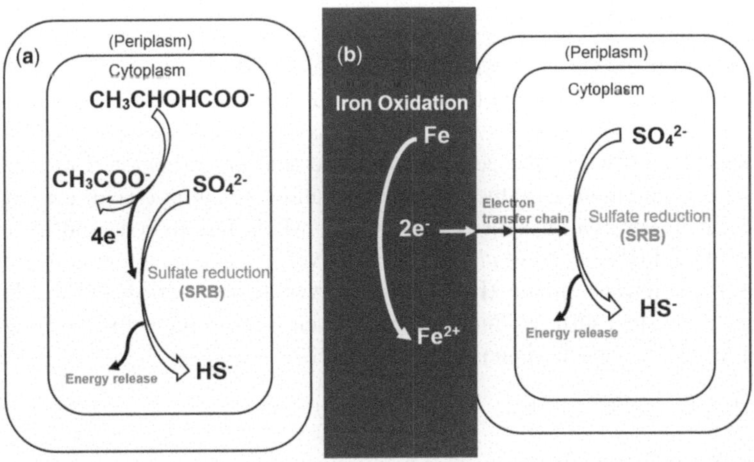

Figure 5.5 Schematic illustration of CMIC (a) and EMIC (b): the electron donors for CMIC and EMIC are organic matter and elemental iron, respectively (adapted based on Li *et al.*, 2018).

According to Enning *et al.* (2012), the reaction leading to the formation of large amounts of corrosion products occurs at a distinct stoichiometry, i.e. Equation 5.16. The oxidization of four moles of elemental Fe to Fe^{2+} is coupled to the reduction of only one mole of sulfate. As a result, H_2S from this reaction precipitates only one mole of FeS, a highly insoluble corrosion product. The remaining three moles of Fe^{2+} precipitate as nonsulfidic iron minerals, e.g. with carbonate, which is abundant in produced waters and other anoxic environments (Usher *et al.*, 2014).

$$4Fe + SO_4^{2-} + 3HCO_3^- + 5H^+ \xrightarrow{SRB} 3FeCO_3 + FeS + 4H_2O \quad (5.16)$$

The electron transfer between a metal surface and the SRB cells can have two routes, i.e. the direct electron transfer (DET) and mediated electron transfer (MET) (Gu, 2012; Jia *et al.*, 2019). In DET, SRB cell walls have a direct contact with a metal surface or they use conductive pili to link cell walls with a metal surface for the electron transfer; while in MET, transfer mediators (electron shuttles) exist which can absorb electrons from metal surfaces and release them to the cells (Gu, 2012; Jia *et al.*, 2019). Enning *et al.* (2012) demonstrated that the bulky black crusts formed through EMIC are electrically conductive and therefore, direct contact between corrosive SRB and the metal is not a necessary condition for corrosion. Iron sulfide minerals can act as a mediator for electrons flowing from the metal to the SRB. Nevertheless, studies also revealed that iron corrosion caused by direct electron uptake seems to be independent of the presence of iron sulfides (Dinh *et al.*, 2004; Mori *et al.*, 2010). Usher *et al.* (2014) claimed nanowires may also play a role in transferring electrons from steel to microorganisms.

5.2.2.4 SOB influenced corrosion

Different sulfur species, e.g. elemental sulfur, thiosulfates, metal sulfides, H_2S, and tetrathionates, can be oxidized to sulfuric acid (Little *et al.*, 2000). Similar to that described in Section 5.2.1.3, various SOB may be involved in the oxidization processes. The attack of the generated sulfuric acid on carbon steel depends on the acid concentration. At rather low concentrations of sulfuric acid, the reaction may occur in accordance with Equation 5.17, where ferrous ions and hydrogen gas are formed (Panossian *et al.*, 2012). However, if the concentration of sulfuric acid is high, ferrous sulfate ($FeSO_4$) is believed to be formed and the $FeSO_4$ adheres to the steel surface which somehow even forms a protective layer against further attack by the acid (Panossian *et al.*, 2012).

$$Fe + 2H^+ \rightarrow Fe^{2+} + H_2 \quad (5.17)$$

$$Fe + H_2SO_4 \rightarrow FeSO_4 + H_2 \quad (5.18)$$

Acid attack has been considered an important cause of corrosion in the presence of microorganisms (Gu, 2009). Studies have shown that drops in pH are typically associated with an increase of Fe release to solution (Barker *et al.*, 1998). Biezma (2001) reported that biogenic acid can increase corrosion rates by changing the local pH through hydrogen permeation, where hydrogen is known to induce cracking or blistering (Woodtli & Kieselbach, 2000). Similar conclusions, e.g. increased acidity, hydrogen penetration and corrosion rates caused by SOB, were also reported by Little *et al.* (2000). It may be of interest to mention that in a study conducted by Dowling (1992), the principal form of attack was actually concluded not due to the generated acid but to the steel oxidization coupled to sulfur reduction through an electron conductive iron sulfide layer.

5.3 MIC OBSERVATIONS

Based on site investigations and laboratory studies, some important aspects related to the MIC phenomena are highlighted and summarized below.

5.3.1 MIC of concrete

5.3.1.1 Corrosion areas

It is noted that corrosion of concrete sewer pipes is often not uniformly distributed. In general, the most corroded areas are at the crown and the waterline level (Figure 5.1). High corrosion rates at the crown have been attributed to temperature differences between concrete walls and the wastewater (US EPA, 1974): an airflow is created inside of the pipes and as a result, the transfer rate of $H_2S_{(g)}$ to sewer walls is maximized at the crown. Migration of acid-containing condensates down the pipe walls has been considered the trigger for the high corrosion rates around the waterline (US EPA, 1974, 1991). Concrete reacts with the condensates, forming deteriorated pasty products at these areas, which are washed away intermittently by the waste stream and subsequently bare and new surfaces are exposed for further attack. It is noted that there are also different opinions with respect to which area is more severely corroded, i.e. at the crown or around the waterline (Mori *et al.*, 1992; Vincke *et al.*, 2001).

5.3.1.2 Corrosion rates

Corrosion rate is a parameter of practical significance in the design of sewer structures (Loganathan *et al.*, 2008; US EPA, 1985; Wells *et al.*, 2009). Many studies have been devoted to measurement and prediction of the corrosion rate. Table 5.1 provides a summary for some of the average corrosion rates reported in the literature based on site observations/measurements (or full-scale experiments simulating site conditions) and laboratory studies (including those using alternative acids rather than biogenic acids).

Table 5.1 shows that the reported corrosion rates have a rather large variation, despite the values being based on site observations or laboratory studies. This is not that surprising as so many factors can affect the results especially under

Table 5.1 Concrete corrosion rates under different test conditions.

$H_2S(g)$ (ppm)	Test Condition (Exposure Medium)	Concrete/Mortar	Corrosion Rate ($mm \cdot year^{-1}$)	Reference
0.1–50	Site conditions	Concrete	2.5–10	US EPA (1991)
N.A.	Site conditions	Concrete	1.3	Morton *et al.* (1991)
N.A.	Site conditions	Concrete (OPC, GGBS and FA cements)	0.8–1.0	van Mechelen and Polder (1991)
5–400	Site conditions	Concrete	1.4, 4.3–4.7	Mori *et al.* (1992)
5–400	Site conditions	Mortar	5.7–32	
25–300	Demonstration plant	Mortar	7.6	
N.A.	Site conditions	Concrete	1.1–1.8	Vincke *et al.* (2002)
40–75	Full-scale experimental sewer	Concrete (OPC + Sil. Ag.)	>7.5	Alexander & Fourie (2011)
		Concrete (OPC + Cal. Ag.)	3.1	
		Concrete (CAC + Sil. Ag.)	1.9	
		Concrete (CAC + Cal. Ag.)	0.6	
0–1000	Pilot reactors	Concrete	3.1	Vollertsen *et al.* (2008)
10–50	Site conditions	Mortar	4	Okabe *et al.* (2007)
10–50		Mortar	3–6.6	Satoh *et al.* (2009)
50–550	Site conditions	Concrete	12	Wells *et al.* (2012)
2		Concrete	1–2	
20		Concrete	5	
1.5–6	Site conditions	Concrete	0.6–6.5	Wells and Melchers (2014a)
80–420		Concrete	3.7–12.7	

40–100	Site conditions	Concrete (Sil. Ag.)	9–12	Wells and Melchers (2014b)
0.1–367	Site conditions	Concrete (OPC (C3A-free) + FA)	1–10	Grengg *et al.* (2015)
250	Lab (artificial solution)	Concrete (OPC or GGBS cement + polymer)	up to 5.7	Vincke *et al.* (1999)
100	Lab (sewage)	Mortar (OPC)	2.1–3.8	Ismail *et al.* (1993)
250	Lab (artificial solution)	Mortar (OPC + polymer or silica fume)	1.6–4.3	Vincke *et al.* (2002)
N.A.	Lab (0.5% H_2SO_4)	Mortar (OPC + polymer or silica fume)	up to 2.7	
400	Lab (sewage) Lab (artificial solution) Lab (distilled water)	Mortar	6.1 3.8 0	Mori *et al.* (1992)
2–4	Lab (sewage)	Concrete (OPC)	10	Æsøy *et al.* (2002)
15–25	Lab (sewage)	Concrete (OPC)	14	
–	Lab (2 M H_2SO_4)	Concrete (OPC)	~10–57 ~57–81	Kawai *et al.* (2005)
	Lab (0.06 M H_2SO_4)	Mortar (OPC)	~2–20 ~16–57	
50–550	Lab (artificial solution)	Concrete (OPC)	0.08–0.21	Gutiérrez-Padilla *et al.* (2010)
0–50	Lab (sewage)	Concrete	0.7–1.2	Jiang *et al.* (2015)

Notes: 1. 'N.A.' indicates information not available/identified; 2. Abbreviations: OPC – ordinary Portland cement; GGBS – ground granulated blast-furnace slag; FA – fly ash; CAC – calcium aluminate cement; Sil. Ag. – silicecus aggregate; Cal. Ag. – calcareous aggregate; C_3A – cement notation for 3 CaO Al_2O_3.

site conditions, e.g. as discussed by Wu *et al.* (2020). Compared with site observations, relevant parameters that can affect the corrosion rates are better known or controlled in small-scale laboratory experiments, e.g. temperature, relative humidity (RH), bacteria, sulfide level (or concentrations of the used sulfur source) and concrete/mortar composition. However, it should also be noted that the test conditions can differ significantly. This will inevitably lead to variation of the obtained results. To enable comparison and evaluation of the results obtained in different laboratories, there is an urgent need to standardize the testing method(s) for MIC of concrete.

5.3.1.3 Cement types

Cement types may have an important impact on the performance of the concrete for sewers subjected to MIC. This section summarizes the results of the studies using different types of cement. The effects of supplementary cementitious materials (SCMs) are also discussed.

- Calcium aluminate cement (CAC)

  CAC contains a high alumina content. In general, the performance of concrete prepared by CAC is better than that prepared by Portland cement when exposed to acidic conditions. Three main reasons for this have been proposed (Saucier & Herisson, 2015; Scrivener & Capmas, 1998): (1) large neutralization capacity (or high alkalinity); (2) precipitation of alumina gel that slows down the attack and (3) bacteriostatic effects. The beneficiary effects provided by CAC have been observed in a number of studies, see e.g. Scrivener and Capmas (1998), Alexander and Fourie (2011) and Saucier and Herisson (2015).
- Sulfate resistant cement

  This is a type of cement designed to mitigate problems for concrete exposed to aggressive sulfate environments. Studies have been performed to investigate if sulfate resistant cement can help improve the performance of concrete exposed to sulfuric acid. However, no significant difference was found for concrete samples prepared by sulfate resistant cement and OPC (Attiogbe & Rizkalla, 1988; Fattuhi & Hughes, 1988). It should be stressed that important differences exist between sulfate attack and sulfuric acid attack (Attiogbe & Rizkalla, 1988). The severity of concrete deterioration caused by sulfuric acid is much more pronounced than that caused by sulfate attack (Monteny *et al.*, 2000).
- Supplementary cementitious materials (SCMs)

  Numerous studies have been conducted to investigate the effects of SCMs, e.g. ground granulated blast-furnace slag (GGBS), fly ash (FA) and silica fume, on the performance of concrete corrosion. Based on most of the tests using mineral acids, the addition of SCMs could have positive effects in reducing the corrosion to some extent, see e.g. Monteny *et al.* (2000) and

Alexander and Fourie (2011). However, when exposed to biological conditions, the addition of SCMs may not always result in superior performance compared to plain OPC samples, see e.g. Sand *et al.* (1994) and Ehrich *et al.* (1999). More studies are needed to clarify the effects of SCMs on concrete corrosion, especially under biological conditions.

- Geopolymer

  Excellent acid resistant properties of geopolymers (or alkali-activated materials) have been reported, see e.g. Shi and Wu (2007), Provis *et al.* (2015) and Grengg *et al.* (2018). Under acid attack, due to the low initial permeability and the typical chemical composition (e.g. low CaO/SiO_2 ratio) of geopolymers, a coherent layer of aluminosilicate gel will be formed even after decalcification (Shi, 2003). As a result, further ingress of acids is hindered (Lloyd *et al.*, 2012; Provis *et al.*, 2015). Nevertheless, it should be mentioned that limited studies can be found with respect to the interactions between the geopolymer and microorganisms (bacteria). In addition, the applications of geopolymers as a structural material has up until now been rather limited, though applications such as sprayed linings for rehabilitation purposes have been reported, see e.g. Saucier and Herisson (2015).

5.3.1.4 Siliceous and calcareous aggregates

Two types of aggregates are normally used in concrete, i.e. the siliceous type (e.g. granite and gabbro, with the main oxide SiO_2) and the calcareous type (e.g. limestone and dolomite, with the main oxide CaO). Aggregates can affect the corrosion rate of concrete influenced by biogenic acids. Calcareous aggregates can slow down corrosion rates of concrete subject to biogenic acids, as demonstrated both under site conditions (or full-scale experiments) (Alexander & Fourie, 2011; Pomeroy, 1969) and in laboratory studies (Hewayde *et al.*, 2007). The effect of the aggregate type on the corrosion rate can be attributed to the 'alkalinity' of the concrete. Concrete with calcareous aggregates has a higher alkalinity compared to concrete with siliceous aggregates (e.g. the higher the CaO, the higher the alkalinity). Essentially high alkalinity provides more substances to react with the generated biogenic acids, which may effectively slow down the corrosion rate.

5.3.2 MIC of carbon steel

The effects of SRB and SOB are similar in MIC of both concrete and metal to some extent, i.e. leading to the formation of sulfides and sulfuric acids (Section 5.2). However, there are also some differences in terms of the corrosion phenomena. In general, the generated sulfides may only cause limited corrosion (i.e. through direct reactions with concrete), while the biogenic sulfuric acid is the main culprit for the significant corrosion reported for concrete (House, 2013; Sun, 2015; Wells *et al.*, 2009). While for MIC of metal, both SRB and SOB can induce significant corrosion (Javaherdashti, 2017; Little & Lee, 2007). Some of the phenomena reported for MIC of carbon steel are summarized below.

5.3.2.1 Corrosion caused by SRB

In the presence of SRB, various iron sulfides can be formed during the corrosion of iron and steel, e.g. mackinawite (FeS), greigite (Fe_3S_4), marcasite (orthorhombic FeS_2) and pyrite (cubic crystal FeS_2) (Little & Lee, 2007). The most commonly found iron sulfides produced along MIC are mackinawite and greigite, while they may be transformed into marcasite and pyrite over time (Jack *et al.*, 1996; Little *et al.*, 2000). It has been reported that marcasite is not stable under natural environmental conditions and therefore, its presence in the corrosion products may indicate that the corrosion is SRB-induced (Little *et al.*, 2000; Vuković *et al.*, 2009).

Considerable attention has been paid to the effects of formed sulfide films on metal corrosion, while the results are rather controversial (Gu, 2012; Little *et al.*, 2000). In theory, it seems reasonable to assume that the formation of adherent iron sulfide films would have some protective effects, e.g. by impeding the diffusion of oxidized iron (ions) from the metal surfaces to the bulk liquid (Enning & Garrelfs, 2014; Sun & Nesic, 2007). Enning and Garrelfs (2014) claimed that the presence of the FeS films may lead to the protection of iron against corrosion, at least temporarily. However, the effectiveness of the sulfide films in preventing metal corrosion in practice has been questioned, e.g. by Little *et al.* (2000) and Xu *et al.* (2011). It was argued that the growth of the films can cause cracking and the formation of loose, porous films (Sun & Nesic, 2007). In addition, several authors revealed that the microbiologically-induced iron sulfides can accelerate the corrosion (Booth *et al.*, 1968; King *et al.*, 1973). It was concluded by Bourdoiseau *et al.* (2011) that the transformation of mackinawite films to greigite or pyrite (in the continued presence of SRB) will lead to enhanced corrosion, as the greigite or pyrite readily flakes from the steel surface, breaking the protective film and allowing high corrosion rates to be re-established. Jack *et al.* (1996) suggested that cathodic protection may fail in the presence of iron sulfides, since the iron sulfides are protected instead of the iron metal. The iron sulfide layer may also be able to adsorb corrosion inhibitors, rendering them less efficient (Sun *et al.*, 2012). In summary, the role of sulfide films in SRB induced corrosion is a complex subject and it is not sufficiently understood. More studies are needed to further advance the understanding in this context.

Stress corrosion cracking and hydrogen embrittlement are two types of damage often reported (De Romero *et al.*, 2005; Koh *et al.*, 2004). Localized corrosion pits on the metal surfaces can be induced by SRB via the CMIC and/or EMIC mechanisms. Under the action of tensile stresses, the electrochemical activity of the metal at the pit tips increases, and accordingly the corrosion rate increases. As a consequence, cracking failures will occur (Wu *et al.*, 2018b). By studying the corrosion behavior of a steel pipeline under the combined action of SRB and external stresses (Wu *et al.*, 2019), a synergistic effect was found on the corrosion of the pipe by the formation of secondary pitting in addition to the primary pitting (i.e. enhanced corrosion). Hydrogen embrittlement, as mentioned in Section 5.2.2.2, is related to the slowing down of the recombination kinetics of hydrogen atoms in the presence of the biogenic H_2S and then the hydrogen atoms diffusing into the metal (Robinson & Kilgallon, 1998; Wu *et al.*, 2018b). But without SRB (or other relevant microorganisms), the dominant failure mode for carbon steel is actually ductile failure (Raman *et al.*, 2005). Stress corrosion cracking and hydrogen embrittlement differ in many ways, e.g. the morphology of metal surfaces after hydrogen embrittlement is apparently different from that after stress corrosion cracking (Biezma, 2001).

Corrosion rate is a parameter of high importance from a practical point of view, as it is related to how fast the metal is destructed. Li *et al.* (2001) reported that corrosion rates can be more than 20 times higher in anaerobic soil in the presence of SRB when compared with abiotic corrosion. According to Enning and Garrelfs (2014), typical corrosion rates of unprotected steel in permanently anoxic environments are 0.2–0.4 mm per year and the main mechanism causing such corrosion can be CMIC. Nevertheless, the corrosion rates caused by CMIC may span at least two orders of magnitude (Enning, 2012). It was also shown that the stimulation of corrosion by the catalytic FeS is generally rather low (Booth *et al.*, 1968; King *et al.*, 1973), i.e. $\leq$0.06 mm per year, and negligible with more compacted crusts (Venzlaff *et al.*, 2013). While in studies designed for EMIC, corrosion rates can be much higher, i.e. up to 0.9 mm pear year (Enning, 2012). This may be partly explained by the fact that the corrosion products produced by CMIC and EMIC differ inherently in terms of the relative amounts of sulfidic and nonsulfidic iron (Enning & Garrelfs, 2014): CMIC produces FeS as the sole mineral product, while FeS accounts for only about 25% of the total iron minerals formed by EMIC (Equation 5.15 vs. Equation 5.16).

5.3.2.2 Corrosion caused by SOB

A variety of SOB exists and they may be aerobic or anaerobic, acidophilic or neutrophilic (Friedrich *et al.*, 2001). As mentioned in Section 5.2.2.4, hydrogen sulfide, metal sulfides and elemental sulfur are among the most reported sulfur species in the oxidization. The generated sulfuric acid, as a very strong acid, can corrode, e.g. underground pipelines or other facilities, by reacting with the metals.

With the coexistence of oxygen and hydrogen sulfide, acidification of seawater occurred in the presence of *thiobacilli* (a type of SOB) and corrosion of mild steel was reported (Eashwar *et al.*, 1993). Sulfide oxidization was considered responsible for the carbon steel deterioration in a dam structure (Mittelman & Danko, 1995). Oxidization of metal sulfides under the influence of SOB has been reported to be the cause for the corrosion of mining equipment, shipping containers and pipelines (Little *et al.*, 2000). Metal monosulfides can react with acid leading to the formation of hydrogen sulfide. When in contact with air, hydrogen sulfide will oxidize rapidly to form sulfur, thiosulfate, sulfite, and, ultimately but more slowly, sulfate (Little *et al.*, 2000). In the presence of elemental sulfur, corrosion rates may be significantly increased. The reasons for this can be attributed to the formation of sulfuric acid and H_2S, which inhibit repassivation promoting crevice corrosion and stress corrosion cracking (Little *et al.*, 2000). Schmitt (1991) provided a summary for the effects of element sulfur on the corrosion activities of several iron and steel products, e.g. carbon steels, ferric steels, austenitic steels and ferritic-austenitic steels.

It is of interest to mention that natural systems are generally much more complex, e.g. different bacteria/microorganisms are often found together and their cooperative activities can lead to increased corrosion rates (Little *et al.*, 2000; Usher *et al.*, 2014). In terms of SOB and SRB, the presence of SOB may encourage the growth of SRB by producing the necessary products for SRB to grow, e.g. sulfate, and 'pods' containing certain bacteria that regenerate the sulfur substrates (Norlund *et al.*, 2009; Usher *et al.*, 2014). Consequently, coexistence of SOB and SRB communities may significantly enhance the corrosion activities.

5.4 MITIGATION AND CONTROL MEASURES

5.4.1 For MIC of concrete

Various measures to control and mitigate MIC have been proposed, see e.g. Zhang *et al.* (2008), Noeiaghaei *et al.* (2017), Grengg *et al.* (2018) and Wang *et al.* (2020). They may be categorized into three groups: (1) improving sewer design features; (2) controlling the causative sewer environment, especially the sulfide level; and (3) improving the performance of concrete. A brief summary of the possible mitigation measures and their principles is given below.

5.4.1.1 Improving sewer design features

Sulfide generation in sewers is affected by hydraulic design features, e.g. flow velocity and retention time of the sewage, inclusion of structures that promote turbulence and anaerobic conditions (US EPA, 1985). Flow velocity is a very pertinent design factor for MIC (see Chapter 4). To some extent, high velocity is beneficial in reducing MIC by enhancing diffusion of atmospheric oxygen into the wastewater (Wang *et al.*, 2020; Wu *et al.*, 2018a). High turbulence is another important factor which can increase the severity of concrete corrosion (US EPA, 1985, 1991). In general, design should be optimized to limit the use of structures that facilitate the occurrence of turbulence and to avoid an excessive retention time (Hewayde, 2005; US EPA, 1991).

Ventilation helps to reduce concrete corrosion in sewer structures (Grengg *et al.*, 2015). The potential of ventilation to reduce MIC has been reported already decades ago (Pomeroy, 1945). However, it was revealed that natural ventilation through vented manhole covers and inlets by the action of wind and pressure differences is rather limited (Pescod & Price, 1982). On the other hand, active ventilation of sewer systems has been considered impractical and costly, and consequently it can only be applied to selected areas (US EPA, 1991).

5.4.1.2 Controlling sulfide in the sewer environment

Control of sulfide formation (and/or buildup) and then the release to the gas phase can be effective in mitigating the concrete corrosion. Typical strategies to control sulfide in the sewer environment include changing redox conditions in wastewater, inhibiting the activity of SRB, and chemical removal of sulfide (Noeiaghaei *et al.*, 2017; Zhang *et al.*, 2008).

- Changing redox conditions

  Injection of air or oxygen has been shown to be effective in controlling the formation of sulfide (Tanaka *et al.*, 2000). Compared with air injection, injection of pure oxygen may be more effective as pure oxygen is five times more concentrated than air (Zhang *et al.*, 2008). Both transient and long-term effects of adding nitrate to sewers in preventing sulfide production have been reported (Poduska & Anderson, 1981; Yang *et al.*, 2005). It was found that nitrate was preferentially reduced over sulfate in oxygen deficient environments, thus reducing the formation of sulfide (Zhang *et al.*, 2008).
- Inhibiting the activities of SRB

  Several methods have been proposed either to destroy the biofilms where SRB reside or inhibit their activities. These include the use of different inhibitors, e.g. biocides (biocide enhancers), free nitrous acid or molybdate (Jayaraman *et al.*, 1999; Jia *et al.*, 2019; Narenkumar *et al.*, 2017; Zhang *et al.*, 2008). In addition, pH elevation of wastewater is also reported, e.g. through feeding sodium hydroxide (NaOH) (Gutierrez *et al.*, 2009; US EPA, 1991).
- Chemical sulfide removal

  Addition of metal ions to wastewater is an important way to remove the dissolved sulfide, i.e. by forming insoluble metal salts followed by precipitation (Nielsen *et al.*, 2005). Among the different metal ions, iron (ferrous and ferric) has been widely used, e.g. as chloride, sulfate or nitrate salts (Ganigué *et al.*, 2018; Nielsen *et al.*, 2005; US EPA, 1974). Strong oxidants which can oxidize sulfide directly have also been proposed, e.g. hydrogen peroxide, chlorine and potassium permanganate (Talaiekhozani *et al.*, 2016; US EPA, 1991).

In addition, biological oxidization of sulfide has been investigated as a means to control sulfide levels in sewers. For instance, nitrate has been used together with nitrate-reducing, sulfur oxidizing bacteria (NR-SOB) to oxidize the H_2S generated by SRB, see e.g. Okabe *et al.* (2003) and Zhang *et al.* (2008). Formaldehyde or paraformaldehyde has been suggested as a possible inhibitor for bacteria in sewers (Zhang *et al.*, 2008). The effects of slow release of solid phase oxygen, by using magnesium peroxide and calcium peroxide, on sulfide emission control have been studied (Chang *et al.*, 2007; Wang *et al.*, 2019). Other emerging possibilities include microbial fuel cells (Allami *et al.*, 2018) and dosing of phages (Akanda *et al.*, 2018).

5.4.1.3 Improving the performance of concrete

A number of methods may be adopted to improve the performance of concrete subjected to MIC. The possible benefits of using supplementary cementitious materials, alternative cement systems (e.g. calcium aluminate cement or geopolymer) and calcareous aggregates have been discussed in Section 5.3.1. The effects of using polymers to modify concrete mixes, e.g. styrene acrylic ester, acrylics, styrene butadiene, vinylcopolymer and polyethylene terephthalate have also been studied (Araghi *et al.*, 2015; Monteny *et al.*, 2001). However, it is worth mentioning that due to different reasons (either technical or economic), Portland cements are still the most used cements for producing structural concrete, especially for cast in-situ structures of large sewers. Apart from the above mentioned methods, the following strategies are highlighted:

- Liners and coatings

  The liners/coatings used to prevent concrete corrosion include PVC liners, cement mortar coatings and coal tar coatings, epoxy, polyurethane and acrylic resins, polyester based polymers and high density polyethylene liners (Berndt, 2011; Loganathan *et al.*, 2008; Olliver & Lockhart, 2017). Due to the better acid resistance, calcium aluminate cement and geopolymer have also been used as a protection layer (liner) for sewer pipes (Wu *et al.*, 2020). However, due to the uncertainties of the long-term performance, liners and coatings alone are often considered not sufficient to protect concrete in large sewers (Loganathan *et al.*, 2008). In many cases, a two-barrier system, i.e. a layer of liner/coating in combination with a sacrificial concrete layer, is required to increase the robustness of the design (Davidson *et al.*, 2008; Loganathan *et al.*, 2008).
- Sacrificial concrete

  Modern (large) sewer tunnel design often requires a sacrificial layer of concrete (Loganathan *et al.*, 2008). As the name denotes, sacrificial concrete has no structural purpose and can be 'sacrificed' (or 'disappear') completely when the lifetime of the structure is secured. One primary purpose of designing sacrificial concrete is to consume the generated biogenic acid in sewers and thus to protect the structural concrete underneath (Goyn, 2008; US EPA, 1974).
- Antimicrobial agents in concrete

  Chemicals with efficient inhibitory effects on the bacterial activities relevant for MIC, or antimicrobial (bactericidal) agents, may be included in concrete mixes to increase the performance. Among the different chemicals considered, (heavy) metal ions appear to be an important group (Grengg *et al.*, 2018). Other chemicals having similar effects include calcium formate, isothiazoline/cabamate and nitrofuran (Noeiaghaei *et al.*, 2017). In addition, antimicrobial nano-particles may be added as normal admixtures during mixing or they can be added into coating systems to improve concrete durability (Noeiaghaei *et al.*, 2017; Woo *et al.*, 2008)

5.4.2 For MIC of carbon steel

Carbon steel pipelines are widely used for the transmission of water, chemicals, and oil and gas. When it comes to MIC studies, considerable attention has been paid to these pipelines, either buried or not buried, as they are among the most affected facilities. Some of the mitigation and control measures are briefly summarized below.

5.4.2.1 Biocides

As a traditional way to mitigate MIC, biocide (including biocide enhancer) treatment has been widely adopted in various industries (see also Section 5.4.1.2). The aim of the treatment is to control or inhibit the activities of microorganisms, e.g. SRB and SOB, to a sufficiently low level. Biocides can be inorganic or organic, oxidizing or non- oxidizing compounds (Jia *et al.*, 2019). To maintain long-term effects, it is normal practice to apply biocide treatments periodically (Kermani & Harrop, 2019). Despite their proven efficiency, one increasing concern with the use of biocides is the potential environmental impact as many of them are generally toxic (Zuo, 2007). To address this concern accordingly, several measures may be considered, including minimizing the volumes of affected materials to be disposed of (e.g. produced water in the oil and gas industry), using more environmentally friendly products or optimizing/limiting the amounts of biocides in use (without compromising the efficiency), see e.g. Kermani and Harrop (2019).

5.4.2.2 Inhibitors

In general, corrosion inhibitors may be categorized into three types based on their working mechanisms. The first type includes oxidizing compounds which encourage the production of low solubility corrosion product films on the metal surfaces, e.g. chromates, nitrates, and molybdates. The second type can be non-oxidizing compounds which form protective films, e.g. carbonates, silicates, and phosphates. The third type are adsorption-type inhibitors which adsorb on metal surfaces and suppress metal dissolution, e.g. large, highly polar, anionic organic molecules (Fontana, 2005; Zuo, 2007). The use of corrosion inhibitors may be considered when facing serious MIC issues (Kermani & Harrop, 2019). However, it should be mentioned that, similar to the biocide treatments discussed above, some of the inhibitors also pose potential risks to human health and the environment (Zuo, 2007).

5.4.2.3 *Biological inhibition*

Biological inhibition concerns the techniques that use particular sorts of microorganisms to battle against MIC (Saji & Umoren, 2020). Zuo (2007) summarized the possible mechanisms of biological inhibition, which include: (1) removal of corrosive cathodic agents through bacterial activities (e.g. oxygen under aerobic conditions); (2) growth inhibition of the bacteria causing corrosion through antimicrobials secreted by certain bacteria; (3) formation of a protective layer on the metal surface, e.g. passive metal oxides or biofilms secreting sticky corrosion inhibitors. The biological techniques as discussed in Section 5.4.1.2, e.g. the use of NR-SOB and phages, may also be employed for MIC mitigation of metals (Jia *et al.*, 2019; Saji & Umoren, 2020). As an emerging methodology, biological inhibition may offer a huge potential to address MIC. However, it is noted that most of the techniques developed so far have rather limited applications. Further studies have to be performed before they can be widely used in the field.

5.4.2.4 *Periodic pigging/assuring cleanliness*

Periodic pigging and other measures to assure cleanliness can be a fundamental strategy of MIC mitigation for pipelines. A mechanical pig is an appliance typically made of hard plastic plates. It is considered as an essential measure to clean garbage from the interior portion of the pipes before a biocide injection (Kermani & Harrop, 2019). This treatment removes (part of the) biofilms and deposited solids, if any, and therefore facilitates quicker and easier action of biocides on the metal surfaces. It is worth mentioning that pigging/cleaning alone may not be a perfect solution to avoid MIC, while the likelihood of controlling MIC can be improved when it is used in combination with biocide injections (Kermani & Harrop, 2019; Saji & Umoren, 2020).

5.4.2.5 *Protective coatings*

When properly installed, protective coatings can protect metals by creating a barrier and isolating them from the external environments. In general, coating materials should possess excellent properties, such as strong tensile bond strength, high density and long term durability. However, it should be mentioned that many microorganisms can degrade common coating materials, which may actually lead to the accelerated corrosion of the underlying steel (Gu, 2009; Usher *et al.*, 2014). Nevertheless, by proper design and when defects are not present in coatings, they in general can perform well, especially when used together with cathodic protection (Section 5.4.2.6).

5.4.2.6 Cathodic protection

Two types of cathodic protection are often adopted, i.e. the galvanic (also known as sacrificial) anode system and the impressed current system (Miyanaga *et al.*, 2007). In the galvanic anode method, the protective current is applied to the target material (e.g. carbon steel) connected to the low-potential galvanic anode, where the anode is corroded while the cathode is protected. For the impressed current method, the protective current is applied from the direct-current power supply and the system can be used for a long term (Miyanaga *et al.*, 2007). The effects of cathodic protection on MIC have been studied decades ago, see e.g. Little *et al.* (1988) and Guezennec (1994). It should be mentioned that numerous discussions are ongoing concerning the mechanisms and interactions between cathodic protection and biofilms (Kermani & Harrop, 2019). Additional work may help to shed more lights on relevant aspects.

REFERENCES

Æsøy A., Østerhus S. and Bentzen G. (2002). Controlled treatment with nitrate in sewers to prevent concrete corrosion. *Water Science and Technology: Water Supply*, **2**, 137–144.

Akanda Z. Z., Taha M. and Abdelbary H. (2018). Current review – the rise of bacteriophage as a unique therapeutic platform in treating peri-prosthetic joint infections. *Journal of Orthopaedic Research*, **36**, 1051–1060.

Alexander M. and Fourie C. (2011). Performance of sewer pipe concrete mixtures with portland and calcium aluminate cements subject to mineral and biogenic acid attack. *Materials and Structures*, **44**, 313–330.

Allami S., Hasan B., Redah M., Hamody H. and Ali Z. D. A. (2018). Using low cost membrane in dual-chamber microbial fuel cells (MFCs) for petroleum refinery wastewater treatment. In: Journal of Physics: Conference Series, IOP Publishing, Karbala, Iraq, p. 012061.

Apgar D., Witherspoon J. and Easter C. (2008). Minimization of Odors and Corrosion in Collection Systems. Water Environment Research Foundation, London, UK.

Araghi H. J., Nikbin I., Reskati S. R., Rahmani E. and Allahyari H. (2015). An experimental investigation on the erosion resistance of concrete containing various pet particles percentages against sulfuric acid attack. *Construction and Building Materials*, **77**, 461–471.

Attiogbe E. K. and Rizkalla S. H. (1988). Response of concrete to sulfuric acid attack. *ACI Materials Journal*, **85**, 481–488.

Ayoub G., Azar N., Fadel M. E. and Hamad B. (2004). Assessment of hydrogen sulphide corrosion of cementitious sewer pipes: a case study. *Urban Water Journal*, **1**, 39–53.

Barker W., Welch S., Chu S. and Banfield J. (1998). Experimental observations of the effects of bacteria on aluminosilicate weathering. *American Mineralogist*, **83**, 1551–1563.

Berndt M. (2011). Evaluation of coatings, mortars and mix design for protection of concrete against sulphur oxidising bacteria. *Construction and Building Materials*, **25**, 3893–3902.

Biezma M. V. (2001). The role of hydrogen in microbiologically influenced corrosion and stress corrosion cracking. *International Journal of Hydrogen Energy*, **26**, 515–520.

Booth G. and Tiller A. (1960). Polarization studies of mild steel in cultures of sulphate-reducing bacteria. *Transactions of the Faraday Society*, **56**, 1689–1696.

Booth G., Elford L. and Wakerley D. (1968). Corrosion of mild steel by sulphate-reducing bacteria: an alternative mechanism. *British Corrosion Journal*, **3**, 242–245.

Bourdoiseau J. A., Jeannin M., Rémazeilles C., Sabot R. and Refait P. (2011). The transformation of mackinawite into greigite studied by raman spectroscopy. *Journal of Raman Spectroscopy*, **42**, 496–504.

Cayford B. I., Dennis P. G., Keller J., Tyson G. W. and Bond P. L. (2012). High-throughput amplicon sequencing reveals distinct communities within a corroding concrete sewer system. *Applied and Environmental Microbiology*, **78**, 7160–7162.

Chang Y. J., Chang Y. T. and Chen H. J. (2007). A method for controlling hydrogen sulfide in water by adding solid phase oxygen. *Bioresource Technology*, **98**, 478–483.

Cord-Ruwisch R. (2000). Microbially influenced corrosion of steel. In: Environmental microbe-metal interactions, D. Lovley (ed.), American Society of Microbiology, Washington, DC, USA, pp. 159–173.

Costello J. (1974). Cathodic depolarization by sulphate-reducing bacteria. *South African Journal of Science*, **70**, 202–204.

Daniels L., Belay N., Rajagopal B. S. and Weimer P. J. (1987). Bacterial methanogenesis and growth from CO_2 with elemental iron as the sole source of electrons. *Science*, **237**, 509–511.

Davidson G., Trim M., Franklin D., Myers J. and Hansen P. et al. (2008). Northern sewerage project-liner selection in a corrosive environment. In: 13th Australian Tunnelling Conference 2008, Engineers Australia and Australasian Institute of Mining and Metallurgy, Melbourne, Australia, p. 115.

De Romero M., Duque Z., Rodríguez L., De Rincón O., Pérez O. and Araujo I. (2005). A study of microbiologically induced corrosion by sulfate-reducing bacteria on carbon steel using hydrogen permeation. *Corrosion*, **61**, 68–75.

Dinh H. T., Kuever J., Mußmann M., Hassel A. W., Stratmann M. and Widdel F. (2004). Iron corrosion by novel anaerobic microorganisms. *Nature*, **427**, 829–832.

Dowling N. (1992). Corrosion of materials used in storage and handling of solid elemental sulphur. Materials Performance: Sulphur and Energy, 31st Annual Conference of Metallurgists of CIM, Edmonton, Australia, pp. 103–115.

Eashwar M., Chandrasekaran P., Subramanian G. and Balakrishnan K. (1993). Microbiologically influenced corrosion of steel during putrefaction of seawater: evidence for a new mechanism. *Corrosion*, **49**, 108–113.

Ehrich S., Helard L., Letourneux R., Willocq J. and Bock E. (1999). Biogenic and chemical sulfuric acid corrosion of mortars. *Journal of Materials in Civil Engineering*, **11**, 340–344.

Elboujdaini M. (2011). Hydrogen-induced cracking and sulfide stress cracking. In: R.W. Revie (ed.), Uhlig's Corrosion Handbook **3**, pp. 183–194.

Enning D. (2012). Bioelectrical corrosion of iron by lithotrophic sulfate-reducing bacteria. Ph.D. thesis.

Enning D. and Garrelfs J. (2014). Corrosion of iron by sulfate-reducing bacteria: new views of an old problem. *Applied and Environmental. Microbiology*, **80**, 1226–1236.

Enning D., Venzlaff H., Garrelfs J., Dinh H. T., Meyer V., Mayrhofer K., Hassel A. W., Stratmann M. and Widdel F. (2012). Marine sulfate-reducing bacteria cause serious corrosion of iron under electroconductive biogenic mineral crust. *Environmental Microbiology*, **14**, 1772–1787.

Fattuhi N. and Hughes B. (1988). The performance of cement paste and concrete subjected to sulphuric acid attack. *Cement and Concrete Research*, **18**, 545–553.

Fontana M. G. (2005). Corrosion Engineering. Tata McGraw-Hill Education, New Delhi, India.

Friedrich C. G., Rother D., Bardischewsky F., Quentmeier A. and Fischer J. (2001). Oxidation of reduced inorganic sulfur compounds by bacteria: emergence of a common mechanism? *Applied and Environmental. Microbiology*, **67**, 2873–2882.

Ganigué R., Jiang G., Liu Y., Sharma K., Wang Y. C., Gonzalez J., Nguyen T. and Yuan Z. (2018). Improved sulfide mitigation in sewers through on-line control of ferrous salt dosing. *Water Research*, **135**, 302–310.

George R. P., Vishwakarma V., Samal S. and Mudali U. K. (2012). Current understanding and future approaches for controlling microbially influenced concrete corrosion: a review. *Concrete Research Letters*, **3**, 491–506.

Goyn A. (2008). Sewer Design Manual. Pipe and Infrastructural Product Division, CMA, Midrand, South Africa.

Grengg C., Mittermayr F., Baldermann A., Böttcher M. E., Leis A., Koraimann G., Grunert P. and Dietzel M. (2015). Microbiologically induced concrete corrosion: A case study from a combined sewer network. *Cement and Concrete Research*, **77**, 16–25.

Grengg C., Mittermayr F., Ukrainczyk N., Koraimann G., Kienesberger S. and Dietzel M. (2018). Advances in concrete materials for sewer systems affected by microbial induced concrete corrosion: A review. *Water Research*, **134**, 341–352.

Gu J. D. (2009). In: Corrosion, Microbial, M. Schaechter (ed.), 3rd edn, Academic Press, Amsterdam, pp. 259–269.

Gu T. (2012). New understandings of biocorrosion mechanisms and their classifications. *Journal of Microbial & Biochemical Technology*, **4**, iii–vi.

Gu J. D., Ford T. E., Berke N. S. and Mitchell R. (1998). Biodeterioration of concrete by the fungus fusarium. *International Biodeterioration & Biodegradation*, **41**, 101–109.

Gu T., Jia R., Unsal T. and Xu D. (2019). Toward a better understanding of microbiologically influenced corrosion caused by sulfate reducing bacteria. *Journal of Materials Science & Technology*, **35**, 631–636.

Guezennec J. (1994). Cathodic protection and microbially induced corrosion. *International Biodeterioration & Biodegradation*, **34**, 275–288.

Gutierrez O., Park D., Sharma K. R. and Yuan Z. (2009). Effects of long-term ph elevation on the sulfate-reducing and methanogenic activities of anaerobic sewer biofilms. *Water Research*, **43**, 2549–2557.

Gutiérrez-Padilla M. G. D., Bielefeldt A., Ovtchinnikov S., Hernandez M. and Silverstein J. (2010). Biogenic sulfuric acid attack on different types of commercially produced concrete sewer pipes. *Cement and Concrete Research*, **40**, 293–301.

Hao O. J., Chen J. M., Huang L. and Buglass R. L. (1996). Sulfate-reducing bacteria. *Critical Reviews in Environmental Science and Technology*, **26**, 155–187.

Hewayde E. (2005). Investigation on degradation of concrete sewer pipes by sulfuric acid attack. Ph.D. thesis. The University of Western Ontario, Canada.

Hewayde E., Nehdi M., Allouche E. and Nakhla G. (2007). Effect of mixture design parameters and wetting-drying cycles on resistance of concrete to sulfuric acid attack. *Journal of Materials in Civil Engineering*, **19**, 155–163.

Hou B., Li X., Ma X., Du C., Zhang D., Zheng M., Xu W., Lu D. and Ma F. (2017). The Cost of Corrosion in China. Npj Materials Degradation, **1**, pp. 1–10.

House M. W. (2013). Using biological and physico-chemical test methods to assess the role of concrete mixture design. In resistance to microbially induced corrosion. Master's thesis. Purdue University.

House M. and Weiss W. (2014). Review of microbially induced corrosion and comments on needs related to testing procedures. In: 4th International Conference on the Durability of Concrete Structures, West Lafayette, USA, pp. 94–103.

Islander R. L., Devinny J. S., Mansfeld F., Postyn A. and Shih H. (1991). Microbial ecology of crown corrosion in sewers. *Journal of Environmental Engineering*, **117**, 751–770.

Ismail N., Nonaka T., Noda S. and Mori T. (1993). Effect of carbonation on microbial corrosion of concretes. *Doboku Gakkai Ronbunshu*, **1993**, 133–138.

Jack T. R., Wilmott M. J., Sutherby R. L. and Worthingham R. G. (1996). External corrosion of line pipe–a summary of research activities. *Materials Performance*, **35**, 18–24.

Japan Sewage Works Agency (2007). Corrosion control measures for sewerage structure: Technologies for corrosion prevention and protection. URL: https://www.jswa.go.jp/e/r_d/r_d2/pdf/009.pdf. (Accessed 02 March 2020).

Javaherdashti R. (2017). Microbiologically Influenced Corrosion (MIC). In: Microbiologically Influenced Corrosion, Springer, Cham, Switzerland, pp. 29–79.

Jayaraman A., Mansfeld F. and Wood T. (1999). Inhibiting sulfate-reducing bacteria in biofilms by expressing the antimicrobial peptides indolicidin and bactenecin. *Journal of Industrial Microbiology and Biotechnology*, **22**, 167–175.

Jensen H. S. (2009). Hydrogen sulfide induced concrete corrosion of sewer networks. Ph.D. thesis. Department of Biotechnology, Chemistry and Environmental Engineering.

Jia R., Unsal T., Xu D., Lekbach Y. and Gu T. (2019). Microbiologically influenced corrosion and current mitigation strategies: A state of the art review. *International Biodeterioration & Biodegradation*, **137**, 42–58.

Jiang G., Sun X., Keller J. and Bond P. L. (2015). Identification of controlling factors for the initiation of corrosion of fresh concrete sewers. *Water Research*, **80**, 30–40.

Joseph A. P., Keller J., Bustamante H. and Bond P. L. (2012). Surface neutralization and H_2S oxidation at early stages of sewer corrosion: influence of temperature, relative humidity and H_2S concentration. *Water Research*, **46**, 4235–4245.

Kawai K., Yamaji S. and Shinmi T. (2005). Concrete deterioration caused by sulfuric acid attack. In: International Conference on Durability of Building Materials and Components, Lyon (France), pp. 17–20.

Kermani B. and Harrop D. (2019). Corrosion and Materials in Hydrocarbon Production: A Compendium of Operational and Engineering Aspects. John Wiley & Sons Ltd, New York, USA.

King R., Miller J. and Smith J. (1973). Corrosion of mild steel by iron sulphides. *British Corrosion Journal*, **8**, 137–141.

Kleinjan W., De Keizer A. and Janssen A. (2003). Biologically Produced Sulfur. Springer Berlin/Heidelberg, pp. 167–188.

Koh S., Kim J., Yang B. and Kim K. (2004). Effect of line pipe steel microstructure on susceptibility to sulfide stress cracking. *Corrosion*, **60**, 244–253.

Li S., Kim Y., Jeon K., Kho Y. and Kang T. (2001). Microbiologically influenced corrosion of carbon steel exposed to anaerobic soil. *Corrosion*, **57**, 815–828.

Li Y., Xu D., Chen C., Li X., Jia R., Zhang D., Sand W., Wang F. and Gu T. (2018). Anaerobic microbiologically influenced corrosion mechanisms interpreted using bioenergetics and bioelectrochemistry: A review. *Journal of Materials Science & Technology*, **34**, 1713–1718.

Little B. J. and Lee J. S. (2007). Microbiologically influenced corrosion. **3**. John Wiley & Sons, New York, USA.

Little B., Wagner P. and Duquette D. (1988). Microbiologically induced increase in corrosion current density of stainless steel under cathodic protection. *Corrosion*, **44**, 270–274.

Little B., Ray R. and Pope R. (2000). Relationship between corrosion and the biological sulfur cycle: a review. *Corrosion*, **56**, 433–443.

Lloyd R. R., Provis J. L. and van Deventer J. S. (2012). Acid resistance of inorganic polymer binders. 1. corrosion rate. *Materials and Structures*, **45**, 1–14.

Loganathan L. N., Flanagan R. F. and Tee T. B. (2008). Optimisation of corrosion protection lining (CPL) thickness for concrete sewer tunnels. In: Underground Facilities for Better Environment and Safety: Proceedings of the World Tunnel Congress, Agra, India, pp. 1063–1070.

Magot M., Ravot G., Campaignolle X., Ollivier B., Patel B., Fardeau M. L., Thomas P., Crolet J. L. and Garcia J. L. (1997). *Dethiosulfovibrio peptidovorans gen. nov., sp. nov.*, a new anaerobic, slightly halophilic, thiosulfate-reducing bacterium from corroding offshore oil wells. *International Journal of Systematic and Evolutionary Microbiology*, **47**, 818–824.

Marchand J., Odler I. and Skalny J. P. (2003). Sulfate Attack on Concrete. CRC Press, London.

Mittelman M. and Danko J. (1995). Corrosion of a concrete dam structure: evidence of microbially influenced corrosion activity. In: 1995 International Conference on Microbial Influenced Corrosion, National Association of Corrosion Engineers Houston, Texas, USA, pp. 15/1–15/7.

Miyanaga K., Terashi R., Kawai H., Unno H. and Tanji Y. (2007). Biocidal effect of cathodic protection on bacterial viability in biofilm attached to carbon steel. *Biotechnology and Bioengineering*, **97**, 850–857.

Monteiro P. and Mehta P. (1993). *Concrete: Structure, Properties, and Materials*. Prentice-Hall, Englewood Cliffs.

Monteny J., Vincke E., Beeldens A., De Belie N., Taerwe L., Van Gemert D. and Verstraete W. (2000). Chemical, microbiological, and in situ test methods for biogenic sulfuric acid corrosion of concrete. *Cement and Concrete Research*, **30**, 623–634.

Monteny J., De Belie N., Vincke E., Verstraete W. and Taerwe L. (2001). Chemical and microbiological tests to simulate sulfuric acid corrosion of polymer-modified concrete. *Cement and Concrete Research*, **31**, 1359–1365.

Mori T., Nonaka T., Tazaki K., Koga M., Hikosaka Y. and Noda S. (1992). Interactions of nutrients, moisture and pH on microbial corrosion of concrete sewer pipes. *Water Research*, **26**, 29–37.

Mori K., Tsurumaru H. and Harayama S. (2010). Iron corrosion activity of anaerobic hydrogen-consuming microorganisms isolated from oil facilities. *Journal of Bioscience and Bioengineering*, **110**, 426–430.

Morton R. L., Yanko W. A., Graham D. W. and Arnold R. G. (1991). Relationships between metal concentrations and crown corrosion in los angeles county sewers. *Research Journal of the Water Pollution Control Federation*, **63**(5), 789–798.

Narenkumar J., Parthipan P., Nanthini A. U. R., Benelli G., Murugan K. and Rajasekar A. (2017). Ginger extract as green biocide to control microbial corrosion of mild steel. *3 Biotech*, **7**, 133.

Newman R., Webster B. and Kelly R. (1991). The electrochemistry of srb corrosion and related inorganic phenomena. *ISIJ International*, **31**, 201–209.

Nica D., Davis J., Kirby L., Zuo G. and Roberts D. (2000). Isolation and characterization of microorganisms involved in the biodeterioration of concrete in sewers. *International Biodeterioration & Biodegradation*, **46**, 61–68.

Nielsen A. H., Lens P., Vollertsen J. and Hvitved-Jacobsen T. (2005). Sulfide–iron interactions in domestic wastewater from a gravity sewer. *Water Research*, **39**, 2747–2755.

Noeiaghaei T., Mukherjee A., Dhami N. and Chae S. R. (2017). Biogenic deterioration of concrete and its mitigation technologies. *Construction and Building Materials*, **149**, 575–586.

Norlund K. L., Southam G., Tyliszczak T., Hu Y., Karunakaran C., Obst M., Hitchcock A. P. and Warren L. A. (2009). Microbial architecture of environmental sulfur processes: A novel syntrophic sulfur-metabolizing consortia. *Environmental Science & Technology*, **43**, 8781–8786.

O'Connell M., McNally C. and Richardson M. G. (2010). Biochemical attack on concrete in wastewater applications: A state of the art review. *Cement and Concrete Composites*, **32**, 479–485.

Okabe S., Santegoeds C. M. and De Beer D. (2003). Effect of nitrite and nitrate on in situ sulfide production in an activated sludge immobilized agar gel film as determined by use of microelectrodes. *Biotechnology and Bioengineering*, **81**, 570–577.

Okabe S., Odagiri M., Ito T. and Satoh H. (2007). Succession of sulfur-oxidizing bacteria in the microbial community on corroding concrete in sewer systems. *Applied and Environmental Microbiology*, **73**, 971–980.

Olliver A. N. and Lockhart T. (2017). Innovative one-pass lining solution for Doha's deep tunnel sewer system. In: Proceedings of the World Tunnel Congress 2017, Bergen, Norway.

Olmstead W. and Hamlin H. (1900). Converting portions of the los angeles outfall sewer into a septic tank. *Engineering News*, **44**, 317–318.

Panossian Z., de Almeida N. L., de Sousa R. M. F., de Souza Pimenta G. and Marques L. B. S. (2012). Corrosion of carbon steel pipes and tanks by concentrated sulfuric acid: a review. *Corrosion Science*, **58**, 1–11.

Pescod M. and Price A. (1982). Major factors in sewer ventilation. *Journal (Water Pollution Control Federation)*, **54**(4), 385–397.

Peyre Lavigne M., Lors C., Valix M., Herisson J., Paul E. and Bertron E. (2016). Microbial induced concrete deterioration in sewers environment: mechanisms and microbial populations. In: RILEM TC 253 MCI Proceedings, Delft, the Netherland, pp. 20–36.

Poduska R. A. and Anderson B. (1981). Successful storage lagoon odor control. *Journal (Water Pollution Control Federation)*, **53**(3), 299–310.

Pomeroy R. (1945). The pros and cons of sewer ventilation. *Sewage Works Journal*, **XVII**, 203–208.

Pomeroy R. D. (1969). Calcareous pipe for sewers. *Journal (Water Pollution Control Federation)*, **4**, 1491–1493.

Provis J. L., Palomo A. and Shi C. (2015). Advances in understanding alkali-activated materials. *Cement and Concrete Research*, **78**, 110–125.

Radkevych O. and Pokhmurs'kyi V. (2001). Influence of hydrogen sulfide on serviceability of materials of gas field equipment. *Materials Science*, **37**, 319–332.

Raman R. S., Javaherdashti R., Panter C. and Pereloma E. V. (2005). Hydrogen embrittlement of a low carbon steel during slow strain testing in chloride solutions containing sulphate reducing bacteria. *Materials Science and Technology*, **21**, 1094–1098.

Roberts D., Nica D., Zuo G. and Davis J. (2002). Quantifying microbially induced deterioration of concrete: initial studies. *International Biodeterioration & Biodegradation*, **49**, 227–234.

Robinson M. and Kilgallon P. (1998). A review of the effects of sulphate reducing bacteria in the marine environment on the corrosion fatigue and hydrogen embrittlement of high strength steels. Health and Safety Executive – Offshore Technology Report.

Saji V. S. and Umoren S. A. (eds). (2020). Corrosion Inhibitors in the Oil and Gas Industry. Wiley, Weinheim, Germany.

Sand W., Dumas T. and Marcdargent S. (1994). Accelerated biogenic sulfuric-acid corrosion test for evaluating the performance of calcium-aluminate based concrete in sewage applications. In: Microbiologically Influenced Corrosion Testing, ASTM International, Volume **1232**, pp. 234–234.

Santo Domingo J. W., Revetta R. P., Iker B., Gomez-Alvarez V., Garcia J., Sullivan J. and Weast J. (2011). Molecular survey of concrete sewer biofilm microbial communities. *Biofouling*, **27**, 993–1001.

Satoh H., Odagiri M., Ito T. and Okabe S. (2009). Microbial community structures and in situ sulfate-reducing and sulfur-oxidizing activities in biofilms developed on mortar specimens in a corroded sewer system. *Water Research*, **43**, 4729–4739.

Saucier F. and Herisson J. (2015). Use of Calcium Aluminate Cement in H_2S Biogenic Environment. In: Institute of Concrete Technology, P. Hewlett, R. Mangabhai, G. Taylor, P. Domone, E. Trout, C. Goodier, I. Berrie, D. Killoran (ed.), Yearbook 2015–2016, Surrey, UK, pp. 67–80.

Schmitt G. (1991). Effect of elemental sulfur on corrosion in sour gas systems. *Corrosion*, **47**, 285–308.

Scrivener K. L. and Capmas A. (1998). Calcium Aluminate Cements. 4th ed. Elsevier Science & Technology, pp. 713–782.

Sharma K. R., Yuan Z., De Haas D., Hamilton G., Corrie S. and Keller J. (2008). Dynamics and dynamic modelling of H_2S production in sewer systems. *Water Research*, **42**, 2527–2538.

Shi C. (2003). Corrosion resistance of alkali-activated slag cement. *Advances in Cement Research*, **15**, 77–81.

Shi H. and Wu M. (2007). The geo-polymerization research status of geopolymeric cement. *Materials Review*, **8**, 88–91. (In Chinese).

Sun X. (2015). Improving the understanding of concrete sewer corrosion through investigations of the gaseous hydrogen sulfide uptake and transformation processes in the corrosion layer. Ph.D. thesis. The University of Queensland: Brisbane, *Queensland.*

Sun W. and Nesic S. (2007). A mechanistic model of H_2S corrosion of mild steel. In: CORROSION 2007, NACE International, Tennessee, pp. 1054–1080.

Sun H., Fang H., Davis J., Hudgins R. et al. (2012). Investigation of effects of iron sulfide on corrosion and inhibition of carbon steel in H_2S containing conditions. In: CORROSION 2012, NACE International, Utah, USA, pp. 50884–5102.

Talaiekhozani A., Bagheri M., Goli A. and Khoozani M. R. T. (2016). An overview of principles of odor production, emission, and control methods in wastewater collection and treatment systems. *Journal of Environmental Management*, **170**, 186–206.

Tanaka N., Hvitved-Jacobsen T., Ochi T. and Sato N. (2000). Aerobic–anaerobic microbial wastewater transformations and re-aeration in an air-injected pressure sewer. *Water Environment Research*, **72**, 665–674.

Tazawa E., Morinaga T. and Kawai K. (1994). Deterioration of concrete derived from metabolites of microorganisms. Proceedings of 3rd CANMET/ACI International Conference on Durability of Concrete, Special Publication 145–59, Nice, France, pp. 1087–1098.

US EPA (1974). Process Design Manual for Sulfide Control in Sanitary Sewerage Systems. US Environmental Protection Agency, Technology Transfer, Washington, DC, USA.

US EPA (1985). Design Manual: Odor and Corrosion Control in Sanitary Sewerage Systems and Treatment Plants. Center for Environmental Research Information, US Environmental Protection Agency, Office of Research and Development, Washington, DC, USA.

US EPA (1991). Hydrogen Sulfide Corrosion in Wastewater Collection and Treatment. Technical Report. US Environmental Protection Agency, Office of Water, Washington DC.

Usher K., Kaksonen A., Cole I. and Marney D. (2014). Critical review: microbially influenced corrosion of buried carbon steel pipes. *International Biodeterioration & Biodegradation*, **93**, 84–106.

Valdebenito-Rolack E. H., Araya T. C., Abarzua L. E., Ruiz-Tagle N. M., Sossa K. E., Aroca G. E. and Urrutia H. E. (2011). Thiosulphate oxidation by thiobacillus thioparus and halothiobacillus neapolitanus strains isolated from the petrochemical industry. *Electronic Journal of Biotechnology*, **14**, 7–8.

van Mechelen A. and Polder R. (1991). Rioleringen (XI), biogene zwavelzuuraantasting, verslag van een vervolgonderzoek in Rotterdam. *Cement*, **2**, 26–30.

Venzlaff H., Enning D., Srinivasan J., Mayrhofer K. J., Hassel A. W., Widdel F. and Stratmann M. (2013). Accelerated cathodic reaction in microbial corrosion of iron due to direct electron uptake by sulfate-reducing bacteria. *Corrosion Science*, **66**, 88–96.

Vincke E., Verstichel S., Monteny J. and Verstraete W. (1999). A new test procedure for biogenic sulfuric acid corrosion of concrete. *Biodegradation*, **10**, 421–428.

Vincke E., Monteny J., Beeldens A., De Belie N., Taerwe L., Van Gemert D. and Verstraete W. (2000). Recent developments in research on biogenic sulfuric acid attack of concrete. In: Environmental Technologies to Treat Sulfur Pollution: Principles and Engineering, P. Lens and L. Pol (eds), London, UK, pp. 515–541.

Vincke E., Boon N. and Verstraete W. (2001). Analysis of the microbial communities on corroded concrete sewer pipes–A case study. *Applied Microbiology and Biotechnology*, **57**, 776–785.

Vincke E., Van Wanseele E., Monteny J., Beeldens A., De Belie N., Taerwe L., Van Gemert D. and Verstraete W. (2002). Influence of polymer addition on biogenic sulfuric acid attack of concrete. *International Biodeterioration & Biodegradation*, **49**, 283–292.

Vollertsen J., Nielsen A. H., Jensen H. S., Wium-Andersen T. and Hvitved-Jacobsen T. (2008). Corrosion of concrete sewers – the kinetics of hydrogen sulfide oxidation. *Science of the Total Environment*, **394**, 162–170.

von Wolzogen Kuehr C. and van der Vlugt L. (1934). The graphitization of cast iron as an electrobiochemical process in anaerobic soils. *Water*, **18**, 147–165.

Vuković M., Cvetkovski V. and Conić V. (2009). Mechanisms of microbiologically induced corrosion of metals in the environments containing sulphate-reducing bacteria. *Corrosion Reviews*, **27**, 1–22.

Wang D., Zhang D., Xu Q., Liu Y., Wang Q., Ni B. J., Yang Q., Li X. and Yang F. (2019). Calcium peroxide promotes hydrogen production from dark fermentation of waste activated sludge. *Chemical Engineering Journal*, **355**, 22–32.

Wang T., Wu K., Kan L. and Wu M. (2020). Current understanding on microbiologically induced corrosion of concrete sewers: a review of the evaluation methods and mitigation measures. *Construction and Building Materials*, **247**, 118539.

Wells P. T. and Melchers R. (2014a). Findings of a 4 year study of concrete sewer pipe corrosion. In: Annual Conference of the Australasian Corrosion Association, pp. 1–12.

Wells T. and Melchers R. (2014b). An observation-based model for corrosion of concrete sewers under aggressive conditions. *Cement and Concrete Research*, **61**, 1–10.

Wells T., Melchers R. E. and Bond P. (2009). Factors involved in the long term corrosion of concrete sewers. Australasian Corrosion Association Proceedings of Corrosion and Prevention, Coffs Harbour, Australia **11**.

Wells T., Melchers R., Joseph A., Bond P., Vitanage D., Bustamante H., De Grazia J., Kuen T., Nazimek J. and Evans T. (2012). A collaborative investigation of the microbial corrosion of concrete sewer pipe in Australia. In: Proceedings of OzWater-12 Australia's National Water Conference and Exhibition, pp. 8–10.

Woo R. S., Zhu H., Chow M. M., Leung C. K. and Kim J. K. (2008). Barrier performance of silane–clay nanocomposite coatings on concrete structure. *Composites Science and Technology*, **68**, 2828–2836.

Woodtli J. and Kieselbach R. (2000). Damage due to hydrogen embrittlement and stress corrosion cracking. *Engineering Failure Analysis* **7**, 427–450.

Wu L., Hu C. and Liu W. V. (2018a). The Sustainability of Concrete in Sewer Tunnel A Narrative Review of Acid Corrosion in the City of Edmonton, Canada. *Sustainability*, **10**, 517.

Wu T., Sun C. and Ke W. (2018b). Interpreting microbiologically assisted cracking with Ee-pH diagrams. *Bioelectrochemistry*, **120**, 57–65.

Wu T., Yan M., Yu L., Zhao H., Sun C., Yin F. and Ke W. (2019). Stress corrosion of pipeline steel under disbonded coating in a SRB-containing environment. *Corrosion Science*, **157**, 518–530.

Wu M., Wang T., Wu K. and Kan L. (2020). Microbiologically induced corrosion of concrete in sewer structures: A review of the mechanisms and phenomena. *Construction and Building Materials*, **239**, 117813.

Xu J., Wang K., Sun C., Wang F., Li X., Yang J. and Yu C. (2011). The effects of sulfate reducing bacteria on corrosion of carbon steel Q235 under simulated disbonded coating by using electrochemical impedance spectroscopy. *Corrosion Science*, **53**, 1554–1562.

Yamanaka T., Aso I., Togashi S., Tanigawa M., Shoji K., Watanabe T., Watanabe N., Maki K. and Suzuki H. (2002). Corrosion by bacteria of concrete in sewerage systems and inhibitory effects of formates on their growth. *Water Research*, **36**, 2636–2642.

Yang W., Vollertsen J. and Hvitved-Jacobsen T. (2005). Anoxic sulfide oxidation in wastewater of sewer networks. *Water Science and Technology*, **52**, 191–199.

Zhang L., De Schryver P., De Gusseme B., De Muynck W., Boon N. and Verstraete W. (2008). Chemical and biological technologies for hydrogen sulfide emission control in sewer systems: a review. *Water Research*, **42**, 1–12.

Zhou E., Li H., Yang C., Wang J., Xu D., Zhang D. and Gu T. (2018). Accelerated corrosion of 2304 duplex stainless steel by marine *Pseudomonas aeruginosa* biofilm. *International Biodeterioration & Biodegradation*, **127**, 1–9.

Zuo R. (2007). Biofilms: strategies for metal corrosion inhibition employing microorganisms. *Applied Microbiology and Biotechnology*, **76**, 1245–1253.

Chapter 6

Biological treatment of organic sulfate-rich wastewaters

Rachel B. Costa, Vincent O'flaherty and Piet N. L. Lens

6.1 INTRODUCTION

Many wastewaters have a high sulfate and organic carbon content, such as those generated by industrial processes that use sulfuric acid (fermentation industry), or sulfate-rich feedstocks (seafood processing industry). The use of sulfurous compounds less oxidized than sulfate in industrial processes, such as sulfide (tanneries, Kraft pulping), sulfite (sulfite pulping), or dithionite (pulp bleaching) also results in the generation of sulfate-rich wastewaters. Besides these organic wastewaters, some sulfate-rich effluents hardly contain any organic matter. These are generated during the leaching of sulfur-rich wastes (mine spoils and landfills, see Chapter 15) or during the scrubbing of sulfur-containing off-gases (see Lagas, 2000). Specific aspects of the treatment of these inorganic wastewaters are presented in Chapter 7.

Aqueous sulfate emissions are generally considered safe. In humans, sulfate concentrations exceeding 500 mg L^{-1} can have laxative effects (US EPA, 1990), but the Drinking Water Standards of the U.S. Public Health Service proposed a maximum of 250 mg L^{-1}, based on aesthetic aspects. In the last decade, the water quality guidelines have included sulfur compounds in the protection of freshwater aquatic life. The Province of British Columbia (Canada), for instance, has set the maximum sulfate concentration between 128–429 mg L^{-1} (BC ME, 2013), depending on the water hardness (Elphick *et al.*, 2011) to ensure aquatic life protection. This range was established based on toxicity assays with different

DOI: 10.2166/9781789060966_0167

freshwater species, representing different families, i.e. a fish, an invertebrate and an acutely sensitive species (Wang *et al.*, 2016). Despite the regulatory changes in different countries, sulfate emission guidelines remain high due to its low toxicity.

Table 6.1 lists some important disadvantages as well as potentials of sulfate reduction in anaerobic reactors. The main concern of the biological treatment of sulfate-rich wastewaters is the application of anaerobic biological treatment of these wastewaters. Under anaerobic conditions, dissimilatory sulfate-reducing bacteria (SRB) use sulfate as a terminal electron acceptor for the degradation of organic compounds and hydrogen (Muyzer & Stams, 2008; Oude Elferink *et al.*, 1994), which results in the generation of sulfide (Figure 6.1). Therefore, the major problem associated with the anaerobic treatment of sulfate-rich wastewaters is the production of sulfide, which can cause both corrosion to the reactor structures (Chapter 5) and toxicity to the bacterial populations involved in the biological wastewater treatment system (see section 6.2.2).

Table 6.1 Effects of sulfide formation in anaerobic reactors.

Disadvantages	Reference
Reduced COD removal efficiency due to the presence H_2S in the effluent	*This chapter*
Potential toxicity	*This chapter*
Corrosion	Chapter 5
Accumulation of inert material in the sludge (metal sulfides)	Utgikar *et al.* (2003)
Less methane formation	*This chapter*
Need for H_2S removal from the biogas	Chapter 11
Deterioration of aerobic post-treatment system (activated sludge bulking, excessive growth of phototrophs)	Eikelboom (2000)
Advantages	**Reference**
Biological sulfate removal	
Organic wastewaters	*This chapter*
Inorganic wastewaters	Chapter 7 and 15
Degradation of xenobiotics	Li *et al.* (2015); Delforno *et al.* (2015)
Heavy metal removal	Chapter 7; Hao (2000)
Methanogenic archaea sulfur requirements	Daniels *et al.* (1986); Allen *et al.* (2015)
Precipitated metal sulfides (e.g. FeS) form good precursors for granulation	Lettinga (1995); Tiwari *et al.* (2006)

COD: Chemical oxygen demand.

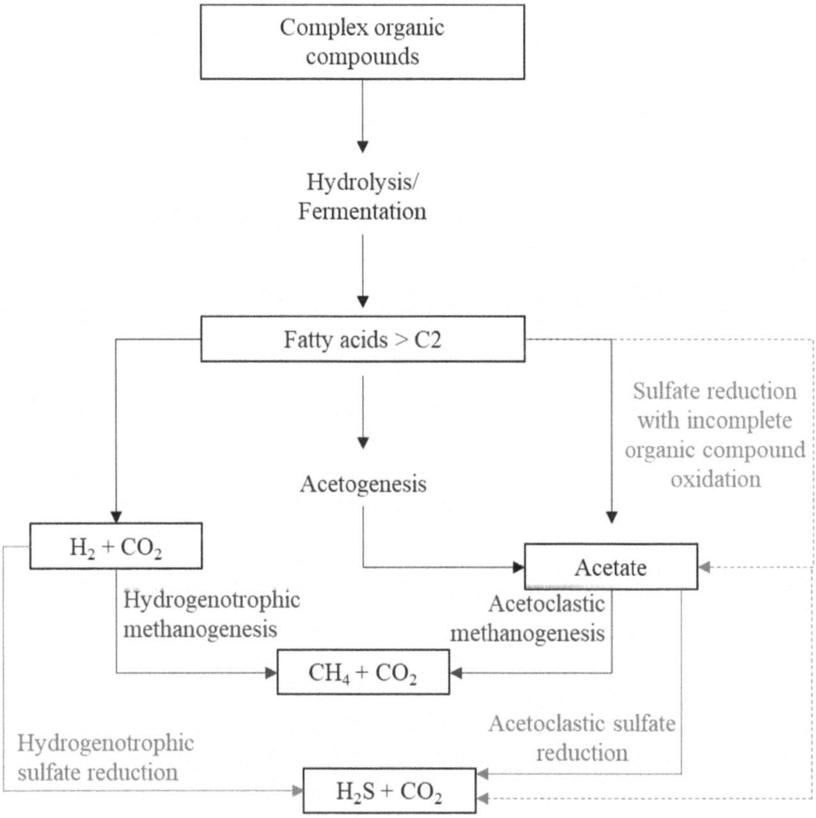

Figure 6.1 Substrate competition between sulfate-reducing and acetogenic bacteria, and methanogenic archaea during the anaerobic digestion of organic matter.

On the other hand, the occurrence of sulfate reduction in anaerobic reactors can be beneficial to some extent. Methanogenesis requires sulfur-containing co-factors (coenzyme M and coenzyme B) (Liu *et al.*, 2012), but most methanogens lack assimilatory sulfate reductases (Daniels *et al.*, 1986). Their sulfur requirements are satisfied by a combination of inorganic sulfide and organic sulfur compounds, such as cysteine or glutathione (Daniels *et al.*, 1986). Microorganisms affiliated with the methanogenic archaea (MA) *Methanocaldococcus jannaschii* are able to biosynthesize homocysteine from H_2S, a trait that seems to be shared with other MA, e.g. *Methanosarcina acetivor* (Allen *et al.*, 2015). The uptake of sulfide by MA to fulfil their sulfur requirements can be related to a stimulation in methanogenic metabolism with addition of sulfide in concentrations between 1 and 2 mM (32–64 mg H_2S L^{-1}) (Daniels *et al.*, 1986). Consequently, the production of sulfide by dissimilatory bacterial species during anaerobic digestion may enhance methanogenesis by satisfying the sulfur growth requirements of methanogens. Furthermore, sulfate reduction can serve as a biological method, together with a sulfide removal step to remove sulfate (or other oxidized sulfur compounds) and SO_2 from sulfur-containing waste streams and off-gases (see also Chapters 7 and 11).

6.2 ANAEROBIC TREATMENT OF SULFATE-RICH WASTEWATERS

6.2.1 Competition between sulfate-reducing bacteria and methanogenic archaea

Microbial activity inhibition due to sulfide relates to two mechanisms: substrate competition and toxicity (section 6.2.2). In biological reactors SRB compete with other organisms for common organic and inorganic substrates (Figure 6.1). Since SRB are ubiquitous and versatile, they compete with several different groups of microorganisms involved in anaerobic bioprocesses (Chen *et al.*, 2008). Much of the research developed on anaerobic digestion focuses on methane production, and the competition between sulfate-reducing bacteria and methanogenic archaea is extensively discussed in the scientific literature (Damianovic & Foresti, 2009; Godoi *et al.*, 2017; Kiyuna *et al.*, 2017; O'Flaherty *et al.*, 1998a, b; Omil *et al.*, 1998).

The main intermediates in the anaerobic mineralization of complex organic matter are hydrogen, acetate, propionate and butyrate (Figure 6.1). From a thermodynamic point of view (Table 6.2), SRB should out-compete the methanogenic consortia during growth on these substrates. Thus, sulfate reduction would theoretically

Table 6.2 Stoichiometry of the anaerobic degradation of propionate, acetate and molecular hydrogen by SRB and MA.

Reactions	$\Delta G°$
Propionate	
$CH_3CH_2COO^- + 3H_2O \rightarrow CH_3COO^- + HCO_3^- + H^+ + 3H_2$	+76.0
$CH_3CH_2COO^- + 0.75\,SO_4^{2-} \rightarrow CH_3COO^- + HCO_3^- + 0.75HS^- + 0.25H^+$	−37.7
$CH_3CH_2COO^- + 1.75\,SO_4^{2-} \rightarrow 3HCO_3^- + 1.75HS^- + 0.5H^+ + 0.25OH^-$	−88.9
Acetate	
$CH_3COO^- + H_2O \rightarrow CH_4 + HCO_3^-$	−31.0
$CH_3COO^- + SO_4^{2-} \rightarrow 2HCO_3^- + HS^-$	−47.6
Hydrogen	
$4H_2 + HCO_3^- + H^+ \rightarrow CH_4 + 3H_2O$	−32.7
$4H_2 + SO_4^{2-} + H^+ \rightarrow HS^- + 4H_2O$	−38.1

Data from Thauer *et al.* (1977) ($\Delta G°$ at 37°C in kJ/reaction).

become the dominant process in anaerobic digesters treating sulfate-rich wastewaters.

It is well established that, provided enough sulfate is present in the wastewater, hydrogenotrophic SRB (H-SRB) outcompete hydrogenotrophic methanogenic archaea (Dar *et al.*, 2008; Visser *et al.*, 1993a; Weijma *et al.*, 2002). This corroborates the process fundamentals, as H-SRB gain more energy from the consumption of molecular hydrogen and have a higher substrate affinity than hydrogenotrophic methanogenic archaea (Oude Elferink *et al.*, 1994). It also explains the rapid inhibition of hydrogenotrophic methanogenic archaea when sulfate-rich wastewater is fed to an anaerobic bioreactor. Furthermore, propionate and butyrate are also degraded faster by SRB than by the syntrophic consortia (Colleran *et al.*, 1995). In contrast, for acetate consumption, conflicting results regarding SRB or MA predominance were reported for several years.

Thermodynamic constants indicate that the use of acetate by acetate-utilizing SRB (A-SRB) is more favourable than its uptake by acetoclastic methanogenic archaea (A-MA) (Table 6.2). Indeed, A-SRB were found to outcompete A-MA in completely stirred chemostats (Gupta *et al.*, 1994; Middleton & Lawrence, 1977), but A-SRB and A-MA can coexist even in excess of sulfate in natural environments (Ozuolmez *et al.*, 2015). In anaerobic reactors treating sulfate-rich wastewaters, however, there is evidence that acetate is completely converted into methane, even in excess of sulfate (Hoeks *et al.*, 1984; Mulder, 1984). Indeed, there are consistent reports on methanogenesis enhancement when acetate is provided as the electron donor in anaerobic reactors treating sulfate-rich water (Damianovic & Foresti, 2009; Omil *et al.*, 1998; Visser *et al.*, 1993a).

SRB are typically divided into complete and incomplete oxidizers. The former mineralize organic compounds to CO_2, whereas the latter partially oxidize organic compounds and generate acetate as a by-product, but are unable to use acetate as an electron donor (Muyzer & Stams, 2008). Incomplete oxidizers account for the majority of SRB (Hao *et al.* 2014; Muyzer & Stams 2008), and are able to use many organic compounds as electron donors, including alcohols (Weijma *et al.*, 2003), amino acids (Stams *et al.*, 1985), aliphatic compounds (Rabus *et al.*, 1993), alkanes (Aeckersberg *et al.*, 1998), and even carbon monoxide (Henstra *et al.*, 2007) and some sugars (Sass *et al.*, 2002).

Organic wastewaters usually contain a complex mixture of several organic compounds, that can be used as electron donors by incomplete oxidizing SRB, but are not available for methanogenic archaea (Figure 6.1). In this scenario, incomplete oxidizing SRB act as acetogens, and the yielded acetate can be utilized by A-MA (Damianovic & Foresti, 2009). Indeed, acetate generation by SRB is reported to stimulate the methanogenesis in several anaerobic systems (Godoi *et al.*, 2015; Kiyuna *et al.*, 2017; Zan & Hao, 2020).

In sulfate-rich organic wastewater treatment, the balance between carbonaceous matter and sulfate is an important parameter to predict establishment of methanogenesis. This balance is conventionally measured as the ratio between the chemical oxygen demand (COD) and sulfate concentration (COD/SO_4^{2-} ratio). It is accepted that methane production will not be inhibited in anaerobic reactors treating wastewaters with a COD/SO_4^{2-} ratio higher than 10 (Lens *et al.*, 1998). However, the carbon source also plays a role. When acetate was provided as the only carbon source, the predominance of A-SRB over A-MA seemed to be achieved only after a long operation time, and after a COD/SO_4^{2-} ratio of 0.5 was adopted (Omil *et al.*, 1998; Figure 6.2).

Moreover, successful methane production has been reported in bioreactors treating waters with low COD/SO_4^{2-} ratios and complex organic substrates (Cetecioglu *et al.*, 2019; Godoi *et al.*, 2015; Zan & Hao, 2020). Godoi *et al.* (2015) operated two structured-bed reactors treating synthetic wastewater mimicking the soluble fraction of sugarcane vinasse. One of the reactors (control)

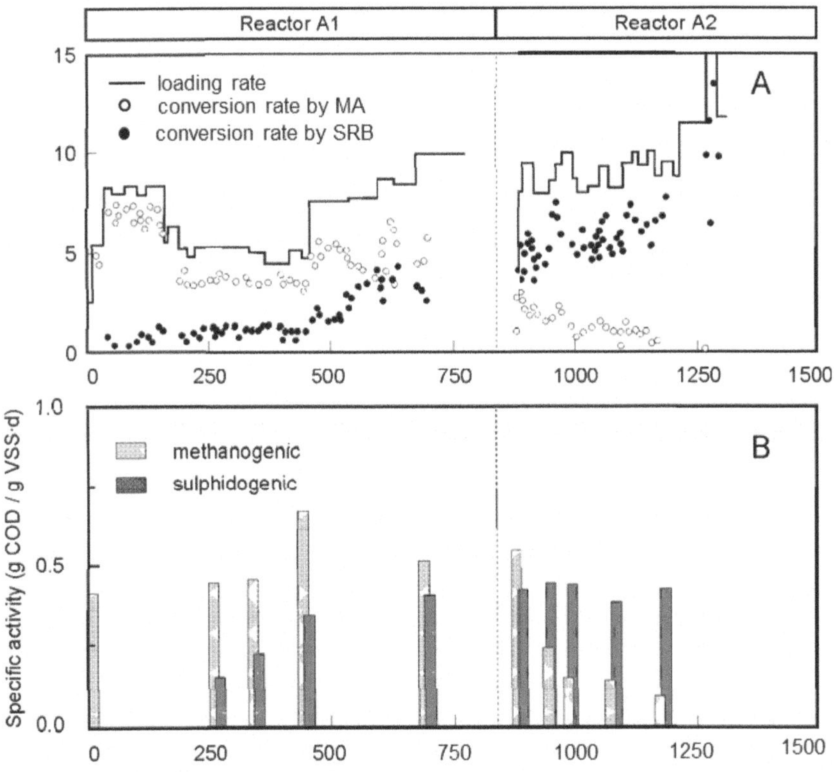

Figure 6.2 Evolution of the performance and sludge activity of two reactors treating acetate as the sole substrate. In this experiment, the COD:sulfate ratio was gradually decreased from 2 to 1 (reactor A1) and finally to 0.5 (reactor A2). (a) Organic loading rate and conversion rates by A-MA and A-SRB and (b) Specific methanogenic and sulfidogenic activities. Adapted from Omil *et al.* (1998).

was fed with a low sulfate-content, while the influent stream of the second reactor (sulfidogenic-methanogenic) contained a COD/SO_4^{2-} ratio of 3. Both reactors presented high COD removal efficiencies (>96%), and similar methane production yields (350 $mLCH_4$ $gCOD^{-1}_{removed}$), indicating that sulfate-reducing activity did not lead to an inhibition of methanogenesis.

Kiyuna *et al.* (2017) added sodium sulfate to real sugarcane vinasse to obtain COD/SO_4^{2-} ratios of 7.5, 10.0 and 12.0 (raw vinasse) in batch reactors operated simultaneously. The authors found that methane production occurred following two different patterns: an initial plateau was reached during the first 20 to 60 days of the experiment (Figure 6.3). After the initial plateau was reached, methane production followed an exponential increase. It is likely that during the initial period (before the first plateau), electron flow was diverted to sulfidogenesis, especially in the condition in which a COD/SO_4^{2-} ratio of 7.5 was applied to the batch reactor.

However, kinetic modelling indicated that the acetate produced by SRB enhanced the specific methane production rate (r'_M) by 50% in the second methane production increase (Kiyuna *et al.*, 2017). Despite the difference in the r'_M, the diversion of the electron flow led to 35% lower final methane production

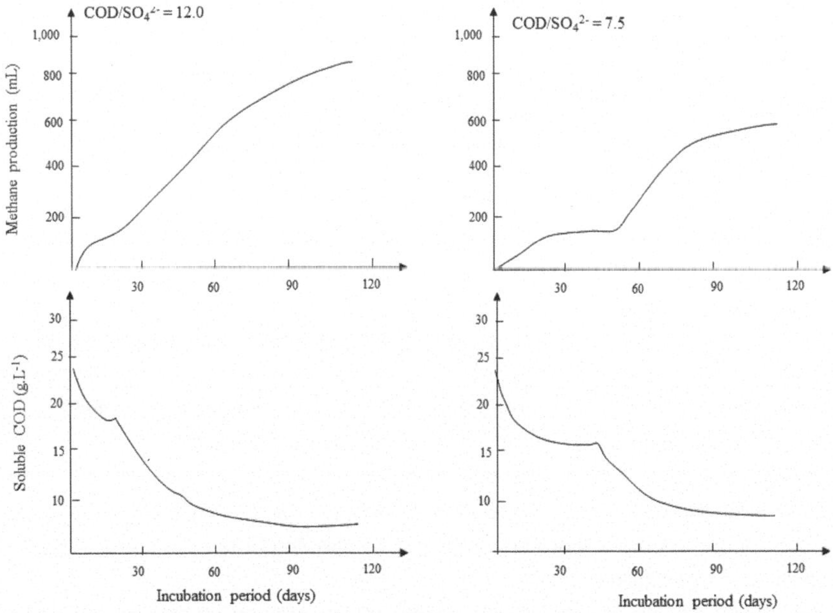

Figure 6.3 Methane production and COD consumption in batch reactors treating sugarcane vinasse with COD/SO_4^{2-} ratios of 12 (raw sugarcane vinasse, left side) and 7.5 (mimicking conditions at the end of the sugarcane harvest, right side). Adapted from Kiyuna *et al.* (2017).

in the conditions with a COD/SO_4^{2-} ratio of 7.5, due to the predominance of sulfidogenesis in the first 20–60 days of the experiment. Kiyuna *et al.* (2017) highlight that the COD removal was similar in all conditions, reinforcing that different metabolic pathways of sulfate reduction and methanogenesis can coexist.

Regardless of the competition between specific groups of SRB and MA, there are several factors that can determine which bioprocess will predominate (Table 6.3). Understanding these factors is a useful tool to adopt feasible engineering strategies in order to optimize the target process. It should be noted that in a complex process such as anaerobic digestion, more than one factor will play a role in determining the outcome of the competition between the two types of metabolism. For instance, Siles *et al.* (2010) attributed lower methane production to the increasing SO_4^{2-} influent concentration in a thermophilic (52°C) continuous-flow stirred tank reactor (CSTR) fed with glucose. According to the authors, biogas production was hampered when the influent stream reached

Table 6.3 Factors determining the outcome of the competition between SRB and methanogenic archaea in high-rate anaerobic reactors (adapted from Hulshoff Pol *et al.*, 1998).

Factor	Reference
Inoculum composition	
Type of seed sludge	McCartney and Oleszkiewicz (1991); Rodriguez and Zaiat (2011)
Bacterial composition	Omil *et al.* (1998); Gao *et al.* (2020)
Attachment properties of bacteria	Isa *et al.* (1986); Silva *et al.* (2006)
Experimental run time	Omil *et al.* (1998); Damianovic and Foresti (2009)
Bio-augmentation	Omil *et al.* (1997b); Mohan *et al.* (2005)
Influent composition	
Type of COD	Rodriguez and Zaiat (2011); van den Brand *et al.* (2014)
Acetate concentration	Yoda *et al.* (1987); Oude Elferink *et al.* (1998)
Sulfate concentration	Dar *et al.* (2008); Siles *et al.* (2010)
COD/SO_4^{2-} ratio	Godoi *et al.* (2015)
Sulfide concentration	Omil *et al.* (1996); Li *et al.* (2015)
Ca^{2+} and Mg^{2+} concentration	De Smul and Verstraete (1999a; Cao *et al.* (2009)
Operational conditions	
pH, mixed liquor	Visser *et al.* (1996); Sousa *et al.* (2015)
Salinity	Vallero *et al.* (2004); Sivakumar *et al.* (2019)
Temperature	Visser *et al.* (1993b); Madden *et al.* (2014); Kiyuna *et al.* (2017)

1400 mg L^{-1} (COD/SO_4^{2-}: 1.6). Nonetheless, Kiyuna *et al.* (2017) reported a lower final volumetric methane production with an initial sulfate concentration of 2900 mg SO_4^{2-} L^{-1} and a COD/SO_4^{2-} ratio of 7.5, based on batch reactor experiments treating carbohydrate-rich wastewater at 55°C.

Unravelling microbial fundaments and metabolic aspects of sulfate-reducing bacteria and their syntrophic partners also represents a scientific frontier (Chapter 3). There are many unknown SRB species that are not yet isolated nor described (Hao *et al.*, 2014; Qian *et al.*, 2019). Moreover, molecular techniques such as metagenomics, metaproteomics and metabolomics have been extensively developed in the last decade (Vanwonterghem *et al.*, 2014). These important tools on linking microbial community structure and interactions have only marginally been used in sulfate-reducing bioreactor systems. It is expected that understanding microbial community links will empower engineers and microbiologists to design more efficient and robust systems (De Vrieze & Verstraete, 2016), resulting in smaller and better performing sulfate-fed bioreactors.

6.2.2 Sulfide toxicity in anaerobic digestion

The sulfide generated as a product of the sulfate reduction can have toxic effects on anaerobic digestion. Sulfide is highly reactive, corrosive and toxic to microorganisms, plants and humans. Hydrogen sulfide dissociates in water according to equations (6.1) and (6.2) (Garrels & Christ, 1965). The main toxic effect is attributed to undissociated sulfide (H_2S), which diffuses through the cell membrane, and denatures proteins through formation of sulfide and disulfide crosslinks between polypeptide chains (Speece, 1983). H_2S may also interfere with coenzymes A and M through the formation of sulfide linkages. This may result in interference with the acetyl coenzyme A pathway for CO_2 fixation which is common to both SRB and methanogens (Widdel, 1988). Another toxicity mechanism is related to the precipitation of metal sulfides, thus decreasing the bioavailability of micronutrients (Gustavsson *et al.*, 2013).

$$H_2S \overset{k_1}{\Longleftrightarrow} H^+ + HS^- \qquad k_1 = 1.0 \times 10^{-7} \tag{6.1}$$

$$HS^- \overset{k_2}{\Longleftrightarrow} H^+ + S^{2-} \qquad k_2 = 1.0 \times 10^{-14} \tag{6.2}$$

The toxicity of sulfide is pH dependent, since hydrogen sulfide dissociation in water is pH-dependent (see Figure 6.4 and Chapter 2). The pK_a value of the dissociation equilibrium of H_2S is about 7.04 at 18°C (Weast, 1981). Above pH 8.0–9.0, virtually all dissolved sulfide is present in its ionized form. At neutral pH, values typical of methanogenic systems, 20–50% of the dissolved sulfide is present as H_2S, depending on the temperature.

Due to the diffusion of H_2S through cell membranes, it was expected that higher undissociated sulfide concentrations would correlate with higher toxicity. However, sulfide inhibition to a range of methanogens, syntrophs and SRB was related to the

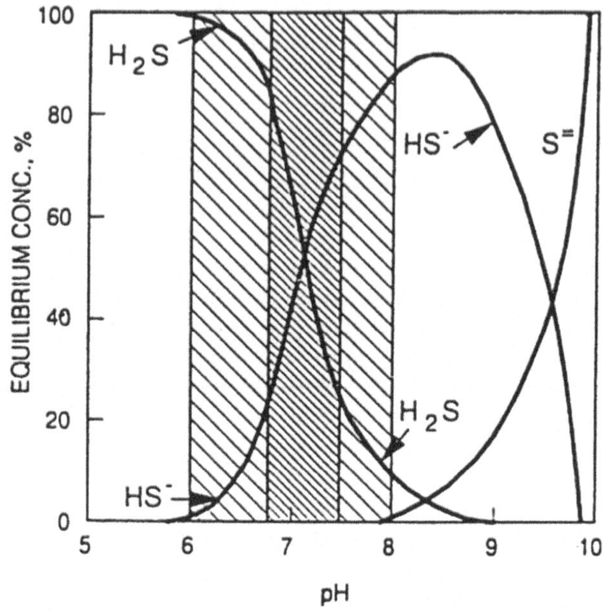

Figure 6.4 Prevalence of sulfide forms at different pH values (▧) pH range of anaerobic digestion and (▩) optimum pH for methanogenic digestion (after Rintala & Puhakka, 1994).

undissociated H_2S concentration between pH 6.8 and 7.2, whereas above pH 7.2 the inhibition was related to the total sulfide concentration (Figure 6.5, Tables 6.4 and 6.5). These observations suggest that H_2S may become more toxic at higher pH levels, possibly because of the development of strong pH gradients across the cell membrane, which may affect the diffusion properties of the H_2S molecule.

Different trophic groups present distinct sulfide toxicity thresholds (Tables 6.4–6.5), and different substrates also impact the toxicity threshold (Maillacheruvu *et al.*, 1993; Oleszkiewicz *et al.*, 1989). Propionate oxidizers are more sensitive to sulfide toxicity than other groups involved in the anaerobic digestion (Lens *et al.*, 2001; Maillacheruvu *et al.*, 1993; O'Flaherty *et al.*, 1999; Oleszkiewicz *et al.*, 1989). It should be noted that organic sulfate-rich wastewaters usually present a blend of different electron donors, and sulfate-reducing processes are more efficiently established in complex matrices containing several carbon sources (Liamleam & Annachhatre, 2007; Waybrant *et al.*, 1998). It is hypothesized that electron donor complexity enables a higher microbial diversity, which might lead to the establishment of multiple pathways of sulfate-reduction (Cao *et al.*, 2012). Indeed, it has been suggested that higher diversity in the microbial community

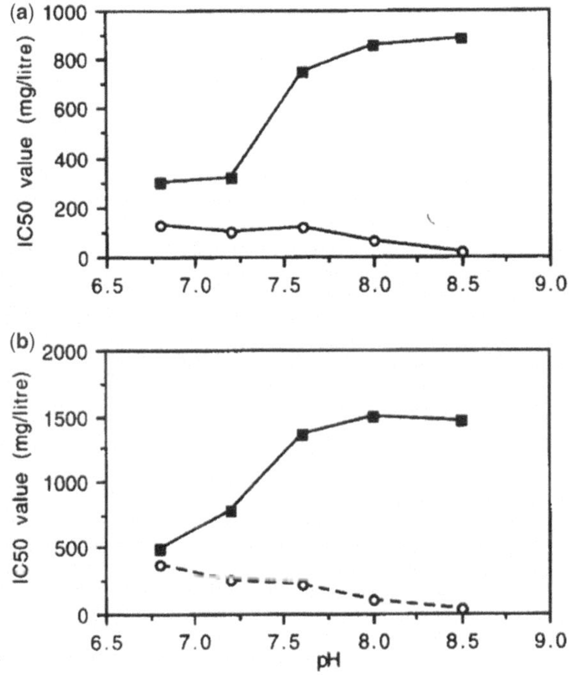

Figure 6.5 IC_{50} values for H_2S and total sulfide inhibition of pure cultures of MA and SRB. (a) *M. soehngenii* and (b) *D. acetoxidans*. H_2S (-O-), total sulfide (-■-) (after O'Flaherty *et al.* 1998b).

enhances stability and robustness of anaerobic systems aiming for biogas production (De Vrieze & Verstraete, 2016). This idea is based on functional redundancy, which implies that many different microbial species perform the same process, e.g. methanogenesis. Hence, if one microbial group is inhibited, it is likely that other populations still perform the same process, ensuring the system response does not change drastically upon environmental changes such as toxic shocks. Nonetheless, there are reports that diverse SRB populations thrive even when only one electron donor is provided (Dar *et al.*, 2007; Kaksonen *et al.*, 2004). The correlation between microbial diversity and robustness in biomethane production has been explored in anaerobic digesters (De Vrieze *et al.*, 2015; Langer *et al.*, 2015; Niu *et al.*, 2015), but there are so far no studies associating functional redundancy and stability of sulfate-reducing or methanogenic reactors submitted to high sulfate loads.

Another system feature that impacts sulfide toxicity thresholds is related to the immobilization of biomass (Lens *et al.*, 2001; Oleszkiewicz *et al.*, 1989). For instance, Isa *et al.* (1986) concluded that SRB growing in a fixed film reactor

Table 6.4 IC_{50} (mg L^{-1}) values in the form of total sulfide concentration for a range of pure cultures of SRB and MA over the pH range 6.8–8.5 (undissociated H_2S concentration in parentheses).

Microorganism	Substrate	IC_{50} (mg L^{-1}) at Varying pH				
		6.8	7.2	7.6	8.0	8.5
D. magnum	Acetate	443	671	660	659	708
		(239)	(218)	(105.6)	(46)	(14)
D. acetoxidans	Acetate	487	775	1360	1500	1450
		(263)	(252)	(218)	(105)	(29)
D. vulgaris	H_2/CO_2	554	840	1343	1499	1400
		(299)	(273)	(215)	(105)	(28)
D. sapovorans	Butyrate	513	796	1133	1170	1215
		(277)	(259)	(181)	(82)	(24.3)
D. postgatei	Acetate	583	926	1248	1119	1290
		(315)	(301)	(200)	(78)	(26)
D. multivorans	Ethanol	498	851	1383	1488	1570
		(269)	(277)	(221)	(105)	(31.5)
D. propionicus	propionate	223	355	500	525	570
		(120.5)	(115)	(80)	(37)	(11.5)
M. barkeri	H_2/CO_2	226	363	681	1114	1000
		(122)	(118)	(109)	(78)	(20)
M. hungatei	H_2/CO_2	278	514	937.5	1042	980
		(150)	(167)	(150)	(73)	(40)
M. mazei	Acetate	130	175	287.5	621	690
		(70)	(57)	(46)	(43.47)	(13.8)
M. soehngenii	Acetate	295	317	744	857	880
		(123)	(103)	(119)	(60)	(17.5)

Values presented are the mean of triplicates (after O'Flaherty *et al.*, 1998b).

were not affected by high levels of sulfide. Immobilized biomass allows the establishment of sulfide concentration profiles, inducing a population stratification along the biofilm depth (Sun *et al.*, 2014). Moreover, sulfate-reducing bacteria predominate in the outer layer of the biofilm, and the sulfide concentration is lower in the inner layers due to diffusion (Santegoeds *et al.*, 1999; Sun *et al.*, 2014). The type of immobilization can also affect the toxicity threshold (Celis-García *et al.*, 2004), which can be related to the different adhesion properties of sulfate-reducing bacteria and methanogens (Isa *et al.*, 1986), or different types of support material (Silva *et al.*, 2006).

Table 6.5 Sulfide toxicity thresholds of various anaerobic systems.

Target process	Biomass	Substrate	°C	pH	H_2S ($mg\ L^{-1}$)	Ref.
Methanogenesis	Suspended sludge	distillery wastewater	37	7.0–7.2	130	Karhadkar *et al.* (1987)
	Suspended sludge	lactate	35	7.0	100	McCartney and Oleszkiewicz (1991)
				8.0	100	
	Suspended sludge	acetate	35	6.5–7.4	125	Oleszkiewicz *et al.* (1989)
				7.7–7.9	100	
	Granular sludge	acetate	30	6.2–6.4	246	Koster *et al.* (1986)
				7.0–7.2	252	
				7.8–8.0	90	
	Granular sludge	acetate	30	7.2–7.4	184	Visser (1995)
				8.1–8.3	38	
	Granular sludge	acetate and propionate	35	7.5	800	Celis-García *et al.* (2004)
Sulfate-reduction	Biofilm				1250	
	Desulfovibrio enriched culture	lactate	37	6.2–6.6	517	Reis *et al.* (1992)
	Granular sludge	acetate and propionate	35	7.5	750	Celis-García *et al.* (2004)
	Biofilm				860	
	Granular sludge	sucrose	55	5.0	95	Lopes *et al.* (2010)

6.2.3 Techniques for quantification of sulfide toxicity on microbial populations involved in anaerobic digestion

There are two principal approaches for analysis of the toxic effects of sulfide on the microbial populations involved in anaerobic digestion:

(1) Sludge activity/toxicity tests
(2) Determination of kinetic growth properties of microbial populations

The main aim of both of these approaches is to determine the effect of various concentrations of a toxicant, i.e. sulfide, on the trophic groups involved in methanogenesis or sulfate reduction. This information will be valuable in predicting the levels of in-reactor sulfide which will be acceptable for successful anaerobic digestion and also in determining the outcome of competition between MA and SRB.

6.2.3.1 Specific methanogenic activity/toxicity tests

Numerous toxicity testing strategies have been developed to monitor the effects of inhibitory compounds on anaerobic microorganisms, and in particular on MA. These tests are based on the effect of inhibitors on sludge or biomass activity compared with a control. Anaerobic toxicity tests can be classified on the basis of their operating conditions. They can be conducted under batch (Colleran *et al.*, 1992; Kiyuna *et al.*, 2017; O'Flaherty *et al.*, 1998a; Owen *et al.*, 1979) or continuous (Godoi *et al.*, 2015; Parkin & Speece, 1983) conditions. Tests can be done directly with reactor sludge (Godoi *et al.*, 2015; Kiyuna *et al.*, 2017), with enrichments from reactor sludge (Paulo *et al.*, 2017; Wang *et al.*, 1991) or with pure cultures of anaerobic bacteria (O'Flaherty *et al.*, 1998b). The parameters that indicate inhibition are:

(1) a drop in methane yield
(2) volatile fatty acid (VFA) accumulation
(3) a decrease in the chemical oxygen demand (COD) removal efficiency

Biogas or methane production can be monitored by water displacement (Veiga *et al.*, 1990; Walker *et al.*, 2009), by the displacement of a plunger of glass syringes (Owen *et al.*, 1979) or by electronic pressure transducers (Braga *et al.*, 2018; Colleran & Pistilli, 1994; Souto *et al.*, 2010). The methane percentage of the biogas produced and the VFA levels are usually monitored by gas chromatography (GC).

Specific sludge activities are typically analysed using batch experiments. Sludge samples are placed in glass serum bottles in the presence of a basal medium which may or may not contain nutrients and a buffering agent (Braga *et al.*, 2018; Kiyuna *et al.*, 2017; O'Flaherty *et al.*, 1998a). Substrates of interest are added in appropriate concentrations to control vials, whereas an inhibitor (sulfide) is included at various concentrations in parallel tests for an assessment of activity as a function of the sulfide concentration. The pH of the vials can be corrected by addition of acid or base (Fajardo *et al.*, 2014). All vials are prepared under anaerobic conditions by flushing with a N_2/CO_2 mixture.

The methanogenic activity can be determined by measuring the methane concentration in the headspace of the serum vials over time (Visser, 1995). Alternatively, the gas pressure increase (or decrease in the case of H_2/CO_2) in the headspace can be measured with a pressure transducer device and the concentration of methane can then be determined by GC (Braga *et al.*, 2018; O'Flaherty *et al.*, 1998a). In both cases, the activity is calculated from the slope of the progress line, either relative methane concentration or millilitres of methane. The volatile suspended solids (VSS) present in the serum vial are determined on completion of the test and the specific methanogenic activity is expressed per unit VSS. The effect of added sulfide concentrations can be expressed as a percentage inhibition of the control activity and thus a 50% inhibition concentration value (IC_{50}) can be calculated (Figure 6.6). The same

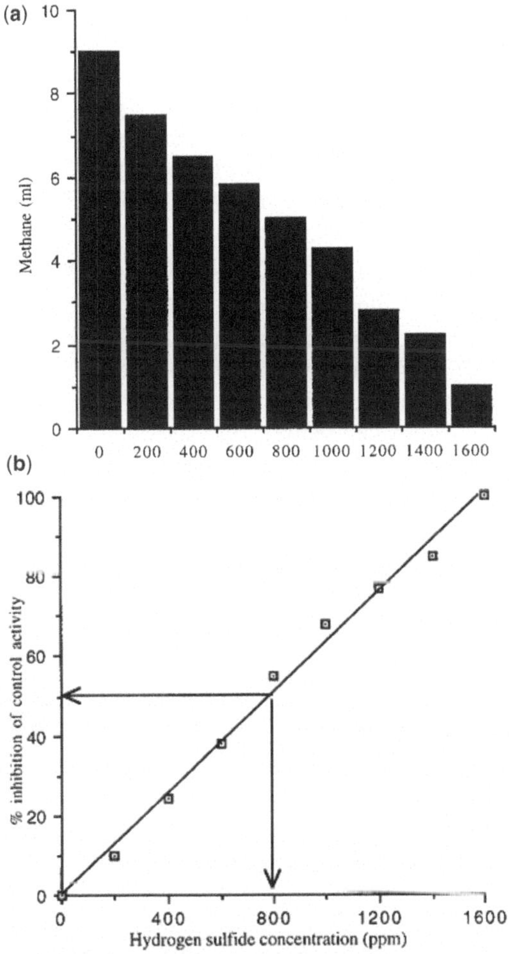

Figure 6.6 Effect of increasing H_2S concentrations on (a) cumulative methane production from acetate after 5 h at pH 8.0 and (b) percentage inhibition by H_2S of control methanogenic activity with acetate at pH 8.0 (O'Flaherty & Colleran, 2000).

procedure can be carried out in the presence of bromoethane sulfonic acid (BES) or sodium molybdate (Mb), specific inhibitors for MA and SRB, respectively. As well as the methane production, the degradation of substrates of interest can also be measured directly, normally by GC, and the effect of added sulfide on the substrate degradation rates can be determined.

In general, batch toxicity tests of the type described above generate acute toxicity data about the effects of a single dose of an inhibitor, such as sulfide, a short time after the administration of the dose. These data can then be used to determine the dose/response relationship and toxicity thresholds. These values are useful when making comparisons between the toxicity of compounds or between the sensitivities of different trophic groups of organisms. There are also several toxicity testing protocols that incorporate both batch and continuous flow strategies to determine IC_{50} values for toxic compounds. These protocols have also been used to monitor the ability of the biological system to recover from, or to acclimatize to, a toxic dose (Soto *et al.*, 1993; Young & Tabac, 1993).

6.2.3.2 Specific sulfidogenic activity/toxicity tests

In this case, the basal medium used in the activity tests is supplemented with sulfate, at an appropriate concentration, so that sulfate will never become limiting (Kiyuna *et al.*, 2017; O'Flaherty *et al.*, 1998a). Methanogenic activity can be eliminated with BES (Oremland & Capone, 1988). The substrate degradation can be measured directly, e.g. by GC analysis (O'Flaherty *et al.*, 1998a; Visser, 1995). The sulfidogenic activity can then be calculated from the slope of substrate degradation and the VSS content (Kiyuna *et al.*, 2017; O'Flaherty *et al.*, 1998a; Visser, 1995). Toxicity thresholds are determined by the dose response relationship, as described in section 6.2.3.1.

6.2.3.3 Determination of kinetic growth properties of microbial populations

The growth and kinetic properties of microorganisms can be determined by simple procedures in batch cultures (O'Flaherty *et al.*, 1998b; Visser, 1995). Typically, the properties of most interest are the specific methanogenic or sulfidogenic growth rates and the affinities for substrate/sulfate. These parameters are of importance in studies to determine the likely outcome of the competition between SRB and other anaerobic bacteria, and also to determine the effect of sulfide on the growth of microbial populations. Normally in these experiments the parameters are determined using substrate (or sulfate) depletion curves (O'Flaherty *et al.*, 1998b; Visser, 1995).

6.2.4 Sulfite toxicity

Sulfite is employed in several industrial processes, as antimicrobial agent in winery or raisin fabrication, or for lignin extraction in the pulp industry (Athanassopoulos *et al.*, 1989; Pokhrel & Viraraghavan, 2004). It has recently been suggested to add sulfite to waste activated sludge to enhance its biodegradability in anaerobic digesters (Zan *et al.*, 2019). However, sulfite damages the structure of proteins, lipids, and DNA, and most living cells possess sulfite-reductases to prevent damage against them (Kappler & Enemark, 2015).

Sulfite inhibits several anaerobic microorganisms, including fermentative (Irwin *et al.*, 2017), and sulfur-reducing bacteria (Florentino *et al.*, In Press), as well as methanogenic archaea either in natural environments (Balderston & Payne,

1976), or in engineered reactors treating high-sulfite wastewater (Athanassopoulos *et al.*, 1989).

In some cases, sulfite induces a lag phase in methane production of variable length. Yang *et al.* (1979) observed lag phases of between 60 hours and 12 days after dosing 25 and 75 mg L^{-1} SO_3^{2-}, respectively, to an acetoclastic methanogenic enrichment culture. The effects of sulfite inhibition appear to be far less severe with adapted sludges, however, probably owing to the presence of SRB which reduces the sulfite to sulfide. Eis *et al.* (1983) reported no lag phase after addition of 100 mg L^{-1} SO_3^{2-} to an adapted sludge. Maaskant and Hobma (1981) found a 50% inhibition of the methanogenic activity at about 150–200 mg L^{-1} SO_3^{2-}, an effect that declined upon repeated dosing of SO_3^{2-} to the sludge.

It is generally considered that sulfite inhibition will be insignificant in continuous cultures and during reactor operation, as in these cases SRB populations that will eliminate the sulfite will develop (Rinzema & Lettinga, 1988; Visser, 1995). It has been shown that wastewaters with sulfite concentrations of up to 800 mg L^{-1} SO_3^{2-} can be treated satisfactorily in anaerobic reactors (Eis *et al.*, 1983; Ferguson *et al.*, 1982; Särner, 1986, 1989).

6.2.5 Cation inhibition in anaerobic digestion

Sulfate-rich wastewater can also contain a high cation (Na^+, Ca^{2+}, Mg^{2+}) content, such as in tanneries (Lefebvre *et al.*, 2006), seafood processing (Omil *et al.*, 1995), petrochemical (Wei *et al.*, 2018), and pharmaceutical (Li *et al.*, 2018) wastewaters. Sewage can also present high sulfate and sodium concentrations in the case of intrusion of brackish groundwater in the collection network or use of seawater in toilet flushing (Van Den Brand *et al.*, 2015). High cation (Na^+, Ca^{2+}, Mg^{2+}) concentrations in these wastewaters can inhibit anaerobic bacteria (Jeison *et al.*, 2008; Muñoz Sierra *et al.*, 2017; Rinzema *et al.*, 1988; Vallero *et al.*, 2003), since they cause osmotic stress to the cells (Pollice *et al.*, 2000; Valentine, 2007), or cause granules disruption followed by biomass washout from the bioreactor (Jeison *et al.*, 2008; Lefebvre *et al.*, 2006).

The sodium concentration reported to inhibit methanogenesis in literature varies, probably due to the history of the sludge, and antagonistic or synergistic effects that may occur related to the different real or synthetic wastewater composition. Activity of acetoclastic methanogens was reduced by 50% at 10 g Na^+ L^{-1} and completely inhibited at 16 g Na^+ L^{-1}. In other studies, the IC_{50} for methanogenesis inhibition was 23–25 g Na^+ L^{-1}, and total inhibition occurred when the sodium concentration was increased to 34 g Na^+ L^{-1} (Jeison *et al.*, 2008; Muñoz Sierra *et al.*, 2017).

In bioreactors treating saline, organic and sulfate-rich wastewaters, SRB are more resistant to toxicity due to salinity than MA. In a mesophilic (30°C) up-flow anaerobic sludge bed (UASB) reactor, sulfate-reducing activity and methanogenic activities decreased to less than 10% of the initial rate when the salinity in the influent stream changed from low salinity to 50 g NaCl L^{-1} and 1 g $MgCl_2$ L^{-1} (Vallero *et al.*, 2004). When the biomass was exposed back to freshwater conditions, sulfate-reducing activity resumed up to 80% of the initial activity, whereas the recovery of methanogenesis was limited to 25% of the original rate (Vallero *et al.*, 2004). At moderate temperature (20°C), the SRB activity rate decreased with increasing salinity, but even with saline water (35 g L^{-1} of aquaria salt), the sequencing batch reactor effluents gave high COD removal efficiencies (van den Brand *et al.*, 2015). SRB were also more resistant to salinity under thermophilic conditions (55°C) (Vallero *et al.*, 2003).

In high salinity wastewaters, trace metals were found to be less bioavailable, and to be washed out from the system (Muñoz Sierra *et al.*, 2017). This could be related to the displacement of trace metals from the biomass matrix by sodium. A COD removal efficiency of 90% was achieved when doubling the trace metals addition (Muñoz Sierra *et al.*, 2017). The addition of potassium to the system can also help to overcome sodium inhibition in biological treatment of high-saline wastewaters, although this was tested for saline organic wastewaters with a low sulfate content only. This strategy is called 'salt in', and is based on balancing the cell osmotic pressure through the accumulation of inorganic ions (Li *et al.*, 2018). The 'salt in' strategy led to an increment of 40% of methane production when 1.74 g L^{-1} of potassium was added to a bioreactor treating heparine sodium production wastewater containing 10 g Na^+ L^{-1} (Li *et al.*, 2018). Maintaining the K^+/Na^+ ratio at 0.05 allowed high COD removal efficiencies (>90%) when treating synthetic industrial wastewater containing 20 g Na^+ L^{-1} and aromatic compounds (0.7–70 mg phenols L^{-1}) (Muñoz Sierra *et al.*, 2018).

6.3 PROCESS TECHNOLOGY OF TREATMENT OF ORGANIC SULFATE-RICH WASTEWATERS

6.3.1 Modelling the effect of sulfide toxicity in anaerobic digestion

The primary difficulty in applying anaerobic treatment technologies to sulfate-containing wastewaters arises from the production of H_2S by SRB and from its toxicity towards the various trophic groups of bacteria involved in the process (section 6.2). The outcome of competitive interactions between SRB and other anaerobic bacteria, such as syntrophs and methanogens, in digester sludges will determine the amount of sulfide in the digester, and ultimately the extent of process failure.

Modelling the biochemical reactions in anaerobic bioreactors is troublesome, since the system is highly complex involving a highly diverse microbial community competing for substrate or cooperating, e.g. by-products released by one group are substrate for another group (Figure 6.1). The first models considered only the competition between MA and SRB, but did not take into account other parameters, such as sulfide inhibition (Gupta *et al.*, 1994; Noguera *et al.*, 1999; Omil *et al.*, 1998).

The development in computing capacity has led to more robust models, such as the Anaerobic Digestion Model No. 1 (ADM1) (Lübken *et al.*, 2007). The use of artificial neural networks has also led to models able to predict the performance of anaerobic reactors treating sulfate-rich wastewaters (Janyasuthiwong *et al.*, 2015; Reyes-Alvarado *et al.*, 2017a; Sinharoy *et al.*, 2019). More robust models often require a great amount of data. The use of online sensors allows for more accurate process monitoring, and the development of sensors has also advanced for a more widespread application (Lamb *et al.*, 2019). Process modelling and control are discussed further in detail in Chapter 9.

6.3.2 Alleviating sulfide toxicity

In principle, all common anaerobic bioreactor configurations can be applied for the treatment of sulfate-rich wastewaters (Table 6.6), either focusing on CH_4 production, or sulfate and COD removal. Regardless of the target process, proper measures to prevent the occurrence of high H_2S concentrations in the liquid or in the gas phase must be taken to either alleviate toxicity or to prevent corrosion (Table 6.7). Corrosion due to H_2S damages structures and pipelines, and it is estimated to cause a US$ 100 M loss per year in man-made structures (Rückert, 2016; Chapter 5).

There are well-established techniques to prevent H_2S generation or toxicity in sulfate-rich wastewater treatment (Figure 6.7 and Table 6.7). Wet gas desulfurization is a common strategy, and is widely applied in full-scale anaerobic digesters (Muyzer & Stams, 2008), but the high cost for disposal of its waste is still not fully addressed (Hao *et al.*, 2014).

In order to avoid toxic effects on methanogenesis, application of a two-stage process has been suggested for a long time (Reis *et al.*, 1988). This strategy has been applied under thermophilic conditions (Gil-Garcia *et al.*, 2018; Lopes *et al.*, 2007) and at various pH values (Demirel & Yenigün, 2002; Lopes *et al.*, 2007), which suggests that it can be applied for different types of wastewaters. Moreover, biogas recirculation in an acidogenic reactor led to an enhancement of 70% of sulfide removal from the liquid phase (Gil-Garcia *et al.*, 2018).

Chemical precipitation with ferric iron can also be used to alleviate toxic effects of sulfide (De Smul & Verstraete, 1999b). This is the basis for one of the biogas desulfurization technologies (see Chapter 11), but it has also been applied to sulfate-reducing bioreactors. A S:Fe ratio ranging from 0.3–0.5 did not impact methane production, whereas acetoclastic methanogenic activity was impaired at higher (0.5–1.0) S:Fe ratios (Yekta *et al.*, 2017). The addition of iron has also successfully avoided toxicity of acetoclastic methanogenesis in a fixed-bed reactor operating at a COD:SO_4^{2-} ratio of 2 (Godoi *et al.*, 2017). The authors aimed to co-digest acidic sulfate-rich wastewater with sugarcane vinasse as an

Table 6.6 Overview of reactor types, sulfate removal efficiencies and extent of sulfate reduction in anaerobic digestion of organic and sulfate-rich wastewaters.

Reactor Type	Applied Target	Electron Donor	Operation Conditions	Process Efficiency		Reference
				$\%SO_4^{2-}$ red	%COD	
UASB	CH_4 production	Synthetic VFA mixture	30°C	70	50–90	Visser *et al.* (1993a)
EGSB	CH_4 production	Synthetic VFA mixture	30°C, 0.5–2.5 g $O_2 \cdot L^{-1}$ (COD), 1.2–4.5 g $SO_4^{2-} \cdot L^{-1}$	35	67–81	Omil *et al.* (1996)
AF		Citric acid	25.8 g $O_2 \cdot L^{-1}$, 3.4 g $SO_4^{2-} \cdot L^{-1}$	93	18	Colleran *et al.* (1994)
CAD		Seafood	10–60 g $O_2 \cdot L^{-1}$ 0.6–2.7 g $SO_4^{2-} \cdot L^{-1}$-	96	3–12	Omil *et al.* (1995)
HAIB	COD and SO_4^{2-} removal	Paper recycling industry wastewater	30°C, 12 h HRT, pH 7.3–8.3,	94.2	92.5	Damianovic *et al.* (2018)
ASTBR	SO_4^{2-} removal and acetate accumulation for subsequent methanogenesis	Ethanol	55°C, 6–12 h HRT	18–57.3	9.8–22.5	Gil-Garcia *et al.* (2018)
ASTBR	CH_4 production	Synthetic wastewater simulating soluble fraction of sugarcane vinasse	30°C, 12 h HRT, 4 g $O_2 \cdot L^{-1}$, 1.3 g $SO_4^{2-} \cdot L^{-1}$	97	96	Godoi *et al.* (2015)

UASB: Up-flow Anaerobic Sludge Bed reactor; EGSB: Expanded Granular Sludge Bed reactor; AF: Anaerobic Filter; CAD: Central Activity Digester; HAIB: Horizontal-flow Anaerobic Immobilized Biomass reactor, ASTBR: Anaerobic Structured-Bed Reactor.

Table 6.7 Measures to reduce the reactor sulfide concentration, thus allowing the integration of methanogenesis and sulfate reduction.

Measure	Reference
A. Dilution of the influent	
Non sulfate containing process water	Fox and Venkatasubbiah (1996)
Recycle of the effluent after a sulfide removal step by	
Sulfide stripping	Jensen and Webb (1995); Glória *et al.* (2016)
Sulfide precipitation	Nielsen *et al.* (2008)
Biological sulfide oxidation to elemental sulfur	
Thiobacillus sp., oxygen	Buisman *et al.* (1990); Valdés *et al.* (2016)
Thiobacillus denitrificans, nitrate	Gommers *et al.* (1988); Beristain-Cardoso *et al.* (2006)
Chlorobium limicola, sunlight	Tang *et al.* (2009); Garcia *et al.* (2017)
Chemical oxidation to elemental sulfur	
Ferric sulfate	Abbott and Eskicioglu (2015); Yekta *et al.* (2017)
B. Decrease of the unionized sulfide concentration	
Elevation of the reactor pH	Omil *et al.* (1996); Gutierrez *et al.* (2009); Yan *et al.* (2018)
Elevation of the reactor temperature	Gil-Garcia *et al.* (2018)
Precipitation of sulfide, e.g. with iron salts	Isa *et al* (1986); Godoi *et al.* (2017)
Stripping of the reactor liquid using	
High mixture degree inside the reactor	Morgan-Sagastume and Noyola (2006)
Recirculation of the biogas after scrubbing	Sarner (1990); Fdz.-Polanco *et al.* (2009)
Other stripping gas (e.g. N_2)	Kent (1973)
Application of electrolysis	Yuan *et al.* (2020)
C. Separation of sulfide production and methanogenesis	
Two stage anaerobic digestion	Reis *et al.* (1988); Lopes *et al.* (2010)
Multi-step process	Sipma *et al.* (1999)

(Continued)

Table 6.7 Measures to reduce the reactor sulfide concentration, thus allowing the integration of methanogenesis and sulfate reduction (*Continued*).

Measure	Reference
D. Selective inhibition of SRB	
Sulfate analogues (e.g. MoO_4^{2-})	Yadav and Archer (1989); Isa and Anderson (2005); Xu *et al.* (2011)
Transition elements (e.g. Co, Ni or Zn)	Clancy *et al.* (1992); Utgikar *et al.* (2003)
Antibiotics	Tanimoto *et al.* (1989)
E. Manipulation of reactor design	
EGSB (upflow velocity 4–6 m/h)	Omil *et al.* (1996)
Baffled reactor	Barber and Stuckey (2000)
Horizontal-flow anaerobic immobilized sludge	Vilela *et al.* (2014)
Structured-bed reactor	Camiloti *et al.* (2013)
F. Manipulation of influent composition	
Increase of acetate concentration	Omil *et al.* (1996); van den Brand *et al.* (2014)
Availability of electron donor	Cao *et al.* (2012); Reyes-Alvarado *et al.* (2017b)
Addition of trace elements	Fermoso *et al.* (2008); Braga *et al.* (2018)
G. Manipulation of the biomass composition	
Bioaugumentation	Omil *et al.* (1997b); Pereyra *et al.* (2012)
Biostimulation	Gonçalves *et al.* (2005)
Exposure to oxygen	Hao *et al.* (2016)
Shock treatment	
Temperature decrease to 15°C	Omil *et al.* (1997a); Hao *et al.* (2016)
Temperature increase to 65°C	Visser *et al.* (1993b); Jung *et al.* (2019)

electron donor, but the predominance of incomplete oxidizing SRB led to acetate accumulation in the bioreactor. Total dissolved sulfide and H_2S concentrations decreased after iron addition to the influent stream, and methanogenesis was restored. The sulfide precipitation removed its inhibitory effects on the MA, allowing for recovery of methanogenesis. It should be noted that different mechanisms for iron-aided sulfide toxicity alleviation have been proposed (Table 6.7): iron can be involved in sulfide oxidation to S^0 (De Smul & Verstraete, 1999b) or in chemical precipitation of iron sulfide (Godoi *et al.*, 2017).

Sulfate reduction is widely reported under thermophilic temperatures (Gil-Garcia *et al.*, 2018; Kiyuna *et al.*, 2017; Lopes *et al.*, 2010; Siles *et al.*, 2010; Visser *et al.*,

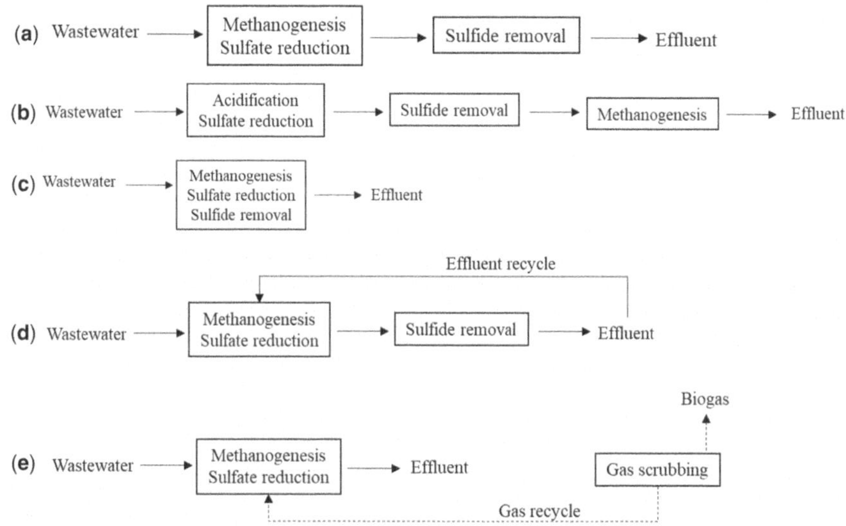

Figure 6.7 Process configurations integrating methanogenesis with sulfate reduction and sulfide removal. Adapted from Visser (1995)

1993b), and SRB respond better than MA when a bioreactor is submitted to temperature shocks (Visser *et al.*, 1993b). Thermophilic conditions also favour sulfide stripping in multi-step anaerobic systems (Gil-Garcia *et al.*, 2018). Nonetheless, H_2S is highly soluble (>1000 mg H_2S L^{-1}) even at 80°C, indicating that at higher temperatures sulfide stripping from the liquid phase is insufficient for complete removal of H_2S (Liu *et al.*, 2018).

The use of selective inhibitors of SRB was one of the first tested strategies to hamper sulfidogenesis (Saleh *et al.*, 1964; Tanimoto *et al.*, 1989). Antibiotics successfully inhibited SRB (Tanimoto *et al.*, 1989), but their application is strongly discouraged since it contributes to increasing antibiotic resistance in environmental biota from several ecosystems (Liu *et al.*, 2018). Transition elements are also reported to inhibit the sulfate reduction process (Clancy *et al.*, 1992; Utgikar *et al.*, 2003), which can be related to their toxic effects (Hao *et al.*, 1994; Hao, 2000; Kiran *et al.*, 2017). Furthermore, addition of transition metals could alleviate sulfide toxicity by precipitating metal sulfides, similar to iron addition (Godoi *et al.*, 2017). Transition metals can also enhance methanogenesis, since some of them are micronutrients for MA (Braga *et al.*, 2018; Fermoso *et al.*, 2008). It is, however, not clear whether supplementation with these transition metals could favour methanogens to an extent that MA can outcompete SRB. The use of sodium tetraborate inhibited the SRB population from swine manure (Yokoyama *et al.*, 2016), but the authors did not report methane production in the study.

6.4 DOWNSTREAM PROCESSES FOR BIOLOGICAL SULFATE-REDUCTION EFFLUENTS

Anaerobic treatment of sulfate-rich organic wastewater has focused on overcoming inhibition of methanogenesis. However, sulfate-reducers are versatile organisms, which can be applied in a wide range of temperature (Gil-Garcia *et al.*, 2018; Lopes *et al.*, 2007; Madden *et al.*, 2014), pH (Sánchez-Andrea *et al.*, 2015; Sousa *et al.*, 2015), or salinity (Wei *et al.*, 2018) conditions, enabling sulfate bio-based processes to be engineered for several purposes. This idea contributed to further research on the linkage of sulfur based biotechnologies with other biogeochemical cycles (Bhattarai *et al.*, 2019; Hao *et al.*, 2014; Muyzer & Stams, 2008; Qian *et al.*, 2019).

6.4.1 Sulfide partial oxidation to elemental sulfur

The association between biological sulfate reduction to sulfide and subsequent conversion to elemental sulfur, which can be later used in sulfuric acid production, was one of the first acknowledged possibilities (Hao *et al.*, 2014; Janssen *et al.*, 1995). Sulfide partial oxidation to S^0 can be achieved through either chemical or biological pathways (Valdés *et al.*, 2016, Chapter 11). The recovery of elemental sulfur through partial sulfide oxidation under anoxic conditions is the basis of biological biogas desulfurization units (Montebello *et al.*, 2012; Ramírez *et al.*, 2009).

Aerobic biological oxidation of sulfide to S^0 is applied at full-scale treating sulfate-rich zinc-smelter wastewater (Muyzer & Stams, 2008). Sulfate-reducing bacteria produce sulfide, that is partially recovered as zinc sulfide (Janssen *et al.*, 2009). The remaining sulfide is recovered as S^0 after partial oxidation. Biological conversion of sulfide to S^0 can be either aerobic (Buisman *et al.*, 1990; Janssen *et al.*, 1995) or anoxic (Di Capua *et al.*, 2019), with the use of nitrate or nitrite as an electron acceptor by colourless sulfur bacteria.

In the case of aerobic oxidation of sulfide to S^0, one of the main difficulties regarding aerobic partial oxidation of sulfide to elemental sulfur is to control aeration, since oxygen in excess can lead to full oxidation to sulfate. To overcome this issue, each part of the process must be kept in separate operation units (Janssen *et al.*, 1997). Another strategy to control the partial sulfide oxidation to S^0 is the use of membranes for aeration (Krayzelova *et al.*, 2015). At lab-scale, biological S^0 recovery has been successfully achieved in sulfate-reducing reactors through the use of a micro-aerobic silicone membrane reactors (Camiloti *et al.*, 2016; Valdés *et al.*, 2016).

6.4.2 Sulfide oxidation using nitrate as electron acceptor

Autotrophic denitrification using sulfide as an electron donor can be applied either for biogas desulfurization (Ramírez *et al.*, 2009; Chapter 13) or post treatment of anaerobic effluents that contain nitrogen oxyanions (Chapter 13). The use of nitrate to prevent H_2S-related issues is particularly widespread in oil and petrochemical wastewaters (Muyzer & Stams, 2008). Furthermore, autotrophic denitrification has been applied at full-scale treating saline sewage (Wang *et al.*, 2009; Wu *et al.*, 2016). The Sulfate reduction, Autotrophic denitrification and Nitrification integrated process (SANI®) addresses the high sulfate content in sewage systems that employ seawater for flushing toilets (Wang *et al.*, 2009).

A further development integrated phosphorous removal in the SANI® system (Liu *et al.*, 2014). The Denitrifying Sulfur cycle-associated Enhanced Biological Phosphorus Removal (DS-EBPR) is based on the establishment of a syntrophic relationship between sulfate-reducing bacteria and sulfur-oxidizing bacteria under intermittent anaerobic/anoxic conditions to achieve COD, nitrogen, and phosphorous removal (Guo *et al.*, 2019). Nitrification occurs in a separate reactor (Wang *et al.*, 2009), and storage polymers such as polyhydroxyalkanoates (PHA), poly-P and poly-S seem to be important to the process (Guo *et al.*, 2019). The integration of sulfate-reducing bacteria, nitrifying and sulfur-oxidizing bacteria is also the basis for treating petrochemical nanofiltration concentrate (NFC), a highly concentrated petrochemical wastewater containing sulfate, nitrogen and hazardous organic compounds (Wei *et al.*, 2018).

Autotrophic denitrification has also been suggested for the post treatment of anaerobic effluents (Souza *et al.*, 2018). Anaerobic systems are suitable for sewage treatment in tropical areas (Chernicharo *et al.*, 2015). These systems, however, are unable to remove nitrogen, and a post-treatment step is required (Foresti *et al.*, 2006). Nonetheless, the conventional denitrification step requires organic compounds, which are no longer usually present in effluents from anaerobic reactors. On the other hand, anaerobic effluents might present endogenous alternative electron donors, such as sulfide. The use of a triple chamber biosystem coupling anaerobic digestion, nitrification and mixotrophic endogenous denitrification allows for organic matter, nitrogen and sulfide removal from domestic sewage (Saia *et al.*, 2016; Souza *et al.*, 2018). An extra advantage in using sulfide as an electron donor for denitrification downstream of an anaerobic reactor is the possibility to recover biogas with a low sulfide content (Souza *et al.*, 2018).

6.5 SRB-BASED BIOREMEDIATION TECHNIQUES

The versatility of SRB regarding operation conditions and use of electron donors has opened promising perspectives for environmental biotechnology. SRB do not require balanced growth with acetogens, which implies less sensitivity to organic overloads. Moreover, SRB are less sensitive to toxicants (section 6.2). Heavy metals are precipitated by sulfide, thus reducing their potential toxic effects (Hao, 2000). SRB can metabolize organic inhibitors such as aromatics (toluene, ethylbenzene), alkanes, chlorinated compounds (chloroform) and long chain fatty acids (Widdel, 1988).

6.5.1 Treatment of inorganic sulfate-rich wastewaters

Reactors based on organic matter removal by sulfate reduction lack one of the major advantages of methanogenic treatment: the recovery of methane from organic compounds. Nonetheless, the predominance of incomplete oxidizers among SRB leads to the accumulation of acetate in sulfate-reducing bioreactors (Bertolino *et al.*, 2014; Gallegos-Garcia *et al.*, 2009), which can be later used as substrate for acetoclastic MA. The acetate produced by SRB leads to residual COD in the bioreactor effluent, restricting its discharge in water bodies. The association between SRB and MA is an alternative to prevent acetate accumulation in SRB-based applications. Godoi *et al.* (2017) suggests that the association between sulfate-reduction and methane production occurs even when treating inorganic wastewater, such as acid mine drainage.

The treatment of sulfate-rich inorganic wastewaters requires addition of an electron donor, the cost of which can impair the system applicability (Liamleam & Annachhatre, 2007). Thus, keeping the COD/SO_4^{2-} ratio low is the more common approach when treating inorganic sulfate-rich wastewater (Cunha *et al.*, 2019; Gallegos-Garcia *et al.*, 2009; Janyasuthiwong *et al.*, 2015; Villa-Gomez *et al.*, 2011). However, the use of a COD/SO_4^{2-} ratio of 2 has allowed for high COD and sulfate removal efficiencies (>90%) and methane production rates (600 mL $CH_4 \cdot L^{-1}\ day^{-1}$) (Godoi *et al.*, 2017). It is not clear, however, whether methane recovery is an economic surplus offset against the additional cost for the electron donor supply to achieve higher COD/SO_4^{2-} ratios.

The high versatility of SRB regarding the use of different electron donors enables the use of several carbon sources in SRB-based bioreactors treating inorganic wastewaters (Table 6.8), including co-digestion with organic wastes. Wastes that have been tested as an electron donor in SRB-based systems include vinasses (Godoi *et al.*, 2017; Nogueira *et al.*, 2019), crude glycerol (Mora *et al.*, 2020), mixtures of different organic wastes (Waybrant *et al.*, 1998), and lignocellulosic materials (Reyes-Alvarado *et al.*, 2018). Although the focus is mainly on sulfate-reducing activities (Chapter 7), methane production was also achieved when using brewery spent grain as an electron donor in an SRB system, reaching nearly 40% biogas under both mesophilic (30°C) and psycrophilic (15°C) conditions (Costa *et al.*, 2020).

6.5.2 Heavy metal removal

Several sulfate-rich wastewaters, e.g. acid mine drainage, spoil leachates and landfill percolates, often are also contaminated with heavy metals (Hao, 2000). The potential of sulfidogenic bioreactors to remove heavy metals from waste streams are presented in detail in Chapter 7.

Table 6.8 Use of different electron donors in SRB systems used for the treatment of inorganic wastewaters.

Electron donor	COD/SO_4^{2-} Ratio	Sulfate-reducing Activity	Reference
Lactate	1.0	11,520 mg $SO_4^{2-} \cdot gVSS^{-1}\ day^{-1}$	Villa-Gomez *et al.* (2011)
Ethanol	4.6$^{(*)}$	211 mg $SO_4^{2-} \cdot gVSS^{-1}\ day^{-1}$	Nevatalo *et al.* (2010)
Ethanol + lactate (3:1)	1.7$^{(*)}$	2016 mg $SO_4^{2-} \cdot gVSS^{-1}\ day^{-1}$	Nevatalo *et al.* (2010)
Glycerol	2.3	167 mg $SO_4^{2-} \cdot gVSS^{-1}\ day^{-1}$	Bertolino *et al.* (2014)
Crude glycerol	8.5	7920 mg $SO_4^{2-}\ L^{-1}\ day^{-1(*)}$	Mora *et al.* (2020)
Synthetic vinasse	2.0	3500 mg $SO_4^{2-}\ L^{-1}\ day^{-1}$	Godoi *et al.* (2017)
Real vinasse	1.5	620 mg $SO_4^{2-}\ L^{-1}\ day^{-1(*)}$	Nogueira *et al.* (2019)
Filter paper	N.R	50–65 mg $SO_4^{2-} \cdot gVSS^{-1}\ day^{-1}$	Reyes-Alvarado *et al.* (2017b)
Scourer	N.R	34 mg $SO_4^{2-} \cdot gVSS^{-1}\ day^{-1}$	Reyes-Alvarado *et al.* (2018)
Cork	N.R	6.1 mg $SO_4^{2-} \cdot gVSS^{-1}\ day^{-1}$	Reyes-Alvarado *et al.* (2018)
Brewery spent grain	N.R	82.8 mg $SO_4^{2-} \cdot gVSS^{-1}\ day^{-1}$	Costa *et al.* (2020)

*Calculated with the data available in the paper; N.R. – not reported.

6.5.3 Biodegradation of xenobiotics

Many persistent xenobiotic compounds in the environment are nitro-substituted aromatics from pesticides, plastics, pharmaceuticals, industrial waste and military use of explosives (Tan & Field, 2000). The use of SRB-based systems is one of the main strategies for emerging organic contaminants removal (Ghattas *et al.*, 2017), due to the ability of SRB in utilizing a wide range of electron donors (Kaksonen & Puhakka, 2007; Liamleam & Annachhatre, 2007).

The use of SRB for degradation of recalcitrant compounds includes nitroaromatic compounds used in explosives (Boopathy *et al.*, 1998), benzene (Abu Laban *et al.*, 2009; Lovley *et al.*, 1995), naphthalene (Bergmann *et al.*, 2011), phenols (Mort & Dean-Ross, 1994), chlorinated compounds (Boopathy, 2002; Häggblom, 1998), and pharmaceutically active compounds, such as caffeine and naproxen (He *et al.*, 2018).

The presence of sulfonated groups in recalcitrant compounds can also be a source of sulfur compounds. The degradation of sulfonated compounds under anaerobic conditions can lead to release of sulfate, thus increasing the sulfate concentration in the effluent stream compared to the influent stream (Li *et al.*, 2015). The reduction of the released sulfate to sulfide can also cause inhibition. Methane production was impaired by sulfide toxicity in a bioreactor treating pharmaceutical wastewater (Li *et al.*, 2015). Sulfide concentrations higher than 20 mg S L^{-1} also impaired linear alkylbenzene sulfonate (LAS) degradation under anaerobic conditions in an Expanded Granular Sludge Bed (EGSB) reactor (Delforno *et al.*, 2015). Lower phenol degradation was attributed to sulfide inhibition in an anaerobic fixed-structured bed reactor operated under mixed sulfidogenic-methanogenic conditions, treating phenol-containing sugarcane vinasse (Godoi *et al.*, 2019). Biomass adaptation led to an enhancement of phenol degradation.

6.5.4 Micro-aerobic treatment of sulfate-rich wastewaters

Micro-aeration of anaerobic digesters enhances process stability (Nguyen & Khanal, 2018), either through boosting hydrolysis (Díaz *et al.*, 2011), leading to higher VFA production (Sawatdeenarunat *et al.*, 2017), allowing for complete

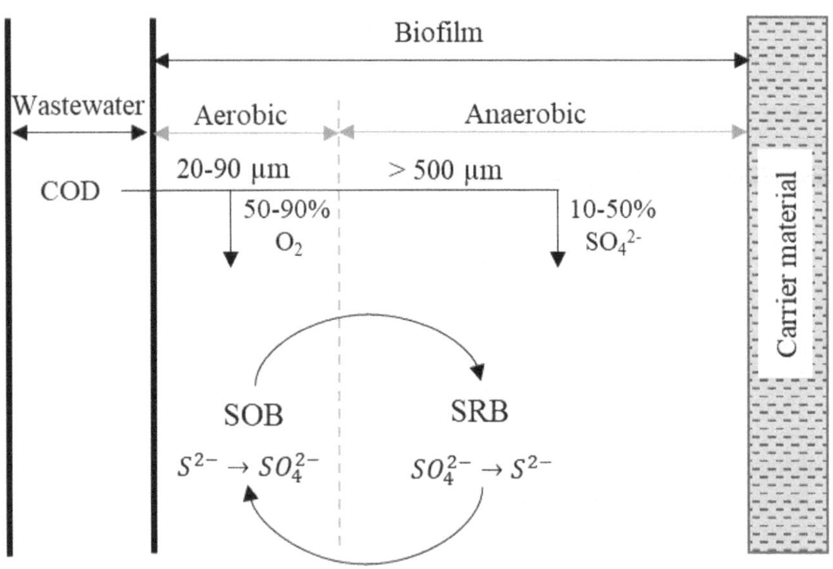

Figure 6.8 Schematic representation of the interaction between aerobic and anaerobic conditions in biofilm reactors (after Lens *et al.* 1995a, b). SOB: Sulfide-oxidizing bacteria, SRB: Sulfate-reducing bacteria.

mineralization of certain xenobiotics (Field *et al.*, 1995), or controlling sulfide (Krayzelova *et al.*, 2015). Indeed, the use of micro-aeration for sulfide control has already been applied in a full-scale anaerobic digester (Jeníček *et al.*, 2017; Chapter 11).

Micro-aerophilic conditions can be created in anaerobic reactors by dosing controlled amounts of oxygen to the sulfate-rich wastewaters. In an anaerobic reactor with immobilized biomass (either self-immobilized in granules or immobilized on a carrier material), biofilms allow for the establishment of a sulfur cycle, the so-called sufureta, comprising of anaerobic and aerobic processes (Figure 6.8).

The close interaction between aerobic and anaerobic conditions also prevails in highly loaded aerobic fixed film reactors (Kuenen *et al.*, 1986). Thus, a sulfur cycle at micro-scale can develop in biofilms growing on aerobic rotating biological contactors (Alleman *et al.*, 1982), trickling filters (Figure 6.8) or flow cells (Santegoeds *et al.*, 1998). Stimulation of the activity of SRB by decreasing the COD:SO_4^{2-} ratio of the wastewater from 20 to 0.5 resulted in a 50% reduction of the waste sludge production (Lens *et al.* 1995b). Practical applications to reduce the waste sludge production of aerobic bioreactors are limited by the lack of sulfate removal by these aerobic fixed film reactors.

6.6 INTEGRATION OF SULFATE REDUCTION IN RESOURCE RECOVERY TECHNOLOGIES

Research and technology development on anaerobic biological processes coupling sulfate-reducing bacteria, methanogenic archaea, and other syntrophic partners focused on wastewater treatment for many years (Colleran *et al.*, 1995; Damianovic & Foresti, 2009; Lens & Kuenen, 2001; Omil *et al.*, 1998; Weijma *et al.*, 2003). The resources depletion has elicited the main focus towards obtaining added-value products and closing cycles to achieve sustainability (Cherubini, 2010). In this context, anaerobic digestion can be designed to produce added-value products other than bio-methane (Kleerebezem *et al.*, 2015). Indeed, the versatility of sulfate-reducing bacteria makes this group central in several applications and the fundamentals established in the initial decades of research on biological sulfate-rich wastewater can be applied in other processes.

6.6.1 Bio-commodities

PHAs are among the added-value products that can be harvested from low-value feedstocks (Qian *et al.*, 2019). These compounds can be used to substitute oil-based plastics, and the integration between wastewater treatment and PHA is suggested as a feasible strategy to overcome the high costs of bioplastic production (Qian *et al.*, 2019; Salehizadeh & van Loosdrecht, 2004). There are several sulfate-reducing bacteria (SRB) that are able to accumulate PHA to an extent comparable to commercial PHA-producing strains (Hai *et al.*, 2004). The use of SRB for PHA in biotechnological applications has, however, not yet been further explored.

Chain-elongation is a process in which short-chain fatty acids and lactate are converted to medium-chain fatty acids (Kleerebezem *et al.*, 2015). These compounds are economically attractive, more easily removed from the aqueous solution due to their increased apolarity, and represent a new opportunity for anaerobic digestion applications (Kleerebezem *et al.*, 2015; Nzeteu *et al.*, 2018). Acetate is one of the main substrates for chain elongation (Grootscholten *et al.*, 2013; Steinbusch *et al.*, 2011), and also the by-product of incomplete sulfate oxidizers, which are the most common SRB (Muyzer & Stams, 2008). For this reason, acetate accumulation is reported in several anaerobic systems, especially when treating sulfate-rich wastewater at low COD/SO_4^{2-} ratios (Damianovic & Foresti, 2009; Godoi *et al.*, 2017). Acetate accumulated in sulfate-rich wastewater treatment could be used in elongation processes, thus allowing the production of value-added products from sulfate-rich wastewaters.

6.6.2 Bio-electricity

Other novel developments include combining anaerobic digestion with bioelectrochemical systems (BESs). The association between the two technologies is deemed as a reasonable strategy in the transition towards a bio-based economy (De Vrieze *et al.*, 2018). However, the H_2S formation during anaerobic digestion can poison the combined heat and power (CHP) unit (Kleerebezem *et al.*, 2015) or the BES (Elzinga *et al.*, 2020). Different approaches have been suggested to prevent these problems, such as sulfate recovery prior to methanogenesis (Zhang & Angelidaki, 2015) or the oxidation of sulfide to elemental sulfur (Sun *et al.*, 2009). More details on the role of SRB on BES performance can be found in Chapter 8.

6.6.3 Biomining and nanoparticles biosynthesis

Novel applications for sulfate reducing-bacteria also include biogenic sulfide production for chemical precipitation of metal sulfides in an urban mining context (Işıldar *et al.*, 2019). In this case, recycling metals from secondary sources (such as electronic waste) can include a precipitation with a biogenic sulfide unit, thus allowing selective recovery of base (Sampaio *et al.*, 2010; Santos & Johnson, 2018) and precious (Nanusha *et al.*, 2019) metals, as well as rare earth elements (Nogueira *et al.*, 2019). Recently, there is also interest in the microbiologically produced sulfide nanoparticles, e.g. CdS, that can be use as photocalyst (Jacob *et al.*, 2019) or to extend the metabolic properties of non-photosyntetic microorganisms (Sakimoto *et al.*, 2016).

REFERENCES

Abbott T. and Eskicioglu C. (2015). Effects of metal salt addition on odor and process stability during the anaerobic digestion of municipal waste sludge. *Waste Management*, **46**, 449–458.

Abu Laban N., Selesi D., Jobelius C. and Meckenstock R. U. (2009). Anaerobic benzene degradation by Gram-positive sulfate-reducing bacteria. *FEMS Microbiology Ecology*, **68**, 300–311.

Alleman J. E., Veil J. A. and Canaday J. T. (1982). Scanning electron microscope evaluation of rotating biological biofilm. *Water Research*, **16**, 543–550.

Allen K. D., Miller D. V., Rauch B. J., Perona J. J. and White R. H. (2015). Homocysteine is biosynthesized from aspartate semialdehyde and hydrogen sulfide in methanogenic archaea. *Biochemistry*, **54**, 3129–3132.

Aeckersberg F., Rainey F. A. and Widdel F. (1998). Growth, natural relationships, cellular fatty acids and metabolic adaptation of sulfate-reducing bacteria that utilize long-chain alkanes under anoxic conditions. *Archives of Microbiology*, **170**, 361–369.

Athanassopoulos N., Kouinis J., Papadimitriou A. and Koutinas A. A. (1989). Effect of high sulfite contents on anaerobic digestion of raisin-finishing wastewater. *Biomedical Wastes*, **30**, 53–60.

Balderston W. L. and Payne W. J. (1976). Inhibition of methanogenesis in salt marsh sediments and whole cell suspensions of methanogenic bacteria by nitrogen oxides. *Applied and Environmental Microbiology*, **32**, 264–269.

Barber W. P. and Stuckey D. C. (2000). Effect of sulfate reduction on Chemical Oxygen Demand removal in an Anaerobic Baffled Reactor. *Water Environment Research*, **72**, 593–601.

Bergmann F. D., Selesi D. and Meckenstock R. U. (2011). Identification of new enzymes potentially involved in anaerobic naphthalene degradation by the sulfate-reducing enrichment culture N47. *Archives of Microbiology*, **193**, 241–250.

Beristain-Cardoso R., Sierra-Alvarez R., Rowlette P., Razo-Flores E., Gómez J. and Field J. (2006). Sulfide oxidation under chemolithoautotrophic denitrifying conditions. *Biotechnology and Bioengineering*, **95**, 1149–1157.

Bertolino S. M., Melgaço L. A., Sá R. G. and Leão V. A. (2014). Comparing lactate and glycerol as a single-electron donor for sulfate reduction in fluidized bed reactors. *Biodegradation*, **25**, 719–733.

Bhattarai S., Cassarini C. and Lens P. N. L. (2019). Physiology and distribution of anaerobic oxidation of methane by archaeal methanotrophs. *Microbiology and Molecular Biology Reviews*, **83**, 9–62.

Boopathy R. (2002). Anaerobic biotransformation of carbon tetrachloride under various electron acceptor conditions. *Bioresource Technology*, **84**, 69–73.

Boopathy R., Kulpa C. F. and Manning J. (1998). Anaerobic biodegradation of explosives and related compounds by sulfate-reducing and methanogenic bacteria: A review. *Bioresource Technology*, **63**, 81–89.

Braga A. F. M., Pereira M. B. O. C., Zaiat M., Silva G. H. R. and Fermoso F. (2018). Screening of trace metal supplementation for black water anaerobic digestion. *Environmental Technology*, **39**, 1776–1785.

British Columbia Ministry of Environment. (2013). Ambient water quality guidelines for sulfate. Victoria, BC, Canada. Available from: http://www.env.gov.bc.ca/wat/wq/BCguidelines/sulphate/pdf/sulphate_final_guideline.pdf

Buisman C. J. N., Geraats B. G., Ijspeert P. and Lettinga G. (1990). Optimization of sulphur production in a biotechnological sulphide-removing reactor. *Biotechnology and Bioengineering*, **35**, 50–56. +

Camiloti P. R., Mockaitis G., Rodrigues J. A. D., Damianovic M. H. Z., Foresti E. and Zaiat M. (2013). Innovative anaerobic bioreactor with fixed-structured bed (ABFSB) for simultaneous sulfate reduction and organic matter removal. *Journal of Chemical Technology and Biotechnology*, **89**, 1044–1050.

Camiloti P. R., Oliveira G. H. D. and Zaiat M. (2016). Sulfur recovery from wastewater using a micro-aerobic external silicone membrane reactor (ESMR). *Water, Air, & Soil Pollution*, **227**, 31.

Cao J., Zhang G., Mao Z., Fang Z., Yang C. and Han B. (2009). Influence of Mg^{2+} on the growth and activity of sulfate reducing bacteria. *Hydrometallurgy*, **95**, 127–134.

Cao J., Zhang G., Mao Z. S., Li Y., Fang Z. and Yang C. (2012). Influence of electron donors on the growth and activity of sulfate-reducing bacteria. *International Journal of Mineral Processing*, **106–109**, 58–64.

Celis-García M. L. B., Ramírez F., Revah S., Razo-Flores E. and Monroy O. (2004). Sulphide and oxygen inhibition over the anaerobic digestion of organic matter: Influence of microbial immobilization type. *Environmental Technology*, **25**, 1265–1275.

Cetecioglu Z., Dolfing J., Taylor J., Purdy K. J. and Eyice Ö. (2019). COD/sulfate ratio does not affect the methane yield and microbial diversity in anaerobic digesters. *Water Research*, **155**, 444–454.

Chen Y., Cheng J. J. and Creamer K. S. (2008). Inhibition of anaerobic digestion process: A review. *Bioresource Technology*, **99**, 4044–4064.

Chernicharo C. A. L., van Lier J. B., Noyola A. and Bressani Ribeiro T. (2015). Anaerobic sewage treatment: state of the art, constraints and challenges. *Reviews in Environmental Science and Biotechnology*, **14**, 649–679.

Cherubini F. (2010). The biorefinery concept: Using biomass instead of oil for producing energy and chemicals. *Energy Conversion and Management*, **51**, 1412–1421.

Clancy P. B., Venkataraman N. and Lynd L. R. (1992). Biochemical inhibition of sulfate reduction in batch and continuous anaerobic digesters. *Water Science and Technology*, **25**, 51–60.

Colleran E. and Pistilli A. (1994). Activity test system for determining the toxicity of xenobiotic chemicals to the methanogenic process. *Ann Microbiol Enzymol*, **44**, 1–20.

Colleran E., Concannon F., Golden T., Geoghan F., Crumlish B., Killilea J. E., Henry M. and Coates J. (1992). Use of methanogenic activity tests to characterise anaerobic sludges, screen for anaerobic biodegradability and determine toxicity thresholds against individual anaerobic trophic groups and species. *Water Science and Technology*, **25**, 31–40.

Colleran E., Finnegan S. and O'Keeffe R. B. (1994). Anaerobic digestion of high sulphate-containing wastewater from the industrial production of citric acid. *Water Science and Technology*, **30**, 263–273.

Colleran E., Finnegan S. and Lens P. N. L. (1995). Anaerobic treatment of sulphate-containing waste streams. *Antonie Van Leeuwenhoek*, **67**, 29–46.

Costa R. B., Bevilaqua D. and Lens P. N. L. (2020). Pre-treatment and temperature effects on the use of slow release electron donor for biological sulphate reduction. *Journal of Environmental Management*, **275**, In Press.

Cunha M. P., Ferraz R. M., Sancinetti G. P. and Rodriguez R. P. (2019). Long-term performance of a UASB reactor treating acid mine drainage: effects of sulfate loading rate, hydraulic retention time, and COD/SO_4^{2-} ratio. *Biodegradation*, **30**, 47–58.

Damianovic M. H. R. Z. and Foresti E. (2009). Dynamics of sulfidogenesis associated to methanogenesis in horizontal-flow anaerobic immobilized biomass reactor. *Process Biochemistry*, **44**, 1050–1054.

Damianovic M. H. R. Z., Godoi L. A. G., Saia F. T. and Foresti E. (2018). Horizontal-flow anaerobic immobilized biomass (HAIB) reactor for organic matter and sulfate removal from paper recycling plant wastewater with simultaneous conversion of sulfide into elemental sulfur. *Journal of Environmental Chemical Engineering*, **6**, 964–969.

Daniels L., Belay N. and Rajagopal B. S. (1986). Assimilatory Reduction of Sulfate and Sulfite by Methanogenic Bacteria. *Applied and Environmental Microbiology*, **51**, 703–709.

Dar S. A., Stams A. J. M., Kuenen J. G. and Muyzer G. (2007). Co-existence of physiologically similar sulfate-reducing bacteria in a full-scale sulfidogenic bioreactor fed with a single organic electron donor. *Applied Microbiology and Biotechnology*, **75**, 1463–1472.

Dar S. A., Kleerebezem R., Stams A. J. M., Kuenen J. G. and Muyzer G. (2008). Competition and coexistence of sulfate-reducing bacteria, acetogens and methanogens in a lab-scale anaerobic bioreactor as affected by changing substrate to sulfate ratio. *Applied Microbiology and Biotechnology*, **78**, 1045–1055.

De Smul A. and Verstraete W. (1999a). Retention of Sulfate-Reducing Bacteria in Expanded Granular-Sludge-Blanket Reactors. *Water Environment Research*, **71**, 427–431.

De Smul A. and Verstraete W. (1999b). The phenomenology and the mathematical modeling of the silicone-supported chemical oxidation of aqueous sulfide to elemental sulfur by ferric sulphate. *Journal of Chemical Technology and Biotechnology*, **74**, 456–466.

Delforno T. P., Moura A. G. L., Okada D. Y., Sakamoto I. K. and Varesche M. B. A. (2015). Microbial diversity and the implications of sulfide levels in an anaerobic reactor used to remove an anionic surfactant from laundry wastewater. *Bioresource Technology*, **192**, 37–45.

Demirel B. and Yenigün O. (2002). Two-phase anaerobic digestion processes: A review. *Journal of Chemical Technology and Biotechnology*, **77**, 743–755.

De Vrieze J., Saunders A. M., He Y., Fang J., Nielsen P. H., Verstraete W. and Boon N. (2015). Ammonia and temperature determine potential clustering in the anaerobic digestion microbiome. *Water Research*, **75**, 312–323.

De Vrieze J. and Verstraete W. (2016). Perspectives for microbial community composition in anaerobic digestion: from abundance and activity to connectivity. *Environmental Microbiology*, **18**, 2797–2809.

De Vrieze J., Arends J. B. A., Verbeeck K., Gildemyn S. and Rabaey K. (2018). Interfacing anaerobic digestion with (bio)electrochemical systems: Potentials and challenges. *Water Research*, **146**, 244–255.

Díaz I., Donoso-Bravo A. and Fdz-Polanco M (2011). Effect of microaerobic conditions on the degradation kinetics of cellulose. *Bioresource Technology*, **102**, 10139–10142.

Di Capua F., Pirozzi F., Lens P. N. L. and Esposito G. (2019). Electron donors for autotrophic denitrification. *Chemical Engineering Journal*, **280**, 643–657.

Eikelboom D. H. (2000). Sulfur-storing bacteria and bulking of activated sludge. In: Environmental technologies to treat sulfur pollution: principles and engineering, P. N. L. Lens and L. Hulshoff Pol (ed.), IWA Publishing, London, pp. 449–466.

Eis B. J., Ferguson J. F. and Benjamin M. M. (1983). The fate and the effect of bisulfate in anaerobic treatment. *Journal Water Pollution Control Federation*, **55**, 1355–1365.

Elphick J. R., Davies M., Gilron G., Canaria E. C., Lo B. and Bailey H. C. (2011). An aquatic toxicological evaluation of sulfate: The case for considering hardness as a modifying factor in setting water quality guidelines. *Environmental Toxicology and Chemistry*, **30**, 247–253.

Elzinga M., Liu D., Klok J. B. M, Roman P., Buisman C. J. N. and Heijne A. (2020). Microbial reduction of organosulfur compounds at cathodes in bioelectrochemical systems. *Environmental Science & Technology*, **1**, 100009.

Fajardo C., Mora M., Fernández I., Mosquera-Coral A., Campos J. L. and Méndez R. (2014). Cross effect of temperature, pH and free ammonia on autotrophic denitrification process with sulphide as electron donor. *Chemosphere*, **97**, 10–15.

Fdz.-Polanco M., Díaz I., Pérez S. I., Lopes A. C. and Fdz-Polanco F. (2009). Hydrogen sulphide removal in the anaerobic digestion of sludge by micro-aerobic processes: Pilot plant experience. *Water Science and Technology*, **60**, 3045–3050.

Ferguson J. F., Eis B. J. and Benjamin M. M. (1982). Neutralisation in anaerobic treatment of an acidic waste. In Preprints presented at the IAWPRC specialised seminar "Anaerobic Treatment of Wastewater in Fixed Film Reactors", Copenhagen, Denmark, 16-18 June, 339–355.

Fermoso F. G., Bartacek J., Chung L. C. and Lens P. N. L. (2008). Supplementation of cobalt to UASB reactors by pulse dosing: $CoCl_2$ versus $CoEDTA^{2-}$ pulses. *Biochemical Engineering Journal*, **42**, 111–119.

Field J. A., Stams A. J. M., Kato M. and Schraa G. (1995). Enhanced biodegradation of aromatic pollutants in cocultures of anaerobic and aerobic bacterial consortia. *Antonie Van Leeuwenhoek*, **67**, 47–77.

Florentino A. P., Costa R. B., Hu Y., O'Flaherty V. and Lens P. N. L. Long chain fatty acid degradation coupled to biological sulfidogenesis: a prospect for enhanced metal recovery. *Frontiers in Bioengineering and Biotechnology*. In Press.

Foresti E., Zaiat M. and Vallero M. (2006). Anaerobic processes as the core technology for sustainable domestic wastewater treatment: Consolidated applications, new trends, perspectives, and challenges. *Reviews in Environmental Science and Biotechnology*, **5**, 3–19.

Fox P. and Venkatasubbiah V. (1996). Coupled anaerobic/aerobic treatment of high-sulfate wastewater with sulfate reduction and biological sulfide oxidation. *Water Science and Technology*, **34**, 359–366.

Gallegos-Garcia M., Celis L. B., Rangel-Mendez R. and Razo-Flores E (2009). Precipitation and recovery of metal sulfides from metal containing acidic wastewater in a sulfidogenic down-flow fluidized bed reactor. *Biotechnology and Bioengineering*, **102**, 91–99.

Gao M., Guo B., Zhang L., Zhang Y., Yu N. and Liu Y. (2020). Biomethane recovery from source-diverted household blackwater: Impacts from feed sulfate. *Process Safety and Environmental Protection*, **136**, 28–38.

Garcia G. P. P, Diniz R. C. O, Bicalho S. K., Franco V., Pereira A. D., Brandt E. F., Etchebehere C., Chernicharo C. A. L. and Araujo J. C. (2017). Microbial community and sulphur behaviour in phototrophic reactors treating UASB effluent under different operational conditions. *International Biodeterioration and Biodegradation*, **119**, 486–498.

Garrels R. M. and Christ C. L. (1965). Solutions, minerals and equilibria, Harper and Row, New York.

Ghattas A. K., Fischer F., Wick A. and Ternes T. A. (2017). Anaerobic biodegradation of (emerging) organic contaminants in the aquatic environment. *Water Research*, **116**, 268–295.

Gil-Garcia C., Godoi L. A. G., Fuess L. T. and Damianovic M. H. R. Z. (2018). Performance improvement of a thermophilic sulfate-reducing bioreactor under acidogenic conditions: Effects of diversified operating strategies. *Journal of Environmental Management*, **207**, 303–312.

Glória R. M., Motta T. M., Silva P. V. O., Costa P., Brandt E. M. F., Souza C. L. and Chernicharo C. A. L. (2016). Stripping and dissipation techniques for the removal of dissolved gases from anaerobic effluents. *Brazilian Journal of Chemical Engineering*, **33**, 713–721.

Godoi L. A. G., Damianovic M. H. R. Z. and Foresti E. (2015). Sulfidogenesis interference on methane production from carbohydrate-rich wastewater. *Water Science and Technology*, **72**, 1644–1652.

Godoi L. A. G., Foresti E. and Damianovic M. H. R. Z. (2017). Down-flow fixed-structured bed reactor: An innovative reactor configuration applied to acid mine drainage treatment and metal recovery. *Journal of Environmental Management*, **197**, 597–604.

Godoi L. A. G., Fuess L. T., Delforno T. P., Foresti E. and Damianovic M. H. R. Z. (2019). Characterizing phenol-removing consortia under methanogenic and sulfate-reducing conditions: potential metabolic pathways. *Environmental Technology*, **40**, 3216–3226.

Gommers P. J. F., Bijleveld W. and Kuenen J. F. (1988). Simultaneous sulfide and acetate oxidation in a denitrifying fluidized bed reactor I – start-up and reactor performance. *Water Research*, **22**, 1075–1083.

Gonçalves M. M. M., Leite S. G. F. and Sant'Anna G. L. (2005). The bio activation procedure for increasing the sulphate-reducing bacteria in a UASB reactor. *Brazilian Journal of Chemical Engineering*, **22**, 565–571

Grootscholten T. I. M., Steinbusch K. J. J., Hamelers H. V. M. and Buisman C. J. N. (2013). Improving medium chain fatty acid productivity using chain elongation by reducing the hydraulic retention time in an upflow anaerobic filter. *Bioresource Technology*, **136**, 735–738.

Guo G., Wu D., Ekama G. A., Ivleva N. P., Hao X., Dai J., Cui Y., Kumar Biswal B. and Chen G. (2019). Investigation of multiple polymers in a denitrifying sulfur conversion-EBPR system: The structural dynamics and storage states. *Water Research*, **156**, 179–187.

Gupta A., Flora J. R. V., Gupta M., Sayles G. D. and Suidan M. T. (1994). Methanogenesis and sulfate reduction in chemostats-I. Kinetic studies and experiments. *Water Research*, **28**, 781–793.

Gustavsson J., Yekta S. S., Karlsson A., Skyllberg U. and Svensson B. H. (2013). Potential bioavailability and chemical forms of Co and Ni in the biogas process-An evaluation based on sequential and acid volatile sulfide extractions. *Engineering in Life Sciences*, **13**, 572–579.

Gutierrez O., Park D., Sharma K. R. and Yuan Z. (2009). Effects of long-term pH elevation on the sulfate-reducing and methanogenic activities of anaerobic sewer biofilms. *Water Research*, **43**, 2549–2557.

Häggblom M. M. (1998). Reductive dechlorination of halogenated phenols by a sulfate-reducing consortium. *FEMS Microbiology Ecology*, **26**, 35–41.

Hai T., Lange D., Rabus R. and Steinbüchel A. (2004). Polyhydroxyalkanoate (PHA) accumulation in sulfate-reducing bacteria and identification of a class III PHA synthase (PhaEC) in *Desulfococcus multivorans*. *Applied and Environmental Microbiology*, **70**, 4440–4448.

Hao O. J. (2000). Metal effects on sulfur cycle bacteria and metal removal by sulfate reducing bacteria. In: Environmental technologies to treat sulfur pollution: principles and engineering, P. N. L. Lens and L. Hulshoff Pol (ed.), IWA Publishing, London, pp. 393–414.

Hao O. J., Huang L., Chen J. M. and Buglass R. L. (1994). Effects of metal additions on sulfate reduction activity in wastewaters. *Toxicological and Environmental Chemistry*, **46**, 197–212.

Hao T. W., Xiang P. Y., Mackey H. R., Chi K., Lu H., Chui H. K., van Loosdrecht M. C. M. and Chen G. H. (2014). A review of biological sulfate conversions in wastewater treatment. *Water Research*, **65**, 1–21.

Hao T., Mackey H. R., Guo G., Liu R. and Chen G. (2016). Resilience of sulfate-reducing granular sludge against temperature, pH, oxygen, nitrite, and free nitrous acid. *Applied Microbiology and Biotechnology*, **100**, 8563–8572.

He Y., Sutton N. B., Rijnaarts H. H. M. and Langenhoff A. A. M. (2018). Pharmaceutical biodegradation under three anaerobic redox conditions evaluated by chemical and toxicological analyses. *Science of the Total Environment*, **618**, 658, 664.

Henstra A. M., Dijkema C. and Stams A. J. M. (2007). *Archaeoglobus fulgidus* couples CO oxidation to sulfate reduction and acetogenesis with transient formate accumulation. Environmental Microbiology, **9**, 1836–1841.

Hoeks F. W. J. M. M., Ten Hoopen H. J. G., Roels J. A. and Kuenen J. G. (1984). Anaerobic treatment of acid water (methane production in a sulfate rich environment). *Progress in Industrial Microbiology*, **20**, 113–119.

Hulshoff Pol L. W., Lens P. N. L., Stams A. J. M. and Lettinga G. (1998). Anaerobic treatment of sulphate-rich wastewaters. *Biodegradation*, **9**, 213–224.

Irwin S. V., Fisher P., Graham E., Malek A. and Robidoux A. (2017). Sulfites inhibit the growth of four species of beneficial gut bacteria at concentrations regarded as safe for food. *PLoS One*, **12**, 1–14.

Isa M. H. and Anderson G. K. (2005). Molybdate inhibition of sulphate reduction in two-phase anaerobic digestion. *Process Biochemistry*, **40**, 2079–2089.

Isa Z., Grusenmeyer S. and Verstraete W. (1986). Sulfate reduction relative to methane production in high-rate anaerobic digestion: Microbiological aspects. *Applied and Environmental Microbiology*, **51**, 572–579.

Işıldar A., van Hullebusch E. D., Lenz M., Du Laing G., Marra A., Cesaro A., Panda S., Akcil A., Kucuker M. A. and Kuchta K. (2019). Biotechnological strategies for the recovery of valuable and critical raw materials from waste electrical and electronic equipment (WEEE) – A review. *Journal of Hazardous Materials*, **362**, 467–481.

Jacob J. M., Rajan R, Tom T. C, Kumar V. S., Kurup G. G., Shanmuganathan R. and Pugazhendhi A. (2019). Biogenic design of ZnS quantum dots – Insights into their in-vitro cytotoxicity, photocatalysis and biosensing properties. *Ceramics International*, **45**, 24193–24201.

Janssen A. J. H., Sleyster R., van der Kaa C., Jochemsen A., Bontsema J. and Lettinga G. (1995). Biological sulphide oxidation in a fed-batch reactor. *Biotechnology and Bioengineering*, **47**, 327–333.

Janssen A. J. H, Ma S. C., Lens P. N. L. and Lettinga G. (1997). Performance of a sulfide-oxidizing expanded-bed reactor supplied with dissolved oxygen. *Biotechnology and Bioengineering*, **53**, 32–40.

Janssen A. J. H., Lens P. N. L., Stams A. J. M., Plugge C. M., Sorokin D. Y., Muyzer G., Dijkman H., Van Zessen E., Luimes P. C. and Buisman J. N. (2009). Application of bacteria involved in the biological sulfur cycle for paper mill effluent purification. *Science of the Total Environment*, **407**(4), 1333–1343.

Janyasuthiwong S., Rene E. R., Esposito G. and Lens P. N. L. (2015). Effect of pH on Cu, Ni and Zn removal by biogenic sulfide precipitation in an inversed fluidized bed bioreactor. *Hydrometallurgy*, **158**, 94–100.

Jeison D., Kremer B. and van Lier J. B. (2008). Application of membrane enhanced biomass retention to the anaerobic treatment of acidified wastewaters under extreme saline conditions. *Separation and Purification Technology*, **64**, 198–205.

Jeníček P., Horejš J., Pokorná-Krayzelová L., Bindzar J. and Bartáček J. (2017). Simple biogas desulfurization by microaeration – Full scale experience. *Anaerobe*, **46**, 41–45.

Jensen A. B. and Webb C. (1995). Treatment of H_2S-containing gases. A review of microbiological alternatives. *Enzyme and Microbial Technology*, **17**, 2–10.

Jung H., Kim J. and Lee C. (2019). Temperature effects on methanogenesis and sulfidogenesis during anaerobic digestion of sulfur-rich macroalgal biomass in sequencing batch reactors. *Microorganisms*, **7**, 682, 1–16. doi:10.3390/microorganisms7120682.

Kaksonen A. H. and Puhakka J. A. (2007). Sulfate reduction based bioprocesses for the treatment of acid mine drainage and the recovery of metals. *Engineering in Life Sciences*, **7**, 541–564.

Kappler U. and Enemark J. H. (2015). Sulfite-oxidizing enzymes. *Journal of Biological Inorganic Chemistry*, **20**, 253–264.

Karhadkar P. P., Audic J. M., Faup G. M. and Khanna P. (1987). Sulfide and sulfate inhibition of methanogenesis. *Water Research*, **21**, 1061–1066.

Kaksonen A. H., Plumb J. J., Franzmann P. D. and Puhakka J. A. (2004). Simple organic electron donors support diverse sulfate-reducing communities in fluidized-bed reactors treating acidic metal- and sulfate-containing wastewater. *FEMS Microbiology Ecology*, **47**, 279–289.

Kent R. D. (1973). Inert gas stripping of contaminated water. US Patent **222**, 480.

Kiran M. G., Pakshirajan K. and Das G. (2017). Heavy metal removal from multicomponent system by sulfate reducing bacteria: Mechanism and cell surface characterization. *Journal of Hazardous Materials*, **324**, 62–70.

Kiyuna L. S. M., Fuess L. T. and Zaiat M. (2017). Unraveling the influence of the COD/sulfate ratio on organic matter removal and methane production from the biodigestion of sugarcane vinasse. *Bioresource Technology*, **232**, 103–112.

Kleerebezem R., Joosse B., Rozendal R. and Van Loosdrecht M. C. M. (2015). Anaerobic digestion without biogas? *Reviews in Environmental Science and Biotechnology*, **14**, 787–801.

Koster I. W., Rinzema A., de Vegt A. L. and Lettinga G. (1986). Sulfide inhibition of the methanogenic activity of granular sludge at various pH-levels. *Water Research*, **20**, 1561–1567.

Krayzelova L., Bartacek J., Díaz I., Jeison D., Volcke E. I. P. and Jenicek P. (2015). Microaeration for hydrogen sulfide removal during anaerobic treatment: a review. *Reviews in Environmental Science and Biotechnology*, **14**, 703–725.

Kuenen J. G., Jorgenson B. B. and Revsbech N. P. (1986). Oxygen microprofiles of trickling filter biofilms. *Water Research*, **20**, 1589–1598.

Lagas, J. (2000). Survey of H_2S and SO_2 removal processes. In: Environmental technologies to treat sulfur pollution: Principles and engineering, P. N. L. Lens and L. W. Hulshoff Pol (eds), IWA Publishing, London, 237–264.

Lamb J. J., Bernard O., Sarker S., Lien K. M. and Hjelme D. R. (2019). Perspectives of optical colourimetric sensors for anaerobic digestion. *Renewable and Sustainable Energy Reviews*, **111**, 87–96.

Langer S. G., Ahmed S., Einfalt D., Bengelsdorf F. R. and Kazda M. (2015). Functionally redundant but dissimilar microbial communities within biogas reactors treating maize silage in co-fermentation with sugar beet silage. *Microbial Biotechnology*, **8**, 828–836.

Lefebvre O., Vasudevan N., Torrijos M., Thanasekaran K. and Moletta R. (2006). Anaerobic digestion of tannery soak liquor with an aerobic post-treatment. *Water Research*, **40**, 1492–1500.

Lens P. N. L. and Kuenen J. G. (2001). The biological sulfur cycle: Novel opportunities for environmental biotechnology. *Water Science and Technology*, **44**, 57–66.

Lens P. N. L., Massone A., Rozzi A. and Verstraete W. (1995a). Effect of sulfate concentration and scraping on aerobic fixed film reactors. *Water Research*, **29**, 857–870.

Lens P. N. L., De Poorter M. P., Cronenberg C. C. and Verstraete W. H. (1995b). Sulfate reducing and methane producing bacteria in aerobic wastewater treatment. *Water Research*, **29**, 871–880.

Lens P. N. L., Visser A., Janssen A. J. H., Hulshoff Pol L. W. and Lettinga G. (1998). Biotechnological Treatment of Sulfate-Rich Wastewaters. *Critical Reviews in Environmental Science and Technology*, **28**, 41–88.

Lens P. N. L., Korthout D., van Lier J. B., Hulshoff Pol L. W. and Lettinga G. (2001). Effect of the liquid upflow velocity on thermophilic sulphate reduction in acidifying granular sludge reactors. *Environmental Technology* (United Kingdom), **22**, 183–193.

Lettinga G. (1995). Anaerobic digestion and wastewater treatment systems. *Antonie Van Leeuwenhoek*, **67**, 3–28.

Li W., Niu Q., Zhang H., Zhe T., Zhang Y., Gao Y., Li Y. Y., Nishimura O. and Yang M. (2015). UASB treatment of chemical synthesis-based pharmaceutical wastewater containing rich organic sulfur compounds and sulfate and associated microbial characteristics. *Chemical Engineering Journal*, **260**, 55–63.

Li J., Shi W., Jiang C., Bai L., Wang T., Yu J. and Ruan W. (2018). Evaluation of potassium as promoter on anaerobic digestion of saline organic wastewater. *Bioresource Technology*, **266**, 68–74.

Liamleam W. and Annachhatre A. P. (2007). Electron donors for biological sulfate reduction. *Biotechnology Advances*, **25**, 452–463.

Liu Y., Beer L. L. and Whitman W. B. (2012). Sulfur metabolism in archaea reveals novel processes. *Environmental Microbiology*, **14**, 2632–2644.

Liu J., Sun F., Wang L., Ju X., Wu W. and Chen Y. (2014). Molecular characterization of a microbial consortium involved in methane oxidation coupled to denitrification under micro-aerobic conditions. *Microbial Biotechnology*, **7**, 64–76.

Liu Z. H., Yin H., Lin Z. and Dang Z. (2018). Sulfate-reducing bacteria in anaerobic bioprocesses: basic properties of pure isolates, molecular quantification, and controlling strategies. *Environmental Technology Reviews*, **7**, 46–72.

Lopes S. I. C., Sulistyawati I., Capela M. I. and Lens P. N. L. (2007). Low pH (6, 5 and 4) sulfate reduction during the acidification of sucrose under thermophilic (55°C) conditions. *Process Biochemistry*, **42**, 580–591.

Lopes S. I. C., Capela M. I. and Lens P. N. L. (2010). Sulfate reduction during the acidification of sucrose at pH 5 under thermophilic (55°C) conditions. II: Effect of sulfide and COD/SO_4^{2-} ratio. *Bioresource Technology*, **101**, 4278–4284.

Lovley D. R., Coates J. D., Woodward J. C. and Phillips E. J. P. (1995). Benzene oxidation coupled to sulfate reduction. *Applied and Environmental Microbiology*, **61**, 953–958.

Lübken M., Wichern M., Schlattmann M., Gronauer A. and Horn H. (2007). Modelling the energy balance of an anaerobic digester fed with cattle manure and renewable energy crops. *Water Research*, **41**, 4085–4096.

Maaskant W. and Hobma S. W. (1981). The influence of sulfite on anaerobic treatment. *H_2O*, **14**(25), 596–598.

Madden P., Al-Raei A. M., Enright A. M., Chinalia F. A., de Beer D., O'Flaherty V. and Collins G. (2014). Effect of sulfate on low-temperature anaerobic digestion. *Frontiers in Microbiology*, **5**, 1–15.

Maillacheruvu K. Y., Parkin G. F., Peng C. Y., Kuo W. C., Oonge Z. I. and Lebduschka V. (1993). Sulfide toxicity in anaerobic systems fed sulfate and various organics. *Water Environment Research*, **65**, 100–109.

McCartney D. M. and Oleszkiewicz J. A. (1991). Sulfide inhibition of anaerobic degradation of lactate and acetate. *Water Research*, **25**, 203–209.

Middleton A. C. and Lawrence W. M. A. (1977). Kinetics of microbial sulfate reduction. *Journal Water Pollution Control Federation*, **49**, 1659–1670.

Mohan S. V, Rao N. C., Prasad K. K and Sarma P. N. (2005). Bioaugmentation of an anaerobic sequencing batch biofilm reactor (AnSBBR) with immobilized sulphate reducing bacteria (SRB) for the treatment of sulphate bearing chemical wastewater. *Process Biochemistry*, **40**, 2849–2857.

Montebello A. M., Fernández M., Almenglo F., Ramírez M., Cantero D., Baeza M. and Gabriel D. (2012). Simultaneous methylmercaptan and hydrogen sulfide removal in the desulfurization of biogas in aerobic and anoxic biotrickling filters. *Chemical Engineering Journal*, **200–202**, 237–246.

Mora M., Lafuente J. and Gabriel D (2020). Influence of crude glycerol load and pH shocks on the granulation and microbial diversity of a sulfidogenic Upflow Anaerobic Sludge Blanket reactor. *Process Safety and Environmental Protection*, **133**, 159–168.

Morgan-Sagastume J. M. and Noyola A. (2006). Hydrogen sulfide removal by compost biofiltration: Effect of mixing the filter media on operational factors. *Bioresource Technology*, **97**, 1546–1553.

Mort S. L. and Dean-Ross D. (1994). Biodegradation of phenolic compounds by sulfate-reducing bacteria from contaminated sediments. *Microbial Ecology*, **28**, 67–77.

Mulder A. (1984). The effects of high sulfate concentrations on the methane fermentation of waste water. *Progress in Industrial Microbiology*, **20**, 133–143.

Muñoz Sierra J. D., Lafita C., Gabaldón C., Spanjers H. and van Lier J. B. (2017). Trace metals supplementation in anaerobic membrane bioreactors treating highly saline phenolic wastewater. *Bioresource Technology*, **234**, 106–114.

Muñoz Sierra J. D., Oosterkamp M. J., Wang W., Spanjers H. and van Lier J. B. (2018). Impact of long-term salinity exposure in anaerobic membrane bioreactors treating phenolic wastewater: Performance robustness and endured microbial community. *Water Research*, **141**, 172–184.

Muyzer G. and Stams A. J. M. (2008). The ecology and biotechnology of sulphate-reducing bacteria. *Nature Reviews – Microbiology*, **6**, 441–454.

Nanusha M. Y., Carlier J. D., Carvalho G. I., Costa M. C. and Paiva A. P. (2019). Separation and recovery of Pd and Fe as nanosized metal sulphides by combining solvent extraction with biological strategies based on the use of sulphate-reducing bacteria. *Separation and Purification Technology*, **212**, 747–756.

Nevatalo L. M., Mäkinen A. E., Kaksonen A. H. and Puhakka J. A. (2010). Biological hydrogen sulfide production in an ethanol-lactate fed fluidized-bed bioreactor. *Bioresource Technology*, **101**, 276–284.

Nguyen D. and Khanal S. K. (2018). A little breath of fresh air into an anaerobic system: How microaeration facilitates anaerobic digestion process. *Biotechnology Advances*, **36**, 1971–1983.

Nielsen A. H., Hvitved-Jacobsen T. and Vollertsen J. (2008). Effects of pH and iron concentrations on sulfide precipitation in wastewater collection systems. *Water Environment Research*, **80**, 380–384.

Niu Q., Kubota K., Qiao W., Jing Z., Zhang Y. and Yu-You L. (2015). Effect of ammonia inhibition on microbial community dynamic and process functional resilience in mesophilic methane fermentation of chicken manure. *Journal of Chemical Technology and Biotechnology*, **90**, 2161–2169.

Nogueira E. W., Licona F. M., Godoi L. A. G., Brucha G. and Damianovic M. H. R. Z. (2019). Biological treatment removal of rare earth elements and yttrium (REY) and metals from actual acid mine drainage. *Water Science and Technology*, **80**, 1485–1493.

Noguera D. R., Pizarro G., Stahl D. A. and Rittman B. E. (1999). Simulation of multispecies biofilm development in three dimensions. *Water Science and Technology*, **39**, 123–130

Nzeteu C. O., Trego A. C., Abram F. and O'Flaherty V. (2018). Reproducible, high-yielding, biological caproate production from food waste using a single-phase anaerobic reactor system. *Biotechnology for Biofuels*, **11**, 1–14.

O'Flaherty V., Lens P. N. L., Leahy B. and Colleran E. (1998a). Long-term competition between sulphate-reducing and methane-producing bacteria during full-scale anaerobic treatment of citric acid production wastewater. *Water Research*, **32**, 815–825.

O'Flaherty V., Mahony T., O'Kennedy R. and Colleran E. (1998b). Effect of pH on growth kinetics and sulphide toxicity thresholds of a range of methanogenic, syntrophic and sulphate-reducing bacteria. *Process Biochemistry*, **33**, 555–569.

O'Flaherty V., Colohan S., Mulkerrins D. and Colleran E. (1999). Effect of sulphate addition on volatile fatty acid and ethanol degradation in an anaerobic hybrid reactor. II: Microbial interactions and toxic effects. *Bioresource Technology*, **68**, 109–120.

O'Flaherty V. and Colleran E. (2000). Sulfur problems in anaerobic digestion. In: Environmental technologies to treat sulfur pollution: Principles and engineering, P. N. L. Lens and L. W. Hulshoff Pol (eds), IWA Publishing, London, pp. 467–489.

Oleszkiewicz J. A., Marstaller T. and McCartney D. M. (1989). Effects of pH on sulfide toxicity to anaerobic processes. *Environmental Technology Letters*, **10**, 815–822.

Omil F., Méndez R. and Lema J. M. (1995). Anaerobic treatment of saline wastewaters under high sulphide and ammonia content. *Bioresource Technology*, **54**, 269–278.

Omil F., Lens P. N. L., Hulshoff Pol L. and Lettinga G. (1996). Effect of upward velocity and sulphide concentration on volatile fatty acid degradation in a sulphidogenic granular sludge reactor. *Process Biochemistry*, **31**, 699–710.

Omil F., Bakker C. D, Hulshoff Pol L. W. and Lettinga G. (1997a). Effect of ph and low temperature shocks on the competition between sulphate reducing bacteria and methane producing bacteria in uasb reactors. *Environmental Technology* (United Kingdom), **18**, 255–264.

Omil F., Oude Elferink S. J. W. H., Lens P. N. L., Hulshoff Pol L. W. and Lettinga G. (1997b). Effect of the inoculation with *Desulforhabdus amnigenus* and pH or O_2 shocks on the competition between sulphate reducing and methanogenic bacteria in an acetate fed UASB reactor. *Bioresource Technology*, **60**, 113–122.

Omil F., Lens P. N. L., Visser A., Hulshoff Pol L. W. and Lettinga G. (1998). Long-term competition between sulfate reducing and methanogenic bacteria in UASB reactors treating volatile fatty acids. *Biotechnology and Bioengineering*, **57**, 676–685.

Oremland R. S. and Capone D. G. (1988). Use of specific inhibitors in biogeochemistry and microbial ecology. In: Advances in Microbial Ecology, K. C. Marshall (ed.), Plenum Press, New York, pp. 295–383.

Oude Elferink S. J. W. H., Visser A., Hulshoff Pol L. W. and and Stams A. J. M. (1994). Sulfate reduction in methanogenic bioreactors. *FEMS Microbiology Reviews*, **15**, 119–136.

Oude Elferink S. J. W. H., Luppens S. B. I., Marcelis C. L. M. and and Stams A. J. M. (1998). Kinetics of acetate oxidation by two sulfate reducers isolated from anaerobic granular sludge. *Applied and Environmental Microbiology*, **64**, 2301–2303.

Owen W. F., Stuckey D. C., Healy J. B., Young L. Y. and McCarty P. L. (1979). Bioassay for monitoring biochemical methane potential and anaerobic toxicity. *Water Research*, **13**, 485–492

Ozuolmez D., Na H., Lever M. A., Kjeldsen K. U., Jørgensen B. B. and Plugge C. M. (2015). Methanogenic archaea and sulfate reducing bacteria co-cultured on acetate: Teamwork or coexistence? *Frontiers in Microbiology*, **6**, 1–12.

Parkin G. F. and Speece R. E. (1983). Attached versus suspended growth anaerobic reactors: Response to toxic substances. *Water Science and Technology*, **15**, 261–289

Paulo L. M., Ramiro-Garcia J., van Mourik S., Stams A. J. M. and Souza D. Z. (2017). Effect of nickel and cobalt on methanogenic enrichment cultures and role of biogenic sulfide in metal toxicity attenuation. *Frontiers in Microbiology*, **8**, 1–12.

Pereyra L. P., Hiibel S. R., Perrault E. M., Reardon K. F. and Pruden A. (2012). Effect of bioaugmentation and biostimulation on sulfate-reducing column startup captured by functional gene profiling. *FEMS Microbiology Ecology*, **82**, 135–147.

Pokhrel D. and Viraraghavan T. (2004). Treatment of pulp and paper mill wastewater – A review. *Science of the Total Environment*, **333**, 37–58.

Pollice A., Rozzi A., Tomei M. C., Di Pinto A. C. and Limoni N. (2000). Monitoring the inhibitory effect of nacl on anaerobic wastewater treatment processes by the rantox biosensor. *Environmental Technology* (United Kingdom), **21**, 535–544.

Qian Z., Tianwei H., Mackey H. R., van Loosdrecht M. C. M. and Guanghao C. (2019). Recent advances in dissimilatory sulfate reduction: From metabolic study to application. *Water Research*, **150**, 162–181.

Rabus R., Nordhaus R., Ludwig W. and Widdel F. (1993). Complete oxidation of toluene under strictly anoxic conditions by a new sulfate-reducing bacterium. *Applied and Environmental Microbiology*, **59**, 1444–1451.

Ramírez M., Gómez J. M., Aroca G. and Cantero D. (2009). Removal of hydrogen sulfide by immobilized *Thiobacillus thioparus* in a biotrickling filter packed with polyurethane foam. *Bioresource Technology*, **100**, 4989–4995.

Reis M. A. M., Goncalves L. M. D. and Carrondo M. J. T. (1988). Sulphate removal in acidogenic phase anaerobic digestion. *Environmental Technology Letters*, **9**, 775–784.

Reis M. A. M., Almeida J. S., Lemos P. C. and Carrondo M. J. T. (1992). Effect of hydrogen sulfide on growth of sulfate reducing bacteria. *Biotechnology and Bioengineering*, **40**, 593–600.

Reyes-Alvarado L. C., Okpalanze N. N., Kankanala D., Rene E. R., Esposito G. and Lens P. N. L. (2017a). Forecasting the effect of feast and famine conditions on biological sulphate reduction in an anaerobic inverse fluidized bed reactor using artificial neural networks. *Process Biochemistry*, **55**, 146–161.

Reyes-Alvarado L. C., Okpalanze N. N., Rene E. R., Rustrian E., Houbron E. and Lens P. N. L. (2017b). Carbohydrate based polymeric materials as slow release electron donors for sulphate removal from wastewater. *Journal of Environmental Management*, **200**, 407–415.

Reyes-Alvarado L. C., Camarillo-Gamboa Á., Rustrian E., Rene E. R., Esposito G., Lens P. N. L. and Houbron E. (2018). Lignocellulosic biowastes as carrier material and slow release electron donor for sulphidogenesis of wastewater in an inverse fluidized bed bioreactor. *Environmental Science and Pollution Research*, **25**, 5115–5128.

Rintala J. A. and Puhakka J. A. (1994). Anaerobic treatment in pulp- and paper-mill waste management: A review. *Bioresource Technology*, **47**, 1–18.

Rinzema A. and Lettinga G. (1988). The effect of sulphide on the anaerobic degradation of propionate. *Environmental Technology Letters*, **9**, 83–88.

Rinzema A., van Lier J. B. and Lettinga G. (1988). Sodium inhibition of acetoclastic methanogens in granular sludge from a UASB reactor. *Enzyme and Microbial Technology*, **10**, 24–32.

Rodriguez R. P. and Zaiat M. (2011). Influence of carbon source and inoculum type on anaerobic biomass adhesion on polyurethane foam in reactors fed with acid mine drainage. *Bioresource Technology*, **102**, 5060–5065.

Rückert C. (2016). Sulfate reduction in microorganisms – recent advances and biotechnological applications. *Current Opinion in Microbiology*, **33**, 140–146.

Saia F. T., Souza T. S. O., Duarte R. T. D., Pozzi E., Fonseca D. and Foresti E. (2016). Microbial community in a pilot-scale bioreactor promoting anaerobic digestion and sulfur-driven denitrification for domestic sewage treatment. *Bioprocess and Biosystems Engineering*, **39**, 341–352.

Sakimoto K. K., Wong A. B. and Yang P. (2016). Self-photosensitization of nonphotosynthetic bacteria for solar-to-chemical production. *Science*, **351,** 74–77.

Saleh A. M., Macpherson R. and Miller J. D. A. (1964). The effect of inhibitors on Sulphate Reducing Bacteria: a compilation. *Journal of Applied Bacteriology*, **27**, 281–293.

Salehizadeh H. and van Loosdrecht M. C. M. (2004). Production of polyhydroxyalkanoates by mixed culture: Recent trends and biotechnological importance. *Biotechnology Advances*, **22**, 261–279.

Sampaio R. M. M., Timmers R. A., Kocks N., André V., Duarte M. T., van Hullesbusch E. D., Farges F. and Lens P. N. L. (2010). Zn-Ni sulfide selective precipitation: The role of supersaturation. *Separation and Purification Technology*, **74**, 108–118.

Sánchez-Andrea I., Stams A. J. M., Hedrich S., Ňancucheo I. and Johnson D. B. (2015). *Desulfosporosinus acididurans sp. nov.*: an acidophilic sulfate-reducing bacterium isolated from acidic sediments. *Extremophiles*, **19**, 39–47.

Santegoeds C. M., Ferdelman T. G., Muyzer G. and de Beer D. (1998). Structural and functional dynamics of sulfate reducing bacterial populations in biofilms. *Applied and Environmental Microbiology*, **64**, 3731–3739.

Santegoeds C. M., Damgaard L. R., Hesselink G., Zopfi J., Lens P. N. L., Muyzer G. and Beer D. (1999). Distribution of sulfate-reducing and methanogenic bacteria in anaerobic aggregates determined by microsensor and molecular analyses. *Applied and Environmental Microbiology*, **65**, 4618–4629.

Santos A. L. and Johnson D. B. (2018). Design and application of a low pH upflow biofilm sulfidogenic bioreactor for recovering transition metals from synthetic waste water at a Brazilian copper mine. *Frontiers in Microbiology*, **9**, 1–11.

Särner E. (1986). Influence and control of H_2S on full scale plants and pilot plant experiments. In Proceedings of the IAWPRC conference "Anaerobic Wastewater Treatment – a grown-up technology", Amsterdam, The Netherlands, 189–204.

Särner E. (1989). Removal of sulfate and sulfite in an anaerobic trickling (antric) filter. In Proceedings of the Conference on Recent Developments: Technological Advances in Biofilm Reactors, Nice, France, 4-6 April, 405–414.

Sarner E. (1990). Removal of sulphate and sulphite in an anaerobic trickling (antric) filter. *Water Science and Technology*, **22**, 395–404.

Sass A., Rütters H., Cypionka H. and Sass H. (2002). *Desulfobulbus mediterraneus sp. nov.,* a sulfate-reducing bacterium growing on mono- and disaccharides. *Archives of Microbiology*, **177**, 468–474.

Sawatdeenarunat C., Sung S. and Khanal S. K. (2017). Enhanced volatile fatty acids production during anaerobic digestion of lignocellulosic biomass via micro-oxygenation. *Bioresource Technology*, **237**, 139–145.

Siles J. A., Brekelmans J., Martín M. A., Chica A. F. and Martín A. (2010). Impact of ammonia and sulphate concentration on thermophilic anaerobic digestion. *Bioresource Technology*, **101**, 9040–9048.

Silva A. J., Hirasawa J. S., Varesche M. B., Foresti E. and Zaiat M. (2006). Evaluation of support materials for the immobilization of sulfate-reducing bacteria and methanogenic archaea. *Anaerobe*, **12**, 93–98.

Sinharoy A., Baskaran D. and Pakshirajan K. (2019). Process integration and artificial neural network modeling of biological sulfate reduction using a carbon monoxide fed gas lift bioreactor. *Chemical Engineering Journal*, 123518.

Sipma J., Lens P. N. L., Vieira A., Miron Y., van Lier J. B., Hulshoff Pol L. W. and Lettinga G. (1999). Thermophilic sulphate reduction in upflow anaerobic sludge bed reactors under acidifying conditions. *Process Biochemistry*, **35**, 509–522.

Sivakumar K., Scarascia G., Zaouri N., Wang T., Kaksonen A. H. and Hong P. Y. (2019). Salinity-mediated increment in sulfate reduction, biofilm formation, and quorum sensing: A potential connection between quorum sensing and sulfate reduction? *Frontiers in Microbiology*, **10**, 1–13.

Soto M., Mendez R. and Lema J. M. (1993). Methanogenic and non-methanogenic activity tests – theoretical basis and experimental set-up. *Water Science and Technology*, **27**, 1361–1376.

Sousa J. A. B., Plugge C. M., Stams A. J. M. and Bijmans M. F. M. (2015). Sulfate reduction in a hydrogen fed bioreactor operated at haloalkaline conditions. *Water Research*, **68**, 67–76.

Souto T. F., Aquino S. F., Silva S. Q. and Chernicharo C. A. L. (2010). Influence of incubation conditions on the specific methanogenic activity test. *Biodegradation*, **21**, 411–424.

Souza T. S. O., Okada D. Y. and Foresti E. (2018). Proof of concept and improvement of a triple chamber biosystem coupling anaerobic digestion, nitrification and mixotrophic endogenous denitrification for organic matter, nitrogen and sulfide removal from domestic sewage. *Bioprocess and Biosystems Engineering*, **41**(12), 1839–1850.

Speece R. E. (1983). Anaerobic biotechnology for industrial wastewater. *Environmental Science and Technology*, **17**, 416A–427A.

Stams A. J. M., Hansen T. A. and Skyring G. W. (1985). Utilization of amino acids as energy substrates by two marine *Desulfovibrio* strains. *FEMS Microbiology Letters*, **31**, 11–15.

Steinbusch K. J. J., Hamelers H. V. M., Plugge C. M. and Buisman C. J. N. (2011). Biological formation of caproate and caprylate from acetate: fuel and chemical production from low grade biomass. *Energy & Environmental Science*, **4**, 216–224.

Sun M., Mu Z. X., Chen Y. P., Sheng G. P., Liu X. W., Chen Y. Z., Zhao Y., Wang H. L., Yu H. Q., Wei L. and Ma F. (2009). Microbe-assisted sulfide oxidation in the anode of a microbial fuel cell. *Environmental Science and Technology*, **43**, 3372–3377.

Sun J., Hu S., Sharma K. R., Ni B. J. and Yuan Z. (2014). Stratified microbial structure and activity in sulfide- and methane- producing anaerobic sewer biofilms. *Applied and Environmental Microbiology*, **80**, 7042–7052.

Tan N. C. G. and Field J. A. (2000). Biodegradation of sulfonated aromatic compounds. In: Environmental technologies to treat sulfur pollution: Principles and engineering, P. N. L. Lens and L. W. Hulshoff Pol (eds), IWA Publishing, London, pp. 377–392.

Tang K., Baskaran V. and Nemati M. (2009). Bacteria of the sulphur cycle: An overview of microbiology, biokinetics and their role in petroleum and mining industries. *Biochemical Engineering Journal*, **44**, 73–94.

Tanimoto Y., Tasaki M., Okamura K., Yamaguchi M. and Minami K. (1989). Screening growth inhibitors of sulfate-reducing bacteria and their effects on methane fermentation. *Journal of Fermentation and Bioengineering*, **68**, 353–359.

Thauer R. K., Jungermann K. and Decker K. (1977). Energy conservation in chemotrophic anaerobic bacteria. *Bacteriological Reviews*, **41**, 100–180.

Tiwari M. K., Guha S., Harendranath C. S. and Tripathi S. (2006). Influence of extrinsic factors on granulation in UASB reactor. *Applied Microbiology and Biotechnology*, **71**, 145–154.

U.S. EPA. (1990). National Primary and Secondary Drinking Water Regulations; Synthetic Organic Chemicals and Inorganic Chemicals. *Federal Register*, **55**(143), 30370.

Utgikar V. P., Tabak H. H., Haines J. R. and Govind R. (2003). Quantification of toxic and inhibitory impact of copper and zinc on mixed cultures of sulfate-reducing bacteria. *Biotechnology and Bioengineering*, **82**, 306–312.

Valdés F., Camiloti P. R., Rodriguez R. P., Delforno T. P., Carrillo-Reyes J., Zaiat M. and Jeison D. (2016). Sulfide-oxidizing bacteria establishment in an innovative microaerobic reactor with an internal silicone membrane for sulfur recovery from wastewater. *Biodegradation*, **27**, 119–130.

Valentine D. L. (2007). Adaptations to energy stress dictate the ecology and evolution of the Archaea. *Nature Reviews. Microbiology*, **5**, 316–323.

Vallero M. V. G., Hulshoff Pol L. W., Lettinga G. and Lens P. N. L. (2003). Effect of NaCl on thermophilic (55°C) methanol degradation in sulfate reducing granular sludge reactors. *Water Research*, **37**, 2269–2280.

Vallero M. V. G., Sipma J., Lettinga G. and Lens P. N. L. (2004). High-rate sulfate reduction at high salinity (up to 90 $mS \cdot cm^{-1}$) in mesophilic UASB reactors. *Biotechnology and Bioengineering*, **86**, 226–235.

van den Brand T. P. H., Roest K., Chen G. H. and van Loosdrecht M. C. M. (2014). Influence of acetate and propionate on sulphate-reducing bacteria activity. *Journal of Applied Microbiology*, **117**, 1839–1847.

van den Brand T. P. H., Roest K., Chen G. H., Brdjanovic D. and van Loosdrecht M. C. M. (2015). Long-term effect of seawater on sulfate reduction in wastewater treatment. *Environmental Engineering Science*, **32**, 622–630.

Vanwonterghem I., Jensen P. D., Ho D. P., Batstone and D. J. and Tyson G. W. (2014). Linking microbial community structure, interactions and function in anaerobic digesters using new molecular techniques. *Current Opinion in Biotechnology*, **27**, 55–64.

Veiga M. C., Soto M., Méndez R. and Lema J. M. (1990). A new device for measurement and control of gas production by bench scale anaerobic digesters. *Water Research*, **24**, 1551–1554.

Vilela R. S., Damianovic M. H. R. Z. and Foresti E. (2014). Removing organic matter from sulfate rich wastewater via sulfidogenic and methanogenic pathways. *Water Science and Technology*, **69**, 1669–1675.

Villa-Gomez D., Ababneh H., Papirio S., Rousseau D. P. L. and Lens P. N. L. (2011). Effect of sulfide concentration on the location of the metal precipitates in inversed fluidized bed reactors. *Journal of Hazardous Materials*, **192**, 200–207.

Visser A. (1995). The anaerobic treatment of sulfate-containing wastewater. *PhD Thesis*, Wageningen Agricultural University, The Netherlands. Available at https://edepot.wur.nl/200073.

Visser A., Beeksma I., van der Zee F., Stams A. J. M. and Lettinga G. (1993a). Anaerobic degradation of volatile fatty acids at different sulphate concentrations. *Applied Microbiology and Biotechnology*, **40**, 549–556.

Visser A., Gao Y. and Lettinga G. (1993b). Effects of short-term temperature increases on the mesophilic anaerobic breakdown of sulfate containing synthetic wastewater. *Water Research*, **27**, 541–550.

Visser A., Hulshoff Pol L. W. and Lettinga G. (1996). Competition of methanogenic and sulfidogenic bacteria. *Water Science and Technology*, **33**, 99–110.

Walker M., Zhang Y., Heaven S. and Banks C. (2009). Potential errors in the quantitative evaluation of biogas production in anaerobic digestion processes. *Bioresource Technology*, **100**, 6339–6346.

Wang Y., Gabbard D. and Pai P. (1991). Inhibition of acetate methanogenesis by phenols. *Journal of Environmental Engineering*, **117**, 487–500.

Wang J., Lu H., Chen G. H., Lau G. N., Tsang W. L. and van Loosdrecht M. C. M. (2009). A novel sulfate reduction, autotrophic denitrification, nitrification integrated (SANI) process for saline wastewater treatment. *Water Research*, **43**, 2363–2372.

Wang N., Dorman R. A., Ingersoll C. G., Hardesty D. K., Brumbaugh W. G., Hammer E. J., Bauer C. R. and Mount D. R. (2016). Acute and chronic toxicity of sodium sulfate to four freshwater organisms in water-only exposures. *Environmental Toxicology and Chemistry*, **35**, 115–127.

Waybrant K. R., Blowes D. W. and Ptacek C. J. (1998). Selection of reactive mixtures for use in permeable reactive walls for treatment of mine drainage. *Environmental Science and Technology*, **32**, 1972–1979.

Weast R. C. (1981). Handbook of Chemistry and Physics, 62th ed., CRC Press Inc., Boca Raton, U.S.A.

Wei C., Wei L., Li C., Wei D. and Zhao Y. (2018). Effects of salinity, C/S ratio, S/N ratio on the BESI process, and treatment of nanofiltration concentrate. *Environmental Science and Pollution Research*, **25**, 5129–5139.

Weijma J., Gubbels F., Hulshoff Pol L. W., Stams A. J. M., Lens P. N. L. and Lettinga G. (2002). Competition for H_2 between sulfate reducers, methanogens and homoacetogens in a gas-lift reactor. *Water Science and Technology*, **45**, 75–80.

Weijma J., Chi T. M., Hulshoff Pol L. W., Stams A. J. M. and Lettinga G. (2003). The effect of sulphate on methanol conversion in mesophilic upflow anaerobic sludge bed reactors. *Process Biochemistry*, **38**, 1259–1266.

Widdel F. (1988). Microbiology and ecology of sulfate-and sulfur reducing bacteria. In: Biology of Anaerobic Microorganisms, A. J. B. Zehnder (ed.), John Wiley and Sons, New York, pp. 69–586.

Wu D., Ekama G. A., Chui H. K., Wang B., Cui Y. X., Hao T. W., van Loosdrecht M. C. M. and Chen G. H. (2016). Large-scale demonstration of the sulfate reduction autotrophic denitrification nitrification integrated (SANI®) process in saline sewage treatment. *Water Research*, **100**, 496–507.

Xu Q., Townsend T. and Bitton G. (2011). Inhibition of hydrogen sulfide generation from disposed gypsum drywall using chemical inhibitors. *Journal of Hazardous Materials*, **191**, 204–211.

Yadav V. K. and Archer D. B. (1989). Sodium molybdate inhibits sulphate reduction in the anaerobic treatment of high-sulphate molasses wastewater. *Applied Microbiology and Biotechnology*, **31**, 103–106.

Yan L., Ye J., Zhang P., Xu D., Wu Y., Liu J., Zhang H., Fang W., Wang B. and Zeng G. (2018). Hydrogen sulfide formation control and microbial competition in batch anaerobic digestion of slaughterhouse wastewater sludge: Effect of initial sludge pH. *Bioresource Technology*, **259**, 67–74.

Yang C. H. J., Parkin G. F. and Speece R. E. (1979). Recovery of anaerobic digestion after exposure to toxicants. Final report to the U.S. Department of Energy, Contract No. EC-77-S-02-4391.

Yekta S., Ziels R. M., Björn A., Skyllberg U., Ejlertsson J., Karlsson A., Svedlund M., Willén M. and Svensson B. H. (2017). Importance of sulfide interaction with iron as regulator of the microbial community in biogas reactors and its effect on methanogenesis, volatile fatty acids turnover, and syntrophic long-chain fatty acids degradation. *Journal of Bioscience and Bioengineering*, **123**, 597–605.

Yoda M., Kitagawa M. and Miyaji Y. (1987). Long term competition between sulfate-reducing and methane-producing bacteria for acetate in anaerobic biofilm. *Water Research*, **21**, 1547–1556.

Yokoyama M. T., Spence C., Hengemuehle S. M., Whitehead T. R., von Bernuth R. and Cotta M. (2016). Sodium Tetraborate Decahydrate Treatment Reduces Hydrogen Sulfide and the Sulfate-Reducing Bacteria Population of Swine Manure. *Journal of Environmental Quality*, **45**, 1838–1846.

Young J. C. and Tabac H. H. (1993). Multilevel protocol for assessing the fate and effect of toxic chemicals in anaerobic treatment processes. *Water Environment Research*, **65**(1), 34–45

Yuan Y., Cheng H., Chen F., Zhang Y., Xu X., Huang C., Chen C., Liu W., Ding C., Li Z., Chen T. and Wang A. (2020). Enhanced methane production by alleviating sulfide inhibition with a microbial electrolysis coupled anaerobic digestion reactor. *Environment International*, **136**, 105503.

Zan F. and Hao T. (2020). Sulfate in anaerobic co-digester accelerates methane production from food waste and waste activated sludge. *Bioresource Technology*, **298**, 122536.

Zan F., Zeng Q., Hao T., Ekama G. A., Hao X. and Chen G. (2019). Achieving methane production enhancement from waste activated sludge with sulfite pretreatment: Feasibility, kinetics and mechanism study. *Water Research*, **158**, 438–448.

Zhang Y. and Angelidaki I. (2015). Recovery of ammonia and sulfate from waste streams and bioenergy production via bipolar bioelectrodialysis. *Water Research*, **85**, 177–184.

Chapter 7

Biological removal of sulfurous compounds and metals from inorganic wastewaters

David Barrie Johnson and Ana Laura Santos

7.1 INTRODUCTION

Sulfur-rich wastewaters that contain relatively little dissolved organic carbon may be generated as a result of several industrial processes: for example, galvanic processes, scrubbing of flue gases at power plants and detoxification of metal-contaminated soils (Table 7.1). However, the mining of heavy metals and coal is, by far and away, responsible for generating the bulk of this type of industrial wastewater (Blowes *et al.*, 2014). Sulfate- and metal-rich effluents produced at the sites of active and derelict mines are frequently referred to as 'acid mine (or rock) drainage' (AMD/ARD). ARD can also be found at sites that have not been impacted by mining, such as gossans in the high Arctic, and upstream of the large and historic mining works at the Rio Tinto, Spain. The pollution of global watercourses due to AMD is immense and occurs in countries that have legacies of historic as well as current mining operations. In the 1990s, it was estimated that there were between 20,000 and 50,000 mines releasing AMD in US Forest Service lands alone (USDA, 1993), impacting many thousands of kilometers of streams and rivers. Flooding of voids left by opencast mining creates 'pit lakes' that have variable chemistries (depending on the ore that was mined, and the local geology) but these are also often acidic and enriched with metals and sulfate (Geller *et al.*, 1998; Sánchez España *et al.*, 2013).

 The chapter is from the book *Environmental Technologies to Treat Sulfur Pollution: Principles and Engineering, 2nd Edition*, Piet N.L. Lens (Ed.).
DOI: 10.2166/9781789060966_0215

Table 7.1 Examples of inorganic sulfate-containing wastewaters produced by industrial activities.

Wastewater Origin	Sulfate (g/L)	Sulfite (g/L)	Other Pollutants	Reference
Industrial activity				
Chemical industry	0.2–50	0–5		Stucki *et al.* (1993)
Mining industry	0.1–50	–	Heavy metals	Banks *et al.* (1997)
Galvanic industry	0.2–50	0–25	Heavy metals	Tichy *et al.* (1998)
Flue-gas scrubbing	1–2	1–2		Dijkman (1995)

7.2 SULFUR-RICH WASTEWATERS ASSOCIATED WITH MINING ACTIVITIES

There is a long history of sulfur-rich wastewater production resulting from mining activities; for example, the Iberian pyrite belt has been associated with mining activities for over 5000 years (López-Archilla *et al.*, 1995). While not all mining activities result in AMD production, operations targeting base and precious metals and coals (lignite and anthracite) have been widely implicated for generating most of this form of environmental pollution. Many AMD streams have low pH, and contain elevated concentrations of sulfate, dissolved transition and other metals and occasionally metalloids (such as arsenic). The chemical compositions of these wastewaters can vary greatly on a regional basis and from site to site within a region (Table 7.2).

7.2.1 Origin of acid mine drainage

The major cause of acid mine drainage pollution is the (microbially) accelerated oxidative dissolution of sulfide minerals and the transportation of the products of these and of secondary reactions in flowing water. Coal contains both inorganic sulfur (principally iron sulfides and sulfate) and organic sulfur (chiefly thiol, thiophene, sulfoxide and sulfone), the relative proportions of which vary with different ranks of coal and degrees of oxidation. The sulfur content of coals is generally between 1 and 10% (Dugan *et al.*, 1991). Many metals of commercial importance, such as copper, cobalt, zinc and lead occur chiefly in the lithosphere as, or are associated with, sulfide minerals. Although the most abundant of all sulfide minerals, pyrite (FeS_2), is no longer mined as a major source of elemental sulfur, it (and other iron sulfide minerals) is a major gangue mineral in many metal ore bodies. Significant amounts of iron sulfides can be present in waste rock dumps and, in particular, in mine tailings produced by froth flotation, where the milling of ores further compounds the reactivity of the waste by reducing the size of particles and increasing their surface areas.

Table 7.2 Examples of mine water chemistries draining from metal and coal mines.

	Parys Cu Mine Wales (Abandoned)	**Sossego Cu Mine Brazil (Active)**	**Curilo Uranium Bulgaria (Abandoned)**	**Maurliden Zn Mine Sweden (Active)**	**Minsterley Brook Pb/Zn Mine England (Abandoned)**	**Bullhouse Coal Mine England (Abandoned)**
pH	2.5	5.0	2.15–4.15	2.3	7.1	5.9
SO_4^{2-} (mg/L)	3100	1505	341–1784		768	
Fe_{total} (mg/l)	650	0.15	73–1072	403	<0.01	61
Al (mg/L)	70	0.98		132		1.2
Mn (mg/L)	10	8.74	0.95–55	49		15
As (mg/L)				1.3		
Cu (mg/L)	60	477–483	0.41–14.9	7.7	<0.01	<1
U (mg/L)			0.28–4.82			
Zn (mg/L)	40	1.02	0.80–24.2	464	58	<1
Reference	Johnson (2006)	Santos and Johnson (2017)	Groudev *et al.* (2008)	Hedrich and Johnson (2014)	Holanda and Johnson (2020)	Johnson (2006)

Sulfide minerals are stable in situations where both oxygen and water are absent or excluded, such as an undisturbed ore body or coal seam. However, in moist aerobic environments they are unstable, and can oxidize spontaneously to form (ultimately) sulfate, the free metal or metalloid (which may be subject to subsequent hydrolysis and precipitation) and, if the net metal:sulfur ratio is <1 and the pH is >1.9, hydronium ions. Several factors determine the rate at which sulfide minerals oxidize, such as the type of mineral and its surface area. Some species of lithotrophic ('rock eating') prokaryotes can, in low pH aerated and moist environments, accelerate the oxidation of sulfide minerals by factors of 10^4 to 10^6 and therefore render the spontaneous chemical oxidation of these minerals of little importance under these conditions. While in the 1950s, only one bacterial species was thought to be able to do this, it is now known that a large number of biodiverse acidophilic bacteria and archaea, acting alone or even more efficiently in consortia, can accelerate the oxidative dissolution of pyrite and other sulfide minerals (Quatrini & Johnson, 2016). The ability of these microorganisms to release or expose base and precious metals from sulfidic ores has been harnessed since the 1960s in a technology referred to as 'biomining', which operates in a variety of engineering configurations (waste dumps, bioheaps and stirred tanks) in countries throughout the world (Rawlings & Johnson, 2007).

The mechanism(s) by which bacteria actually solubilize these hard, dense minerals has been the subject of much debate. While molecular oxygen can act as the primary oxidant of sulfides, ferric iron is thought to have a more significant role, especially at low pH (<3) where this form of iron becomes increasingly soluble (Evangelou, 1995; Vera *et al.*, 2013). Ferric iron is reduced to ferrous when it reacts with sulfide minerals. In the case of pyrite, six ferric ions are required to oxidize one 'FeS_2', with the resultant partial oxidation of one of the sulfur atoms in the sulfide (S^{-1}) moiety to thiosulfate ($S_2O_3^{2-}$, where the sulfur atoms have oxidation states of $+5$ and -1; Steudel, 2000).

Since spontaneous chemical oxidation of ferrous iron proceeds very slowly at low pH, microbiological oxidation is of critical importance, and iron-oxidizing acidophiles are therefore the primary agents involved in the oxidative dissolution of sulfidic minerals at low pH. The thiosulfate generated is unstable in low pH liquors and transforms to elemental sulfur and a variety of sulfur oxy-anions, all of which can be oxidized by sulfur-oxidizing prokaryotes (Quatrini & Johnson, 2016), forming sulfuric acid, which helps to maintain the pH conducive for the iron-oxidizers. A third group of acidophiles also contributes indirectly to the process by catabolizing organic materials arising as exudates or lysate from the (mostly) carbon-fixing iron- and sulfur-oxidizers. A schematic outlining how consortia of these three groups of acidophiles interact during the oxidative dissolution of sulfide minerals is shown in Figure 7.1.

The complete chemical oxidation of pyrite that generates soluble ferric iron and sulfate (and/or bisulfate) can generate, or consume, proton (hydronium ion) acidity, depending on solution pH, since the pK_a of sulfate/bisulfate is 1.9

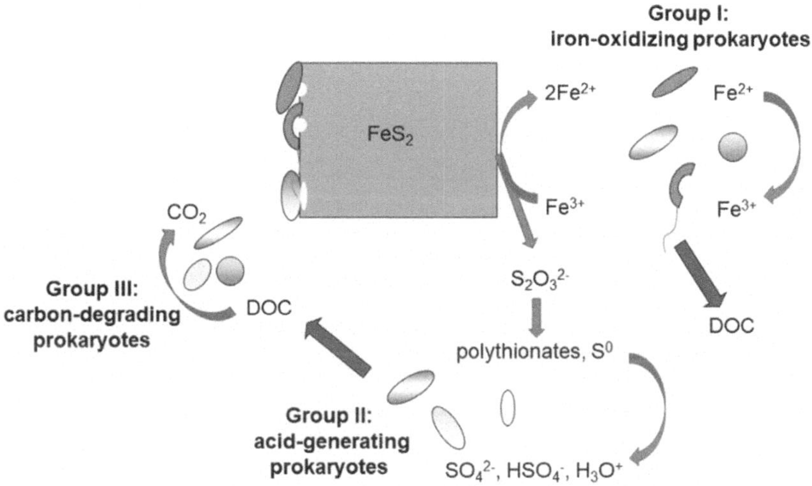

Figure 7.1 Interactive roles of acidophilic microbial consortia in the oxidative dissolution of pyrite. DOC, dissolved organic carbon.

(reactions (7.1) and (7.2)):

$$FeS_2 + 3.75\,O_2 + 0.5\,H_2O \rightarrow Fe^{3+} + 2\,SO_4{}^{2-} + H^+ \quad (7.1)$$

$$FeS_2 + 3.75\,O_2 + H^+ + 0.5\,H_2O \rightarrow Fe^{3+} + 2\,HSO_4{}^- \quad (7.2)$$

However, some or all of the ferric iron generated from pyrite oxidation can undergo mineralization reactions to form secondary oxidized minerals such as jarosites, ferrihydrite or (as shown in reaction (7.3)) schwertmannite:

$$16\,FeS_2 + 60\,O_2 + 60\,H_2O \rightarrow Fe_8O_8(SO_4)(OH)_6 + 8\,Fe^{3+} + 38\,H_3O^+ + 31\,SO_4{}^{2-} \quad (7.3)$$

This reaction is strongly acid-generating, which both limits the extent of secondary mineralization to form schwertmannite and provides a strong pH buffer to acidic iron-rich water bodies, helping to maintain their pH often within the range 2.2–2.8. This is most dramatically demonstrated in the Rio Tinto in south-west Spain, which remains within this pH range for most of its 60 km length (Garcia-Moyano *et al.*, 2012).

Figure 7.2 illustrates the major point sources of AMD at a hypothetical abandoned deep mine where waste rocks have been deposited in heap dumps and mill wastes in tailings deposits. During the active life of mines, water tables are maintained at artificially low levels by pumping water to the surface. However, when mines are abandoned this is discontinued, and water levels rebound at

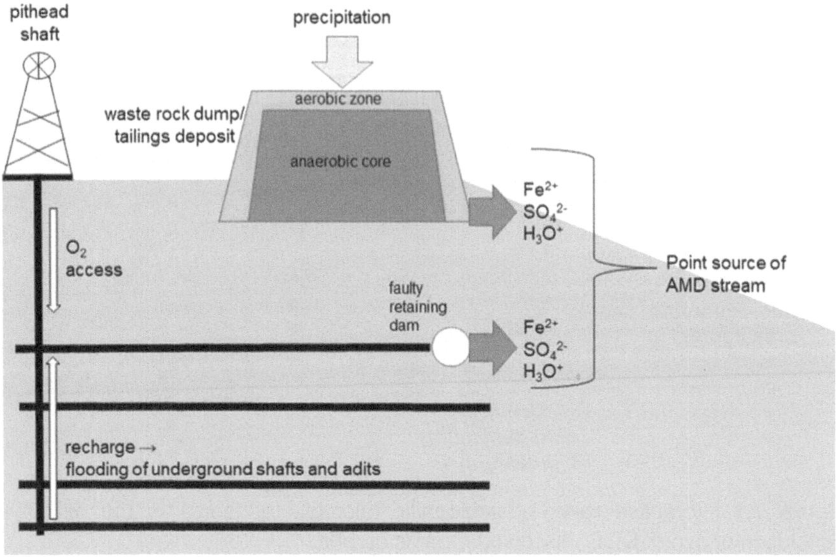

Figure 7.2 Sources of discharge of acid mine drainage at a hypothetical abandoned deep mine and associated mine waste deposits.

varying rates. The humid aerated atmosphere in the drained mine provides favorable conditions for sulfide mineral oxidation, and the up-welling waters cause iron, sulfate and other oxidation products to come into solution, so that the initial breakout of water tends to be the most acidic and polluting. Flooding mines limits oxygen diffusion rates and therefore sulfide oxidation, though this may be modified by fluctuating water tables. In the case of waste rock dumps and tailings deposits, aerated surface layers frequently overlie anaerobic deeper layers. In the aerobic region, sulfide mineral oxidation may proceed rapidly due to the microbially-accelerated reactions outlined above. However, the oxidation of sulfidic minerals can also occur within the anaerobic regions of waste dumps and tailings deposits due to the anoxic attack of soluble ferric iron present in waters flowing from the aerobic regions (7.4):

$$FeS_2 + 6\,Fe^{3+} + 3\,H_2O \rightarrow 7\,Fe^{2+} + S_2O_3{}^{2-} + 6\,H^+ \quad (7.4)$$

Waters emanating from these deposits are, therefore, often initially dominated by ferrous rather than ferric iron at the point of discharge. Aeration of these drainage waters facilitates the microbiological oxidation of the iron, giving rise to the characteristic deposition of orange-brown ‘ochre’ (or ‘yellow boy’) in affected streams and rivers (Figure 7.3).

Figure 7.3 Acidic, iron- and sulfate-rich water draining a waste rock dump at the abandoned San Telmo mine (Spain).

7.2.2 Chemical characteristics of AMD

Although the term 'acid mine drainage' is often used generically for iron- and sulfate-rich mine-impacted waters, their pH values may be close to neutral, particularly at the point of discharge. The total acidity of AMD comprises their measured hydronium ion concentrations (pH; 'proton acidity') and the hydronium ions that derive from the hydrolysis of dissolved metals, most notably iron, manganese and aluminum ('mineral acidity'). As noted above, ferrous iron is often the dominant ionic form of this metal in oxygen-depleted waters emanating directly from underground mines or mine spoils and, unlike ferric iron, this species is soluble at circum-neutral pH. Oxidation of ferrous iron to ferric consumes one hydronium ion/iron oxidized, while the subsequent hydrolysis of iron (7.5) generates three hydronium ions/iron, so the net balance is two hydronium ions for the oxidation and subsequent precipitation of iron.

$$Fe^{3+} + 6\,H_2O \rightarrow Fe(OH)_3 + 3\,H_3O^+ \qquad (7.5)$$

Similarly, the oxidation and hydrolysis of manganese is net productive of proton acidity, while in the case of aluminum (which exists only in the +3 oxidation state), net hydronium production is 3/aluminum hydrolyzed.

Mine waters that have pH values above 4.5 are considered to contain alkalinity due to the presence of (predominantly) dissolved bicarbonate. This can counter hydronium ion acidity, as shown in reaction (7.6):

$$HCO_3^- + H_3O^+ \rightarrow CO_2 + 2\,H_2O \qquad (7.6)$$

The total acidity of discharged AMD can be calculated from measuring the concentrations of iron, manganese and aluminum. If this exceeds the bicarbonate alkalinity, the waters are assumed to be net acidic.

Besides acidity, other components of AMD are highly variable, as illustrated in Table 7.2. Elevated concentrations of iron and sulfate (compared to non-polluted surface waters) are their most characteristic feature, but other metals and metalloids may also be present at relatively high concentrations. The latter may also derive from the oxidation of sulfide minerals (e.g. copper from chalcopyrite; arsenic from arsenopyrite) or from accelerated weathering of non-sulfide minerals (e.g. aluminum from aluminosilicates). The solubility of most (cationic) metals, including some of the more toxic metals such as cadmium and lead, is far greater in acidic than in neutral and alkaline waters. The more acidic mine streams may therefore act as conduits for distributing metals to watercourses (including groundwater) considerable distances away from their sites of origin.

Redox potential measurements are a particularly useful physico-chemical index of AMD chemistry, since their E_h values are dictated by one of two major redox systems that occur in this type of polluted water: the ferrous/ferric couple and the sulfide/sulfate couple (Nordstrom *et al.*, 1979). Their E^o values are affected by pH, both becoming more electronegative as pH increases. The E^o value of the ferrous/ferric couple in acidic, sulfate-rich waters is about +710 mV (Johnson *et al.*, 2017), so that values above this potential indicate the predominance of ferric iron over ferrous, and values below this potential the opposite.

Concentrations of dissolved inorganic carbon (DIC) in low pH (<3.5) AMD waters tend to be very low (<0.5 mg/L) due to the dissociation of carbonic acid and the low solubility of CO_2 in acidic waters. In acidic mining lakes, DIC concentrations in the hypolimnion may greatly exceed those in the epilimnion. Concentrations of dissolved organic carbon (DOC) also tend to be very small (generally <10 mg/L) and are comparable with those in neutral/non-acidic oligotrophic waters. Some organic carbon may originate from the degradation of, for example, wooden pit props in underground mines. Primary production in mine waters themselves is mediated either by acid-tolerant micro-algae or by chemolithotrophic bacteria (Johnson & Aguilera, 2019).

The dominant form of nitrogen in AMD is usually ammonium, though significant concentrations of nitrate may occur where explosives are used (Blowes *et al.*, 2014). Levels of total nitrogen in AMD tend to be well in excess of total phosphorus, mostly due to the adsorption of the latter onto ferric iron precipitates (Blowes *et al.*, 2014). This macronutrient is therefore often considered to limit primary production in mine-impacted waters.

7.2.3 Impact of AMD on the biosphere

Acid mine drainage waters have a deleterious impact on most aquatic organisms, though these effects may be immediate or protracted. The major effects are due to: (i) presence of toxic metals and metalloids; (ii) low pH values of the more severely acidic waters; (iii) precipitation of iron-rich ochre deposits on stream sediments, and (iv) osmotic stress. AMD impacted watercourses tend to be devoid of fish (due to mortality, and also because of loss of spawning gravel), have limited biodiversity in planktonic and benthic organisms, and display lower rates of primary production, compared to non-polluted surface waters.

Toxic metals and metalloids in AMD are more often encountered in waters draining metal mines than in those draining coal sites (Table 7.2). In aquatic ecosystems, metal distribution, speciation, and bioavailability in sediments and the water column can contribute to AMD toxicity. Many of these metals and metalloids can also be hazardous to human and animal health, if exposed to high concentrations, due to their ability to persist in the environment for an extended period and to accumulate in different levels of the food chain, which may result in acute or chronic diseases (Simate & Ndlovu, 2014). For example, exposure to high concentrations of chromium can disrupt vital metabolic functions and cause damage to the nervous system and some types of cancer (Singh *et al.*, 2011).

In extremely acidic mine waters, proton acidity can have a direct toxic effect (e.g. to fish, promoting formation of mucus in gills) or an indirect toxic effect, by enhancing the solubility of metals. The deposition of ochre from ferruginous mine drainage resulting from oxidation and hydrolysis of iron can result in long stretches of downstream waters being impacted. These iron-rich precipitates smother stream and river sediments, impeding oxygen diffusion and killing the majority of benthic organisms. This may have a dramatic effect on the food chain in impacted streams and rivers, with salmonid species being particularly sensitive to this type of pollution (Thoreau, 2002). In faster flowing waters, the ochre and other metal precipitates may remain in suspension rather than settle out. In that case, the turbidity of the water can have a major impact on light penetration, and thus negatively affect algal biomass and biodiversity (Robb, 1994).

Numbers and biodiversity of phytoplankton in AMD lakes follow the same general trends as described above, though exceptions to this pattern have been reported. Some macrophytes may establish in AMD waters, though mechanisms of metal tolerance and uptake are still not fully understood. Sheoran & Sheoran (2006) found that some species of macrophytes, such as *Typha angustata*, *Desmostachya bipinnata* and *Saccharum bengalense* could tolerate high concentrations of several metals with no growth impairment. Elsewhere, Nixdorf *et al.* (1998) showed that primary production by phytoplankton in acidic mine lakes in Germany was likely to be more limited by the availability of light, inorganic carbon and phosphorus, rather than by the low pH.

Most microorganisms are also severely damaged or killed upon exposure to AMD. Microbial communities that develop in acid polluted streams are very different from those that are found in local non-polluted streams (Mills & Mallory, 1987). However, it is now appreciated that the biodiversity of obligately acidophilic microorganisms active in highly acidic (pH <3) metal-rich waters, such as AMD, is much greater than was previously recognized, and includes chemolithotrophic (iron- and/or sulfur-oxidizing) and heterotrophic prokaryotes (bacteria and archaea) as well as eukaryotes, such as fungi, yeasts, algae and protozoa (Johnson & Aguilera, 2019). These microorganisms living in extreme acidity sometimes form communities with others of their own species, but also different species where a variety of interactions may occur. Most types of interactions between the indigenous microbiota in non-polluted ecosystems have also been observed in extremely acidic environments.

7.3 PREVENTION, CONTAINMENT AND TREATMENT OF AMD

In recent years, environmental agencies throughout the world have been implementing more strict regulations regarding disposal and discharge levels of industrial effluents. Several mechanical, chemical and biological approaches have been used to preclude the formation of AMD or, where this is not feasible, to remediate AMD-impacted waters. 'Source control' measures are designed to prevent or minimize AMD production, while 'migration control' is aimed at restricting the movement of contaminated waters.

7.3.1 Non-biological prevention and remediation systems

Most source control approaches aim to minimize or even eliminate AMD production by isolating potentially acid forming materials from exposure to either oxygen or water. Underground mine backfilling is one of the management techniques used to prevent the contact between air and sulfide minerals, therefore limiting their oxidation. The backfills (e.g. soil, sand materials, cement, and paste backfill) are placed into mine shafts, which not only limits air infiltration, but may also provide sufficient alkaline to neutralize acidity, which in turn improves underground conditions (Villain *et al.*, 2013). However, it is appropriate if they can be securely sealed, though this is often a more risk-laden strategy and is not feasible where underground passages are intricately connected, or where detailed and accurate knowledge of mine workings is unavailable. Underwater storage of mine tailings, in which water acts as a diffusion barrier to oxygen, may be effective, particularly when the tailings are covered with a layer of sediment to further limit oxygen penetration and reduce surface perturbations (Blowes *et al.*, 2014). Land based waste heaps may be sealed with covers of clay, plastic liners or organic materials (e.g. wood wastes), both to reduce oxygen access and (in the case of plastic liners and clay skins) to limit water percolation (Olds *et al.*, 2012).

Not all abandoned mines and mine spoils will produce acidic effluents, since some may be virtually devoid of acid forming minerals, and others may contain sufficient quantities of neutralizing materials (e.g. carbonates). Blending of potentially acid generating with acid consuming materials has been used to produce environmentally benign composites (e.g. Miller *et al.*, 2010). Other techniques involve the formation of highly insoluble iron phosphates on the surfaces of oxidizing pyritic minerals by phosphate amendment (Mauric & Lottermoser, 2011). Anionic surfactants have been demonstrated to inhibit sulfide mineral-oxidizing bacteria, though the effectiveness of controlling AMD production by applying these chemicals tends to be highly variable (Loos *et al.*, 1989).

Techniques developed to control the formation of AMD have mostly proven to be costly, impractical or inefficient, and therefore studies have been directed towards migration control. The most widespread method used to remediate AMD is to aerate and add alkaline chemicals to neutralize pH and remove metals, a process generally described as 'active' treatment. Different neutralizing agents may be used, such as limestone, soda ash ($NaCO_3$), ammonium and sodium hydroxide. The major objective is to accelerate the oxidation of ferrous iron (for which active aeration is also often necessary) and to precipitate iron and other metals as their hydroxides and/or carbonates. Some removal of sulfate (as gypsum) is achieved when calcium-containing neutralizing materials are used. Although active treatment can provide effective remediation of AMD, it has disadvantages of high infrastructure and operating costs, and problems with adequate disposal of the bulky sludge produced. The latter tends to be highly voluminous, typically containing only 2–4% (by weight) of solids and may contain highly toxic metals (and arsenic) that can be remobilized and released again into the environment (Veloso *et al.*, 2012).

Alternative approaches for adding alkalinity to AMD include the use of anoxic limestone drains (ALDs), open limestone channels (OLCs), diversion wells and others (Skousen *et al.*, 2017). With ALDs, waters containing little or no dissolved oxygen (<1 mg/L) percolate through trenches or pits containing a layer of limestone and a plastic liner both underneath and also covering the limestone bed to avoid O_2 influx and CO_2 escape. Iron oxidation must be avoided with ALDs, as ferric iron deposits inactivate the limestone, a process known as 'armoring'. Anoxic limestone drains require low maintenance, however metal removal must occur downstream of the drains or in association with biological treatment (e.g. constructed wetlands) especially in the cases of Al^{3+} and Fe^{3+}-rich AMD. Open limestone channels are streams lined with limestone rocks usually in areas with steep slopes into which contaminated water flows. Mine water is neutralized and oxidized by the OLC causing metal hydroxides precipitation. Precipitates may coat (or armor) the limestone surface, thus dissolution and acid neutralization may occur at lower rates than for uncoated limestone (Skousen *et al.*, 2017). These systems are more efficient when applied up-front in an integrated passive remediation system, where the AMD is most acidic.

Recent research has highlighted the potential use of 'naturally available' resources, such as crushed seashell, lignite, bentonite and attapulgite, for remediating AMD. Falayi and Ntuli (2014) reported the use of inactivated attapulgite (a magnesium aluminum phyllosilicate) as an adsorbent for the removal of trace metals from a drainage of a gold mine. Over 95% of copper, nickel and ferrous iron were removed after a reaction time of 4 h.

7.3.2 Biological remediation systems

There are a number of biological processes that generate alkalinity (or consume acidity) and which therefore have potential use in remediating AMD. These include oxygenic photosynthesis, ammonification, and several reductive processes (methanogenesis, and iron and sulfate reduction). Of these, iron and sulfate reduction are obvious candidates for AMD bioremediation, given the chemical nature of AMD. Acid consumption resulting from the reduction of ferric iron depends on whether the latter is present in soluble or mineral form. Whilst the reduction of soluble ferric iron does not generate alkalinity, the reduction of amorphous and soluble ferric iron minerals does (7.7):

$$4\,Fe(OH)_3 + CH_2O \rightarrow 4\,Fe^{2+} + H_2CO_3 + 2\,H_2O + 8\,OH^- \tag{7.7}$$

The reduction of sulfate to sulfide consumes protons in acidic solutions by transforming a strong acid (sulfuric) into a weak acid (hydrogen sulfide), as:

$$2\,H^+ + SO_4^{2-} + CH_2O \rightarrow H_2S + 2\,H_2CO_3 \tag{7.8}$$

Both reactions require electron donors, which are generally organic ('CH_2O' in reactions (7.7) and (7.8)). An important additional feature of biological sulfate (as opposed to iron) reduction is that its product, sulfide, forms highly insoluble products with many toxic heavy metals that occur in AMD. The very low solubility products of many metal sulfides mean that these metals are removed highly effectively in the presence of only trace amounts of soluble sulfide (Diaz *et al.*, 1997). Dissimilatory reduction of sulfate can therefore fulfill three critical roles in remediating acidic sulfurous wastewaters: (i) increasing water pH, (ii) lowering concentrations of sulfate, and (iii) facilitating the (selective) precipitation of metals. In contrast, the dissimilatory reduction of elemental (or zero-valent) sulfur (ZVS) can mitigate only the last of these.

7.4 SULFATE REDUCTION IN MINE DRAINAGE WATERS AND OTHER EXTREMELY ACIDIC ENVIRONMENTS

The microbial ecology of metal-rich acidic environments, such as AMD, has been widely studied using both traditional microbiological and biomolecular approaches, sometimes in tandem. These have, however, mostly focused more on pure cultures and consortia responsible for accelerating the oxidative dissolution of sulfide minerals and thereby helping generate acidic, metal-rich wastewaters. Bacteria and archaea that are metabolically active in anaerobic low pH environments have received less attention, though it is known that many species of acidophiles are able to grow via ferric iron respiration in oxygen-free or depleted zones (Coupland & Johnson, 2008). Fermentative and methanogenic prokaryotes are rare or unknown in extremely acidic environments, the former possibly because of the enhanced toxicity of their metabolic products (small molecular weight aliphatic acids) at low pH. Waters draining coal and metal mines often contain very elevated sulfate concentrations, and therefore it would be reasonable to expect that anaerobic sites (mostly sediments) in AMD streams and similar environments (e.g. mill tailings deposits, which are usually stored under water) would be populated with prokaryotes (mostly bacteria) that couple the oxidation of electron donors to the dissimilatory reduction of sulfate, generating hydrogen sulfide. However, it is interesting to note that up until the late 1990s, the only sulfate-reducing bacteria that had been isolated from acidic mineral sediments were neutrophilic strains that were highly sensitive to even moderate acidity. It was thought that these bacteria were inhabiting micro-sites that had much higher pH than implied by bulk measurements, and that they were active in maintaining the more (for them) conducive pH in these sites, as dissimilatory sulfate reduction is a proton-consuming reaction below ~ pH 6.

These early studies had been prompted by observations of hydrogen sulfide formation in mine impacted water bodies. The first such report was by Tuttle *et al.* (1969) who described a situation where a mine water stream (pH 2.84) was constrained to flow through a porous dam containing wood dust into a receiving pond that was slightly less acidic (pH 3.38) and contained ~30% less sulfate. These chemical changes were ascribed to the activities of sulfate-reducing bacteria that were detected ($<10^3$/mL) in the pond but not in the stream water, the implication being that these bacteria had been stimulated by organic compounds derived from the sawdust. Mixed bacterial populations reduced sulfate in the laboratory using sawdust as a carbon and energy source at pH 3.0, but none of the pure cultures of SRB they obtained were active below pH 5.5. Gyure *et al.* (1990) detected reduction of $^{35}SO_4^{2-}$ in sediment slurries in a pH 3.8 mine lake, though the optimum pH for sulfate reduction in sediment samples was found to be pH 5. Herlihy and Mills (1985) reported seasonal sulfate reduction occurring in freshwater sediments in lakes impacted by acid mine drainage (pH 2.5–3.5). There was no apparent point source of organic carbon, and dissolved carbon concentrations were not recorded, though water was described as 'eutrophic'. Fortin *et al.* (1996) and Fortin and Beveridge (1997) noted that sulfate reduction occurred in acidic (pH 3–4) mine tailings and isolated SRB from these environments, though again these were unable to grow in low pH (<5.5) media in the laboratory. Sulfate reduction was subsequently identified and quantified in extremely acidic (pH ≤3) sediments in a lake impacted by volcanism in Argentina by Koschorreck *et al.* (2003), acidic mine tailings in Canada (Praharaj & Fortin, 2004), Chile (Diaby *et al.*, 2007) and Siberia (Karnachuk *et al.*, 2005), moderately acidic (pH ~4) sediments around flooded mine workings in the USA (Church *et al.*, 2007) and sediments in the Rio Tinto, Spain (Sánchez-Andrea *et al.*, 2011).

Growths of 'acid streamers' have frequently been reported in flowing mine waters (Johnson & Aguilera, 2019). These are assemblages of bacteria and archaea, many of which oxidize ferrous iron, enmeshed in extracellular polymeric substances. Acid streamers are usually light cream/orange colored, though blackened growths may occasionally be found in sub-surface streamers. Johnson *et al.* (1993) added glycerol to cream-colored streamers and water, taken from an extremely acidic (pH 2.3–2.4) subterranean stream within an abandoned pyrite mine, and incubated them in the laboratory. An initial phase of iron reduction (marked by bleaching of the orange color of the mine water) was followed by a second phase of sulfate reduction (marked by the blackening of the cultures and the evolution of H_2S). This simple experiment inferred that acidophilic SRB were present in the streamer communities, but that their activities were restricted by the small amounts of organic compounds that could act as electron donors for sulfate reduction. This suggestion was supported by Rowe *et al.* (2007) who described a stream draining a small abandoned copper mine in the south of Spain that was being partially remediated by 'natural attenuation'. The low pH and metal-rich drainage stream was rapidly colonized by a consortium of eukaryotic and prokaryotic acidophiles when it emerged from underground. These formed a thick multi-layered microbial mat, the upper layer of which was green colored due to the dominance of eukaryotic algae, while the lower zone was dark grey/black (due to accumulation of copper sulfide) and smelled strongly of hydrogen sulfide. Concentrations of dissolved organic carbon increased over three fold when the mine water passed over the surface algal growths, and were orders of magnitude greater in pore waters extracted from the lower (black) mat layers. Organic carbon derived from the surface algae streamers (as exudates and lysates) were thought to sustain the enhanced level of sulfate reduction in these macroscopic growths in comparison with the subterranean streamers where algae were absent.

7.4.1 Physiological constraints on sulfate- and sulfur-reduction

Sulfate is unique among the electron acceptors used by microorganisms in its necessity to be activated before it can be reduced (Hamilton, 1998). Sulfate activation consumes two adenosine triphosphate (ATP) molecules/sulfate, forming adenosine phosphosulfate which, in a two-electron reaction, is first reduced to sulfite, prior to a further six electron reduction to hydrogen sulfide (HS^-, and/or H_2S, depending on pH). In addition to this, transport of sulfate (in co-transport of H^+ or Na^+) into the cell is also energy-dependent. The redox potential of the sulfate/hydrogen sulfide couple ($E_h = -220$ mV, at pH 7) is much lower than many potential alternative electron acceptors, such as nitrate and ferric iron, though it is interesting to note that this E_h value actually becomes more electro-positive as the pH falls (by 74 mV/unit pH fall) and is +75 mV at pH 3. Compared to aerobic microorganisms, and many anaerobes that use electron acceptors other than sulfate, SRB are severely disadvantaged with regard to the energy yield that is achievable from catabolism of organic substrates, which is further lowered in the case of SRB species that mediate incomplete oxidation of organic substrates and excrete small molecular mass organic metabolites, such as acetate. Standard redox potentials tend not to be identical to values experienced in actual metabolic situations as, under actual conditions, thermodynamic parameters might be modified by, for example, removal of reaction products in a linked series of reactions (i.e. syntrophism), resulting in dis-equilibrium (Hamilton, 1998).

In contrast to sulfate, which requires eight electrons for its complete reduction, ZVS requires only two electrons to be reduced to sulfide. Sulfur is produced in large quantities as a waste product and can be an attractive alternative to sulfate where the major or sole objective is to generate hydrogen sulfide. However, where the removal of sulfate from wastewaters is of primary concern, the use of ZVS is redundant. In addition, the insolubility of ZVS in water is a negative attribute, though acidophilic bacteria that use this solid phase material as either an electron donor or electron acceptor are able to facilitate direct contact with it and convert hydrophobic ZVS to a hydrophilic ('wetted') form (Dopson & Johnson, 2012).

The nature and amounts of suitable electron donors are also critical in determining the activities of acidophilic sulfate- and sulfur-reducing prokaryotes. While SRB as a group can use a variety of substrates as electron donors, this is generally much more restricted in terms of individual species. The most widely used organic electron donors are small molecular weight aliphatic acids and alcohols (Rabus *et al.*, 2013), though the former are problematic for acidophilic SRB, as discussed below. Bulky and complex organic substrates (e.g. straw and animal manures) are used in compost 'bioreactors' because of their (often) local availability and low cost (Neculita *et al.*, 2007). SRB are unable to degrade and utilize many complex (e.g. lignin and cellulose) biopolymers directly, and this is carried out by other microorganisms (eukaryotes and prokaryotes) also present in these constructed environments, generating organic materials that can be fermented (e.g. glucose produced by cellulose hydrolysis), and the fermentation products are thought to be the major electron donors used by the SRB in these systems. Apart from organic electron donors, hydrogen is widely used by SRB (Widdel & Hansen, 1992). This is a non-toxic, highly reduced and energetic substrate and is present in many autotrophic environments since it can be generated abiotically as well as biologically. Oxidation of hydrogen coupled to the reduction of sulfate generates only water and hydrogen sulfide as metabolic end products, and so no inhibition related to organic compounds would be anticipated. Hydrogen has been used to sustain full-scale sulfidogenic bioreactors (e.g. at the Nyrstar zinc refinery in the Netherlands; Meulepas *et al.*, 2009), though technical issues related to its flammability and mass transfer into aqueous phases are significant.

Organic acids, such as lactate, have widespread use as carbon and energy sources, and are used frequently to enrich, and to cultivate, SRB. However, acidophilic bacteria tend to be sensitive to organic acids, with concentrations of <1 mM often proving inhibitory or lethal. Many small molecular weight aliphatic acids exist predominantly as undissociated (and lipophilic) molecules in acidic waters such as AMD, and are able to pass freely through bacterial cell membranes. The internal pH of acidophiles tends to be circum-neutral which causes the organic acids to dissociate, and therefore a dis-equilibrium between internal and external concentrations of the (undissociated) acid, causing further influx. The resulting accumulation of protons within the bacterial cells can exceed the buffering capacity of cell cytoplasm and cause severe acidification. Gyure *et al.* (1990) found that 5 mM concentrations of organic acids inhibited sulfate reduction at pH 3.8, but concentrations of around 0.1 mM stimulated activity, while Fortin *et al.* (1996) noted that SRB isolated from minerals tailings were able use 1 mM acetate and formate as electron donors in medium adjusted to pH 7.5, but not that poised at pH 5.5 and below. In natural situations, concentrations of aliphatic acids in extremely acidic environments probably rarely exceed toxic levels, allowing them to be metabolized as with neutrophilic SRB. The situation is very different, however, in laboratories and constructed environments such as bioreactors, and different strategies need to be used to cultivate and sustain acidophilic SRB (Ñancucheo *et al.*, 2016).

Besides the nature and availabilities of electron donors and acceptors, other physio-chemical factors determine the activities of sulfidogenic SRB in natural and constructed environments. Temperature constraints impact the efficiencies of SRB bioreactors used for treating inorganic wastewaters. Daily and seasonal temperature fluctuations may affect the efficiency of biological sulfide production. In some areas, most notably in the high latitudes and at elevation, low temperatures (<10°C) may prevail, which will limit *in situ* SRB activity. While psychrophilic SRB with temperature optima at around 20°C and which are active at <10°C have been isolated (Isaksen & Teske, 1996), all characterized acidophilic sulfidogenic bacteria are mesophilic or thermo-tolerant and acidophilic sulfidogenic archaea are mostly extreme thermophiles, as described below. Another factor that has been observed to affect the growth and activity of SRB is the availability of suitable surfaces for bacterial attachment and colonization (e.g. Bass *et al.*, 1996).

7.4.2 Acidophilic sulfate- and sulfur-reducing prokaryotes

Currently there are only four fully described and validated species of acidophilic SRB, though several others have been reported that appear to represent novel taxa. Sen and Johnson (1999) obtained the first documented pure culture acidophilic bacterial isolate (strain M1) from the White River draining the Soufriere Hills in Monserrat (WI) and a second, very closely related, isolate (strain P1) from an abandoned copper mine in Wales. Sixteen years later, strain M1 was fully described and named as the type strain of *Desulfosporosinus acididurans* by Sánchez-Andrea *et al.* (2015), together with two related strains that had been isolated from the Rio Tinto (Spain). During this time, two other species of acidophilic SRB were described: *Thermodesulfobium narugense* which was isolated from a hot spring in Japan (Mori *et al.*, 2003), and *Desulfosporosinus acidiphilus* which was isolated from acid rock drainage in France (Alazard *et al.*, 2010). More recently, the novel species *Thermodesulfobium acidiphilum* (isolated from geothermally heated soil in Kamchatka, Russia) has been described (Frolov *et al.*, 2017). One of the characteristics of all acidophilic SRB described to date is the fact that they are moderate rather than extreme acidophiles (which have pH optima for growth at or below pH 3). The four species do, however, display different responses to temperature. *Td. narugense* is a moderate thermophile and grows between pH 4 and 6.5. *D. acidiphilus* grows between pH 3.6 and 5.5, and *D. acididurans* between pH 3.8 and 7. Both *Desulfosporosinus* (Gram-positive bacteria of the *Peptococcaceae* family and *Firmicutes* phylum, and characterized by endospore formation) species are mesophiles, and oxidize glycerol incompletely, generating acetic acid as a metabolic waste product. *Td. acidiphilum* grows at 37 65°C but has a similar pH range to the mesophilic isolates (3.7–6.5), though a closely related isolate (strain 3baa) has been reported to grow between pH 2.6 and 6.6 (Rüffel *et al.*, 2018).

Other *Peptococcaceae* strains of acidophilic SRB have been described, though not fully characterized. These include strains (Ñancucheo *et al.*, 2012) isolated from the Cantareras mine in Spain (Rowe *et al.*, 2007) and others isolated from the Rio Tinto (Sánchez-Andrea *et al.*, 2013). These are more closely related to the genus *Desulfitobacterium*, species of which reduce sulfite but not sulfate. However, the moderately acidophilic isolates can reduce sulfate and, in contrast to acidophilic *Desulfosporosinus* spp., oxidize glycerol completely to carbon dioxide. A novel genus ('*Desulfobacillus*') has been proposed for these isolates (Dopson & Johnson, 2012; Willis *et al.*, 2019).

While all moderately acidophilic SRB characterized to date are *Firmicutes*, there is circumstantial evidence to suggest that bacteria having this trait are much more diverse, phylogenetically. Molecular signatures (mostly 16S rRNA genes) of sulfate-reducing bacteria of the genera *Desulfatirhabdium, Desulfobulbus, Desulfobacterium, Desulfobacca, Desulfomonile* and *Desulfatirhabdium* have been detected in a variety of low pH sulfidogenic bioreactors (Montoya *et al.*, 2013; Sánchez-Andrea *et al.*, 2014). However, until these bacteria are isolated and the tolerance to pH tested in axenic cultures in the laboratory, it remains unclear whether they are genuine acidophiles or (like many SRB in environmental situations) occupy and sustain higher pH micro-niches within these bioreactors.

Many of the moderately acidophilic SRB that have been described can also catalyze the dissimilatory reduction of ZVS to hydrogen sulfide. Some species of acidophiles can also grow under anaerobic conditions via sulfur respiration, but do not reduce sulfate. Interestingly, most of these (exclusively archaea) were characterized long before the first acidophilic SRB, and are extreme acidophiles. Both *Thermoplasma acidophilum* (Darland *et al.*, 1970) and *Thermoplasma volcanium* (Segerer *et al.*, 1988) are moderately thermophilic facultative anaerobic euryarchaeotes that couple the oxidation of organic carbon to the reduction of either oxygen or elemental sulfur. Other ZVS-reducing acidophilic archaea are crenarchaeotes, and include *Acidianus brierleyi* and *Acidianus infernus*, both of which can oxidize as well as reduce ZVS, depending on oxygen availability (Segerer *et al.*, 1986), *Sulfurisphaera ohwakuensis* (Kurosawa *et al.*, 1998) and the obligately anaerobic thermo-acidophile *Stygiolobus azoricus* (Segerer *et al.*, 1991). The only currently classified species of ZVS (non-sulfate)-reducing acidophilic bacteria is *Desulfurella amilsii* (Florentino *et al.*, 2016). This is a thermo-tolerant (growth between 20° and 52°C), acid-tolerant (pH range of 3 to 7, with growth optimum between 6 and 6.5) bacterium that can also reduce thiosulfate and mediate the disproportionation of ZVS. Another ZVS-reducing acidophilic bacterium described though not yet classified is strain I2511 (Holanda & Johnson, 2020). This isolate is a mesophilic *Firmicute*, distantly related to *Alicyclobacillus* spp. that can grow by ZVS- or iron-respiration at an optimum and minimum pH of ~3.5 and ~2.5, respectively.

7.5 BIOENGINEERING APPROACHES FOR REMEDIATING SULFATE-RICH MINE WATERS

7.5.1 Constructed wetlands

Most biological processes used for remediation of mine water came about as result of observations of these processes in the natural environment. Constructed wetlands, for example, mimic their natural counterparts by creating an engineered ecosystem, which provides required redox, acid neutralization and precipitate settling functions (Skousen *et al.*, 2017). In the late 1970s, Huntsman *et al.* (1978) reported that outflows from naturally dominated *Sphagnum* bogs displayed improved water chemistry compared with inflowing AMD from a coal mine in Ohio, USA. The increased pH of the outflow was accompanied by decreased concentrations of sulfate, iron and other metals. Wetlands are highly complex ecosystems, and modifications of water chemistry may occur due to several mechanisms, including dilution, precipitation (by oxidative and reductive mechanisms), adsorption, and uptake by biomass.

There are three main types of constructed wetland, which operate on radically opposed biogeochemical activities based on either oxidative (aerobic wetlands) or reductive (anaerobic and vertical flow wetlands/reducing and alkalinity producing systems (RAPS)) processes. The main mechanisms involved in these systems include: (i) biological metal oxidation, precipitation and capture,

(ii) biological sulfate reduction and precipitation of metal sulfides in an organic matter layer, (iii) complexation with organic matter, (iv) sorption, and (v) direct uptake by plant roots (Skousen *et al.*, 2017). The main objective of aerobic wetlands is to promote iron oxidation, hydrolysis and metal hydroxide precipitation. Aerobic wetlands are relatively shallow systems, within which acid- and metal-tolerant macrophytes are planted in soil or other substrate, to control the water flow and to filter and stabilize iron precipitates. These systems remove metals by slowing the water flow and allowing for oxidation and may be used in tandem at a treatment site, often in association with anoxic limestone drains. *Sphagnum* mosses have proved to be somewhat sensitive to changes in AMD chemistry and accumulation of iron within constructed wetlands, and therefore cattails (*Typha latifolia* and *Typha orientalis*) are generally the preferred vegetation. The oxidation and precipitation of iron can result in significant production of proton acidity as AMD flows through aerobic wetlands, which slows down the rate of spontaneous (chemical) oxidation.

'Compost bioreactors' (anaerobic wetlands) function by actively generating alkalinity as a result of microbiological reductive activities, and are therefore commonly used to treat net-acidic waters. Since organic electron donors fuel these processes, various organic materials, such as cow and horse manure, spent mushroom compost, peat, sawdust and wood chips, have been used as substrates in compost wetlands. The choice of organic materials used varies according to their effectiveness and local availability. In some cases, anaerobic wetland substrates are mixed with limestone, which contributes to continuous alkalinity supply since there is neither Fe^{2+} oxidation nor Fe^{3+} precipitation to coat the limestone surface. The main mechanisms involved in anaerobic wetland systems are the formation and precipitation of metal sulfides, removal of sulfate and alkalinity generation by sulfate reduction reactions and continuous generation of carbonate alkalinity. Dissolution of basic minerals within the compost mix will also contribute to acid neutralization. The reduction of both iron and sulfate are important processes in anaerobic wetlands. In some systems, cattails and other emergent vegetation provide a continuous supply of carbon to the system. However, root penetration may introduce oxygen into the anaerobic zone and reduce the effectiveness of the reductive processes. In some systems, surface plants are excluded, and the anaerobic 'cell' is so constructed as to limit any oxygen diffusion. Microbial populations that develop in anaerobic wetlands are, at least in the initial phases, the indigenous populations of the composting materials, and direct inoculation tends not to be used. Microorganisms, including obligate and facultative acidophilic and acid-tolerant bacteria present in inflowing AMD, may also establish in these constructed environments.

Vertical flow wetlands/RAPS (Figure 7.4) were developed in the late 1980s (Younger *et al.*, 2002). These vertical ponds have perforated pipes embedded in a limestone layer at the bottom of the system and a layer of organic matter substrate at the top. The AMD flows downwards through the organic substrate

Figure 7.4 A reducing and alkalinity producing (RAP) wetland constructed to treat acidic, iron-rich water draining an abandoned coal mine (Tan-y-Garn, South Wales).

layer, where O_2 is consumed, generating anoxic conditions. Sulfate reduction contributes to acid neutralization, and ferric iron reduction eliminates potential coating of the underlying limestone, which provides further acid neutralization. Aerobic wetlands or settling ponds may be used to precipitate iron and manganese downstream of compost-based wetlands.

7.5.2 Bioreactor systems

While it is generally accepted that biological processes (mostly driven by microorganisms) are the most significant involved in remediating AMD in natural and constructed wetlands, the fact that these ecosystems do not lend themselves to ready control and manipulation means that their performance is difficult to predict and is subject to seasonal and other variations (Johnson & Hallberg, 2002). Biological treatment of AMD and similar wastewaters has several major advantages over active (chemical) treatment. Operating costs and footprint of the latter are high, metal removal efficiencies tend to be relatively low, bulky sludges requiring dewatering and adequate disposal are produced, and all potentially valuable metals which accumulate in the sludge are eventually lost or remobilized. In contrast, bioreactor systems facilitate the recovery of chalcophilic metals (and of sulfur, when integrated with elemental sulfur production), the sludge produced is far denser (some 6–10 fold), less voluminous and more stable (Neculita *et al.*, 2007), and sulfate concentrations can be lowered to potable water levels. On the other hand, they may require fairly sophisticated and expensive engineering systems, though this can be partially offset by the commercial value of the metals recovered. In the past few decades, various biotechnologies have been developed based on isolating 'useful' microorganisms from the environment and engineering a controlled system to promote their growth and activities. Generally, these involve the use of bioreactors in order to neutralize AMD, precipitate metals as oxy-hydroxide phases (mostly iron) or as sulfides (such as copper and zinc), recover potentially valuable metals and reduce the amount of waste that requires disposal since water and metals may be recycled and reused (Johnson & Sánchez-Andrea, 2019).

Biological sulfate reduction has been widely applied for remediating metal-, metalloid- and sulfate-rich mine waters using laboratory-scale systems. This technology harnesses the metabolic diversity and versatility of sulfate- and sulfur-reducing bacteria in order to design bioremediation processes. Biogenic or chemical sulfide not only remove potentially toxic metals from wastewaters but can also provide a selective metal precipitation (Ñancucheo *et al.*, 2012; Santos and Johnson, 2018). The metal removal is based on the reaction of hydrogen sulfide with some metal cations (usually metals with +2 oxidation states) to form insoluble metal sulfide precipitates, such as Zn, Co, Cu, Ni, and Cd (Lewis, 2010). There are two commercial-scale processes based on microbial mediated sulfate reduction using bioreactor systems for the treatment of acidic wastewaters from metal mines and related situations: the *BioSulphide*®, developed by BioteQ Environmental Technologies Inc., Canada and the *SULFATEQ*™ by Paques, the Netherlands. In both, the central process is the generation of sulfide by neutrophilic SRB with the main objective being to lower the levels of toxic metals to environmentally acceptable concentrations. In addition, these bioreactor systems offer the opportunity to lower sulfate concentrations, precipitate metals selectively and minimize/eliminate the production of waste sludges.

The *BioSulphide*® process essentially utilizes biogenic H_2S generated by the reduction of ZVS by sulfur-reducing bacteria (biological stage) to remove transition metals from acid mine drainage (chemical stage). In this case, biogenic H_2S is delivered to a contactor stirred tank containing AMD (Figure 7.5). By carefully controlling pH and sulfide concentrations, the selective precipitation of a particular metal can be achieved due to differences in the solubility product of metal sulfides, e.g. copper in a waste stream containing a mixture of iron, zinc and copper (Ashe *et al.*, 2008). The sulfide concentrate is recovered in a clarifier and subsequently dewatered using a filter press. The waste stream can then proceed to a second precipitation tank where further metal precipitation (e.g. zinc) can be induced by controlling pH and flow rates. In cases where more than one metal product is to be recovered, multiple contactors and clarifiers can be used.

In 2001, the first commercial *BioSulphide*® process plant was established at the Caribou Mine in New Brunswick (Canada) designed to remove Cu, Zn, Cd and Pb from AMD upstream of the existing lime plant. Due to the success of the plant in demonstrating its technical feasibility, other full-scale commercial plants were commissioned, for example, in North America at the Copper Queen Mine in

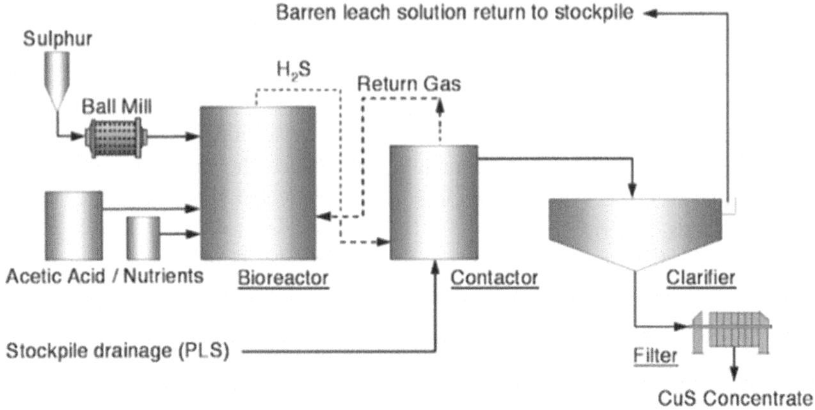

Figure 7.5 Simplified flowsheet of the *BioSulphide®* process at the Copper Queen Mine. (Ashe *et al.*, 2008, with permission). PLS, pregnant leach solution.

Bisbee, Arizona and at the Raglan Nickel mine in northern Quebec (Canada). The Bisbee Project started in 2004 producing a high-grade copper concentrate, typically containing >40% copper. About 974,000 m^3 wastewater were treated using this technology (with >99% copper recovery) and 492,000 pounds of copper were produced in the year of 2012. The operations at the Bisbee plant were suspended in 2013 and the plant was decommissioned in 2015. The Raglan Project, also established in 2004, has been operated as a seasonal water treatment plant at an active nickel mine in the Canadian Arctic region. The nickel sulfide produced in the process is up to 40% (dry weight) and has been successfully incorporated with the conventional flotation concentrate produced at the Raglan mine (Bratty *et al.*, 2006). The operation contract of this water treatment plant has been extended until 2020.

The *SULFATEQ*™ process utilizes two different biological processes, in contrast to the single bioprocess in the *BioSulphide®* technology. As before, sulfide is generated in a sulfate-reducing bioreactor in the first biological stage of the operation (again generating net alkalinity and promoting the precipitation of chalcophilic metals in the wastewater). The H_2S generated is directed to a contactor where zinc is precipitated as sulfide, which is collected and reused in smelters. Using this technology, 95% of the dry sludge generated is zinc sulfide. The second biological stage involves the oxidation of the excess of hydrogen sulfide, present as HS^-, to ZVS under oxygen-controlled conditions by sulfide-oxidizing bacteria in a S^0-generating bioreactor (Boonstra *et al.*, 1999). Two useful products are therefore produced: metal sulfides, which may be processed to recover the metal(s) concerned, and hydrophilic sulfur, which may be used to produce sulfuric acid or as an agrochemical by-product. This technology has been demonstrated at a number of pilot-scale operations and, since 1992, the first full-scale plant has been in operation to remediate acidic metal- and sulfate-rich groundwater at the Nyrstar zinc refinery in Budel-Dorplein (The Netherlands). Over 100 years of zinc refining at the site had produced severe acid and metal contamination of groundwater and soils adjacent to the site, and there was concern about the pollution plume spreading to nearby aquifers, which were sourced for drinking water. A hydrogeological containment system was designed, in which the contaminated groundwater is pumped to the surface, processed so that its quality falls within the limits set by the Dutch authorities, and then discharged into a nearby river. Various options were considered, and the decision was made, following extensive laboratory- and pilot-scale testing, to install a biological treatment facility. The system installed utilized an upflow anaerobic sludge blanket (UASB) reactor for the sulfate reduction step, and a submerged fixed film reactor for the aerobic conversion of sulfide to elemental sulfur. In the processed groundwater, typical effluent concentrations of zinc, cadmium and sulfate are, respectively, <0.05, <0.001 and <200 mg/L. Annually, around 300 tons of zinc are recovered using the *SULFATEQ*™ biotechnology at the Nyrstar zinc refinery.

Both of the above processes use bacteria that are sensitive to even moderate acidity and relatively small concentrations of some transition metals, and therefore these bacteria need to be protected from direct exposure to AMD. As a result, the engineering complexity of the reactor systems can cause financial implications for both their capital and operating expenditures. The adaptation of neutrophilic species of SRB to moderately acidic conditions has been reported as an alternative approach, though studies have shown that incrementally lowering the bioreactor pH may negatively impact both bacterial activity and biomass yield (Bijmans *et al.*, 2008). More recently, the development of novel biological sulfate reduction systems using species of acidophilic/acid-tolerant sulfate-reducing bacteria have been described (Ñancucheo & Johnson 2012, 2014; Santos & Johnson, 2017). In an acidophilic sulfate-reducing bioreactor contact between AMD and the bacterial consortium is possible, and therefore a single bioreactor can be used simultaneously to grow biomass and to precipitate target metals. Since the reaction catalyzed by these bacteria consumes protons, the bioreactors can be operated as continuous flow systems where pH is maintained by balancing the consumption of protons by the inflow of AMD. The system operates with minimal control and input of reagents and is robust throughout continuous testing using laboratory modules and different water chemistries (Ñancucheo *et al.*, 2012; Ñancucheo & Johnson, 2014; Santos & Johnson, 2018), and can therefore be considered as an attractive alternative to existing technologies.

Laboratory-scale systems have shown that biosulfidogenesis is feasible either as a stand-alone technology or as part of integrated systems for AMD treatment and the removal of metals can occur both in-line (precipitation within the reactor vessel) and off-line. However, the design and configuration of the system will depend mainly on the chemistry of the wastewater being treated. Ñancucheo and Johnson (2014) used a sulfidogenic bioreactor, maintained at low pH, to lower concentrations of sulfate from two extremely low pH mine waters that contained small concentrations of iron, but no other chalcophilic metals. Over 98% of sulfate was removed from one and 50–60% from the other. The hydrogen sulfide generated in this process was converted to colloidal sulfur in an off-line vessel. Santos and Johnson (2018) described the application of a single sulfidogenic bioreactor housing a consortium of acidophilic SRB and other acidophiles, operated at low pH (4–5) used to remove transition metals (copper, nickel, cobalt, and zinc) from a moderately acidic (pH 5) drainage stream at an operating copper mine in northern Brazil. Copper was the most abundant transition metal in the mine water (7.5 mM) while concentrations of the other transition metals were <0.25 mM. The treatment involved two stages: (i) off-line copper recovery via a stream of H_2S-containing gas generated by the bioreactor and (ii) in-line removal of nickel, zinc and cobalt from the partially processed water, which was amended with glycerol, yeast extract and basal salts and fed directly into the bioreactor. All four metals were successfully removed with recovery of a high-grade CuS (Figure 7.6).

Hedrich and Johnson (2014) proposed an integrated system designed to combine remediation of and metal recovery from AMD at the Maurliden mine in Sweden. This mine water contained ~7 mM zinc and iron and smaller concentrations of arsenic, cadmium, aluminum and manganese. This integrated system comprised of four main steps. First, soluble arsenic was removed from the mine water upstream of the other modular bioreactors by adsorbing the negatively charged arsenate onto the positively charged schwertmannite in order to avoid the co-precipitation of this toxic metalloid in the following modules and contamination of the metal sulfides recovered. The second step involved a bioreactor used to oxidize ferrous iron in the AMD to ferric, and to produce schwertmannite, a small portion of which (~11%) was used to remove As in the previous step. Soluble copper and cadmium were removed in the next stage by reacting with biogenic H_2S in an off-line reactor vessel. Lastly, the Cu/Cd-free mine water was then amended with glycerol and fed into an acidophilic sulfidogenic bioreactor. Most of the zinc was removed by maintaining the pH of the unit at 4.0, which also avoided co-precipitation of aluminum and manganese.

Elsewhere, Falagán *et al.* (2017) described the use of a low pH sulfidogenic modular system to remove soluble aluminum from synthetic acidic Al-rich mine waters. The precipitation of aluminum hydroxysulfate minerals (hydrobasaluminite and felsőbányaite) from soluble aluminum was mediated by sulfate-reducing bacteria, which utilized the protons released during the formation of these aluminum hydroxysulfates as well as those present in the feed liquor in the sulfate reduction process (equation (7.8)), thereby facilitating the continuous hydrolysis and precipitation of aluminum. The bioreactor was dominated by *D. acididurans* and two species of facultative anaerobic non-sulfidogens *Acidocella aromatica* and *Acidithiobacillus ferrooxidans*. Recently, a modified version of the acidophilic sulfidogenic bioreactors was designed to combine the

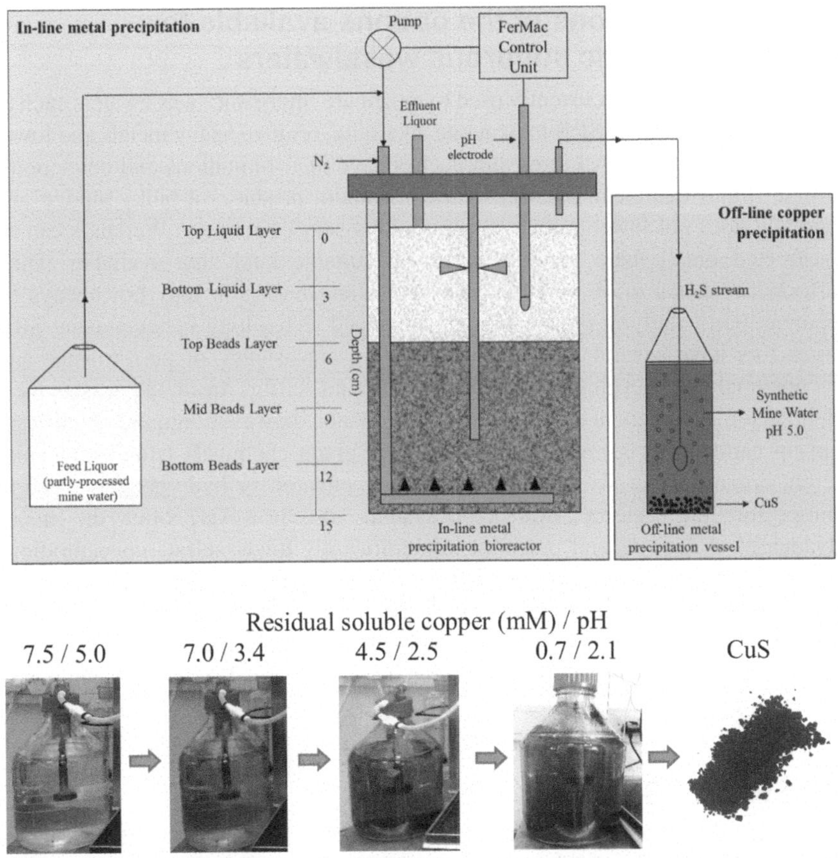

Figure 7.6 Top: Schematic representation of the low pH sulfidogenic bioreactor. Excess hydrogen sulfide produced in the bioreactor was delivered to an off-line vessel, containing synthetic mine water, in order to mediate selective removal of copper. Following copper precipitation, the more acidic partly-processed water was used as feed liquor for the bioreactor. Bottom: Time-lapse images of different stages of copper mineralization in the off-line vessel. X-ray diffraction analysis of the solid phase product confirmed that it was covellite (CuS; Santos & Johnson, 2018).

use of sulfate and ZVS as electron acceptors. The 'hybrid' sulfidogenic bioreactor (HSB) was populated with acidophilic sulfidogens (*D. acididurans* and *Peptococcaceae* CEB3, which can reduce sulfate as well as ZVS), other acidophilic bacteria (*Actinobacterium* AR3, *At. ferrooxidans* and *Ac. aromatica*) and also a novel *Firmicute* that does not reduce sulfate, but reduces ZVS. The HSB effectively removed zinc from circum-neutral pH mine-impacted waters from abandoned Zn/Pb mines in the UK (Holanda & Johnson, 2020).

7.5.3 Pros and cons of the options available for remediating acidic sulfurous wastewaters

Of the three approaches currently used to remediate 'inorganic' wastewaters such as AMD, two make use of SRB to generate alkalinity, remove heavy metals and lower sulfate concentrations. All three approaches have their limitations and detractions. Abiotic (lime) treatment is expensive to maintain, produces a bulky sludge and generally does not lower sulfate concentrations very effectively. Wetlands can be constructed only where there is sufficient suitable land area available. Their performance tends to be variable, and net sulfate reduction may not always be apparent, particularly in older systems. Advanced technology systems using SRB bioreactors have several advantages, including consistency of performance and the potential for recycling materials (metals and sulfur) from the wastewaters which have commercial value. Bioreactor plants, however, require significant start-up capital and current systems use pure-grade chemicals (small molecular weight organic materials such as ethanol or methanol, or hydrogen) as electron donors for the bacteria. Bioreactor systems do, however, offer the major advantages of control, and potential to significantly lower sulfate concentrations and to selectively precipitate metals. By being unique among the available options in generating end products that can be re-used and recycled (e.g. ZVS and transition metal sulfides), this biotechnology would seemingly be far better suited to a future where resource recovery combined with environmental protection is critical.

REFERENCES

Alazard D., Joseph M., Battaglia-Brunet F., Cayol J. L. and Ollivier B. (2010). *Desulfosporosinus acidiphilus* sp. nov.: a moderately acidophilic sulfate-reducing bacterium isolated from acid mining drainage sediments. *Extremophiles*, **14**, 305–312. doi:10.1007/s00792-010-0309-4.

Ashe N. L., McLean I. and Nodwell M. (2008). Review of operations of the biosulphide process plant at the copper Queen mine, Bisbee, Arizona. Hydrometallurgy 2008: Proceedings of the 6th International Symposium on Hydrometallurgy, Society for Mining, Metallurgy and Exploration Inc., Phoenix, Arizona, pp. 98–107.

Banks D., Younger P. L., Arnesen R.-T., Iversen E. R. and Banks S. B. (1997). Mine-water chemistry: the good, the bad and the ugly. *Environmental Geology*, **32**, 157–174. doi:10.1007/s002540050204.

Bass C. J., Webb J. S., Sanders P. F. and Lappin-Scott H. M. (1996). Influence of surfaces on sulfidogenic bacteria. *Biofouling*, **10**, 95–109.

Bijmans M. F. M, Dopson M., Ennin F., Lens P. N. and Buisman C. J. (2008). Effect of sulfide removal on sulfate reduction at pH 5 in a hydrogen fed gas-lift bioreactor. *Microbiology and Biotechnology*, **18**, 1809–1818. doi:10.4014/jmb.0800.109.

Blowes D. W., Ptacek C. J., Jambor J. L., Weisener C. G., Paktunc D., Gould W. D. and Johnson D. B. (2014). The geochemistry of acid mine drainage. *Treatise on Geochemistry*, **9**, 149–204. doi:10.1016/B978-0-08-095975-7.00905-0.

Boonstra J., van Lier R., Janssen G., Dijkman H. and Buisman C. J. N. (1999). Biological treatment of acid mine drainage. *Process Metallurgy*, **9**, 559–567. doi:10.1016/S1572-4409(99)80145-X.

Bratty M., Lawrence R., Kratochvil D. and Marchant B. (2006). Applications of biological H2S production from elemental sulfur in the treatment of heavy metal pollution including acid rock drainage. Proceedings of the 7th International Conference on Acid Rock Drainage (ICARD), American Society of Mining and Reclamation (ASMR), St. Louis, Missouri, pp. 271–281.

Church C. D., Wilkin R. T., Alpers C. N., Rye R. O. and McCleskey R. B. (2007). Microbial sulfate reduction and metal attenuation in pH 4 acid mine water. *Geochemical Transactions*, **8**, 10. doi:10.1186/1467-4866-8-10.

Coupland K. and Johnson D. B. (2008). Evidence that the potential for dissimilatory ferric iron reduction is widespread among acidophilic heterotrophic bacteria. *FEMS Microbiology Letters*, **279**, 30–35. doi: 10.1111/j.1574-6968.2007.00998.x.

Darland G., Brock T. D., Samsonoff W. and Conti S. F. (1970). A thermophilic, acidophilic mycoplasma isolated from a coal refuse pile. *Science*, **170**, 1416–1418. doi:10.1126/science.170.3965.1416.

Diaby N., Dold B., Pfeifer H. R., Holliger C., Johnson D. B. and Hallberg K. B. (2007). Microbial communities in a porphyry copper tailings impoundment and their impact on the geochemical dynamics of the mine waste. *Environmental Microbiology*, **9**, 298–307. doi:10.1111/j.1462-2920.2006.01138.x.

Diaz M. A., Monhemius A. J. and Narayanan A. (1997). Consecutive hydroxide – sulfide precipitation treatment of acid mine drainage. Proceedings of the 4th International Conference on Acid Rock Drainage, MEND, Vancouver, BC, Canada, pp. 1179–1194.

Dijkman H. (1995). Biological gas desulfurization. Mededelingen van de Faculteit Landbouwwetenschappen. *University of Gent*, **60**, 2677–2684.

Dopson M. and Johnson D. B. (2012). Biodiversity, metabolism and applications of acidophilic sulfur- metabolizing microorganisms. *Environmental Microbiology*, **14**, 2620–2631. doi:10.1111/j.1462-2920.2012.02749.x.

Dugan P. R., Quigley D. R. and Attia Y. A. (eds) (1991). Processing and Utilization of High Sulfur Coals. Elsevier, Amsterdam.

Evangelou V. P. (1995). Pyrite Oxidation and its Control. CRC Press, New York.

Falagán C., Grail B. M. and Johnson D. B. (2017). New approaches for extracting and recovering metals from mine tailings. *Minerals Engineering*, **106**, 71–78. doi:10.1016/j.mineng.2016.10.008.

Falayi T. and Ntuli F. (2014). Removal of heavy metals and neutralisation of acid mine drainage with un-activated attapulgite. *Journal of Industrial and Engineering Chemistry*, **20**, 1285–1292. doi:10.1016/j.jiec.2013.07.007.

Florentino A. P., Brienza C., Stams A. J. and Sánchez-Andrea I. (2016). *Desulfurella amilsii* sp. nov., a novel acidotolerant sulfur-respiring bacterium isolated from acidic river sediments. *International Journal of Systematic and Evolutionaty Microbiology*, **66**, 1249–1253. doi:10.1099/ijsem.0.000866.

Fortin D. and Beveridge T. J. (1997). Microbial sulfate reduction within sulfidic mine tailings: formation of diagenetic Fe sulfides. *Geomicrobiology Journal*, **14**, 1–21. doi: 10.1080/01490459709378030.

Fortin D., Davis B. and Beveridge T. J. (1996). Role of *Thiobacillus ferrooxidans* and sulfate-reducing bacteria in iron biocycling in oxic and acidic mine tailings. *FEMS Microbiology Ecology*, **21**, 11–24. doi:10.1111/j.1574-6941.1996.tb00329.x.

Frolov E. N., Kublanov I. V., Toshchakov S. V., Samarov N. I., Novikov A. A., Lebedinsky A. V., Bonch-Osmolovskaya E. A. and Chernyh N. A. (2017). *Thermodesulfobium acidiphilum* sp. nov., a thermoacidophilic, sulfate-reducing, chemoautotrophic bacterium from a thermal site. *International journal of Systematic and Evolutionary Microbiology*, **67**, 1482–1485. doi:10.1099/ijsem.0.001745.

García-Moyano A., González-Toril E., Aguilera Á. and Amils R. (2012). Comparative microbial ecology study of the sediments and the water column of the Río Tinto, an extreme acidic environment. *FEMS Microbiology Ecology*, **81**, 303–14. doi:10.1111/j.1574-6941.2012.01346.x.

Geller W., Klapper H. and Salomons W. (1998). Acidic Mining Lakes. Springer-Verlag, Berlin.

Groudev S., Georgiev P., Spasova I. and Nicolova M. (2008). Bioremediation of acid mine drainage in a uranium deposit. *Hydrometallurgy*, **94**, 93–99. doi:10.1016/j.hydromet.2008.05.023.

Gyure R. A., Konopka A., Brooks A. and Doemel W. (1990). Microbial sulfate reduction in acidic (pH 3) strip-mine lakes. *FEMS Microbiology Ecology*, **73**, 193–202. doi:10.1016/0378–1097(90)90730-E.

Hamilton W. A. (1998). Sulfate-reducing bacteria: physiology determines their environmental impact. *Geomicrobiology Journal*, **15**, 19–28. doi:10.1080/01490459809378059.

Hedrich S. and Johnson D. B. (2014). Remediation and selective recovery of metals from acidic mine waters using novel modular bioreactors. *Environmental Science and Technology*, **48**, 12206–12212. doi:10.1021/es5030367.

Herlihy A. T. and Mills A. L. (1985). Sulfate reduction in freshwater sediments receiving acid mine drainage. *Applied Environmental Microbiology*, **49**, 179–186.

Holanda R. and Johnson D. B. (2020). Removal of zinc from circum-neutral pH mine-impacted waters using a novel 'hybrid' low pH sulfidogenic bioreactor. *Frontiers in Environmental Science*, **8**, 22. doi:10.3389/fenvs.2020.00022.

Huntsman B. E., Solch J. G. and Porter M. O. (1978). Utilization of Sphagnum species dominated bog for coal acid mine drainage abatement. Proceedings of the 91st annual meeting of the Geological Society of America, Toronto, ON, Canada, pp 322–327.

Isaksen M. F. and Teske A. (1996). *Desulforhopalus vacuolatus* gen. nov., sp. nov., a new moderately psychrophilic sulfate-reducing bacteria with gas vacuoles isolated from a temperate estuary. Archives of Microbiology, **166**, 160–168. doi: 10.1007/s002030050371.

Johnson D. B. (2006). Biohydrometallurgy and the environment: Intimate and important interplay. *Hydrometallurgy*, **83**, 153–166. doi:10.1016/j.hydromet.2006.03.051.

Johnson, D. B. and Aguilera, A. (2019). Extremophiles and acidic environments. In: Encyclopedia of Microbiology, Schmidt, Thomas M. (ed.), 4th edn. **2**, Elsevier, UK, pp. 206–227. https://doi.org/10.1016/B978-0-12-809633-8.90687-3

Johnson D. B. and Hallberg K. B. (2002). Pitfalls of passive mine drainage. *Re/Views in Environmental Biotechnology*, **1**, 335–343. doi:10.1023/A:1023219300286.

Johnson D. B. and Sánchez-Andrea I. (2019) Dissimilatory reduction of sulfate and zero-valent sulfur at low pH and its significance for bioremediation and metal recovery. *Advances in Microbial Physiology*, **75**, 205–231. doi:10.1016/bs.ampbs. 2019.07.002.

Johnson D. B., Ghauri M. A. and McGinness S. (1993). Biogeochemical cycling of iron and sulfur in leaching environments. *FEMS Microbiology Review*, **11**, 63–70. doi:10.1111/j.1574-6976.1993.tb00268.x.

Johnson D. B., Hedrich S. and Pakostova E. (2017). Indirect redox transformations of iron, copper and chromium catalyzed by extremely acidophilic bacteria. *Frontiers in Microbiology*, **8**, 211. doi: 10.3389/fmicb.2017.00211.

Karnachuk O. V., Pimenov N. V., Yusupov S. K., Frank Y. A., Kaksonen A. H., Puhakka J. A., Ivanov M. V., Lindström E. B. and Tuovinen O. H. (2005). Sulfate Reduction Potential in Sediments in the Norilsk Mining Area, Northern Siberia. *Geomicrobiology Journal*, **22**, 11–25. doi:10.1080/01490450590922523.

Koschorreck M., Wendt-Potthoff K. and Geller W. (2003). Microbial sulfate reduction at low pH in sediments of an acidic lake in Argentina. *Environmental Science and Technology*, **37**, 1159–1162. doi:10.1021/es0259584.

Kurosawa N., Itoh Y. H., Iwai T., Sugai A., Uda I., Kimura N., Horiuchi T. and Itoh T. (1998). *Sulfurisphaera ohwakuensis* gen. nov., sp. nov., a novel extremely thermophilic acidophile of the order *Sulfolobales*. International Journal of Systematic Bacteriology, **48**, 451–456. doi:10.1099/00207713-48-2-451.

Lewis A. E. (2010). Review in metal sulphide precipitation. *Hydrometallurgy*, **104**, 222–234. doi:10.1016/j.hydromet.2010.06.010.

Loos M. A., Bosch C., Maré J., Immelman E. and Sanderson R. D. (1989). Evaluation of sodium lauryl sulfate, sodium benzoate and sorbic acid as inhibitors of acidification of South African coal waste. Groundwater and Mining. Proceedings of the 5th Biennial symposium of the groundwater division of the Geological Survey of South Africa Randberg, Transvaal, Pretoria, South Africa, pp. 193–200.

López-Archilla A. I., Marin I. and Amils R. (1995). Microbial ecology of an acidic river: biotechnological applications. In: Biohydrometallurgical Processing, T. Vargas, C. A. Jerez, J. V. Wiertz and H. Toledo (eds), University of Chile, Santiago, Chile, pp. 63–74.

Mauric A. and Lottermoser B. G. (2011). Phosphate amendment of metalliferous waste rocks, Century Pb–Zn mine, Australia: Laboratory and field trials. *Applied Geochemistry*, **26**, 45–56. doi:10.1016/j.apgeochem.2010.11.002.

Meulepas R. J., Jagersma C. G., Gieteling J., Buisman C. J., Stams A. J. and Lens P. N. (2009). Enrichment of anaerobic methanotrophs in sulfate-reducing membrane bioreactors. *Biotechnology and Bioengineering*, **104**, 458–470. doi:10.1002/bit.22412.

Miller S. D., Stewart W. S., Rusdinar Y., Schumannoseph R. E., Ciccarelli J. M., Li J. and Smart R. St. C. (2010). Methods for estimation of long-term non-carbonate neutralisation of acid rock drainage. *Science of the Total Environment*, **408**, 2129–2135. doi:10.1016/j.scitotenv.2010.01.011.

Mills A. L. and Mallory L. M. (1987). The community structure of sessile heterotrophic bacteria stressed by acid mine drainage. *Microbial Ecology*, **14**, 219–232. doi:10.1007/BF02012942.

Montoya L., Celis L. B., Gallegos-García M., Razo-Flores E. and Alpuche-Solís A. G. (2013). Consortium diversity of a sulfate-reducing biofilm developed at acidic pH influent conditions in a down-flow fluidized bed reactor. *Engineering in Life Sciences*, **13**, 302–311. doi:10.1002/elsc.201200047.

Mori K., Kim H., Kakegawa T. and Hanada S. (2003). A novel lineage of sulfate-reducing microorganisms: Thermodesulfobiaceae fam. nov., *Thermodesulfobium narugense*, gen. nov., sp. nov., a new thermophilic isolate from a hot spring. *Extremophiles*, **7**, 283–290. doi:10.1007/s00792-003-0320-0.

Ñancucheo, I. and Johnson, D. B. (2012). Selective removal of transition metals from acidic mine waters by novel consortia of acidophilic sulfidogenic bacteria. *Microbial Biotechnology*, **5**, 34–44. doi:10.1111/j.1751-7915.2011.00285.x

Ñancucheo I. and Johnson D. B. (2014). Removal of sulfate from extremely acidic mine waters using low pH sulfidogenic bioreactors. *Hydrometallurgy*, **150**, 222–226. doi:10.1016/j.hydromet.2014.04.025.

Ñancucheo I., Hedrich S. and Johnson D. B. (2012). New microbiological strategies that enable the selective recovery and recycling of metals from acid mine drainage and mine process waters. *Mineralogical Magazine*, **76**, 2683–2692. doi:10.1180/minmag.2012.076.7.04.

Ñancucheo I., Rowe O. F., Hedrich S. and Johnson D. B. (2016). Solid and liquid media for isolating and cultivating acidophilic and acid-tolerant sulfate-reducing bacteria. *FEMS Microbiology Letters*, **363**, 1–6. doi:10.1093/femsle/fnw083.

Neculita C. M., Zagury G. J. and Bussiere B. (2007). Passive treatment of acid mine drainage in bioreactors using sulfate-reducing bacteria: critical review and research needs. *Journal of Environmental Quality*, **36**, 1–16. doi:10.2134/jeq2006.0066.

Nixdorf B., Wollmann K. and Deneke R. (1998). Ecological potential for planktonic development in extremely acidic mining lakes in Lusatia. In: Acidic Mining Lakes, W. Geller, H. Klapper and W. Salomons (eds), Springer-Verlag, Berlin, Germany, pp. 147–168.

Nordstrom D. K., Jenne E. A. and Ball J. W. (1979). Redox equilibria of iron in acid mine waters. In: Chemical Modeling in Aqueous Systems, E. A. Jenne (ed.), ACS Symposium Series, American Chemical Society, Washington D.C, pp. 51–79.

Olds W., Weber P. and Pizey M. (2012). Alkalinity producing covers for minimisation of acid mine drainage generation in waste rock dumps. In: Proceedings of the 8th Australian workshop on acid and metalliferous drainage, H. Miller and L. Preuss (eds), pp. 253–262.

Praharaj T. and Fortin D. (2004). Indicators of microbial sulfate reduction in acidic sulfide-rich mine tailings. *Journal of Geomicrobiology*, **21**, 457–467. doi:10.1080/01490450490505428.

Quatrini R., Ossandon F. J. and Rawlings D. E. (2016). The flexible genome of acidophilic prokaryotes. In: Acidophiles: Life in Extremely Acidic Environments, R. Quatrini and D. B. Johnson (eds), Caister Academic Press, Poole, UK, pp. 199–220. doi:10.21775/9781910190333.12

Rabus R., Hansen T. A. and Widdel F. (2013). Dissimilatory sulfate-and sulfur-reducing prokaryotes. In: The Prokaryotes: Prokaryotic Physiology and Biochemistry, E. Rosenberg, E. F. DeLong, S. Lory, E. Stackebrandt and F. Thompson (eds), Springer, Heidelberg, pp. 309–404. doi:10.1007/978-3-642-30141-4-70.

Rawlings D. E. and Johnson D. B. (2007). The microbiology of biomining: development and optimization of mineral-oxidizing microbial consortia. *Microbiology*, **153**, 315–324. doi:10.1099/mic.0.2006/001206-0.

Robb G. A. (1994). Environmental consequences of coal mine closure. *The Geographical Journal*, **160**, 33–40. doi:10.2307/3060139.

Rowe O. F., Sánchez-España J., Hallberg K. B. and Johnson D. B. (2007). Microbial communities and geochemical dynamics in an extremely acidic, metal-rich stream at an abandoned sulfide mine (Huelva, Spain) underpinned by two functional primary production systems. *Environmental Microbiology*, **9**, 1761–1771. doi:10.1111/j.1462-2920.2007.01294.x.

Rüffel V., Maar M., Dammbrück M. N., Hauröder B., Neu T. R. and Meier J. (2018). *Thermodesulfobium* sp. strain 3baa, an acidophilic sulfate reducing bacterium forming biofilms triggered by mineral precipitation. *Environmental Microbiology*, **20**, 3717–3731. doi:10.1111/1462-2920.14374.

Sánchez-Andrea I., Rodríguez N., Amils R. and Sanz J. L. (2011). Microbial Diversity in Anaerobic sediments at Río Tinto, a naturally acidic environment with a high heavy metal content. *Applied and Environmental Microbiology*, **77**, 6085–6093. doi:10.1128/AEM.00654-11

Sánchez-Andrea I., Stams A. J., Amils R. and Sanz J. L. (2013). Enrichment and isolation of acidophilic sulfate-reducing bacteria from Tinto River sediments. *Environmental Microbiology Reports*, **5**(5), 672–678. doi:10.1111/1758-2229.

Sánchez-Andrea I., Sanz J. L., Bijmans M. F.M. and Stams A. J.M. (2014). Sulfate reduction at low pH to remediate acid mine drainage. *Journal of Hazardous Materials*, **269**, 98–109. doi:10.1016/j.jhazmat.2013.12.032.

Sánchez-Andrea I., Stams A. J. M., Hedrich S., Ñancucheo I. and Johnson D. B. (2015). *Desulfosporosinus acididurans* sp. nov., an acidophilic sulfate-reducing bacterium isolated from acidic sediments. *Extremophiles*, **19**, 39–47. doi:10.1007/s00792-014-0701-6.

Sánchez-España J., Diez M. and Santofimia E. (2013). Mine pit lakes of the Iberian Pyrite Belt: Some basic limnological, hydrogeochemical, and microbiological considerations. In: Acidic Pit Lakes: The Legacy of Coal and Metal Surface Mines, W. Geller, H. Klapper and W. Salomons (eds), Springer-Verlag, Berlin, Germany, pp. 315–342.

Santos A. L. and Johnson D. B. (2017). The effects of temperature and pH on the kinetics of an acidophilic sulfidogenic bioreactor and indigenous microbial communities. *Hydrometallurgy*, **168**, 116–120. doi:10.1016/j.hydromet.2016.07.018.

Santos A. L. and Johnson D. B. (2018). Design and application of a low pH upflow biofilm sulfidogenic bioreactor for recovering transition metals from synthetic waste water at a Brazilian copper mine. *Frontiers in Microbiology*, **9**, 2051. doi:10.3389/fmicb.2018.02051.

Segerer A., Neuner A. M., Kristjansson J. K. and Stetter K. O. (1986). *Acidianus infernus* gen. nov., sp. nov., and *Acidianus brierleyi* comb. nov.: facultatively aerobic, extremely acidophilic thermophilic sulfur-metabolizing archaebacteria. *International Journal of Systematic Bacteriology*, **36**, 559–564. doi:10.1099/00207713-36-4-559.

Segerer A., Langworthy T. A. and Stetter K. O. (1988). *Thermoplasma acidophilum* and *Thermoplasma volcanium* sp. nov. from Solfatara Fields. *Systematic and Applied Microbiology*, **10**, 161–171. doi:10.1016/S0723-2020(88)80031-6.

Segerer A., Trincone A., Gahrtz M. and Stetter K. O. (1991). *Stygiolobus azoricus* gen. nov., sp. nov. represents a novel genus of anaerobic, extremely thermoacidophilic archaebacteria of the order *Sulfolobales*. *International Journal of Systematic Bacteriology*, **41**, 495–501. doi:10.1099/00207713-41-4-495.

Sen A. M. and Johnson D. B. (1999). Acidophilic sulfate-reducing bacteria: candidates for bioremediation of acid mine drainage. *Process Metallurgy*, **9**, 709–718. doi:10.1016/S1572-4409(99)80073-X.

Sheoran A. S. and Sheoran V. (2006). Heavy metal removal mechanism of acid mine drainage in wetlands: a critical review. *Minerals Engineering*, **19**, 105–116. doi:10.1016/j.mineng.2005.08.006.

Simate G. S. and Ndlovu S. (2014). Acid mine drainage: Challenges and opportunities. *Journal of Environmental Chemical Engineering*, **2**, 1785–1803. doi:10.1016/j.jece.2014.07.021.

Singh R., Gautam N., Mishra A. and Gupta R. (2011). Heavy metals and living systems: an overview. *Indian Journal of Pharmacology*, **43**, 246–253. doi:10.4103/0253-7613.81505.

Skousen J., Zipper C. E., Rose A., Ziemkiewicz P. F., Nairn R., McDonald L. M. and Kleinmann R. L. (2017). Review of passive systems for acid mine drainage treatment. *Mine Water Environment*, **36**, 133–153. doi:10.1007/s10230-016-0417-1.

Steudel R. (2000). The chemical sulfur cycle. In: Environmental Technologies to Treat Sulfur Pollution: Principles and Engineering, P. Lens and L. Hulshoff Pol (eds), International Association on Water Quality, London, pp. 1–31.

Stucki G., Hanselmann K. W. and Hurzeler A. (1993). Biological sulfuric acid transformation: reactor design and process optimization. *Biotechnology and Bioengineering*, **41**, 303–315. doi:10.1002/bit.260410304.

Thoreau H. D. (2002). Effects of acid rain on aquatic species. http://www.rst2.edu/ties/acidrain/PDF/3effects/ef6.pdf.

Tichý R., Lens P., Grotenhuis J. T. C. and Bos P. (1998). Solid-state reduced sulfur compounds: environmental aspects and bioremediation. *Critical Reviews in Environmental Science and Technology*, **28**, 1–40. doi:10.1080/10643389891254188.

Tuttle J. H., Dugan P. R. and Randles C. I. (1969). Microbial dissimilatory sulfur cycle in acid mine water. *Journal of Bacteriology*, **97**, 594–602.

USDA (1993). Acid drainage from mines in the National Forest: a management challenge. U. S. Department of Agriculture, Forest Service. *Program Aid*, **1505**, 1–12. https://archive.org/details/CAT31108485/page/n1.

Veloso R. W., Glasauer S., Mello J. W.V. and Cardoso L. P. (2012). Comparison between two sequential extraction procedures to assess Arsenic mobility. *Annual Activity Report* 2011–2012 INTC-Acqua, **1**, 88–89.

Vera M., Sand W. and Shippers A. (2013). Progress in bioleaching: Fundamentals and mechanisms of bacterial metal sulfide oxidation-Part A. *Applied Microbiology and Biotechnology*, **97**, 7529–7541. doi:10.1007/s00253-013-4954-2.

Villain L., Alakangas L. and Öhlander B. (2013). The effects of backfilling and sealing the waste rock on water quality at the Kimheden open-pit mine, northern Sweden. *Journal of Geochemical Exploration*, **134**, 99–110. doi: 10.1016/j.gexplo.2013.08.003.

Widdel F. and Hansen T. A. (1992). The dissimilatory sulfate- and sulfur-reducing bacteria. In: The Prokaryotes, A. Balows, H. G. Trüper, M. Dworkin, W. Harder and K. H. Schleifer (eds), Springer-Verlag, Berlin, Germany, pp. 583–624.

Willis G., Nancucheo I., Hedrich S., Giaveno A. G., Donati E. and Johnson D. B. (2019) Enrichment and isolation of acid-tolerant sulfate-reducing microorganisms from anoxic, acidic hot spring sediments in the Copahue volcano, Argentina. *FEMS Microbiology Ecology*, **95**(12), fiz175.

Younger P. L., Banwart S. A. and Hedin R. S. (2002). Mine Water: Hydrology, Pollution, Remediation. Springer Science and Business Media, Berlin.

Chapter 8

Electrochemical removal of sulfur pollution

Eleftheria Ntagia, Antonin Prévoteau and Korneel Rabaey

8.1 INTRODUCTION

Sulfur compounds are common contaminants in gaseous and aqueous waste streams such as sour gases, sewage and industrial wastewaters. The broad range of sulfur compounds contained in the wastewater, as well as the total sulfur concentration, the pH range and the presence of additional contaminants, e.g. organic compounds or trace metals, will define the selection of the treatment strategies. Sulfur pollution treatment today comprises a wide range of biological (Alcántara *et al.*, 2004; Buisman *et al.*, 1989, 1999; Janssen *et al.*, 2009; Lens *et al.*, 1998) and physicochemical (Eow, 2002; Pandey & Malhotra, 1999; Zhang *et al.*, 2008) processes. These treatment strategies target the removal of different sulfur forms, ranging from the toxic hydrogen sulfide gas (H_2S) to sulfur oxyanions and organosulfur compounds. The majority of the physicochemical approaches are chemical and energy intensive. Conversely, biological processes are environmentally benign in comparison, but usually suffer from lower conversion or removal rates. Furthermore, biological processes fail to accommodate the treatment of higher strength industrial wastewaters without intensive pre-treatment steps to reach biocompatibility.

In recent years, increasing attention has been given to electrochemical treatment. It has been proposed as a more sustainable solution as the oxidation and reduction

DOI: 10.2166/9781789060966_0247

reactions required are driven by a clean reagent that can be produced in a sustainable manner: the electron (Katsounaros *et al.*, 2014; Radjenovic & Sedlak, 2015). Additionally, electrochemical treatment allows for the recovery of potentially valuable chemicals, as well as, in certain applications, electrical power generation (Feng *et al.*, 2016; Sun *et al.*, 2016).

In this chapter, we first provide an introduction to the field of environmental electrochemistry and the basics of electrochemical engineering. Secondly, the electrochemical properties of the sulfur components possibly involved in electrochemical treatment are outlined. Finally, the different concepts of applied electrochemistry for the treatment of sulfur-containing wastewaters are overviewed for both abiotic and biotic methods.

8.2 ENVIRONMENTAL ELECTROCHEMISTRY TO TREAT SULFUR POLLUTION

8.2.1 Brief introduction to environmental electrochemistry

Environmental electrochemistry emerged around 30 years ago, when it was suggested that the industrially efficient electrochemistry could be used for environmental protection in a more sustainable manner (Comninellis & Chen, 2010; Feng *et al.*, 2016; Janssen & Koene, 2002). Already in the 1970s, Bockris gave, in the introduction of his book "Electrochemistry of Cleaner Environments" (Bockris, 1972), the perspectives of an electrochemical future, which included a range of applications of electrochemistry for the treatment of sewage. In his book an optimistic closed-loop approach was introduced, starting from the electrochemical oxidation of organic matter, to the electroreduction of CO_2 to formaldehyde and from there to the production of protein. Nowadays electrochemical methods are proposed for organic pollutants removal (Martínez-Huitle *et al.*, 2015), microbial electrosynthesis of fine chemicals (Nevin *et al.*, 2010; Rabaey & Rozendal, 2010), as well as abiotic CO_2 capture and reuse (Luna *et al.*, 2019).

The main principle of environmental electrochemistry is that the driving force for environmental remediation is an electron flux generally generated by an external power source (Bockris, 1972; Feng *et al.*, 2016). The electrooxidation and electroreduction reactions that are used in physicochemical waste treatment are conducted in an electrochemical cell. The promising advantages of electrochemical technologies arise mainly from the use of electrical energy rather than more conventional chemical treatments. The ability to easily control the parameters (current or voltage) controlling the performance of the cell with a virtually non-limited electron source is considerably advantageous when compared to the safety, environmental and logistic issues generally induced by chemical-intensive processes. This can result in high operational and energy efficiencies, especially when applied to concentrated streams, besides ease of automation, amenability to versatile operating conditions, cost-effectiveness and sustainability when renewable energy resources and remote areas of applications are considered (Feng *et al.*, 2016; Martínez-Huitle *et al.*, 2015; Radjenovic & Sedlak, 2015). A major advantage is that electrochemical processes generally do not require excessive use of chemicals that could end up as residual salts in effluent streams, requiring further treatment.

The versatility of electrochemistry has now led to many different approaches to treat sulfur-containing streams, ranging from, direct or indirect, abiotic electrooxidation or electroreduction of the sulfur compounds, electrodialysis, electrocoagulation and electroflotation to microbial fuel cells and microbial electrolysis cells. In the following section, we will explain some of the basic principles governing electrochemistry.

8.2.2 Basics of electrochemical engineering for environmental applications

Three key aspects need to be considered when assessing electrochemical processes: (i) the thermodynamics of the electrochemical reactions possibly involved, (ii) the electron transfer kinetics at the electrode/electrolyte interface, (iii) the design of the electrochemical cell or electrochemical reactor, and (iv) the performance of the processes such as current efficiency and electric power investment.

8.2.2.1 The electrochemical cell

An electrochemical cell is usually comprised of: (1) two electrodes: an anode and a cathode on which oxidation and reduction reactions are conducted, respectively; (2) one or several electrolytes, which are conductive solutions containing the chemical reagents taking part in the reactions as well as supporting (i.e. non-electroactive) ions allowing sufficient conductivity; (3) one or several solid separators, which are typically ion-exchange membranes which can be permselective or not, depending on the process (Figure 8.1).

Electrochemical cells are divided into galvanic and electrolytic cells. The cell is considered as galvanic when the overall cell reaction (i.e. considering both anodic and cathodic reactions as a single redox process) is thermodynamically favourable ($\Delta G < 0$), allowing for spontaneous reactions to occur on both electrodes and possibly for generation of electric energy, if the cell is not under short circuit (i.e. inducing a non-zero voltage). The anode of a galvanic cell is the negative electrode, whereas the cathode is positive. Conversely, an electrolytic cell requires the addition of an external electric power source to drive a non-spontaneous reaction ($\Delta G \geq 0$), i.e. to translate electric energy into chemical energy (Ciobanu *et al.*, 2007; Pourbaix, 1966). The corresponding electrode polarity, applied by the external power source, is opposite to the galvanic cell, with a positive anode and a negative cathode.

The electrodes are connected to a power supply or potentiostat/galvanostat through an electrical circuit, that enables electron flow from the anode to the

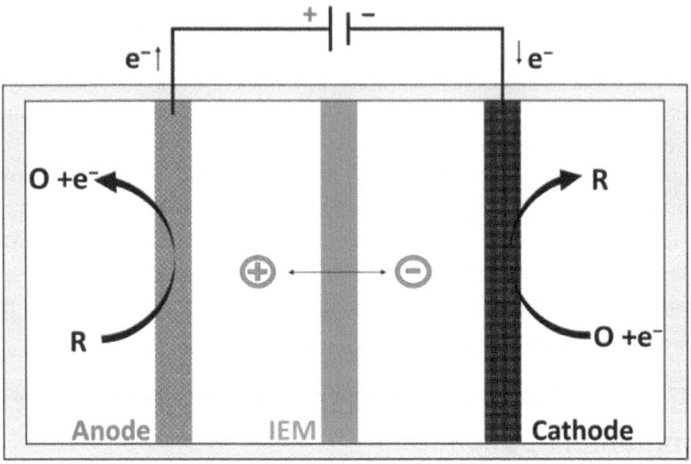

Figure 8.1 Schematic drawing of a two-compartment electrochemical cell controlled with an external power source that enables electron flow from the anode to the cathode. The two electrodes, anode and cathode, are separated with an ion-exchange membrane (IEM), that allows for (selective) passage of positively or negatively charged ions. An oxidation reaction is conducted at the anode (positively charged electrode) and a reduction reaction is conducted at the cathode (negatively charged electrode).

cathode. Simultaneously, an ion flow takes place in the electrolyte, where anions migrate towards the anode and cations towards the cathode to ensure charge balance. The ion-exchange membrane allows for selective passage of negatively (anion-exchange membrane) or positively (cation-exchange membrane) charged ions. Using a membrane results in the formation of two electrolyte chambers (Jiang *et al.*, 2012). This separation provides specific advantages, such as product recovery in the separate electrolyte chambers, pH gradient, and separation of toxic products in the case of bioelectrochemical systems.

During the operation of an electrochemical cell the experimental conditions might vary and defining the absolute potential of an electrode is an almost impossible achievement. Therefore, a third electrode with a well-defined equilibrium potential is adjusted close to the electrode of interest (anode or cathode) in the electrochemical cell, known as a reference electrode. The reference electrode is an electrode that is immersed in a defined electrolyte (usually 3 M or saturated KCl) and maintains a constant potential during the experimental period (Smith & Stevenson, 2007). This enables continuous monitoring of the electrode potentials, relative to the reference electrode (Rabaey, 2010). A number of different reference electrodes exist and their conversion to the standard hydrogen electrode (SHE) at 25°C is given in Table 8.1.

Table 8.1 Common aqueous reference electrodes (RE) and their potentials (V) versus the Standard Hydrogen Electrode (SHE), at 25°C, depending on the composition of their inner electrolyte and headspace partial pressure when applicable, adjusted from Smith and Stevenson (2007).

Electrode Description		Electrolyte Composition	V vs SHE ($T = 25°C$)
Hydrogen	NHE or SHE	$a_{H}^{+} = 1$; $p_{H2} = 1$ atm	0
Calomel	SCE	Saturated KCl	+0.241
	SSCE	Saturated NaCl	+0.236
	NCE	1 M KCl	+0.280
Mercury/Mercurous Sulfate	MSRE	Saturated K_2SO_4	+0.640
Mercury/Mercury Oxide		1 M NaOH	+0.098
Silver/Silver Chloride		Saturated KCl	+0.197
		3 M KCl/NaCl	+0.209
Copper/Copper Sulfate		Saturated $CuSO_4$	+0.316

NHE: Normal Hydrogen Electrode, SHE: Standard Hydrogen Electrode, SCE: Standard Calomel Electrode, SSCE: Saturated Salt Calomel Electrode, NCE: Normal Calomel Electrode, MSRE: Mercury Sulfate Reference Electrode.

8.2.2.2 Thermodynamics of electrochemical reactions and the electrode potential

Electrochemical reactions differ from homogeneous chemical reactions because they occur at the interface between the electrodes and their respective electrolyte, stressing the dramatic impact of electrode surface area and mass transfer parameters on the overall rate of reactions. An electrochemical reaction can be seen as a "delocalized" redox reaction, physically separating the two half-reactions that are the oxidation (at the anode/electrolyte interface) and the reduction (at the cathode) (Pourbaix, 1966). As such, electrochemical studies generally focus on those half-reactions separately, considering they are specific, amongst others, to the redox couple(s) involved, the electrode material and the nature of the electrolyte.

In electrochemistry, the basic form of the Nernst equation characterizes the equilibrium for an electrochemical reaction involving an inert electrode and a single dissolved redox couple O/R (where O represents the oxidized form and R the reduced form of the redox couple). The Nernst equation links the electrode potential with the ratio of the interfacial concentration of the oxidized and reduced forms of the redox couple when the system is at equilibrium (Bard & Faulkner, 2001). For an electrochemical reaction involving a single redox couple O/R (without a proton coupled to the electron transfer):

$$O + ne \rightleftarrows R$$

the Nernst equation is:

$$E = E^0 + \frac{RT}{nF}\ln\frac{a_O}{a_R} = E^{0'} + \frac{RT}{nF}\ln\frac{C_O}{C_R}$$

where E is the electrode potential (V), E^0 is the standard potential of the considered redox couple (V), $E^{0'}$ is the formal potential, R is the universal gas constant (8.314 J mol^{-1} K^{-1}), T is the temperature (K), F is the Faraday constant (96,485 C mol^{-1}), n is the number of electrons (dimensionless) involved in the reaction, α_O and α_R are the activities (dimensionless) and C_O and C_R are the interfacial concentrations (mol L^{-1}) of, respectively, the oxidized and reduced form of the considered redox couple. Whereas E^0 is a thermodynamic constant for a specific redox couple at a specific T, $E^{0'}$ varies with the activity coefficients of the redox compounds, and therefore with their concentration as well as the ionic strength of the medium (Bard & Faulkner, 2001). The IUPAC (International Union of Pure and Applied Chemistry) convention hereby used, considers a positive electrode potential as the one favouring oxidation reactions (low energy electrons on electrode surface) and a negative electrode potential as the one favouring reductions (high energy electrons on electrode surface).

It is often assumed that the redox compounds are at sufficiently low concentrations, thus the latter are considered to be equal to the activities, and the Nernst equation is commonly used with the standard potential and the concentrations (Ciobanu *et al.*, 2007). The standard potential is used with reference to the Standard Hydrogen Electrode (SHE), which corresponds to the standard potential of the H^+/H_2 couple, i.e. the corresponding equilibrium potential between the redox couple and an inert electrode for all reagent activities equal to 1 (pH 0 and 1 atm partial pressure of H_2).

A chemical equilibrium is the state at which no change in the properties of the system occurs. For an electrochemical system involving a single redox couple, the chemical equilibrium corresponds to a state at which the oxidation and reduction rates are strictly identical at the electrode/electrolyte interface, implying a zero net reaction and the absence of electric current ($I = 0$). Conversely, one must keep in mind that an electrochemical cell (or a single electrode) in open circuit conditions (i.e. $I = 0$) is not *necessarily* at equilibrium, but can be the result of oxidation and reduction processes of identical rate but involving different redox couples. A typical example is the corrosion process where the metal is oxidized and an oxidizing agent such as O_2 is reduced at the same rate.

The Nernst equation is valid for electrochemical equilibria, but also for non-equilibrium conditions when a so called *reversible* electron transfer occurs at the electrode/electrolyte interface. A system is *reversible* when the intrinsic kinetics of electron transfer at the interface are particularly fast with respect to the kinetics of mass transfer of the reagent(s) and product(s) involved in the reaction. (The full concept(s) of electrochemical *reversibility* is complex and out of the scope of this chapter. An interested reader could refer to the dedicated section "2.1.1 Reversibility" of the seminal electrochemistry textbook from Bard and Faulkner (Bard & Faulkner, 2001)). The electrons can be transferred so fast that there is virtually no delay for the interfacial concentrations of the redox species to obey the surface thermodynamics, i.e. to adapt to the electrode potential with respect to the Nernst equation. As such, the Nernst equation is also valid when a non-zero current goes across an electrode where a reversible electron transfer takes place.

When wastewater is considered as an electrolyte in electrochemical treatment, it is expected that the pH will vary substantially from one case to another. To calculate the electrode potential E at different operational pH values, one can write an electrochemical reaction to include proton exchange coupled to electron transfer:

$$aA + cH_2O + ne^- \rightleftharpoons bB + mH^+$$

where A is the acid form and B the alkaline form of a compound. E can then be calculated based on the Nernst equation taking into account the influence of the pH:

$$E = E^0 - \frac{0.0591 \times m}{n} \times \text{pH} + \frac{0.0591}{n} \times \log\frac{(A)^a}{(B)^b}$$

with E and E^0 in V (Pourbaix, 1966). This is a useful and fast tool to calculate the expected theoretical electrode potential in an electrochemical reactor, operated at pH $\neq 0$.

The free energy change of the overall electrochemical reaction conducted in the cell is related to the cell voltage (electrode potential difference) according to:

$$\Delta G_r^0 = -nFE^0$$

where ΔG_r^0 is the standard Gibbs free energy of the overall cell reaction (J mol^{-1}) (Ciobanu *et al.*, 2007; Comninellis & Chen, 2010).

8.2.2.3 Overpotential and ohmic resistance

The electrochemical reactions are called heterogeneous since they occur at the electrode-electrolyte interface. The rate of a specific electrochemical reaction directly depends on the current (flux of charges) and the coulombic efficiency (*vide infra*) associated with this reaction. The relationship between current density and electrode potential is essential to define an electrochemical system. For a specific combination of electrode and redox couple, the reaction rate depends upon mass transfer to the electrode, kinetic variables and surface effects (Bard & Faulkner, 2001). Assuming all the electrons exchanged at the electrode interface are dedicated to a single electrochemical reaction, the relationship between the surface reaction rate (v) and the electric current (i) is described by

Faraday's law of electrolysis:

$$v = \frac{i}{n \times F \times A} = \frac{j}{n \times F}$$

where v is given in mol s^{-1} m^{-2} and j is the current density, which is the current applied per surface area (A) of the electrode (usually the projected or geometric surface area is reported in electrochemical engineering applications).

When current is applied in an electrochemical cell, previously at equilibrium, both anode and cathode potentials are shifted from their equilibrium value. This shift in the electrode potential results in what is called overpotential (η), which is the difference between an electrode potential under a current density j and its equilibrium potential ($j = 0$):

$$\eta_{(j)} = E_{(j)} - E_{\text{eq.}} = E_{(j)} - E_{(j=0)}$$

where $E_{(j)}$ is the potential that the electrode acquires upon application of a specific current density (j). Reaction overpotentials are, therefore, a loss with respect to the thermodynamic potential and should be minimized in absolute value when possible. The overpotential at a specific current density depends on a series of different parameters, including the nature of the redox reaction and the reagents and products concentration, the mass transfer parameters at the bulk electrolyte and the electrode interface, the nature of the electrode material, the electrolyte composition and the temperature.

An electrochemical cell necessarily exhibits an internal ohmic resistance (R_{int}) which is the sum of ionic resistances (electrolytes and ion exchange membrane (s)) and electronic resistances (solid electrode, external circuit and the contact resistance from their respective connections). Consequently, a current flowing through the electrochemical cell causes an additional energy loss with a corresponding ohmic drop which increases the voltage of an electrolysis cell, or decreases the voltage of an electrochemical power source (fuel cell) (Bard & Faulkner, 2001). This ohmic drop (in Ω) is the product of the applied current and the internal resistance of the cell:

$$\text{Ohmic drop} = I \times R_{\text{int}}$$

The magnitude of the ohmic drop is particularly important since it can substantially impact the voltage and, therefore, the energy efficiency of the system. An internal resistance that is too high can considerably impede the development of an electrochemical process (Prévoteau *et al.*, 2020). It is therefore important to quantify internal resistances, for example via the current interrupt method (Bard & Faulkner, 2001). Since the ionic resistance often represents a substantial fraction of R_{int}, the latter can be decreased by (i) decreasing the distance between the anode and the cathode, (ii) using more conductive membranes or (iii) using more conductive electrolytes. Whereas higher current densities allow for higher removal rates, they necessarily increase the ohmic drop proportionally (assuming R_{int} is invariant within the range of current considered).

The ohmic drop is also an important concept to consider when studying electrochemical reactions on a single electrode, i.e. when focusing on a half-cell of the electrochemical setup. The applied, or recorded, potential (E_{rec}) of the working electrode will include an ohmic drop contribution that is proportional to the uncompensated resistance R_u present between the working electrode and the reference electrode (due to the electrolyte present between them). Thus, the recorded potential can be corrected to obtain the real working electrode potential (E_{real}):

$$E_{real} = E_{rec} - i \times R_u$$

where the current follows the aforementioned convention, i.e. positive for an anode ($E_{real} \leq E_{rec}$) and negative for a cathode ($E_{real} \geq E_{rec}$). It is recommended to limit the amplitude of R_u by positioning the tip of a reference electrode in close vicinity to the working electrode, for example by using a Luggin capillary, as long as the mass transfer properties from and towards the electrode surface are not substantially impacted.

8.2.2.4 Efficiencies of the electrochemical process

The efficiency of an electrochemical process is based on the efficiency with which charge is transported through the electrochemical cell and is facilitating the electrochemical reactions. This is described by Faraday's law, that relates the amount of electrochemical reactions conducted at an electrode to an amount of charges exchanged through the electrochemical cell (Bard & Faulkner, 2001). Based on Faraday's law of electrolysis, the coulombic efficiency (also called "faradaic efficiency", or "current efficiency") can be defined as the ratio of output charges of interest to the input charges. For electrolysis, it corresponds to the amount of charge actually involved in the electrochemical reaction(s) of interest divided by the amount of charges that actually crossed the electrolysis cell:

$$CE\ (\%) = \frac{n \times F \times \Delta m}{\int_0^t I(t) \times dt}$$

where CE (%) is the coulombic efficiency, Δm is the change in the number of moles (mol) of the redox compound of interest during the electrolysis time Δt (s) and I is the current ($1\ A = 1\ C\ s^{-1}$).

Conversely, the coulombic efficiency for an electrochemical power source (fuel cell) is the amount of charge that actually crossed the electric circuit (i.e. generating the electric energy output) divided by the theoretical amount of charge that could be exchanged by the amount of electroactive species consumed (Comninellis & Chen, 2010; Logan *et al.*, 2006; Rabaey *et al.*, 2006).

In the same way, in an electrochemical cell the energy efficiency is used to describe the energy output of the system compared to the energy input. The energy efficiency of the process will highly depend on the internal resistance of the cell, as explained in section 8.2.2.3, as the higher the losses in the electrochemical cell, the lower the energy efficiency of the process will be. The energy efficiency simply describes how much of the energy input, in an electrolysis cell for example, ends up in the desired reaction, rather than side reactions or ohmic losses in the system. The energy efficiency will thus be the product of the voltage and current efficiencies:

$$EE = \frac{E_{eq}}{E_{cell}} \times CE,$$

Where E_{eq} is the cell voltage at equilibrium and E_{cell} the cell voltage obtained under operation at a specific current density (Klans Jüttner, 2007).

The operational performance of the electrochemical cells is usually evaluated based on the removal rate (with respect to an electrode surface or a reactor volume), the removal efficiency or product recovery as well as on the power consumption by an electrolysis cell or the power output of a fuel cell. In both cases, the power input/output is given by a combination of Joule's law with Ohm's law (Bard & Faulkner, 2001; Logan *et al.*, 2006):

$$P = I \times E_{cell}$$

where P is the power (W) and E_{cell} is the cell voltage (V). The operational costs of electrochemical treatment are directly related to the power consumption of the electrochemical cell. It is, therefore, of utmost importance to minimize this power consumption. In an electrolysis cell, this can be done by reducing the cell voltage, by decreasing the distance between the two electrodes, or by increasing the electrolyte conductivity, among other measures.

8.2.3 Oxidation and reduction reactions involving sulfur species

Extensive research on electrochemical equilibria and electrochemical oxidative and reductive processes of sulfur species has taken place in the last 40 years, mainly because of the complexity of the sulfur chemistry (Levillain *et al.*, 2006). Considerable inconsistency exists in the sulfur literature regarding the final products of sulfide oxidation and S^0 reduction, due to a lack of suitable analytical techniques to study many of the different sulfur forms in existence (Levillain *et al.*, 2006). The most common sulfur compounds found in electrochemical sulfur pollution treatment are presented in Table 8.2, with their respective oxidation numbers.

In order to study and predict the thermodynamic stability area of aqueous and solid sulfur compounds, one can use the Pourbaix diagrams. The Pourbaix

Table 8.2 Sulfur compounds considered in electrochemical sulfur pollution treatment with their respective oxidation numbers (Pourbaix, 1966).

Compound Name	Formula	Oxidation Number
Hydrogen sulfide	H_2S	−2
Bisulfide (or hydrogen sulfide ion)	HS^-	−2
Elemental sulfur	S^0	0
Polysulfides	S_n^{2-}	−1 to −0.40
Thiosulfate	$S_2O_3^{2-}$	+2
Sulfite	SO_3^{2-}	+4
Sulfate	SO_4^{2-}	+6

diagrams, or potential (Eh)-pH equilibrium diagrams, indicate which redox species are predominant at thermodynamic equilibrium, expressed as functions of the pH and the redox potential of the solution (Pourbaix, 1966). These, however, only describe the situation under equilibrium conditions, the ultimate understanding and prediction of electrochemical reactions require knowledge of kinetics as well (Bouroushian, 2010).

The potential-pH diagram for aqueous sulfur compounds is given in Figure 8.2, excluding thiosulfate ($S_2O_3^{2-}$), sulfite (SO_3^{2-}), dithionate ($S_2O_4^{2-}$) and polythionates ($S_n(SO_3)_2^{2-}$), which are thermodynamically unstable and tend to decompose in aqueous solutions (Pourbaix, 1966). Only H_2S, HS^-, S, HSO_4^- and SO_4^{2-} will be found in solution in equilibrium (Dutta *et al.*, 2010; Pourbaix, 1966). Zero-valent sulfur is stable in the presence of water and acid solutions, but unstable in alkaline regions, where it disproportionates to HS^-, polysulfides and SO_4^{2-}. Sulfate is stable in the presence of water and aqueous solutions at all pH values and the same applies for sulfide (Pourbaix, 1966). Table 8.3 overviews the most common reversible electrochemical reactions which take place in electrochemical cells treating sulfur containing electrolytes.

8.3 ABIOTIC ELECTROCHEMICAL TREATMENT OF SULFUR SPECIES

8.3.1 Introduction

Electrochemical sulfur removal has mainly focused on the removal of sulfide (Pikaar *et al.*, 2015), which is the corrosive (see Chapter 4) and toxic (see Chapter 6) form of sulfur pollution. For sulfide removal, the electrochemical processes follow a similar approach as any of the other electrochemical wastewater treatment technologies: a combination of oxidation processes that can be direct (i.e. at the electrode/electrolyte interface) and indirect (homogeneous

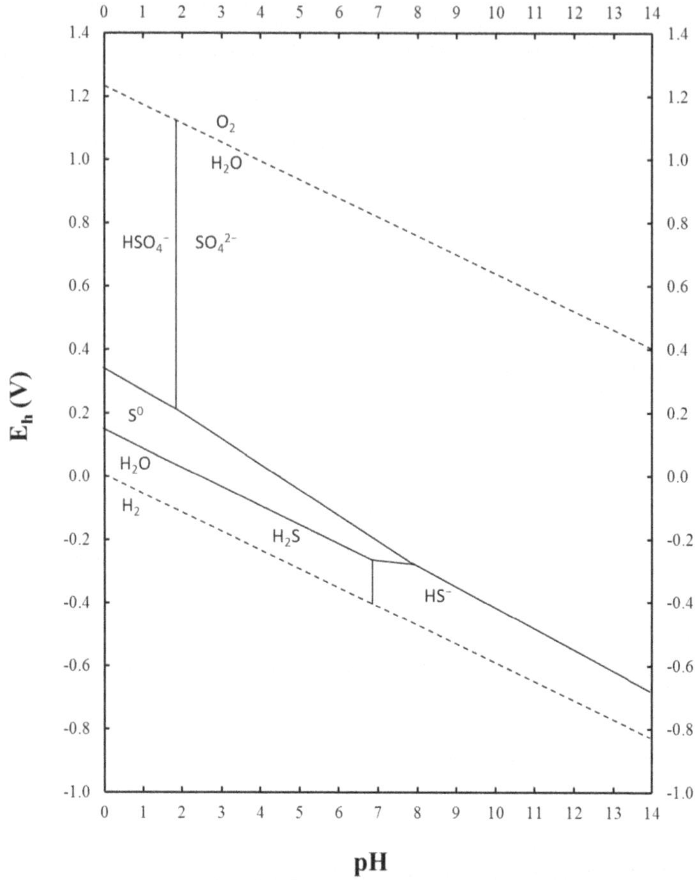

Figure 8.2 Potential (Eh)/pH diagram for the system sulfur-water, including predominance regions at equilibrium for H_2O, O_2, H_2, S^0, H_2S, HS^-, HSO_4^- and SO_4^{2-} at standard temperature and pressure (STP) and [S] = 1 M, produced based on Pourbaix (1966) and Dutta *et al.* (2010). Potentials are expressed with respect to SHE.

redox process in solution). In the first case, the sulfide is electrochemically oxidized at the electrode surface. In the second case, the oxidation proceeds through anodic production of oxidative intermediates such as oxygen or chlorine which can further oxidize sulfide in solution. Electrochemical sulfide removal mainly comprises electrochemical membrane processes, such as membrane divided electrolysis (Ntagia *et al.*, 2020; Pikaar *et al.*, 2011a; Vaiopoulou *et al.*, 2016), or electrochemically induced sulfide removal with "sacrificial anodes" (Hariz *et al.*, 2013; Lin *et al.*, 2017; Murugananthan *et al.*, 2004).

Table 8.3 Standard redox potentials (E^0, pH 0) and apparent standard redox potentials at pH 7 (E^0_{app}) of the most common sulfur involving half-reactions (Bouroushian, 2010; Dutta *et al.*, 2010; Levillain *et al.*, 2006; Mao *et al.*, 1991; Pourbaix, 1966).

Half-reaction	E^0 (V vs SHE, at 25°C)	E^0_{app} (V vs SHE, at 25°C and pH = 7)
$nS(s) + 2e^- \rightarrow S_n^{2-}, 4 \leq n \leq 5$	−0.340 to − 0.315	−0.340 to − 0.315
$S(s) + H^+ + 2e^- \rightarrow HS^-$	−0.065	−0.272
$S_n^{2-} + nH^+ + (2n-2)e^- \rightarrow nHS^-, 2 \leq n \leq 5$	+0.007 to + 0.298	−0.251 to − 0.116
$S(s) + 2H^+ + 2e^- \rightarrow H_2S(aq)$	+0.142	−0.272
$S_2O_3^{2-} + 8H^+ + 8e^- \rightarrow 2HS^- + 3H_2O$	+0.200	−0.214
$SO_4^{2-} + 9H^+ + 8e^- \rightarrow HS^- + 4H_2O$	+0.252	−0.214
$SO_4^{2-} + 8H^+ + 6e^- \rightarrow S(s) + 4H_2O$	+0.357	−0.194

8.3.2 Sulfide removal with electrolysis cells

Studies on the effect of sulfide oxidation on different electrode materials, and vice versa, initiated the field of direct electrochemical desulfurization around 50 years ago (Hamilton & Woods, 1983; Ramasubramanian, 1975). Since then electrochemical sulfide oxidation has been studied extensively, mainly due to the fact that sulfide oxidation to S^0 readily occurs at a low potential ($E_{eq} = -0.419$ at pH 12, Table 8.3), possibly allowing for low operational voltages, i.e. energetically attractive processes. In addition, electrochemical treatment can allow for on-site recovery of sulfur compounds (e.g. elemental sulfur (S^0) and sulfuric acid (H_2SO_4)), or electrolysis side products (H_2 gas, caustic soda (NaOH), see Figure 8.3) (Dutta *et al.*, 2008; Rabaey *et al.*, 2006; Vaiopoulou *et al.*, 2016), ultimately resulting in the minimization of the operational costs (Chatzisymeon *et al.*, 2013; Pikaar *et al.*, 2015; Rankin *et al.*, 2010).

Lab-scale electrochemical desulfurization systems have been tested for a wide range of liquid waste streams, including tannery wastewater (Rajalo & Petrovskaya, 1996; Szpyrkowicz *et al.*, 2005), pulp and paper processed water and alkaline sulfidic streams (Kim & Han, 2014; Vaiopoulou *et al.*, 2016; Wang *et al.*, 2015), geothermal brines (Ateya & Al-Kharafi, 2002; Ateya *et al.*, 2003; El-Sherif *et al.*, 2010; Rankin *et al.*, 2010), domestic wastewater (Pikaar *et al.*, 2011a, 2011b, 2012) and mixed liquor of anaerobic digesters (Lin *et al.*, 2016).

Another method proposed is electrochemically induced precipitation of sulfide, using "sacrificial anodes". In this case, a membraneless electrochemical cell is employed, consisting of two parallel identical metal plates under galvanic control. Under current, the anode usually generates oxidized metal ions, for example iron (II) and iron (III) species, that precipitate dissolved sulfide as metal sulfides

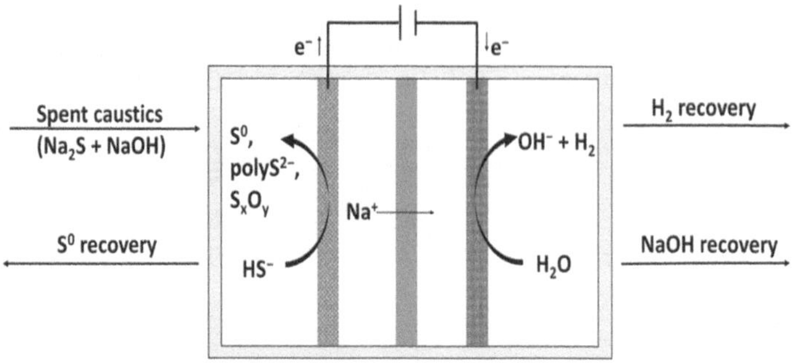

Figure 8.3 Schematic drawing of a two-compartment electrochemical cell for the treatment of spent caustic streams containing sulfide (Ntagia *et al.*, 2020). The stream enters the anolyte compartment where sulfide is electrochemically oxidized, generating different oxidation products that may be recovered. In addition, the current simultaneously drives (i) the migration of sodium ions (Na^+) from the anolyte to the catholyte across a cation exchange membrane (CEM), and (ii) the generation of hydroxide ions (OH^-) and H_2 at the cathode, allowing for NaOH and H_2 recovery.

(Hariz *et al.*, 2013; Lin *et al.*, 2017; Murugananthan *et al.*, 2004). Table 8.4 lists the most recent studies on direct electrochemical desulfurization, as well as a few studies conducted on electrochemically induced sulfide metal precipitation, showing a wide range of cell configurations, wastewater types and electrode materials that have been tested.

Extensive research has been dedicated to the electrochemical treatment of alkaline sulfidic streams, widely known as spent caustic streams (SCS), as they have a remarkably high sulfide content (0.5–4 wt% sulfide-S) and pose a financial burden for petrochemical refineries, chemical manufacturing and the pulp and paper industry (Alnaizy, 2008; Pokhrel & Viraraghavan, 2004). Due to the high sulfidic content and the highly alkaline nature (pH > 13) of these streams, biological treatment is hindered, while the currently employed chemical methods require extreme operational conditions ($125 < T < 320°C$ and $0.5 < P < 20$ MPa), such as for wet oxidation (Hawari *et al.*, 2015; Sheu & Weng, 2001; Zermeño-Montante *et al.*, 2011). Moreover, the key component, caustic soda, is not recovered with the conventional methods. Instead, electrochemical treatment performs efficiently at ambient conditions, allows for recovery of chemicals and is highly adjustable to variable operational conditions (Chaplin, 2019). Finally, an additional motivation for the electrochemical treatment of industrial streams is their typically high conductivity (a few mS cm^{-1}), that makes them an attractive electrolyte, as it reduces the ohmic losses in the electrolysis cell and thus decreases the overall power input required to drive the process (Ateya *et al.*, 2003).

Table 8.4 Electrochemical sulfide removal performance when treating sulfur-contaminated wastewaters, with different anode materials, including sacrificial anodes (updated from Dutta *et al.*, 2010).

Cell Configuration	Anode Material	Electrolyte (T/pH Electrolysis)	Applied Current Density or Potential/E_{AN} (vs SHE)	Reference
Two chamber (separated by CEM)	Ta/Ir coated Ti	Industrial spent caustic streams (SCS) (25°C/14)	300 A m^{-2}	Ntagia *et al.* (2020)
Rotating cylinder reactor (separated by CEM*)	Pb coated Cu	Synthetic alkaline medium (80°C/10)	−0.6 – +1.0 V	Fornés and Bisang (2017)
Two chamber (separated by CEM)	Ta/Ir coated Ti	Synthetic alkaline spent caustic streams (SCS) (25°C/14)	100 – 300 A m^{-2}	Vaiopoulou *et al.* (2016)
Two chamber (separated by microporous membrane)	Ti/RuO_2	Synthetic alkaline medium (20°C/13)	250 A m^{-2}	Wang *et al.* (2015)
Two chamber (separated by CEM)	Ta/Ir, Ru/Ir, Pt/Ir, PbO_2 and SnO_2 coated Ti	Domestic wastewater (~25°C/7.5)	100 A m^{-2}	Pikaar *et al.* (2011b)
Undivided polarization cell	Polycrystalline Pt	Synthetic (25°C/9–12)	−0.95 – +0.80 V	Al-Kharafi *et al.* (2010)
Pyrex beaker	Ti_4O_7 (Ebonex®)	Synthetic sour brines (n.a./~13)	100 A m^{-2}	El-Sherif *et al.* (2010)
Two chamber (separated by CEM)	Carbon fibre brush	Synthetic (21°C/7)	11.9 A m^{-2}	Dutta *et al.* (2009b)

(*Continued*)

Table 8.4 Electrochemical sulfide removal performance when treating sulfur-contaminated wastewaters, with different anode materials, including sacrificial anodes (updated from Dutta *et al.*, 2010) (*Continued*).

Cell Configuration	Anode Material	Electrolyte (T/pH Electrolysis)	Applied Current Density or Potential/E_{AN} (vs SHE)	Reference
Pyrex beaker and glass cell	Boron Doped Diamond (BDD)	Synthetic sour brines (n.a./~13)	160–1300 A m^{-2}	Waterston *et al.* (2007)
Undivided glass cell	Ti/RhO_x–TiO_2, Ti/PdO–Co_3O_4, Ti/PbO_2, Ti/Pt–Ir	Tannery wastewater (n.a./~8)	200–400 A m^{-2}	Szpyrkowicz *et al.* (2005)
Undivided polarization cell	Graphite disk	Synthetic geothermal brines (25°C/12)	−0.8 – +1.4 V	Ateya *et al.* (2003)
Single compartment container*	Fe plates	Domestic wastewater (~21°C/~7.5)	20 A m^{-2}	Lin *et al.* (2017)
Single compartment beaker*	Fe and Al plates	Industrial spent caustic streams (SCS) (~25°C/8–12)	106–367 A m^{-2}	Hariz *et al.* (2013)
Single compartment electroflotation cell*	Fe, Al, Ti plates	Tannery wastewater	120–620 A m^{-2}	Murugananthan *et al.* (2004)

*Works including sacrificial anodes, inducing removal of sulfide mainly as metal sulfides
CEM: Cation Exchange Membrane

8.3.3 Electrochemical sulfide oxidation

8.3.3.1 Direct versus indirect sulfide oxidation

Electrochemical oxidation of sulfide proceeds either directly at the anode or indirectly in solution via electrogenerated oxidizing agents (O_2, Cl_2). Indirect oxidation is usually promoted to avoid the negative effects of direct oxidation, i.e. passivation of the electrode with insulating S^0 (Bockris, 1972; Feng *et al.*, 2016). Passivation is the phenomenon whereby the products of direct oxidation, such as polymeric substances in the case of organics oxidation or elemental sulfur (S^0) in the case of sulfide oxidation, form an insulating layer on the surface of the electrode, covering the electroactive sites (Chen & Miller, 2004; Feng *et al.*, 2016; Miller & Chen, 2006). Indirect oxidation also enables maintaining a higher current density, as the mass transfer of sulfide to the anode surface may be diffusion-limited even under well-mixed conditions. However, O_2 and Cl_2 are electrogenerated at higher potentials compared to sulfide oxidation, resulting ultimately in a higher cell voltage and thus a lower energy efficiency. In addition, indirect oxidation can result in the formation of halogenated toxic by-products, that can limit the application of this method for treatment of complex wastewaters.

8.3.3.2 Sulfide oxidation products

In an electrolysis cell, sulfide oxidation proceeds via both direct and indirect sulfide oxidation reactions that can take place simultaneously under a wide potential range (Table 8.3). The electrode material holds a crucial role in selecting for direct or indirect oxidation. The distribution of the final sulfide oxidation products depends on the nature of the electrode and the electric input (current density or potential), the sulfide concentration, the convection, the pH and composition of the anolyte (Song *et al.*, 2008; Vaiopoulou *et al.*, 2016).

In the case of direct sulfide oxidation, S^0 is produced and is deposited on the anode surface. This is both desirable and undesirable, as it provides a potentially recoverable product with low solubility that can be removed continuously in a continuously operated reactor but, on the other hand, S^0 deposition can lead to the aforementioned passivation phenomenon. When a constant current is applied, passivation typically induces an increase of the anode potential. The decrease of the active surface area results in a decrease of the mass transfer limited current for sulfur species oxidation. When the current cannot be sustained by only the direct oxidation of sulfur compounds, oxygen evolution co-occurs, which is exhibited by a steep increase in anode potential (Figure 8.4). It has been observed that O_2 evolution could favour partial deflaking of the passivating S^0, in this way prolonging the lifetime of the electrode (Ntagia *et al.*, 2019). Another plausible scenario is further oxidation of the electrodeposited S^0 to sulfur oxyanions (Behm & Simonsson, 1997). Behm and Simonsson (1997) described a process where the outer particles of the S^0 layer are released in the electrolyte, while the inner particles are oxidized to sulfate and thiosulfate.

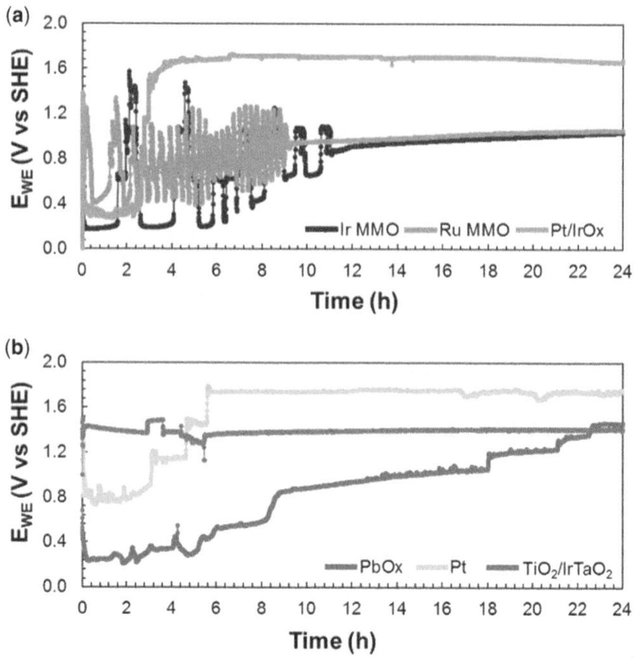

Figure 8.4 Evolution of the ohmic drop corrected anode potential in the electrode materials (A: Ir mixed metal oxide (MMO), Ru MMO, Pt/IrOx and B: PbOx, Pt, $TiO_2/IrTaO_2$) during electrochemical activity tests at constant current density of $j = 50\ A\ m^{-2}$ (Ntagia *et al.*, 2019).

In indirect oxidation, intermediate oxidants ($OH^{\bullet}$, O_2, Cl_2), produced anodically, oxidize sulfide to a mixture of S^0, polysulfides, SO_4^{2-} and $S_2O_3^{2-}$ (Mao *et al.*, 1991; Miller & Chen, 2005; Pikaar *et al.*, 2011b).

8.3.3.3 Impact of electrode material on sulfide oxidation

The deterioration of the anode material as a result of sulfide oxidation is a common phenomenon in desulfurization (Al-Kharafi *et al.*, 2010; Behm & Simonsson, 1999; Miller & Chen, 2006). Therefore, extensive research has been dedicated to selecting electrode materials that would exhibit both electrocatalytic activity towards sulfide oxidation (low overpotential), and high stability. Table 8.4 lists studies performed with various anode materials, ranging from carbon- to metallic-based.

Among the first materials tested where carbon-based electrodes, due to their high selectivity for direct sulfide oxidation towards S^0 and their low cost (Anani *et al.*, 1990; Ateya *et al.*, 2005). Nevertheless, these were soon found non-suitable, especially for industrial wastewater desulfurization, as they not only suffer from passivation (Dutta *et al.*, 2009b), but they can also be oxidized already at low potentials ($E^0_{CO2/C} = 0.207$ vs SHE), leading to electrode dissolution for potentials typically above 1.1 V vs SHE (Fabbri *et al.*, 2014; Yi *et al.*, 2017). Electrocatalytic surface loss and poisoning, as well as subsequent decrease of sulfide oxidation kinetics have also been reported for boron doped diamond (BDD) electrodes (Lawrence *et al.*, 2002; Waterston *et al.*, 2007) as well as for non-carbon-based anodes, such as platinum (Al-Kharafi *et al.*, 2010; Ramasubramanian, 1975), nickel (Behm & Simonsson, 1999) and titanium-oxide electrodes (El-Sherif *et al.*, 2010).

More recently, the focus in electrochemical sulfide oxidation has been set on using titanium-based electrodes with a thin active layer of mixed metal oxides (MMO), as they have shown good activity towards sulfide oxidation and sustain a long lifetime (Cherevko *et al.*, 2016; Pikaar *et al.*, 2011b; Vaiopoulou *et al.*, 2016). This highlights their suitability for industrial wastewater applications, in comparison with the aforementioned materials which have not yet been shown to be stable in alkaline electrolytes (Katsounaros *et al.*, 2014; Song *et al.*, 2008).

Szpyrkowicz *et al.* (2005) tested Pt/Ir, PbO_2, PdO/Co_3O_4 and RhO_x/TiO_2 coated titanium-based electrodes for electrolytic treatment of tannery wastewater containing up to 10 mM sulfide. They achieved more than 90% removal in a short electrolysis time and current densities of 200 and 400 A m^{-2}. Similarly, Pikaar *et al.* (2011b) tested five different MMO coated Ti-based electrodes, Ta/Ir, Ru/Ir, Pt/Ir, SnO_2 and PbO_2 for the desulfurization of sewage (~0.3 mM sulfide) at 100 A m^{-2} applied current density and reported higher than 70% removal and coulombic efficiencies. Ntagia *et al.* (2019) investigated the impact of catalyst choice (Ir MMO, Ru MMO, Pt/IrOx, Pt, PbOx and $TiO_2/IrTaO_2$) on the activity towards sulfide oxidation and the stability under longer-term operation at 200 A m^{-2} in alkaline solutions (pH > 12). The results of the study showed a high activity of Ir and Ru MMO, with greater than 60% coulombic efficiency for direct sulfide oxidation and an anode potential lower than 1 V vs SHE. The stability of all the anodes, with the exception of Ir MMO, was affected after 305 h of continuous operation, which was a result of partial or total (in the case of Pt/IrOx) loss of the catalytic layer.

8.4 BIOELECTROCHEMICAL TREATMENT OF SULFUR SPECIES

8.4.1 Introduction

In microbial electrochemical technologies, microorganisms capable of exchanging electrons with conductive solid are used as the "catalysts" for the electrochemical reactions. Possible advantages are (i) increasing the reaction rate of a sluggish abiotic electron transfer, (ii) the use of less expensive catalytic materials for the electrodes, resulting in lower total operational costs and (iii) the ability of the living biocatalyst to self-regenerate by replication (Logan *et al.*, 2006, Rabaey,

2010). In comparison with abiotic systems, microbial electrochemical systems are limited by the biocatalytic activity as well as the low-conductivity electrolytes used, usually sewage. Among others, limitations include high ohmic drops and thus low energy efficiency, limited conversion or production rates, long start-up periods and high sensitivity to changes in operational conditions (Clauwaert *et al.*, 2008; Prévoteau *et al.*, 2020; Zheng *et al.*, 2020).

Two key types of bioelectrochemical systems (BES) employing bacteria as catalysts exist: microbial fuel cells (MFCs) when electric power can be generated from microbially-catalysed electrooxidation of organics, or microbial electrolysis cells (MECs) when a current or voltage is applied via an external power source to drive non-spontaneous reactions (Figure 8.5). The principle of this technology is that electroactive bacteria oxidize an electron donor (or reduce an electron acceptor) and they deliver the electrons released to the anode (or receive electrons from the cathode), which then enter the electrical circuit and are transferred to the cathode, to be used for the reduction of an electron acceptor. This process is analogous to microbial metabolism, as bacteria gain energy by transferring electrons from electron donors to electron acceptors at higher redox potentials (Rabaey *et al.*, 2006).

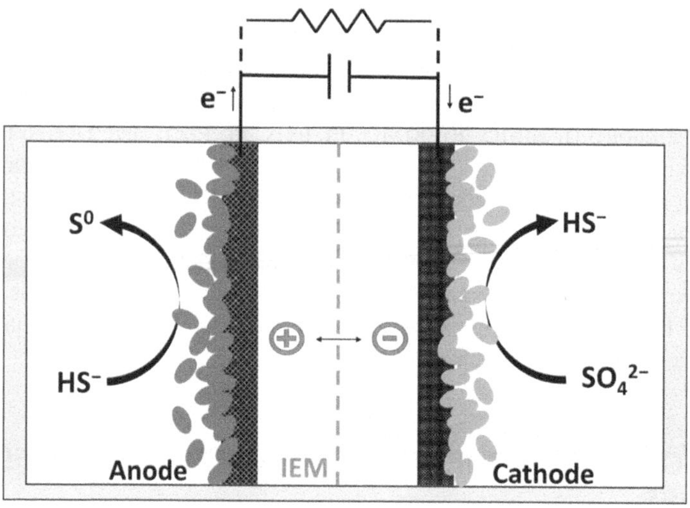

Figure 8.5 Schematic drawing of a bioelectrochemical cell (BES) for the treatment of sulfur pollution. A resistor connected in the electrical circuit between anode and cathode allows for power extraction (microbial fuel cell (MFC)), or else an external power source is needed to drive the anodic and cathodic reactions (microbial electrolysis cell (MEC)). The BES can be separated by an ion exchange membrane (IEM) into two separate electrolyte chambers. Different microbial consortia can catalyze the sulfide oxidation reaction to S^0 and transfer electrons to the anode, or receive electrons from the cathode and reduce SO_4^{2-} to sulfide.

Many industrial wastewaters, including those from the pulp and paper industry, food processing and tanneries, are sulfate-rich and the physicochemical processes applied for their treatment are energy intensive. To tackle this, biological treatment has been proposed as a cost-efficient solution. Biological sulfate removal includes two sequential steps: first the reduction of sulfate to sulfide by sulfate-reducing bacteria (SRB) and second the oxidation of sulfide to S^0, by sulfide-oxidizing bacteria (SOB) (Lens *et al.*, 1998). The same two-step principle is applied in bioelectrochemical treatment, although in this case, bacteria can donate or receive electrons to solid electron acceptors/donors (the electrodes) and in addition, power can be extracted in the case of an MFC (Rabaey *et al.*, 2006).

8.4.2 Bioanodic sulfide removal

Bioelectrochemical sulfate removal was investigated in the 1990s, when Habermann and Pommer (1991) introduced a biological fuel cell with capacity for sulfide storage. In their study, heterotrophic sulfate reduction was conducted in an electrochemical cell, where bacteria were transferring electrons from a carbon source, to sulfate, thereby reducing the latter to hydrogen sulfide. Subsequently, sulfide was oxidized at a graphite anode, thus regenerating sulfate. The electrons released were transferred through the electrical circuit to the cathode to be used for oxygen reduction. Follow-up studies were based on the same working principle, aiming to investigate the microbial communities involved in bioelectrochemical sulfate and sulfide removal, select appropriate catalytic electrode materials for sulfide oxidation, and maximize the current output of the system (Cooney *et al.*, 1996; Dutta *et al.*, 2009a; Liang *et al.*, 2013; Rabaey *et al.*, 2006; Zhao *et al.*, 2008).

Desulfovibrio and *Desulfobacter* sp. were commonly employed in current density production of BES tested for sulfate removal (Cooney *et al.*, 1996; Habermann & Pommer, 1991; Liang *et al.*, 2013), as these genera are associated with sulfate reduction, but also organic carbon oxidation. *Paracoccus* sp., which are capable of oxidizing sulfide with nitrate as the electron acceptor, have been isolated from the anode of an MFC in the study of Rabaey *et al.* (2006). The authors proposed, as a plausible scenario, a switch of electron acceptor from nitrate to the anode, however, no abiotic controls were conducted. Therefore, abiotic oxidation of part or all of the available sulfide at the anode surface cannot be excluded.

The anode materials tested were mainly porous carbon-based materials, as they enable biofilm attachment and promote sulfide oxidation to S^0 (Dutta *et al.*, 2009a; Zhao *et al.*, 2008). Additionally, Dutta *et al.* (2009a) suggested that electrodeposited S^0 on the anode surface could serve as an electron acceptor for microbially-catalysed acetate oxidation in an MFC, because in open circuit potential conditions, acetate oxidation was considerably higher in the presence than in the absence of S^0.

More recently, Daghio *et al.* (2018) have studied the removal of hydrocarbons from oil-contaminated sediments, focusing on the selection of the anode potential to increase sulfide oxidation with concomitant power production. The mechanism proposed included a first step of hydrocarbon microbial oxidation coupled to sulfate reduction to sulfide, followed by abiotic anodic sulfide oxidation to S^0 and subsequent bioelectrochemical oxidation of S^0 to SO_4^{2-}. The final step of the proposed mechanism was observed upon inoculation of the anode with *Desulfobulbus propionicus*. More positive anode potentials (+195 and +300 mV vs Ag/AgCl) were found to promote sulfide removal, and thus S^0 deposition, and power production, whereas more negative potentials (−205 mV vs Ag/AgCl) favoured *D. propionicus* and thus, back oxidation of S^0 to SO_4^{2-}. Yet, there was not enough evidence for direct electron transfer (DET) between the microbes and the electrode.

In conclusion, to the best of our knowledge, there is no rigorous evidence to date for a hypothetical DET between SOB and the anode of an MFC. The mechanism most commonly suggested is heterotrophic sulfate bioreduction to sulfide, that is later abiotically oxidized to S^0 at the anode (Dutta *et al.*, 2009a; Liang *et al.*, 2013). Nevertheless, the advantage of bioelectrochemical cells for SO_4^{2-} removal over the conventional biological SO_4^{2-} reduction lies in the fact that the produced sulfide is readily oxidized to S^0, thus limiting the inhibitory effects of sulfide accumulation on microbial metabolism (Zhao *et al.*, 2008).

8.4.3 Biocathodic sulfate removal

Earlier attempts for sulfate-reducing bioelectrochemical cells focused on MFCs for power harvesting, but recent studies put the focus on microbial electrolysis cells, where the aim is to reduce sulfate to sulfide in the cathodic compartment of a bioelectrochemical cell (Gacitúa *et al.*, 2018; Luo *et al.*, 2014; Su *et al.*, 2012). The suggested mechanism included either DET from the cathode to the bacteria performing sulfate reduction (Su *et al.*, 2012) or cathodic production of dissolved H_2 ("indirect electron transfer") as an electron donor for autotrophic sulfate reduction (Blázquez *et al.*, 2016; Gacitúa *et al.*, 2018; Luo *et al.*, 2014). Both approaches could be advantageous for sulfate-rich industrial wastewaters lacking biodegradable organic compounds, as the cathode can serve as a surrogate for providing electrons to the SRB.

Earlier studies on cathodic bioelectrochemical sulfate reduction aimed at controlling the potential at values where solely sulfate reduction will take place, avoiding hydrogen evolution or other parasitic reactions that might produce toxic by-products (Gacitúa *et al.*, 2018). Su *et al.* (2012) constructed a two-chamber electrochemical cell, with a carbon felt anode and cathode and inoculated their cathode with a mixed culture from a wastewater treatment plant. They controlled the cathode potential at −200 mV vs SHE and observed sulfate depletion and sulfide accumulation, reaching ~236 mA m^{-2} maximum current density. Biofilms dominated by *Desulfobulbus propionicus* and *Geobacter* spp. were present on the cathodes. The authors claimed that the mechanism involved was sulfate reduction via microbial DET, as no H_2 could be detected in their abiotic control tests. However, the electron transport mechanisms were not elucidated in their study. One cannot exclude a very slow H_2 generation even at this relatively high cathode potential, because it is thermodynamically feasible at low H_2 partial pressure at the electrode interface, e.g. if the bacteria readily scavenge the electrogenerated H_2 (Philips, 2020).

Since this initial report suggesting that cathodic sulfate reduction occurred via microbial DET, most studies described H_2-mediated sulfate reduction as the most plausible scenario. Luo *et al.* (2014) observed higher sulfate removal rates, as well as traces of H_2 in the headspace of the MEC, at cathode potentials between −0.6 and −0.2 V vs SHE, which suggested that the main mechanism for sulfate removal was hydrogenotrophic sulfate reduction. Gacitúa *et al.* (2018) observed complete sulfate reduction only when the cathode was controlled at −0.9 V vs SHE, which most presumably confirms that there was not direct, but rather H_2- mediated cathodic sulfate reduction.

An innovative system that combined cathodic sulfate reduction with sulfide oxidation to S^0 in a two-compartment BES (Figure 8.6) was recently proposed by Blázquez *et al.* (2017, 2016). This concept involves cathodically producing H_2 that is used by SRB to reduce sulfate to sulfide, with the latter being subsequently

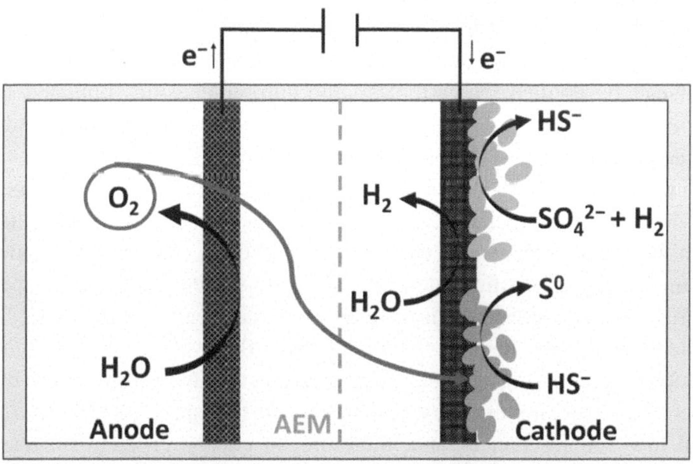

Figure 8.6 Schematic drawing of the two-compartment BES separated by an anion exchange membrane (AEM) proposed for simultaneous cathodic SO_4^{2-} removal with concomitant S^0 production, adapted by Blázquez *et al.* (2016). Hydrogenotrophic sulfate-reducing bacteria conduct the reduction of SO_4^{2-} to HS^-/H_2S. The latter is oxidized to S^0 by sulfide-oxidizing bacteria that use O_2 diffusing from the anodic to the cathodic compartment as the electron acceptor.

oxidized to S^0 by SOB also present in the catholyte. The electron acceptor for the latter process was oxygen, which was produced electrochemically in the anodic compartment and was diffusing through the ion-exchange membrane towards the cathodic compartment. This led to microaerophilic conditions, which favoured partial sulfide oxidation to elemental sulfur. Nevertheless, oxygen diffusion through the membrane cannot be well controlled, which is a bottleneck that needs to be resolved before proceeding to any putative application.

In conclusion, it is commonly agreed that the presence of a bacterial culture at the cathode of a BES is essential to perform sulfate reduction. However, it is not clear yet whether DET from the solid electrode to the bacteria occurs and, in addition, what is the percentage of the biological contribution to the total bioelectrochemical sulfate reduction (Blázquez *et al.*, 2016; Gacitúa *et al.*, 2018; Luo *et al.*, 2014; Su *et al.*, 2012). Finally, if a comparison were to be made between MECs and MFCs for sulfate removal, MECs take advantage of autotrophic sulfate reduction, thus minimizing the need for external electron donor addition, while MFCs alleviate sulfide toxicity and could provide recovery of S^0 and electric power.

8.5 OUTLOOK

The applicability of electrochemistry for treating sulfur pollution in waste streams lies within the versatility of this technology. The diversity of available concepts allows for electrochemical treatment tailoring to specific waste streams. When sulfate-containing waste streams are considered and the overall cost of electrochemical treatment will restrict its application, bioelectrochemical treatment can be applied. Nevertheless, the current densities obtained from the systems tested so far were in the range of a few mA cm^{-2}, far below industrially relevant densities.

When industrial, sulfide loaded wastewaters outside the biological range need to be treated, abiotic electrochemical treatment is recommended. A combination of both can be applied as well, where each of them can be used either as a pre-treatment or as a polishing step for the other. A clear path for the application of electrochemical sulfide removal is the treatment of industrial alkaline wastewaters, as electrochemical treatment can selectively remove sulfide while simultaneously recovering chemicals, such as H_2 and NaOH, to be used on-site. Moreover, these streams are very conductive (in the range of mS cm^{-1}), which results in low cell resistances and, therefore, in energy efficient processes. The anodic effluent of such an electrochemical process could be provided to a BES for polishing and removal of organic constituents.

In order to achieve a wider application of the electrochemical technologies for treating sulfur pollution, challenges need to be overcome. Anode costs need to decrease and their operational lifetime needs to be ensured beyond years. Depending on the nature of the electrolytes, ion-exchange membranes need to be made more robust against aggressive sulfur species, extreme pH conditions, scaling (precipitation of water-insoluble salts), or fouling by organics typically present in wastewaters. Furthermore, relevant recovery possibilities should be considered, such as chemicals, electric power and heat. The larger the electrochemical cell and the higher the current densities applied, the greater the heat recovery that can be achieved (Jiang *et al.*, 2012). Finally, a significant boost to electrochemical applications is expected to be given by a future shift towards decentralized treatment, due to their ease for remote control but also their potential independency from the power grid (Radjenovic & Sedlak, 2015), as well as by the availability of cheap and renewable electricity.

REFERENCES

Alcántara S., Velasco A., Muñoz A., Cid J., Revah S. and Razo-Flores E. (2004). Hydrogen sulfide oxidation by a microbial consortium in a recirculation reactor system: Sulfur formation under oxygen limitation and removal of phenols. *Environmental Science & Technology*, **38**(3), 918–923.

Al-Kharafi F. M., Saad A. Y., Ateya B. G. and Ghayad I. M. (2010). Electrochemical oxidation of sulfide ions on platinum electrodes. *Modern Applied Science*, **4**(3), 2–11.

Alnaizy R. (2008). Economic analysis for wet oxidation processes for the treatment of mixed refinery spent caustic. *Environmental Progress*, **27**(3), 295–301.

Anani A. A., Mao Z., White R. E., Srinivasan S. and Appleby A. J. (1990). Electrochemical production of hydrogen and sulfur by low-temperature decomposition of hydrogen sulfide in an aqueous alkaline solution. *Journal of Electrochemical Society*, **137**(9), 2703–2709.

Ateya B. G. and Al-Kharafi F. M. (2002). Anodic oxidation of sulfide ions from chloride brines. *Electrochemistry*, **4**, 231–238.

Ateya B. G., Alkharafi F. M. and Al-Azab A. S. (2003). Electrodeposition of sulfur from sulfide contaminated brines. *Electrochemical and Solid-State Letters*, **9**(6), 137–140.

Ateya B. G., Al-Kharafi F. M., Abdallah R. M. and Al-Azab A. S. (2005). Electrochemical removal of hydrogen sulfide from polluted brines using porous flow through electrodes. *Journal of Applied Electrochemistry*, **35**, 297–303.

Bard A. J. and Faulkner L. R. (2001). In: Electrochemical Methods: Fundamentals and Applications, I. John Wiley, Sons (ed.), 2nd ed., John Wiley & Sons, Inc, New York, USA.

Behm M. and Simonsson D. (1997). Electrochemical production of polysulfides and sodium hydroxide from white liquor Part I: experiments with rotating disc and ring-disc electrodes. *Journal of Applied Electrochemistry*, **27**, 507–518.

Behm M. and Simonsson D. (1999). Nickel as anode material for the electrochemical production of polysulfides in white liquor. *Journal of New Materials for Electrochemical Systems*, **2**, 11–18.

Blázquez E., Gabriel D., Baeza J. A. and Guisasola A. (2016). Treatment of high-strength sulfate wastewater using an autotrophic biocathode in view of elemental sulfur recovery. *Water Research*, **105**, 395–405.

Blázquez E., Gabriel D., Baeza J. A. and Guisasola A. (2017). Evaluation of key parameters on simultaneous sulfate reduction and sulfide oxidation in an autotrophic biocathode. *Water Research*, **123**.

Bockris J. O. (1972). Electrochemistry of Cleaner Environments. Plenum Press, New York.

Bouroushian M. (2010). Electrochemistry of the Chalcogens. In: Electrochemistry of Metal Chalcogenides, Monographs in Electrochemistry, F. Scholz (ed.), Springer-Verlag Berlin and Heidelberg GmbH & Co. KG, Berlin, Germany, pp. 57–76.

Buisman C., Post R., Ijspeert P., Geraats G. and Lettinga G. (1989). Biotechnological process for sulphide removal with sulphur reclamation. *Acta Biotechnologica*, **9**, 255–267.

Buisman C. J. N., Vellinga S. H. J., Janssen G. H. R. and Dijkman H. (1999). Biological sulfide production for metal recovery. Proceedings of the Symposium on Fundamentals of Lead and Zinc Extraction and Recycling, held at the 1999 TMS Annual Meeting.

Chaplin B. P. (2019). The prospect of electrochemical technologies advancing worldwide water treatment. *Accounts of Chemical Research*, **3**(52), 592–604.

Chatzisymeon E., Foteinis S., Mantzavinos D. and Tsoutsos T. (2013). Life cycle assessment of advanced oxidation processes for olive mill wastewater treatment. *Journal of Cleaner Production*, **54**, 229–234.

Chen A. and Miller B. (2004). Potential oscillations during the electrocatalytic oxidation of sulfide on a microstructured Ti/Ta_2O_5-IrO_2 electrode. *J. Phys. Chem. B*, **108**, 2245–2251.

Cherevko S., Geiger S., Kasian O., Kulyk N., Grote J., Savan A., Shrestha B. R., Merzlikin S., Breitbach B., Ludwig A. and Mayrhofer K. J. J. (2016). Oxygen and hydrogen evolution reactions on Ru, RuO_2, Ir, and IrO_2 thin film electrodes in acidic and alkaline electrolytes: a comparative study on activity and stability. *Catalysis today*, **262**, 170–180.

Ciobanu M., Wilburn J. P., Krim M. L. and Kliffel D. E. (2007). Fundamentals. In: Handbook of Electrochemistry, C. G. Zoski (ed.), Elsevier.

Clauwaert P., Aelterman P., Pham T. H., De Schamphelaire L., Carballa M., Rabaey K. and Verstraete W. (2008). Minimizing losses in bio-electrochemical systems: the road to applications. *Applied Microbiology and Biotechnology*, **79**(6), 901–913.

Comninellis C. and Chen G. (2010). Electrochemistry for the Environment, 1st edn. Springer, New York, USA.

Cooney M. J., Roschi E., Marison I. W., Comninellis C. and von Stockar U. (1996). Physiologic studies with the sulfate-reducing bacterium *Desulfovibrio desulfuricans*: evaluation for use in a biofuel cell. *Enzyme and Microbial Technology*, **18**, 358–365.

Daghio M., Vaiopoulou E., Aulenta F., Sherry A., Head I., Franzetti A. and Rabaey K. (2018). Anode potential selection for sulfide removal in contaminated marine sediments. *Journal of Hazardous Materials*, **360**(April), 498–503.

Dutta P. K., Rabaey K., Yuan Z. and Keller J. (2008). Spontaneous electrochemical removal of aqueous sulfide. *Water Research*, **42**, 4965–4975.

Dutta P. K., Keller J., Yuan Z., Rozendal R. A. and Rabaey K. (2009a). Role of sulfur during acetate oxidation in biological anodes. *Environmental Science & Technology*, **43**(10), 3839–3845.

Dutta P. K., Rozendal R. A., Yuan Z., Rabaey K. and Keller J. (2009b). Electrochemical regeneration of sulfur loaded electrodes. *Electrochemistry Communications*, **11**(7), 1437–1440.

Dutta P. K., Keller J., Yuan Z., Rozendal R. A. and Rabaey K. (2010). Conversion of sulfur species in bioelectrochemical systems. In: Bioelectrochemical Systems: From Extracellular Electron Transfer to Biotechnological Application, K. Rabaey, L. Angenent, U. Schröder and J. Keller (eds), IWA Publishing, London, UK, pp. 243–262.

El-Sherif S., Bejan D. and Bunce N. J. (2010). Electrochemical oxidation of sulfide ion in synthetic sour brines using periodic polarity reversal at Ebonex electrodes. *Canadian Journal of Chemistry*, **88**, 928–936.

Eow J. S. (2002). Recovery of sulfur from sour acid gas: a review of the technology. *Environmental Progress*, **21**(October), 143–162.

Fabbri E., Habereder A., Waltar K., Kötz R. and Schmidt T. J. (2014). Developments and perspectives of oxide-based catalysts for the oxygen evolution reaction. *Catalysis Science & Technology*, **4**, 3800–3821.

Feng Y., Yang L., Liu J. and Logan B. E. (2016). Electrochemical technologies for wastewater treatment and resource reclamation. *Environmental Science: Water Research and Technology*, **2**, 800–831.

Fornés J. P. and Bisang J. M. (2017). Electrochemical production of colloidal sulphur by oxidation of sulphide ion at lead coated-2- and -3-dimensional rotating cylinder anode surfaces. *Electrochimica Acta*, **243**, 90–97.

Gacitúa M. A., Muñoz E. and González B. (2018). Bioelectrochemical sulphate reduction on batch reactors: effect of inoculum-type and applied potential on sulphate consumption and pH. *Bioelectrochemistry*, **119**, 26–32.

Habermann W. and Pommer E.-H. (1991). Biological fuel cells with sulphide storage capacity. *Applied Microbiology and Biotechnology*, **35**, 128–133.

Hamilton I. C. and Woods R. (1983). An investigation of the deposition and reactions of sulphur on gold electrodes. *Journal of Applied Electrochemistry*, **13**, 783–794.

Hariz I. B., Halleb A., Adhoum N. and Monser L. (2013). Treatment of petroleum refinery sulfidic spent caustic wastes by electrocoagulation. *Separation and Purification Technology*, **107**, 150–157.

Hawari A., Ramadan H., Abu-reesh I. and Ouederni M. (2015). A comparative study of the treatment of ethylene plant spent caustic by neutralization and classical and advanced oxidation. *Journal of Environmental Management*, **151**, 105–112.

Janssen L. J. J. and Koene L. (2002). The role of electrochemistry and electrochemical technology in environmental protection. *Chemical Engineering Journal*, **85**, 137–146.

Janssen A. J. H., Lens P. N. L., Stams A. J. M., Plugge C. M., Sorokin D. Y., Muyzer G., Dijkman H., Van Zessen E., Luimes P. and Buisman C. J. N. (2009). Application of bacteria involved in the biological sulfur cycle for paper mill effluent purification. *Science of the Total Environment*, **407**(4), 1333–1343.

Jiang J., Hu J., Cui M. and Tian H. (2012). Integration of hydrogen production and waste heat recovery in electrochemical wastewater treatment. *Renewable Energy*, **43**, 179–182.

Katsounaros I., Cherevko S., Zeradjanin A. R. and Mayrhofer K. J. J. (2014). Oxygen electrochemistry as a cornerstone for sustainable energy conversion. Angewandte Chemie International Edition, **53**, 102–121.

Kim K. and Han J. I. (2014). Performance of direct alkaline sulfide fuel cell without sulfur deposition on anode. *International Journal of Hydrogen Energy*, **39**(13), 7142–7146.

Klans Jüttner. (2007). Technical Scale of Electrochemistry. In: Encyclopedia of Electrochemistry, P. S. A. J. Bard, M. Stratmann, D. McDonald (eds), Electrochemical Engineering, Vol. **5**.

Lawrence N. S., Thompson M., Prado C., Jiang L., Jones T. G. J. and Compton R. G. (2002). Amperometric detection of sulfide at a boron doped diamond electrode: the electrocatalytic reaction of sulfide with ferricyanide in aqueous solution. *Electroanalysis*, **14**, 499–504.

Lens P. N. L., Visser A., Janssen A. J. H., Pol L. W. H. and Lettinga G. (1998). Biotechnological treatment of sulfate-rich wastewaters. *Critical Reviews in Environmental Science and Technology*, **28**(1), 41–88.

Levillain E., Demortier A. and Lelieur J. P. (2006). Sulfur. In: Encyclopedia of Electrochemistry, Inorganic Electrochemistry, Colume 7, W. E. Geiger and C. Pickett (eds), Wiley, New York, USA, pp. 253–271.

Liang F., Xiao Y. and Zhao F. (2013). Effect of pH on sulfate removal from wastewater using a bioelectrochemical system. *Chemical Engineering Journal*, **218**, 147–153.

Lin H., Williams N., King A. and Hu B. (2016). Electrochemical sulfide removal by low-cost electrode materials in anaerobic digestion, **297**, 180–192.

Lin H.-W., Kustermans C., Vaiopoulou E., Prévoteau A., Rabaey K., Yuan Z. and Pikaar I. (2017). Electrochemical oxidation of iron and alkalinity generation for efficient sulfide control in sewers. *Water Research*, **118**, 114–120.

Logan B. E., Hamelers B., Rozendal R., Schröder U., Keller J., Freguia S., Aelterman P., Verstraete W. and Rabaey K. (2006). Microbial fuel cells: methodology and technology. *Environmental Science & Technology*, **40**(17), 5181–5192.

De Luna P., Hahn C., Higgins D., Jaffer S. A., Jaramillo T. F. and Sargent E. H. (2019). What would it take for renewably powered electrosynthesis to displace petrochemical processes? *Science*, **364**(eaav3506).

Luo H., Fu S., Liu G., Zhang R., Bai Y. and Luo X. (2014). Autotrophic biocathode for high efficient sulfate reduction in microbial electrolysis cells. *Bioresource Technology*, **167**, 462–468.

Mao Z., Anani A., White R. E., Srinivasan S. and Appleby A. J. (1991). A modified electrochemical process for the decomposition of hydrogen sulfide in an aqueous alkaline solution. *J. Electrochem. Soc.*, **138**(5), 1299–1303.

Martínez-Huitle C. A., Rodrigo M. A., Sirés I. and Scialdone O. (2015). Single and coupled electrochemical processes and reactors for the abatement of organic water pollutants: a critical review. *Chemical Reviews*, **115**, 13362–13407.

Miller B. and Chen A. (2005). Effect of concentration and temperature on electrochemical oscillations during sulfide oxidation on $Ti/Ta_2O5–IrO_2$ electrodes. *Electrochimica Acta*, **50**, 2203–2212.

Miller B. and Chen A. (2006). Oscillatory instabilities during the electrochemical oxidation of sulfide on a Pt electrode. *Journal of Electroanalytical Chemistry*, **588**, 314–323.

Murugananthan M., Raju G. B. and Prabhakar S. (2004). Removal of sulfide, sulfate and sulfite ions by electro coagulation. *Journal of Hazardous Materials*, **109**(December 2003), 37–44.

Nevin K. P., Woodard T. L., Franks A. E., Summers Z. M. and Lovley D. R. (2010). Microbial electrosynthesis: feeding microbes electricity to convert carbon dioxide and water to multicarbon extracellular organic compounds. *mBio*, **1**(2), e00103–10.

Ntagia E., Fiset E., da Silva Lima L., Pikaar I., Zhang X., Jeremiasse A. W., Prévoteau A. and Rabaey K. (2019). Anode materials for sulfide oxidation in alkaline wastewater: an activity and stability performance comparison. *Water Research*, **149**, 111–119.

Ntagia E., Fiset E., Truong Cong Hong L., Vaiopoulou E. and Rabaey K. (2020). Electrochemical treatment of industrial sulfidic spent caustic streams for sulfide removal and caustic recovery. *Journal of Hazardous Materials*, **388**(November 2019), 121770.

Pandey R. A. and Malhotra S. (1999). Desulfurization of gaseous fuels with recovery of elemental sulfur: an overview. *Critical Reviews in Environmental Science and Technology*, **29**(3), 229–268.

Philips J. (2020). Extracellular electron uptake by acetogenic bacteria: does H_2 consumption favor the H_2 evolution reaction on a cathode or metallic iron ? *Frontiers in Microbiology*, **10**(January), 1–13.

Pikaar I., Rozendal A. R., Yuan Z., Keller J. and Rabaey K. (2011a). Electrochemical sulfide removal from synthetic and real domestic wastewater at high current densities. *Water Research*, **45**, 2281–2289.

Pikaar I., Rozendal R. A., Yuan Z., Keller J. and Rabaey K. (2011b). Electrochemical sulfide oxidation from domestic wastewater using mixed metal-coated titanium electrodes. *Water Research*, **45**, 5381–5388.

Pikaar I., Li E., Rozendal R. A., Yuan Z., Keller J. and Rabaey K. (2012). Long-term field test of an electrochemical method for sulfide removal from sewage. *Water Research*, **46**, 3085–93.

Pikaar I., Likosova E. M., Freguia S., Keller J., Rabaey K. and Yuan Z. (2015). Electrochemical abatement of hydrogen sulfide from waste streams. *Critical Reviews in Environmental Science and Technology*, **45**, 1555–1578.

Pokhrel D. and Viraraghavan T. (2004). Treatment of pulp and paper mill wastewater – a review. *Science of the Total Environment*, **333**, 37–58.

Pourbaix M. (1966). Atlas of electrochemical equilibria in aqueous solutions. Pergamon, Oxford.

Prévoteau A., Carvajal-Arroyo J. M., Ganigué R. and Rabaey K. (2020). Microbial electrosynthesis from CO_2: forever a promise? *Current Opinion in Biotechnology*, **62**, 48–57.

Rabaey K. (2010). Bioelectrochemical systems: A new approach towards environmental and industrial biotechnology. In: Bioelectrochemical Systems: From Extracellular Electron Transfer to Biotechnological Application, K. Rabaey, L. Angenent, U. Schröder and J. Keller (eds), IWA publisher, London, UK.

Rabaey K. and Rozendal R. A. (2010). Microbial electrosynthesis – revisiting the electrical route for microbial production. *Nature reviews. Microbiology*, **8**(10), 706–716.

Rabaey K., Van de Sompel K., Maignien L., Boon N., Aelterman P., Clauwaert P., De Schamphelaire L., Pham H. T., Vermeulen J., Verhaege M., Lens P. and Verstraete W. (2006). Microbial fuel cells for sulfide removal. *Environmental Science & Technology*, **40**(17), 5218–5224.

Radjenovic J. and Sedlak D. L. (2015). Challenges and opportunities for electrochemical processes as next- generation technologies for the treatment of contaminated water. *Environmental Science and Technology*, **49**(19), 11292–11302.

Rajalo G. and Petrovskaya T. (1996). Selective electrochemical oxidation of sulphides in tannery wastewater. *Environmental Technology*, **17**, 605–612.

Ramasubramanian N. (1975). Anodic behavior of platinum electrodes in sulfide solutions and the formation of platinum sulfide. *Journal of Electroanalytical Chemistry*, **64**, 21–37.

Rankin K., Bejan D. and Bunce N. J. (2010). Electrochemical oxidation of the sulfide ion in synthetic geothermal brines in batch cells using coke electrodes. *Industrial and Engineering Chemistry Research*, **49**, 6261–6266.

Sheu S.-H. and Weng H.-S. (2001). Treatment of olefin plant spent caustic by combination of neutralization and fenton reaction. *Water Research*, **35**(8), 2017–2021.

Smith T. J. and Stevenson K. J. (2007). Reference electrodes. In: Handbook of Electrochemistry, C. G. Zoski (ed.), Elsevier, Amsterdam, The Netherlands.

Song S., Zhang H., Ma X., Shao Z., Baker R. T. and Yi B. (2008). Electrochemical investigation of electrocatalysts for the oxygen evolution reaction in PEM water electrolyzers. *International Journal of Hydrogen Energy*, **33**, 4955–4961.

Su W., Zhang L., Tao Y., Zhan G., Li D. and Li D. (2012). Sulfate reduction with electrons directly derived from electrodes in bioelectrochemical systems. *Electrochemistry Communications*, **22**, 37–40.

Sun M., Zhai L.-F., Li W.-W. and Yu H.-Q. (2016). Harvest and utilization of chemical energy in wastes by microbial fuel cells. *Chemical Society Reviews*, **45**, 2847–2870.

Szpyrkowicz L., Kaul S. N., Neti R. N. and Satyanarayan S. (2005). Influence of anode material on electrochemical oxidation for the treatment of tannery wastewater. *Water Research*, **39**, 1601–1613.

Vaiopoulou E., Provijn T., Prévoteau A., Pikaar I. and Rabaey K. (2016). Electrochemical sulfide removal and caustic recovery from spent caustic streams. *Water Research*, **92**, 38–43.

Wang L., Lu P., Liu C. and Wang L. (2015). Electro-oxidation of sulfide on Ti/ RuO_2 electrode in an aqueous alkaline solution. *International Journal of Electrochemical Science*, **10**, 8374–8384.

Waterston K., Bejan D. and Bunce N. J. (2007). Electrochemical oxidation of sulfide ion at a boron-doped diamond anode. *Journal of Applied Electrochemistry*, **37**(3), 367–373.

Yi Y., Weinberg G., Prenzel M., Greiner M., Becker S., Schlögl R. and Heumann S. (2017). Electrochemical corrosion of a glassy carbon electrode. *Catalysis Today*, **295**, 32–40.

Zermeño-Montante I., Nieto-Delgado C., Sagredo-Puente R. D., Cárdenas-Galindo M. G. and Handy B. E. (2011). Catalytic wet air oxidation of sodium sulfide solutions. Effect of the metal-support and acidity of the catalysts. *Top Catal*, **54**, 579–586.

Zhang L., De Schryver P., De Gusseme B., De Muynck W., Boon N. and Verstraete W. (2008). Chemical and biological technologies for hydrogen sulfide emission control in sewer systems: a review. *Water Research*, **42**, 1–12.

Zhao F., Rahunen N., Varcoe J. R., Chandra A., Avignone-Rossa C., Thumser A. E. and Slade R. C. T. (2008). Activated carbon cloth as anode for sulfate removal in a microbial fuel cell. *Environmental Science & Technology*, **42**(13), 4971–4976.

Zheng T., Li J., Ji Y., Zhang W., Fang Y., Xin F., Dong W., Wei P., Ma J. and Jiang M. (2020). Progress and prospects of bioelectrochemical systems: electron transfer and its applications in the microbial metabolism. *Frontiers in Bioengineering and Biotechnology*, **8** (January), 1–10.

Chapter 9

Anaerobic treatment of sulfate-rich wastewaters: process modeling and control

A. Robles, S. Vinardell, J. Serralta, N. Bernet, P. N. L. Lens, J. P. Steyer and S. Astals

9.1 INTRODUCTION

A variety of anthropogenic activities, especially at industrial level, generate wastewater streams rich in sulfate which, if improperly treated, can cause health, social and environmental problems (Serrano *et al.*, 2020). These industrial activities include food production (e.g. edible oil, seafood-processing), fermentation industry, tanneries, coal-burning power plants, paper and pulp industry, mining and metallurgical processes. Additionally, in some regions (e.g. Hong Kong), the use of seawater for toilet flushing to alleviate the pressure on potable water has increased the sulfate concentration in municipal sewage (Li *et al.*, 2018; Wu *et al.*, 2016).

Anaerobic digestion (AD) is a well-established technology to transform municipal and industrial organic-rich wastewaters into renewable energy in the form of methane-rich biogas. Besides energy recovery, AD presents other important advantages compared to aerobic treatment such as low sludge production and low energy requirements (Puyol *et al.*, 2017). AD is also a suitable technology to treat organic-rich municipal and industrial wastewaters containing sulfate. However, a high sulfate content can affect the AD process performance and feasibility due to (i) the competition for easily biodegradable organic matter (e.g. short chain fatty acids, hydrogen) between sulfate reducers and anaerobic microorganisms (Cassidy *et al.*, 2015; Kalyuzhnyi *et al.*, 1998), (ii) the generation of H_2S, the product of sulfate

DOI: 10.2166/9781789060959_0277

reduction, which is malodorous, toxic, and corrosive. H_2S can damage the combined heat and power (CHP) unit and metal parts of the infrastructure (e.g. gas storage tanks, compressors, and pipelines) by itself or through SO_2 formation from combustion (H_2SO_4 when combined with the biogas moisture), which requires equipment and instrumentation resistant to corrosion (Angelidaki *et al.*, 2018), and (iii) the inhibitory impact of H_2S on anaerobic microorganisms since H_2S is a lipophilic compound that can diffuse through the microbial cell membrane and damage cellular metabolism (Chen *et al.*, 2008; Li & Lancaster, 2013). The detrimental impact of H_2S on the AD process and infrastructure requires the effective prediction and control of the H_2S concentration in the biogas and in the digestion liquor.

Several mathematical models have been developed to model the sulfur cycling in anaerobic biotechnologies, primarily for continuous stirred tank reactor (CSTR) and upflow anaerobic sludge blanket (UASB) systems, but also for batch and plug flow systems (Cassidy *et al.*, 2015; Knobel & Lewis, 2002). The sulfur cycling is quite complex since it is metabolically diverse and it is connected to the carbon, nitrogen, phosphorus and iron cycles among others. Most models have focused on the substrate competition between sulfate-reducing bacteria and anaerobic microorganisms (including acetogenic bacteria and methanogenic archaea) and the inhibition of anaerobic microorganisms by hydrogen sulfide. Some models have also incorporated sulfide precipitation (Flores-Alsina *et al.*, 2016; Gupta *et al.*, 1994).

Anaerobic reactors treating sulfate-rich streams can be very unstable (see Chapter 6). Therefore, implementing an efficient monitoring and control system is required to keep the process running adequately. The monitoring and control system is particularly important in the presence of a fluctuating input pollution and/or when the influent has a low biological oxygen demand (BOD) to sulfate ratio. Different types of sensors are available for providing the on-line data required for sulfate reduction process control, such as chemical sensors and analyzers, microsensors or biosensors (Cassidy *et al.*, 2015). The combination of these sensors with molecular techniques would allow monitoring of microbial activity and microbial ecology in bioreactors where sulfate reduction occurs. This would be useful for development of a control strategy considering sulfide production as a control parameter.

This chapter reviews the current state-of-the-art models, monitoring techniques and control strategies applied to anaerobic processes for sulfate-rich wastewater treatment.

9.2 MODELS

Dynamic mathematical models are powerful tools to improve the understanding, monitoring, and control of wastewater treatment plants and anaerobic digesters

(Batstone *et al.*, 2015; Jimenez *et al.*, 2015; Olsson *et al.*, 2014; Solon *et al.*, 2017). The power of mathematical models lies in their capacity to reproduce an empirical behavior on a computer, in a clear and quantifiable manner, where the mathematical equations simulate the physical, chemical, and biological processes. Mathematical models can be used to analyze, predict and optimize the performance of a process or plant under different operating conditions which is a useful tool to complement and improve lab- and pilot-scale studies (Fernández-Arévalo *et al.*, 2017; Lizarralde *et al.*, 2019; Seco *et al.*, 2020). Mathematical models can also help to understand the process from a global point of view and to train process operators. For instance, Kazadi Mbamba *et al.* (2019) analyzed the plant-wide impact of different iron dosages for chemical phosphorus removal, Martí *et al.* (2017) modeled different sludge management strategies in a wastewater treatment plant (WWTP) to enhance phosphorus recovery, and Arnell *et al.* (2016) modeled the plant-wide impact of different co-digestion strategies. Last but not least, simplified mathematical models are used in the control loop to improve process operation.

The dynamic modeling of AD and/or sulfate reduction processes has been largely reported in the literature due to the importance of these biotechnologies in the development of the circular economy (Batstone *et al.*, 2002, 2006; Fedorovich *et al.*, 2003; Lauwers *et al.*, 2013). A comprehensive review of the kinetic parameters related to the sulfur cycle in these models (e.g. maximum specific uptake rate, half-saturation constant, microbial growth yield, decay rate, inhibition function and constants, Henry's law coefficient, p*K*a dissociation value) can be found in Barrera *et al.* (2013) and Cassidy *et al.* (2015). In a recent publication, Ahmed and Rodríguez (2018) carried out a comparative study of five different structures of sulfate reduction models for AD to find a trade-off between model complexity and model accuracy.

9.2.1 Models at reactor level

Pioneering publications dealing with sulfate reduction were published in the 1990s (Gupta *et al.*, 1994; Vavilin *et al.*, 1994), and publications dealing with this topic have been steadily published ever since. The publication of the Anaerobic Digestion Model N°1 (ADM1) (Batstone *et al.*, 2002) represented a breakthrough for anaerobic biotechnologies modeling since it created a common basis for further model development and validation studies. ADM1 also aimed to make model outcomes more comparable and compatible, as well as to assist a technology transfer from research to industry (Batstone *et al.*, 2002). The original ADM1 only included organic and inorganic carbon and nitrogen reactions and did not include sulfur transformations due to the complexity of the sulfur cycle. However, publications including sulfur transformation with the ADM1 model appeared soon after its publication (Batstone *et al.*, 2006; Fedorovich *et al.*, 2003).

Since then, several models have been developed to describe sulfate reduction, either through standalone models or by incorporating sulfate reduction into ADM1 (Ahmed & Rodríguez, 2018). The sulfur cycle involves several biochemical and physical processes which have been incorporated in the models to different degrees depending on the assumptions and the framework of the study. The processes associated with sulfur include: (i) heterotrophic and/or autotrophic sulfate reduction to sulfide, which is linked to sulfate-reducing bacteria (SRB) biomass growth, (ii) SRB biomass decay, (iii) hydrogen sulfide inhibition to anaerobic microorganisms, (iv) sulfide speciation in the liquid phase (i.e. acid-base chemistry), (v) hydrogen sulfide liquid-gas mass transfer, (vi) precipitation of metal sulfides, and (vii) the uptake and release of inorganic sulfur due to biomass growth and decay. The latter has not been included in any model due to the relatively small sulfur concentration in organic molecules. Consequently, it will not be discussed in this chapter.

9.2.1.1 Stoichiometry and kinetics of sulfate reduction

Sulfate-reducing bacteria (SRB) can utilize a wide variety of compounds as electron donors such as alcohols (e.g. methanol and ethanol), organic acids (e.g. formate, lactate, acetate, propionate, butyrate, pyruvate and malate), aromatic compounds (e.g. benzoate and phenol), and hydrogen (Muyzer & Stams, 2008). Thus, sulfate reduction can be carried out by heterotrophic and autotrophic SRB. Heterotrophic SRB use organic compounds as an electron donor and carbon source, while autotrophic SRB use H_2 as an electron donor and CO_2 as a carbon source (Lens & Kuenen, 2001; Muyzer & Stams, 2008). Figure 9.1 shows a schematic representation of the biochemical processes included in the model ADM1, including the sulfate-reducing processes modeled by different authors at different complexity levels. Acetate, propionate and butyrate are the most common organic substrates for SRB in these models since they are intermediate compounds of the AD process. Lactate has also been incorporated in a few models (D'Acunto *et al.*, 2011; Knobel & Lewis, 2002). However, lactate concentrations in anaerobic digesters are low compared to volatile fatty acids concentrations and, consequently, it has been generally omitted. The degradation of butyrate and propionate by SRB can be complete (the carbon source is oxidized to CO_2) or incomplete (the carbon source is oxidized to acetate). The incomplete oxidation of alcohols, lactate and fatty acids by SRB appears as the most common degradation pathway (Muyzer & Stams, 2008) and, therefore, this is the utilized pathway to achieve a more realistic and more flexible model (Barrera *et al.*, 2013; Fedorovich *et al.*, 2003). However, some models have also incorporated the complete oxidation of these compounds (Cassidy *et al.*, 2015; D'Acunto *et al.*, 2011; Gupta *et al.*, 1994; Knobel & Lewis, 2002). Autotrophic sulfate reduction has been included in most models. The stoichiometry of the

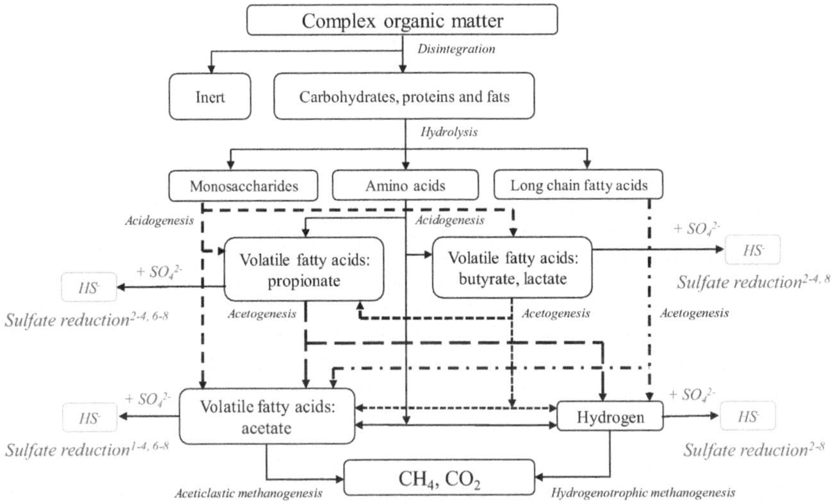

Figure 9.1 Schematic representation of the biochemical processes included in the model ADM1, including sulfate-reducing processes modeled by different authors (adapted from Robles *et al.* (2018)). [1]Fomichev and Vavilin (1997); [2]Kalyuzhnyi *et al.* (1998); [3]Knobel and Lewis (2002); [4]Fedorovich *et al.* (2003); [5]Frunzo *et al.* (2012); [6]Durán (2013); Durán *et al.* (2017); [7]Barrera *et al.* (2015); [8]Ahmed and Rodríguez (2018).

SRB reactions including incomplete heterotrophic and autotrophic sulfate reduction are:

- Butyric acid incomplete oxidation

$$2CH_3CH_2CH_2COOH + H_2SO_4 \rightarrow 4CH_3COOH + H_2S$$

- Propionic acid incomplete oxidation

$$4CH_3CH_2COOH + 3H_2SO_4 \rightarrow 4CH_3COOH + 4CO_2 + 4H_2O + 3H_2S$$

- Lactic acid incomplete oxidation

$$2CH_3CHOHCOOH + H_2SO_4 \rightarrow 2CH_3COOH + 2CO_2 + 2H_2O + H_2S$$

- Acetic acid oxidation

$$CH_3CH_2COOH + H_2SO_4 \rightarrow 2CO_2 + 2H_2O + H_2S$$

- Hydrogen oxidation

$$4H_2 + H_2SO_4 \rightarrow 4H_2O + H_2S$$

Sulfate reduction kinetics are commonly modeled by dual-substrate Monod kinetics, representing either the SRB growth rate (Equation 9.1) or the SRB substrate uptake rate (Equation 9.2). These equations are linked since the maximum specific uptake rate (k_m) is obtained by dividing the maximum specific growth rate (μ_{max}) by the biomass yield (Y) (i.e. $k_m = \mu_{max}/Y$) (Barrera *et al.*, 2013). Both approaches have been used to model sulfate reduction, however, anaerobic models normally focus on catabolism (i.e. substrate consumption, Equation 9.2) rather than on anabolism (i.e. biomass growth, Equation 9.1). Note that the activated sludge models used to model carbon, nitrogen and phosphorus removal under aerobic and anoxic conditions focus on anabolism due to the higher biomass yield under these conditions (Henze *et al.*, 2006). The dual-substrate Monod equations (Equation 9.1 and 9.2) have one term for the electron donor (organic carbon or hydrogen) and one term for the electron acceptor (sulfate). Each of these terms act as a switching function, which "on/off" the reaction depending on the electron donor or electron acceptor availability. Sulfate reduction models incorporate a dual-term Monod equation for each niche of SRB (iSRB), based on the electron donor utilized to carry out sulfate reduction:

$$\rho_{\text{growth},i} = \mu_{\text{max,iSRB}} \cdot \frac{S_i}{K_{S,\text{iSRB}} + S_i} \cdot \frac{S_{\text{SO}_4}}{K_{S,\text{SO}_4,\text{iSRB}} + S_{\text{SO}_4}} \cdot X_{\text{iSRB}} \tag{9.1}$$

$$\rho_{\text{uptake},i} = k_{m,\text{iSRB}} \cdot \frac{S_i}{K_{S,\text{iSRB}} + S_i} \cdot \frac{S_{\text{SO}_4}}{K_{S,\text{SO}_4,\text{iSRB}} + S_{\text{SO}_4}} \cdot X_{\text{iSRB}} \tag{9.2}$$

In Equation 9.2, $k_{m,\text{iSRB}}$ is the maximum specific uptake rate of iSRB biomass, S_i is the electron donor i concentration, $K_{S,\text{iSRB}}$ is the half-saturation constant of iSRB biomass towards the electron donor, S_{SO4} is the sulfate concentration, $K_{S,\text{SO4,iSRB}}$ is the half-saturation constant of iSRB biomass towards sulfate, and X_{iSRB} is the iSRB biomass concentration.

The number of SRB in existing models varies from two (Harerimana *et al.*, 2013; Vavilin *et al.*, 1994) to five (Knobel & Lewis, 2002), although three and four SRB types are the most common approaches, including butyrate-utilizing SRB (bSRB), propionate-utilizing SRB (pSRB), acetate-utilizing SRB (aSRB) and hydrogen-utilizing SRB (hSRB) (Fedorovich *et al.*, 2003; Flores-Alsina *et al.*, 2016). Acetate- and hydrogen-utilizing SRB have been incorporated in most models, which is expected considering that acetate and hydrogen are the final intermediates before methanogenesis and that, under stable conditions, fermentation reactions are relatively fast compared to methanogenesis (Ahmed & Rodríguez, 2018; Peces *et al.*, 2018). Accordingly, some researchers omitted butyrate-utilizing SRB in their models (Barrera *et al.*, 2015; Poinapen & Ekama, 2010). However, propionate-utilizing SRB should not be omitted, since propionate degradation is the least thermodynamically favorable reaction among short chain fatty acids fermentation. Propionate degradation has been reported to be influenced by changes in hydrogen concentration, microbial inhibition and process disturbances (Ito *et al.*, 2012; Yang *et al.*, 2015). Regarding hydrogen-utilizing SRB, Batstone *et al.* (2006) pointed out that for influents with low sulfate concentrations, hydrogen-utilizing SRB would be the dominant sulfate reduction pathway and, to reduce the model complexity, it should be the only sulfate reduction reaction considered in the model. However, at higher influent sulfate concentrations, volatile-fatty-acid-utilizing SRB could be significant and, therefore, they need to be included in the model.

9.2.1.2 Endogenous respiration of sulfate reduction bacteria

Endogenous respiration (i.e. biomass decay) has been included in most models. The first-order function (Equation 9.3) has been used to model endogenous respiration of biomass since pioneering models were developed (Gupta *et al.*, 1994). Later on, it was stablished as the default equation to model biomass decay in the ADM1 (Batstone *et al.*, 2002):

$$\rho_{\text{decay,iSRB}} = k_{d,\text{iSRB}} \cdot X_{\text{iSRB}} \tag{9.3}$$

where $k_{d,\text{iSRB}}$ is the first-order decay rate constant of iSRB biomass and X_{iSRB} is the concentration of iSRB biomass.

9.2.1.3 Inhibition by hydrogen sulfide

Substrate competition between sulfate-reducing bacteria, fermenters, and methanogenic microorganisms is controlled by the maximum specific uptake rate (k_m) and the half-saturation constant (K_S) of the microorganisms involved. However, the sensitivity of the different microorganisms towards inhibition can have a strong impact on the overall process performance by decreasing the substrate uptake rate (Cassidy *et al.*, 2015). The ADM1 included pH inhibition for all microorganisms, hydrogen inhibition from acetogenic microorganisms (i. e. valerate, butyrate and propionate degraders) and free ammonia inhibition from acetoclastic methanogens. H_2S inhibition was not included, since sulfate reduction was not included (Batstone *et al.*, 2002). However, hydrogen sulfide inhibition has been included in most models including sulfate reduction (Ahmed & Rodríguez, 2018; Fedorovich *et al.*, 2003). Inhibition is implemented in the model by including an inhibition term for each inhibitor to the uninhibited Monod equation. For instance, the inclusion of an inhibition term for H_2S inhibition expands Equation 9.2 to Equation 9.4, where $I_{\text{h2s,iSRB}}$ varies from 1 (no-inhibition) to 0 (complete inhibition) as the inhibitor concentration increases:

$$\rho_{\text{uptake},i} = k_{m,\text{iSRB}} \cdot \frac{S_i}{K_{S,\text{iSRB}} + S_i} \cdot \frac{S_{\text{SO}_4}}{K_{S,\text{SO}_4,\text{iSRB}} + S_{\text{SO}_4}} \cdot X_{\text{iSRB}} \cdot I_{\text{h2s,iSRB}} \tag{9.4}$$

The inhibition term ($I_{h2s,iSRB}$) has an inhibition function that determines the relationship between the inhibitor concentration and the decrease in the specific substrate utilization rate. The non-competitive inhibition function is the most utilized function to describe H_2S inhibition in these models (Equation 9.5); likely because it was the default inhibition function in the ADM1 (Barrera *et al.*, 2015; Flores-Alsina *et al.*, 2016; Knobel & Lewis, 2002; Sun *et al.*, 2016). Nonetheless, other inhibition functions have been used as well (Equations 9.5–9.7). Inhibition functions used to model H_2S inhibition include:

- Non-competitive inhibition function:

$$I_{h2s,iSRB} = \frac{K_{I,h2s,iSRB}}{K_{I,h2s,iSRB} + S_{h2s}} \tag{9.5}$$

  where $K_{I,h2s,iSRB}$ is the 50% inhibitory constant of H_2S on biomass iSRB, and S_{h2s} is the H_2S concentration in the liquid phase.
- First-order inhibition function (Fedorovich *et al.*, 2003; Kalyuzhnyi *et al.*, 1998):

$$I_{h2s,iSRB} = \begin{cases} \left(\frac{K_{I,h2s,iSRB} - S_{h2s}}{K_{I,h2s,iSRB}}\right)_{S_{h2S} \leq K_I} \\ 0 \quad {}_{S_{h2S} > K_I} \end{cases} \tag{9.6}$$

  where $K_{I,h2s,iSRB}$ represents the H_2S concentration at which iSRB is 100% inhibited, and S_{h2s} is the H_2S concentration in the liquid phase.
- Modified non-competitive inhibition function (so-called 2×2 constants) (Vavilin *et al.*, 1995):

$$I_{h2s,iSRB} = \frac{1}{1 + \left(\frac{S_{h2s}}{K_{100}}\right)^{\frac{\ln 99}{\ln\left(\frac{K_2}{K_{100}}\right)}}} \tag{9.7}$$

  where K_2 represents the H_2S concentration at which the rate of iSRB is decreased by half, K_{100} represents the H_2S concentration at which the rate of iSRB is decreased one hundred times, and S_{h2s} is the H_2S concentration in the liquid phase.
- Poinapen and Ekama inhibition function (Poinapen & Ekama, 2010):

$$I_{h2s,iSBR} = \exp\left[-\left(\frac{S_{h2s}}{1.20112 \cdot K_{I,h2s,iSBR}}\right)^2\right] \tag{9.8}$$

  where $K_{I,h2s,iSRB}$ is the 50% inhibitory concentration of H_2S on biomass iSRB and S_{h2s} is the H_2S concentration in the liquid phase.

Although each inhibition function has a distinctive profile (see Figure 9.2), most publications did not justify the selection of a specific inhibition function. The key difference between the non-competitive inhibition function (Equation 9.5) and the other inhibition functions (Equations 9.6–9.8) is that in the non-competitive inhibition functions, inhibition is not complete ($I_{h2s,iSRB} = 0$) even at extremely high inhibitor concentrations ($I_{h2s,iSRB} = 0.09$ at 100 mg H_2S-S · L^{-1} for a $K_{I,h2s,iSRB}$ of 10 mg H_2S-S · L^{-1}). On the other hand, there is a high uniformity regarding the microorganisms inhibited by H_2S in these models, where H_2S inhibition has been applied to all AD microorganisms that compete with SRB for the electron donors (e.g. butyrate-degraders, propionate-degraders, lactate-degraders, acetoclastic methanogens, and hydrogenotrophic methanogens) as well as all SRB groups included in the model. Ahmed and Rodríguez (2018) compared the inhibition parameters reported in several publications and found notable consistency among them. Differences between inhibition parameters are approximately two-fold, which can be considered a low variability considering the different experimental conditions and model structures from which these parameters were obtained. However, it should be noted that these parameters were obtained from different equations, which limits transportability among studies.

9.2.1.4 Acid-base equilibria

The inhibition function (Equations 9.5–9.8) uses the free H_2S concentration as the inhibitory agent, since H_2S has been reported to be more inhibitory than HS^- and

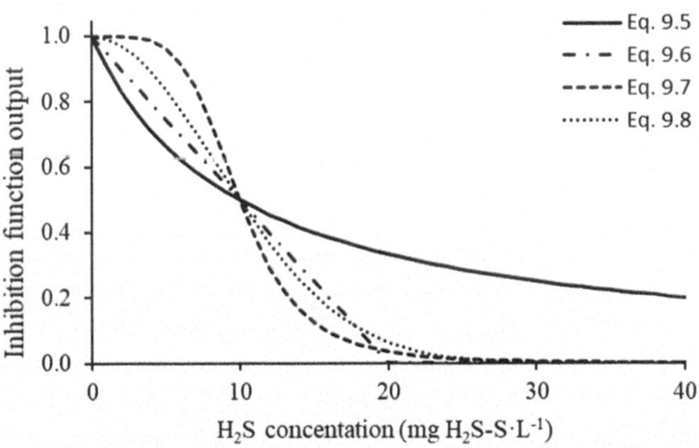

Figure 9.2 Profile of the different inhibition functions utilized to model H_2S inhibition for a 50% inhibitory concentration of 10 mg H_2S-S · L^{-1}. Solid line represents the non-competitive inhibition function. Dash-dotted line represents the first-order inhibition function. Dashed line represents Vavilin *et al.* (1995) inhibition function [K_{100} was set at 26 mg H_2S-S · L^{-1} for comparison purposes since this was the concentration that reduced the rate 100 times in Equation 9.7]. Dotted line represents the Poinapen and Ekama (2010) inhibition function.

S^{2-} (Chen *et al.*, 2008). Therefore, models need to include sulfide acid-base chemistry to resolve sulfide speciation in the liquid phase and to properly describe H_2S inhibition. Nonetheless, sulfide speciation is required to model the pH and the ionic strength of the liquid phase, H_2S liquid-gas mass transfer, and ion pairing for precipitation (Flores-Alsina *et al.*, 2016; Knobel & Lewis, 2002; Solon *et al.*, 2015).

$$H_2S_{(aq)} \underset{pK_{a1}}{\longleftrightarrow} HS^{\mp}H^{+} \underset{pK_{a2}}{\longleftrightarrow} S^{2-} + 2H^{+} \tag{9.9}$$

H_2S is a diprotic acid since it has two hydrogen atoms that can dissociate in its molecule (Equation 9.9). H_2S is a weak acid with a pK_{a1} of 7.01 and pK_{a2} of 13.8 at 25°C (Millero, 1986). However, there is some uncertainty in the literature regarding these values, since reported pK_{a1} values at 25°C range from 6.97 to 7.06, while pK_{a2} values vary from 12.2 to 15.0 (Li & Lancaster, 2013; Millero, 1986). Biological reactors are operated around neutral or slightly acidic pH. Consequently, in most models only the H_2S and HS^- concentration have been included since the S^{2-} concentration was assumed negligible (Barrera *et al.*, 2013; Fedorovich *et al.*, 2003). Sulfide speciation can be calculated using Equations 9.10–9.12:

$$S_{h2s} = \frac{(S_{h+})^2 \cdot S_{IS}}{(S_{h+})^2 + S_{h+} \cdot K_{a1} + K_{a1} \cdot K_{a2}} \approx \frac{S_{h+} \cdot S_{IS}}{K_{a1} + S_{h+}} \tag{9.10}$$

$$S_{hs-} = \frac{S_{h+} \cdot K_{a1} \cdot S_{IS}}{(S_{h+})^2 + S_{h+} \cdot K_{a1} + K_{a1} \cdot K_{a2}} \approx \frac{K_{a1} \cdot S_{IS}}{K_{a1} + S_{h+}} \tag{9.11}$$

$$S_{s2-} = \frac{K_{a1} \cdot K_{a2} \cdot S_{IS}}{(S_{h+})^2 + S_{h+} \cdot K_{a1} + K_{a1} \cdot K_{a2}} \approx 0 \tag{9.12}$$

where S_{h+} is the protons concentration, S_{IS} is the total sulfur concentration (IS stands for inorganic sulfur), K_{a1} is the first acid-base equilibrium constant of H_2S, and K_{a2} is the second acid-base equilibrium constant of H_2S.

Anaerobic digesters and sulfate reducing reactors are frequently operated at mesophilic conditions (35°C) (Serrano *et al.*, 2020), which reduces the pK_{a1} values from 7.0 to 6.8 (Hughes *et al.*, 2009; Millero, 1986). The impact of temperature on the acid dissociation constant has not been considered in most models, however, at mesophilic conditions (~35°C) the calculated free H_2S concentration in the liquid phase is 20–30% lower (depending on the digester pH) than at 25°C. The van 't Hoff equation (Equation 9.13) is the function recommended in ADM1 to describe the variation of equilibrium constants with temperature (where the $pKa1$ was 7.04 at 298 K and θ was 0.029), which provides similar pK_{a1} values to the empirical equations in Millero (1986):

$$K_{a,T_2} = K_{a,T_1} \cdot e^{\theta(T_2 - T_1)} \tag{9.13}$$

Finally, when the inorganic sulfur species in the model are solved using ODE (ordinary differential equation) instead of the algebraic equations (Equations 9.10 and 9.11), one equation is needed to model the inorganic sulfur dynamics and another equation is needed to model either H_2S or HS^- (Batstone *et al.*, 2002; Jeppsson & Rosen, 2006). The other compound is calculated as the total inorganic sulfur minus the calculated compound (e.g. $S_{h2s} = S_{IS} - S_{hs-}$). When utilizing ODE, all chemical species are included in the state vector and thus stoichiometry and kinetics are defined in terms of species (Lizarralde *et al.*, 2015). This modification also requires the inclusion of an additional rate equation for the acid-base reaction (Equation 9.14), where $\rho_{A/B,h2s}$ is the product rate of H_2S from HS^- (i.e. base to acid) and $k_{A/B,h2s}$ is the acid-base kinetic parameter ($1 \times 10^{10}\ M^{-1} \cdot day^{-1}$) (Jeppsson & Rosen, 2006):

$$\rho_{\frac{A}{B},h2S} = k_{\frac{A}{B},h2S} \cdot (S_{hs-}(S_{h+} - K_{a1}) - K_{a1} S_{IS}) \tag{9.14}$$

9.2.1.5 Gas-liquid H_2S mass-transfer

To calculate the concentration of H_2S in the biogas and in the liquid phase, it is necessary to consider the H_2S mass-transfer between the liquid and the gas phase. Henry's law, which expresses the concentration in the liquid phase due to a gas partial pressure, has been used to describe the equilibrium between both the liquid and the gas phase (Cassidy *et al.*, 2015; Kalyuzhnyi *et al.*, 1998). Equation 9.15 is the single-film mass-transfer equation to dynamically model the exchange of H_2S between both phases:

$$\rho_{T,h2S} = k_L a \cdot (S_{h2s} - 64 \cdot K_{H,h2s} \cdot p_{h2S,gas}) \tag{9.15}$$

where $\rho_{T,h2S}$ is the specific mass-transfer rate of H_2S from the liquid to the gas phase, $k_L a$ is the gas-liquid transfer coefficient (200 day^{-1}, as most common default value although highly variable depending on mixing, temperature, etc.), S_{h2s} is the hydrogen sulfide concentration in the liquid phase, $K_{H,h2s}$ is the Henry's law coefficient for H_2S (0.001 $M \cdot bar^{-1}$ (Batstone *et al.*, 2002)), 64 is a conversion factor to convert the Henry's law coefficient from molar to chemical oxygen demand (COD) basis, and $p_{h2s,gas}$ is the partial pressure of H_2S in the headspace (bar). The impact of temperature on the Henry's law coefficient should be also considered in the model (Majer *et al.*, 2008).

9.2.1.6 Metal precipitation

Metal precipitation modeling has received less attention than the physical and biochemical reactions, likely due to the difficulty in implementing and validating the model as well as the inexistence of a standardized model structure. However, metal precipitation is important since (i) sulfate reduction technology is used to produce H_2S which in turn is used to remove/recover metals as metal sulfide precipitates, and (ii) ferrous and ferric chloride are used to precipitate sulfides in anaerobic digesters (Hussain *et al.*, 2016; Kaksonen & Puhakka, 2007; Lewis, 2010).

There are two main methods to model chemical precipitation in anaerobic biotechnologies: (i) an equilibrium-based model solved using algebraic equations, and (ii) a kinetic-based model solved using ODE. The equilibrium-based model assumes that precipitation reactions are fast and, therefore, there is enough reaction time for ions to form the precipitates and reach equilibrium (Barat *et al.*, 2013; Lewis *et al.*, 2003; Lizarralde *et al.*, 2015). The equilibrium-based model precipitates the ions present in a supersaturated solution based on the minerals solubility product constant (K_{sp}) values until the equilibrium is reached. Gupta *et al.* (1994) used an equilibrium-based model to simulate metal sulfide and carbonate precipitation. The main advantages of an equilibrium-based model are the lower computational demand and the fact that it relies on well-known electrolyte solution thermodynamics (Kazadi Mbamba, 2016). However, in these models, the precipitate with the lowest K_{sp} prevails over other mineral phases. On the other hand, the kinetic-based precipitation model allows control of the rate at which the precipitation reaction progresses towards the equilibrium. Therefore, at the end of an integration step the aqueous solution may still be supersaturated or undersaturated (Kazadi Mbamba, 2016). Note that if the precipitation rate is fast enough, the kinetic-based and the equilibrium-based models should achieve identical results. Nonetheless, since there is no standardized model, there are several kinetic-based precipitation models available in the literature utilizing different structures and equations (Barat *et al.*, 2013; Kazadi Mbamba *et al.*, 2015; Kralj *et al.*, 1997; Lewis, 2010; Lizarralde *et al.*, 2015).

A pioneering approach towards a kinetic-based precipitation model for precipitation in wastewater streams was developed for the Activated Sludge Model No. 2d (ASM2d) by Henze *et al.* (1999). In the ASM2d model, iron phosphate ($FePO_4$) precipitation and redissolution were modeled using Equations 9.16 and 9.17, where iron phosphate (X_{FePO4}) precipitates from the reaction between phosphate (S_{PO4}) and ferric hydroxide ($X_{Fe(OH)3}$):

$$\rho_{pre,FePO_4} = k_{cryst,FePO4} \cdot S_{PO_4} \cdot X_{Fe(OH)_3} \tag{9.16}$$

$$\rho_{red,FePO_4} = k_{diss,FePO4} \cdot X_{FePO_4} \tag{9.17}$$

In Equation 9.16 and 9.17, $k_{cryst,FePO4}$ is the iron phosphate precipitation rate, and $k_{diss,FePO4}$ is the iron phosphate redissolution rate. The main limitation of the ADM2d precipitation model is that it does not approach equilibrium conditions, since precipitation occurs because soluble phosphate and iron hydroxide are present in the reaction media (Kazadi Mbamba, 2016). A much more comprehensive precipitation model was proposed by Koutsoukos *et al.* (1980) (Equation 9.18), which was later on modified by Musvoto *et al.* (2000), Barat *et al.* (2013) and Lizarralde *et al.* (2015). Although none of these studies included metal sulfides precipitation, the proposed model is a generic model that can be easily adapted to model different mineral precipitations.

$$\rho_{\text{pre,MA}} = k_{\text{cryst,MA}} \cdot s \cdot \left[([M^{\text{m+}}]^{\upsilon+}[A^{\text{a}-}]^{\upsilon-})^{\frac{1}{\upsilon}} - ([M^{\text{m+}}]_0^{\upsilon+}[A^{\text{a}-}]_0^{\upsilon-})^{\frac{1}{\upsilon}} \right]^n \tag{9.18}$$

where $[M^{\text{m+}}]$, $[A^{\text{a}-}]$ and $[M^{\text{m+}}]_0$, $[A^{\text{a}-}]_0$ are the concentrations of the ions in solution at a given time and at equilibrium, respectively. $k_{\text{cryst,MA}}$ is the apparent precipitation rate constant of MA, s is a factor proportional to the number of growth sites available, υ^+ is the number of cations involved in the reaction, υ^- is the number of anions involved, and $\upsilon = \upsilon^+ + \upsilon^-$. The exponent n is a number determined experimentally and is equal to 2 for most precipitation reactions. Note that the product between $[M^{\text{m+}}]_0^{\upsilon+}$ and $[A^{\text{a}-}]_0^{\upsilon-}$ equals the solubility product of the mineral ($K_{\text{sp,MA}}$). Musvoto *et al.* (2000) suggested that, when no seed material is added, the precipitation rate ($k_{\text{cryst,MA}}$) and the growth site factor (s) could be combined in a single precipitation rate constant ($k_{T,\text{cryst,MA}}$). Barat *et al.* (2013) included a function based on the saturation index in the kinetic expressions related to the precipitation and dissolution processes, accounting therefore for undersaturated and oversaturated operating conditions. Lizarralde *et al.* (2015) expanded the model by including three different Monod-style functions to model the development of the supersaturation of a solution, the nucleation process, and the growth of the precipitates.

An alternative approach to model mineral precipitation is the one described in Kazadi Mbamba *et al.* (2015b, 2015a). The authors propose a semi-empirical first-order kinetic precipitation model in which precipitation rate depends on the concentration of a mineral in the aqueous solution (Equation 9.19). Non-zero seed experimental conditions were used ($X_{\text{MA},0} = 1 \times 10^{-6}$ M). The model also includes a relative supersaturation term (σ), representing the magnitude of the driving force for precipitation, i.e. how far the aqueous solution is from equilibrium (Equation 9.20). As for Koutsoukos *et al.* (1980), the exponent n is the order of the precipitation reaction with respect to supersaturation. In Kazadi Mbamba *et al.* (2015b, 2015a), the exponent n was 2 for all minerals, except for struvite where n was 3.

$$\rho_{\text{pre,MA}} = k_{\text{cryst,MA}} \cdot X_{\text{MA}} \cdot \sigma^n \tag{9.19}$$

$$\sigma = \left(\frac{[M^{\text{m+}}]^{\upsilon+}[A^{\text{a}-}]^{\upsilon-}}{K_{\text{sp}}} \right)^{\frac{1}{\upsilon}} - 1 \tag{9.20}$$

The Kazadi Mbamba *et al.* (2015b, 2015a) model was implemented in the ADM1 by Flores-Alsina *et al.* (2016), who successfully modeled multiple mineral precipitation reactions, including iron sulfide precipitation in anaerobic digesters. It is important to highlight that Flores-Alsina *et al.* (2016) considered non-ideality, including ion complexation/pairing and ion activities instead of molar ion concentrations. Knobel and Lewis (2002) also considered that the aqueous phase is a non-ideal solution. Indeed, few authors have considered non-ideality in their calculations, which represents a source of inaccuracy (Capson-Tojo *et al.*, 2020; Patón *et al.*, 2018; Solon *et al.*, 2015).

9.2.2 Models at biofilm level

Biological sulfate reduction in high-rate granular bioreactors has been largely researched at lab-scale (Hao *et al.*, 2014; Liamleam & Annachhatre, 2007). High-rate bioreactors are also the main bioreactor configuration in the full-scale sulfate reduction technologies by Paques BV, e.g. Thioteq™ and Sulfateq™ (Hedrich *et al.*, 2018; Huisman *et al.*, 2006). The main advantage of granular bioreactors (e.g. UASB and expanded granular sludge bed (EGSB)) is the superior capacity to retain microorganisms in the system than flocculent bioreactors (e.g. CSTR and plug flow), which allows improvement of the system efficiency and reduction of the reactor volume (Hao *et al.*, 2014; van Lier *et al.*, 2015). Additionally, the granule structure provides layered microenvironments and niches for the different microorganisms involved in the process (Feldman *et al.*, 2019; Hao *et al.*, 2014; Sun *et al.*, 2016). Biofilm modeling enables better understanding of the mass transport of substrates and their microbial conversion in the biofilm as well as the dynamics of the microbial community.

The main difference between suspended biomass models (see Section 9.2.1) and biofilm models is that biofilm models need to include additional terms to model the diffusion of the reactants and products within the biofilm. For that purpose, biofilm models have used Fick's second law in spherical coordinates (Equation 9.21), which allows modeling of the diffusion of the compounds in the biofilm under unsteady-state conditions, i.e. the model is able to model changes of concentration gradient over time (Devinny & Ramesh, 2005; Sun *et al.*, 2016). Considering that the diffusion is axisymmetric, the equation is typically simplified to Equation 9.22:

$$\frac{\partial S_i}{\partial r} = \frac{D_f}{r^2}\left(\frac{\partial}{\partial r}\left(r^2\frac{\partial S_i}{\partial r}\right) + \frac{1}{\sin\theta}\frac{\partial}{\partial\theta}\left(\sin\theta\frac{\partial S_i}{\partial\theta}\right) + \frac{1}{\sin^2\theta}\left(\frac{\partial^2 S_i}{\partial^2\phi}\right)\right) \tag{9.21}$$

$$\frac{\partial S_i}{\partial r} = D_f\left(\frac{\partial^2 S_i}{\partial^2 r}\right) \tag{9.22}$$

The diffusion coefficients in the biofilm need to be estimated or fitted using experimental data. Molecular diffusion constants for water are available for most compounds. However, diffusion coefficients within a biofilm may differ since the presence of microorganisms restricts the diffusion of compounds (Devinny & Ramesh, 2005). Fan *et al.* (1990) published an empirical formula that relates the diffusion coefficient measured in water with the one in the biofilm considering the biomass density in the biofilm. However, many other equations to estimate biofilm diffusion coefficients are available in the literature (Stewart, 1998).

A simpler approach was used by Sun *et al.* (2016), who set the diffusivity coefficients of the different compounds at 0.8-fold of the values in water.

Noguera *et al.* (1999) developed a mathematical model able to simulate the three-dimensional growth of two species in an anaerobic biofilm (i.e. SRB *Desulfovibrio vulgaris* and the methanogenic *Methanobacterium formicicum*), which used the kinetic and thermodynamic mechanisms developed in their previous publication (Noguera *et al.*, 1998). The model was able to predict different biofilm structures in the presence and in the absence of sulfate as well as when both microorganisms co-existed. The research by Noguera *et al.* (1999, 1998) complemented the theoretical study conducted by Overmeire *et al.* (1994), who showed the importance of sulfate mass transfer as a factor to comprehend the competition between sulfate reducers and methanogens in the anaerobic granules. The results of Overmeire *et al.* (1994) showed that sulfate reduction can be limited in UASB reactors by the sulfate mass transfer. However, the model did not include the competition between SRB and methanogens for the electron donor, nor the utilization of SRB products by methanogens (i.e. acetate from the incomplete oxidation of the carbon source). The model used by Overmeire *et al.* (1994) was a second-order differential (Equation 9.22) one, with one boundary condition for the center of the granule ($r = 0$, Equation 9.23) and one boundary condition for the granule interface ($r = R$, Equation 9.24):

$$\frac{1}{r^2}\frac{\partial}{\partial r}\left(D_f \cdot r^2 \frac{\partial S_{g,\mathrm{SO4}}}{\partial r}\right) = k_{m,\mathrm{SRB}} \frac{S_{g,\mathrm{SO4}}}{K_{S,\mathrm{SRB}} + S_{g,\mathrm{SO4}}} X_{\mathrm{SRB}} \tag{9.22}$$

$$\frac{\partial S_{g,\mathrm{SO4}}}{\partial r} = 0 \tag{9.23}$$

$$D_f\left(\frac{\partial S_{g,\mathrm{SO4}}}{\partial r}\right) = k_1 \cdot (S_{\mathrm{aq,SO4}} - S_{g,\mathrm{SO4}}) \tag{9.24}$$

where D_f is the diffusion coefficient of sulfate in the biofilm, $S_{g,\mathrm{SO4}}$ is the sulfate concentration in the granule, $k_{m,\mathrm{SRB}}$ is the maximum uptake rate of sulfate, $K_{S,\mathrm{SRB}}$ is the sulfate half-saturation constant, X_{SRB} is the SRB concentration in the granules, R is the radius of the granule, $S_{\mathrm{aq,SO4}}$ is the sulfate concentration in the liquid, and k_1 is the mass transport coefficient of sulfate in the stagnant liquid film between the liquid bulk and the granule surface.

The biological, chemical, and physical processes occurring in a methanogenic and sulfate-reducing granular reactor under dynamic conditions have been modeled by D'Acunto *et al.* (2011) and by Sun *et al.* (2016). D'Acunto *et al.* (2011) developed a convection-diffusion model, where a convection term controlled the biofilm growth and a diffusion term controlled the substrate and product gradients within the biofilm. The model included three distinct microbial groups (i.e. complete oxidizing SRB, incomplete oxidizing SRB, and methanogens) and three reactions (one reaction for each microbial group)

D'Acunto *et al.* (2011) predicted that SRB dominate the outmost layer of the granule while methanogens dominate the inner layers of the granule. D'Acunto *et al.* (2011) also found that the methanogenic layer gains dominance in the granule as the influent BOD:SO_4^{2-} ratio decreases. The microbial distribution obtained by D'Acunto *et al.* (2011) within the granule is in agreement with the one obtained by Sun *et al.* (2016). Sun *et al.* (2016) showed that SRB and fermentative bacteria dominate in the granule outer layer, while methanogens dominate in the granule inner layer. In Sun *et al.* (2016), a mathematical model based on growth kinetics of microorganisms and substrate transportation through biofilms was developed to describe methane production and sulfate reduction with ethanol as the key electron donor. The model was calibrated and validated using experimental data from two case studies conducted in a UASB reactor. The developed model could satisfactorily describe methane and sulfide production as well as ethanol and sulfate removal in both systems. The model outputs revealed a stratified distribution of methanogenic archaea, sulfate-reducing bacteria and fermentative bacteria in the anaerobic granular sludge and the relative abundances of these microorganisms vary with substrate concentration. It also revealed that sulfate-reducing bacteria outcompeted fermentative bacteria for ethanol utilization when the COD/SO_4^{2-} ratio was equal to or above 0.5. Model simulations suggested an optimal granule size for efficient methane production, while the sulfate reduction was not significantly affected by variation in granule size. Sun *et al.* (2016) also provided comprehensive insights by revealing that the methane production and sulfate reduction were affected by ethanol and sulfate loading rates as well as by the development stage of the microbial community. The model of Sun *et al.* (2016) was based on ADM1 and included six microbial groups, i.e. ethanol fermenters, ethanol-degraders SRB, acetate-degraders SRB, autotrophic SRB, acetoclastic methanogens, and hydrogenotrophic methanogens. Most importantly, Sun *et al.* (2016) were able to properly describe the methane production and sulfate reduction of a UASB reactor operated for over 180 days.

9.3 ON-LINE MONITORING

9.3.1 Sensors for sulfurous compounds

Different off-line techniques based on analytical measurements can be applied for monitoring AD processes due to their great reproducibility and precision. These techniques are, however, time consuming and require extensive manual handling. Thus, the choice of appropriate on-line monitoring techniques is crucial for efficiently controlling biological processes of this type, which are complex – due to their strong non-linearity and non-stationarity – and thus very difficult to control. For instance, a simple method for the determination of dissolved sulfide in colored complex media was developed at the end of the 1990s by Percheron *et al.* (1996) using ion exchange chromatography. Its principle was based on the complete oxidation of sulfide into sulfate through the strong oxidant hydrogen peroxide. The difference in sulfate concentration gives the total dissolved sulfide concentration. Sulfide from continuous and discontinuous digestion of sulfate-rich wastewater has been successfully assayed by this technique. However, autoxidation and stripping of H_2S must be avoided, thus basification, refrigeration and/or other handling actions must be performed immediately after sampling, without bacterial separation.

On-line monitoring remains the weakest part of real-time AD systems process control. Accordingly, adequate selection of monitoring equipment is required to achieve accurate actions. Among the available sensors, the choice of a measurement device to be used for on-line control depends on several factors, such as its sensitivity to variations and its response time to a disturbance. Additional criteria should also be considered at the industrial scale such as low cost and low maintenance. For example, Jimenez *et al.* (2015) and Wu *et al.* (2019) presented comprehensive overviews of different monitoring techniques used in AD processes. Moreover, a comprehensive overview of real-time monitoring techniques to measure sulfide, electron donors, sulfate, and biomass composition can be found in Cassidy *et al.* (2015).

Within the anaerobic treatment of sulfate-rich wastewaters, sensors to probe sulfurous compounds are of the greatest importance. Classical on-line and off-line sulfide concentration measurements based on sensors consist of using an ion selective electrode (Bandekar *et al.*, 1995), a redox electrode (Janssen *et al.*, 1998), a titrimetric method (Zancato *et al.*, 1990), or a spectrophotometric method (Bandekar *et al.*, 1995). Nonetheless, the most widely implemented monitoring method to indirectly measure the overall activity of the process consists of analyzing the output gas composition and flow rate.

Different techniques are available for measuring the composition of the biogas including H_2S, such as Infrared (IR), Fourier transform infrared (FTIR), non-dispersive infrared (NDIR), ultraviolet (UV) and ultraviolet-visible (UV/VIS) photometers; diode laser sensing, specific analyzers based on electrochemical cells, thermal conductivity analyzers, and moisture sensors (Jimenez *et al.*, 2015; Redondo *et al.*, 2008; Wu *et al.*, 2019). From these on-line measurements, a soft sensor (also known as a virtual sensor) based on a theoretical evaluation from the Henry's law and the sulfide dissociation equilibrium can be used to calculate the concentration of dissolved compounds. Figure 9.3 shows a general

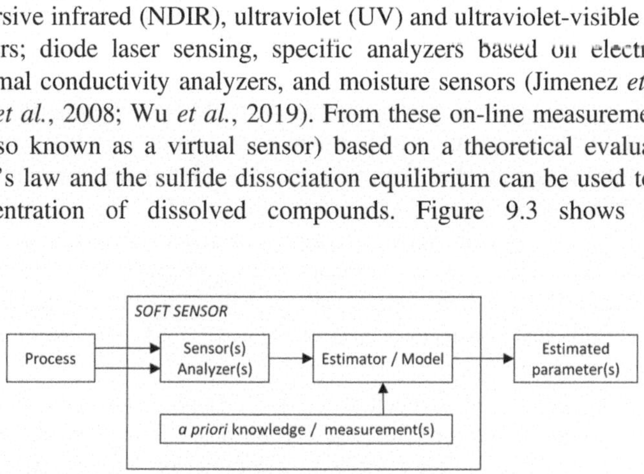

Figure 9.3 Schematic view of a soft sensor.

representation of a soft sensor, which can be defined as a numerical algorithm that combines signals from available hardware sensors within a given mathematical framework. They estimate an unknown parameter, taking advantage of the *a priori* knowledge available about a process (Bastin & Dochain, 1990).

9.3.2 In-situ sulfate sensors

Different electrodes can be used for sulfate monitoring, such as ion-selective electrodes (ISEs), which are usually chosen for routine applications since they have many advantages over other methods for anion concentration determination. Different sulfate-selective compounds were proposed at the end of the 1990s based on the Nernst potential for sulfate determination in liquid streams (see e.g. Berrocal *et al.*, 2000; Li *et al.*, 1999; Nishizawa *et al.*, 1998; Shamsipur *et al.*, 2001). The selectivity of an ISE towards sulfate is of great importance and must be taken into account when choosing an appropriate ISE sensor. The response of an ISE should, in theory, be in accordance with the Nernst equation (Cassidy *et al.*, 2015):

$$E = A - 2.303\frac{R \cdot T}{z_i \cdot F} \cdot \log\left(a_i\right) \tag{9.25}$$

where E is the electrode potential, A is a constant, R is the universal gas constant, T is the temperature, F is the Faraday constant, z_i is the ionic charge, and a_i is the activity of the ion.

Morigi *et al.* (2001) presented a sulfate-ISE based on a dispersion of hydrotalcite particles into a polymeric membrane using coated-wire (CW) configuration, testing membranes based on polyvinyl chloride (PVC) and polydimethylsiloxane (PDMS). Morigi *et al.* (2001) observed that both membranes exhibited strong and selective interactions with sulfate, despite better performances being obtained when using PDMS rather than PVC, with response times below 60 seconds, while the potentiometric response was practically constant in the pH range 4–7. Ganjali *et al.* (2002) presented a novel sulfate ion-selective polymeric membrane electrode based on a derivative of pyrilium perchlorate. This ISE showed good selectivity for the sulfate ion with respect to common organic and inorganic anions, exhibiting good linear responses within a pH range of 4–9. Another sulfate-selective PVC-based ISE was tested by Mazloum-Ardakani *et al.* (2012). These authors investigated the performance of this electrode using potentiometric and electrochemical impedance spectroscopy (EIS). The results showed that the prepared electrode had a detection limit of 6.3×10^{-7} M and a response time of less than 15 s.

Nezamzadeh-Ejhieh and Esmaeilian (2012) proposed a sulfate-ISE based on surfactant-modified zeolite particles into carbon paste. The electrode exhibited a linear response range to sulfate species in the range of 10^{-6} to 10^{-3} M with a detection limit of 2×10^{-6} M and a good sensitivity based on the Nernstian slope within the pH range 4–10. Lomako *et al.* (2006) also evaluated a sulfate-ISE based on the composition of a neutral carrier and a quaternary ammonium salt. The ISE was used for the determination of sulfate in seawater, mineral water and urine by a direct potentiometric method and potentiometric titration, resulting in relative errors below 7%. Mazloum-Ardakani *et al.* (2006) presented a sulfate-ISE based on a complex of copper as a membrane carrier. The electrode exhibited good sensitivity and linearity within the range of 10^{-7}–10^{-1} M. The probe resulted in a fast response time of 10 seconds and had a lifetime of more than three months within the pH range 3.5–8.0. Fibbioli *et al.* (2000) investigated a polymeric membrane ISE based on an anionic chloro-borane cluster and a dihydrochloride analogue for phosphate and sulfate monitoring. The best sulfate selectivity was found for ISE membranes based on the dihydrochloride, whereas those with the zwitterion analogue were shown to possess a reasonably good selectivity for monohydrogen phosphate. More recently, Gołębiewski *et al.* (2019) studied an electrochemical sensor comprising a cyclopeptide connected to the surface of a gold electrode via a bis(dipyrromethene) Cu(II) or Co(II) complex for the detection of sulfate in water. The sensor displayed high selectivity and sensitivity, rendering it potentially useful for environmental monitoring.

Although further research is needed to develop industrially-feasible ISEs, these solutions are of great interest for real-time controlling sulfate-rich processes. On-line dissolved sulfate measurements are very important when treating sulfate-rich wastewaters since the influent BOD:SO_4^{2-}-S ratio strongly affects the feasibility of AD systems due to the competition of SRB and methanogens for the available substrate. Thus, these ISE sensors would enable the development of control strategies aimed to optimize methane production or to minimize H_2S inhibition, among others.

9.3.3 In-situ sulfurous compounds sensors

Regarding H_2S, Pandey *et al.* (2012) broadly discussed the use of different sensors for H_2S measurement in the gas phase, such as the following:

- Semiconducting metal-oxide sensors in the form of thin or thick films, based on metal-oxide semiconductors (e.g. tin oxide). These sensors present a small size, simple construction, low weight, low power consumption, and low cost (Fine *et al.*, 2010). The heated metal-oxide sensor is the most commonly used sensor for monitoring gases.
- Electrochemical sensors based on liquid and solid electrolyte sensors, with solid electrolyte being the most commonly used electrochemical sensors for H_2S-gas analysis. Amperometric-based electrochemical sensors generate a current signal that is related to the H_2S concentration by Faraday's law and the laws of mass transport. Potentiometric-based electrochemical sensors measure electrical potential (voltage).

- Optical sensors use an optical sensing technique, based on the attenuation of light waves, commonly employing optical transduction techniques to yield H_2S information (McDonagh *et al.*, 2008). Optical chemical sensors can generally be categorized into two types, i.e. direct and reagent mediated. In direct sensing, H_2S is detected via its intrinsic optical property (e.g. absorption or luminescence). In reagent-mediated sensors, a change in the optical response of an intermediate agent is used for measurement. For instance, Cho *et al.* (2008) proposed a wireless electronic nose system for real-time quantitative analysis of gas mixtures using a gas sensor micro-array and neuro-fuzzy network, showing good performance in the concentration range of 0.15–1.5 ppm H_2S with a high sensibility and response time (0.2 s).
- Conducting-polymer sensors are widely used for gas characterization due to their easy fabrication (see e.g. Virji *et al.*, 2005), high reproducibility, rapid reaction rate, and low cost compared to other techniques. The basic idea is to mix conducting materials with polymers selected for the target gas to form a conducting-polymer-composite sensing material.
- Piezoelectric sensors are configured as mass-change-sensing devices. Generally, these devices are a resonating polymer-coated disk with metal electrodes, each side of which is connected to a lead wire. These sensors resonate at a characteristic frequency via excitation by a specific oscillating signal. Another type of piezoelectric device is the surface acoustic wave device, which is based on the principle that a Ray-Leigh wave travels over the surface of the device instead of through its volume (Tigli & Zaghloul, 2007).

On the other hand, on-line dissolved sulfide monitoring is also of great importance when treating sulfate-rich wastewaters. ISEs can also be used for this purpose. Moreover, from the measured activity of free sulfide ions with an ISE, the analytical concentration of the total dissolved sulfide (TDS) can be calculated if the protonation constants of the sulfide ion (K_1 and K_2) and the pH of the sample solution are known (Cassidy *et al.*, 2015), as shown in Equation 9.26, creating a virtual sensor based on ISE, pH and equilibrium calculations:

$$S^{2-} = \frac{\text{TDS}}{1 + \frac{H^+}{K_1} + \frac{(H^+)^2}{K_2}} \tag{9.26}$$

For instance, Grootscholten *et al.* (2008) designed an on-line estimator based on Equation 9.26 for the sulfide and zinc concentrations in a precipitation reactor. This virtual sensor allowed estimation of ZnS concentrations and precipitation rates in a CSTR based on the pH and the activity of the S^{2-} measured through an ISE. An appropriate performance was observed on the basis of experimental data, offering good possibilities for application in other metal removal processes.

Different sulfide-ISEs were developed in the 1980s and 1990s for sulfur determination in liquid (see e.g. Atta *et al.*, 1998; Guterman *et al.*, 1983; Tóth *et al.*, 1988) and even solid (García-Calzada *et al.*, 1999) samples. However, results on sulfide-ISEs seem to be less accurate than those obtained for sulfate-ISEs due to different inferences on the measuring principle. For instance, Balasubramanian and Pugalenthi (2000) critically analyzed ISEs and iodimetry for the determination of sulfide in tannery wastewater, showing significant deviations of ISEs from reference values. In this case, the accuracy of the values obtained through ISEs depended on the different effluent streams from which the samples were collected. Villa-Gomez *et al.* (2014) used a sulfide-ISE that was shown to be informative for applications in sulfate-reducing bioreactors and was successfully used for control purposes. However, the response of the probe showed a time-varying behavior due to sulfide accumulation or utilization of substrate sources that were not accounted for. Therefore, the recorded sulfide values needed to be corrected for pH variations and high sulfide concentrations ($>200 \text{ mg} \cdot L^{-1}$). On the other hand, Nezamzadeh-Ejhieh and Afshari (2012) modified a PVC-membrane electrode with the surfactant zeolite for potentiometric determination of sulfide, showing successful sulfide determinations in waste samples from a sugar company. This electrode exhibited a linear response range to sulfide in the range of 10^{-7} to $10^{-1} \text{ mol} \cdot L^{-1}$ with good sensitivity to potential changes and a detection limit of $6.3 \times 10^{-8} \text{ mol} \cdot L^{-1}$ within the pH range 3–10.

Regarding combined systems, Redondo *et al.* (2008) designed an automated H_2S on-line analyzer to assess the composition of the liquid and gas phases. The analyzer consisted of an Ag_2S ISE to detect S^{2-} in the liquid phase and a continuous flow analyzer with a gaseous diffusion step for detecting H_2S in the gas phase. High Nernstian slopes were achieved.

9.3.4 Biosensors

Cassidy *et al.* (2015) reviewed the possibilities of combining in-situ sensors with molecular techniques to get better insights into the processes prevailing in a bioreactor, such as combining microelectrodes with specific oligonucleotide probes (Ramsing *et al.*, 1993), or combining molecular techniques (DGGE: denaturing gradient gel electrophoresis, PCR: polymerase chain reaction and FISH: fluorescence in situ hybridization) with microsensors for H_2S and CH_4 (Santegoeds *et al.*, 1999). However, these techniques for biomass and activity characterization are invasive, destructive, expensive, and provide off-line data. Hence, further research is required to quantify microbial activity in real-time. In this respect, impedimetric biosensors would allow quantification of different biological-related processes, such as the following: monitoring microbial populations using an impedimetric immunosensor (Wan *et al.*, 2009), antibody recognition through direct immobilization on a bioinspired architecture as a sensing platform (Wan *et al.*, 2011), evaluation of redox indicators under dechlorinating conditions (Jones & Ingle, 2005), amplification of responses of vancomycin-functionalized magnetic nanoparticles using a modified quartz crystal microbalance (Wan *et al.*, 2010a), potentiometric stripping analysis for the detection of *Desulforibrio caledoiensis* (Wan *et al.*, 2010b), or monitoring the treatment of acid mine drainage using conductivity (Lyew & Sheppard, 2001).

9.4 CONTROL

9.4.1 Process control in AD

AD must be optimized to enhance resource recovery with minimum energy input. However, AD optimization is not straightforward due to the inherent complexity involved in anaerobic processes. In particular, sensitivity to process overloads and other disturbances such as acidification, inhibition and toxicant exposure, rheology, foaming, stirring and mixing problems, as well as low/high C:N ratios (i.e. ammonia inhibition and lack of macro- and micro-nutrients) are among the main situations that can be faced in practice. These factors may lead to different operational problems including biomass washout. Over the years, research on AD has advanced, allowing for widening the application field of AD: from the side-stream treatment of sewage and biological sludge to the mainstream treatment of wastewater (industrial and municipal/urban), and to the (co-) digestion of the organic fraction of municipal solid waste, food waste or different kinds of agricultural residues. Therefore, control targets have also evolved, moving from the classical regulation of key variables to the prediction of overall AD process performance, including feeding optimization (e.g. blending of co-substrates).

AD is a mature technology that relies on a synergistic effort of a diverse group of microbial communities to metabolize diverse organic substrates (see Figure 9.1). However, AD is sensitive to process disturbances, and thus it is advantageous to use on-line monitoring and process control techniques to efficiently operate AD processes: Section 9.3 overviewed the range of electrochemical, chromatographic and spectroscopic devices that can be implemented for on-line AD monitoring and control. Complexity of the control strategy ranges from a feedback control to advanced control systems.

Different operating and control strategies have been proposed, from open-loop and off-line control approaches to advanced automatic (closed-loop) control systems. Several examples of control strategies applied to different AD systems can be found elsewhere, mostly using continuous in situ monitoring and associated feedback procedures to routinely deliver continuous and optimal performance (Gaida *et al.*, 2017; Jimenez *et al.*, 2015; Nguyen *et al.*, 2015; Olsson *et al.*, 2014; Robles *et al.*, 2018; Ward *et al.*, 2008). Over the last 40 years many different control methodologies for substrate feed control of AD processes have been developed to improve plant efficiency and sustainable long-term energy production (Gaida *et al.*, 2017). Regardless of the application, such control aims to find a compromise between maximizing economic yield, minimizing the ecological footprint and minimizing the risk of process failure (Gaida *et al.*, 2017).

9.4.2 Advanced closed-loop control of AD processes

Besides classical proportional-integral-derivative (PID) control and derivatives (e.g. PI – proportional integral), different advanced control approaches have been applied to different AD processes, such as disturbance monitoring control, model predictive control, adaptive control, robust control, fuzzy control, artificial neural networks, and fault detection and isolation, among others. A summary of the main characteristics of these controllers is presented in this chapter.

9.4.2.1 Disturbance monitoring control

The control of wastewater treatment processes in general – and for the treatment of waters polluted by sulfurous compounds in particular – can be achieved through a careful analysis of the disturbances affecting the main variables. The purpose of the "disturbance monitoring" is to develop a control strategy that, compared to conventional controllers, tackles the problem of process overloading very differently (Steyer *et al.*, 1999). The basic idea of this control strategy is to add a disturbance on the input liquid flow rate. The response of simple parameters, that can be chosen because of their heavy use at the industrial scale (e.g. pH and the output gas flow rate), can be analyzed to determine if it is possible to increase the loading rate of the reactor. A negative effect of a positive disturbance (e.g. an increase of the input liquid flow rate) can be observed by a pH decrease or by an increase of the output gas flow rate lower than theoretically expected. In such a case, the reactor is assumed to be overloaded and the organic loading rate is automatically decreased. In the case of no negative effect of the disturbance, the reactor is assumed to operate safely (i.e. without being overloaded) and the loading rate is increased. Thus, the organic loading rate can be maintained as high as possible despite influent variability and the organic matter concentration of the treated effluent can be kept low and stable (Steyer *et al.*, 1999).

9.4.2.2 Model-based and linearizing control

The model-based control principle is largely used in the process industry. These controllers take advantage of available mathematical models for optimization purposes. Model-based control could allow efficient control of the running of a digester while achieving more precise actions than those controllers based on the difference between measurements and set points. For instance, linearizing control is based on the linearization of a non-linear system in order to achieve linear closed-loop dynamics. However, due to the strong non-linearity of AD processes, linearizing controllers only attain proper results when the process dynamics are bound by a defined linear zone. This kind of controller is built in a two-step procedure (Kurtz & Henson, 1997). Firstly, a non-linear model is used in order to synthesize the non-linear state feedback controller that linearizes the map between a "new" manipulated input and the controlled output. In the second step, a linear pole placement controller is designed for the feedback linearized system.

9.4.2.3 Adaptive control

The key idea of an adaptive control design is to take advantage of the well-known parameters associated with the dynamics of bioprocesses (basically, the reaction pathways and mass balances summarized in a dynamic mass balance model) while accounting for the model uncertainty (mainly the kinetics). Since the model is generally non-linear, the model-based control design might result in a linearizing control structure, in which the on-line estimation of the unknown variables (concentration of pollutant and its intermediates) and parameters (reaction rates and yield coefficients) are incorporated.

Adaptive control schemes can be applied for AD processes because of their good disturbance rejection ability (Renard *et al.*, 1988). In such cases, the parameters of the controller are continuously adapted to follow – and sometimes to estimate – the changes in the operating conditions (Dochain & Perrier, 1993).

9.4.2.4 Robust control

One of the key issues to be addressed in controlling biological wastewater treatment processes is to reject the disturbances that can destabilize the reactor. It is worth mentioning that a clear distinction must be made between disturbance rejection and insensitivity to unmodeled phenomena or parameter variations. Robust control strategies represent a promising option to tackle these two problems since this control strategy includes significant improvements over conventional PID controllers (Flores-Estrella *et al.*, 2019; Méndez-Acosta *et al.*, 2010). Regarding adaptive control, robust control approaches strongly rely on mathematical models such as those depicted in Section 9.2. However, the main difference between robust and adaptive control schemes is that the latter continuously adapt their parameters based on dynamic estimates of the process evolution whereas, in the former scheme, the uncertainty is explicitly accounted for in the building of the controller. The adaptive control scheme can thus achieve a better performance whereas the robust control scheme can handle more severe operating conditions (Steyer *et al.*, 2006).

9.4.2.5 Fuzzy control

AD is a complex process which presents important challenges to be controlled with classic PID methods. However, this imperfect knowledge of the process can be minimized through expertise obtained with the operation of pilot units. Since fuzzy logic naturally captures the acquired experience of human operators, it appears to be a powerful tool for controlling this type of process (Driankov *et al.*, 1996; Zadeh, 1965). In addition, fuzzy logic can incorporate semi-quantitative information into simplified models and it can put subjective information into a form usable by computers.

Fuzzy control allows efficient management of a wastewater treatment plant in general – and an AD process in particular (Estaben *et al.*, 1997). In addition, fuzzy logic has very interesting properties since human operators and/or control engineers generally find it difficult to decide upon adequate parameters to use when dealing with a full-scale wastewater treatment plant. The fuzzy control theory can be used to implement effective control strategies even when suitable models or correct parameters are unknown. Last but not least, this method can be extended to many different applications in the wastewater treatment field since a fuzzy controller can be easily tuned (Driankov *et al.*, 1996).

9.4.2.6 Artificial neural networks

Section 9.4.2 has so far outlined several difficulties which arise upon the implementation of control systems for wastewater treatment and the AD process. Firstly, the behavior of biological processes and the models describing these systems (see Section 9.2) are often highly non-linear. Additionally, these models are complex and their parameters are often difficult to determine. The dynamics of biological processes also tend to vary because of variations in metabolism and environmental parameters.

Application of artificial neural networks (ANNs) overcomes all these problems since ANNs can approximate any non-linear mapping arbitrarily and their learning abilities together with their parallel structure make them suitable for adaptive on-line application (Funahashi, 1989; Hornik *et al.*, 1989). It is thus clear that artificial neural networks are promising for the modeling and control of biological and AD processes (Premier *et al.*, 1999; Wilcox *et al.*, 1995).

9.4.3 Process control of sulfate-reducing bioprocesses

The number of closed-loop control applications in sulfate-reducing bioprocesses is still scarce. Nonetheless, there is a good number of open-loop strategies that may help to improve the performance of these systems. Moreover, mathematical models (see Section 9.2) may serve as support for the design of enhanced operating and control strategies when there are available monitoring techniques (see Section 9.3).

Many operating strategies have been used to control the conversion of sulfate into sulfide and the underlying problems related to corrosive sulfide formation (see Chapter 6). Sulfate-reducing process control has mainly focused on regulating the competition among microbial groups, optimizing the input ratio of electron donors to acceptors, and/or minimizing sulfide production. Specifically, the competition for the use of available substrates between SRB and methanogens strongly reduces the energy recovery from waste, especially in the case of treating wastewaters with low BOD:SO_4^{2-}-S ratios such as urban wastewater.

Moreover, the generated H_2S has a higher impact on methanogenic than on SRB activity (Chen *et al.*, 2008). However, the control of the influent sulfate concentration is generally considered as an unrealistic option (Cirne *et al.*, 2008; Jimenez *et al.*, 2015), thus control of sulfate-reducing processes has been generally focused on four main alternatives conceived to control the negative effects of sulfide: (i) prevent sulfide formation with the addition of SRB inhibitory compounds (e.g. molybdate, divalent transition metals, antibiotics); (ii) addition of small amounts of oxygen in the anaerobic digester to oxidize sulfide to sulfur or sulfate (microaeration); (iii) addition of organic and inorganic salts to precipitate sulfur; and (iv) biogas desulfurization in an external gas treatment unit (Cirne *et al.*, 2008; Jimenez *et al.*, 2015). Specifically, microaeration and salts addition for metal precipitation are two strategies that can be efficiently optimized through the implementation of an automated control system. Chemical technologies to mitigate H_2S inhibition (i.e. to oxidize the sulfide) include precipitation by addition of metal salts and addition of oxidizing reagents like chlorine, hydrogen peroxide, or potassium permanganate. Biological methods include biological oxidation of H_2S by sulfur-oxidizing bacteria and inhibition of SRB activity (Liu *et al.*, 2015).

9.4.3.1 Control of H_2S emissions

9.4.3.1.1 H_2S oxidation in bioreactor effluent

McFarland and Jewell (1989) presented one of the first studies dealing with control algorithms applied to AD processes treating sulfate-rich wastewaters. They recommended pH adjustment for biogas sulfide control under conditions in which the influent sulfur level is below sulfide inhibitory concentrations, showing that control of gaseous sulfide levels through insoluble iron (Fe^{3+}) phosphate addition is an efficient control strategy with no adverse effects on digester performance. Later, Camp and Sublette (1992) used another system – a PC-based machine vision system – to continuously monitor changes in biomass concentration and to control the undesirable production of colloidal elemental sulfur (a reactor upset condition due to an inhibitory sulfide concentration) in a bioreactor containing *Thiobacillus denitrificans*. A video camera was established which produced regions of different background lighting. Mean values of the distribution of red, green and blue intensity components with corresponding regions of a digital image captured from the camera were used to monitor color changes associated with changes in biomass concentration and to determine if the reactor was under suboptimal operational conditions. They showed that the ratio of red to blue intensity was an important parameter in detecting the formation of an elemental sulfur precipitant. Using a stepper motor-driven pressure regulator, process control was then performed by altering the hydrogen sulfide flow rate setpoint based on the vision system measurements. At the end of the 1990s, Janssen *et al.* (1998) showed that the formation of elemental sulfur from the biological oxidation of sulfide could be optimized by controlling the redox state of the solution. They successfully applied a PI control of the redox potential so that nearly stoichiometric amounts of oxygen were supplied to the system. In this way, the supplied oxygen was sufficient for oxidizing sulfide to sulfur, although about 10% of the sulfide was still oxidized to sulfate.

9.4.3.1.2 H_2S oxidation in biogas

Microaeration is considered a highly efficient, simple, reliable, and economically feasible technique for removing hydrogen sulfide from biogas (Cirne *et al.*, 2008; Khoshnevisan *et al.*, 2017; Krayzelova *et al.*, 2015). During microaeration, the oxygen added to the anaerobic digester promotes sulfide oxidization to elemental sulfur through chemical oxidation or by the so-called sulfide-oxidizing bacteria (SOB). However, microaeration is commonly conducted as an open-loop strategy, and only a few automated applications are available. In any case, precise dosing of O_2 is required to maximize the efficiency of anaerobic digestion. The appropriate oxygen dosing depends on the microaeration target: preventing volatile fatty acids (VFA) accumulation or removing hydrogen sulfide (Nguyen & Khanal, 2018).

The optimization of the O_2:S^{2-} ratio or the dissolved oxygen (DO) concentration (i.e. working under oxygen limiting conditions: DO concentrations $<0.1\ mg \cdot L^{-1}$) enables elemental sulfur to be obtained as a main end-product, which is the most desirable option from an economic point of view (Janssen *et al.*, 1995). Besides, optimizing the amount of required oxygen for H_2S oxidation can lead to the potential recovery of elemental sulfur. Hydrogen sulfide removal efficiencies higher than 90% can be achieved with microaeration (Fdz.-Polanco *et al.*, 2009; Jenicek *et al.*, 2010; Ramos *et al.*, 2014). Another factor affecting the efficiency of microaeration that could be used as a manipulated variable in a closed-loop controller is the residence time of biogas in the headspace volume, especially when air is dosed into the headspace. Removal efficiencies close to 100% have been reported (Díaz *et al.*, 2011; Ramos & Fdz-Polanco, 2014) with biogas residence times over 5 hours. These high efficiencies are due to the fact that SOB usually grow on the walls of the headspace.

Different control parameters have been used for maintaining an appropriate oxygen injection rate in dynamic systems in order to cope with variations in hydrogen sulfide production. Ramos and Fdz-Polanco (2014) developed a PID controller to control the air flow rate according to the H_2S concentration in biogas. Other authors (Khanal & Huang, 2006; Nghiem *et al.*, 2014) modified the air dosing rate in order to maintain the set point established for the oxidation reduction potential (ORP). Microaeration operated at a constant dosing rate could indeed result in oxygen overdosing thus decreasing methane production (Kobayashi *et al.*, 2012). The strategy of intermittent microaeration has also been applied in different studies. Nguyen *et al.* (2019) avoided VFA accumulation by means of an intermittent (every 24 h) ORP-controlled microaeration strategy (at ORP of −470 mV).

Nonetheless, there are different factors to take into account in order to adequately establish control boundaries in terms of maximum oxygen dose, such as:

- Oxygen toxicity to methanogens: although oxygen is a well-known inhibitor for the activity of methanogens, the majority of microaeration studies have not revealed a significant decrease in the methanogenic activity. According to different authors (Estrada-Vázquez *et al.*, 2003; Shen & Guiot, 1996), methanogens are protected by facultative anaerobic bacteria in both granular and suspended sludge.
- Explosion risks of methane/oxygen mixtures: the flammable range for oxygen/methane mixtures are 85–95% air and 5–15% methane by volume (Appels *et al.*, 2008). The oxygen required for hydrogen sulfide removal is very small and not close to this range. However, the explosion risk should always be considered.
- Partial oxidation of soluble organic matter: the oxygen dosing rate required for microaerobic H_2S removal (0.001–0.01 · $kg\ m^{-3} \cdot day^{-1}$) is three orders of magnitude lower than the organic loading rate (OLR, 1–10 kg COD · $m^{-3} \cdot day^{-1}$). Therefore, the amount of oxidized substrate can be considered negligible (Krayzelova *et al.*, 2014).
- Dilution of biogas by nitrogen from air: when air is used for microaeration, nitrogen can significantly reduce the percentage of methane in the biogas, making its use difficult in cogeneration units. Celis (2012) reported that the percentage of nitrogen in the biogas increased up to 20% when treating biogas with high H_2S concentrations (around 12,000 ppm). Pure oxygen could be used in such cases, however, it increases the operational costs.
- An innovative method of microaeration is the application of water electrolysis within UASB reactors so that O2 is produced directly at the anode in the reactor (Tartakovsky *et al.*, 2011). A feedback control of the applied voltage can be used to control the oxygen production rate.

Automatic process control of microaeration still remains a great challenge for its implementation in full-scale applications (Chen *et al.*, 2020). Mathematical models may serve as support for the design of control strategies when there are available monitoring techniques. In this respect, Pokorna-Krayzelova *et al.* (2017) proposed the model ADM1-S:O, an ADM1 extension including the processes related to sulfate reduction and sulfide oxidation to elemental sulfur.

9.4.3.2 Metal precipitation

9.4.3.2.1 Metal addition for H_2S control

Another open-loop strategy commonly conducted for H_2S control is metal addition. Nonetheless, few automated applications are yet available. Closed-loop control for metal precipitation would allow optimization of the dose of metal to be used for precipitation, optimizing the economics of the process.

Metal addition can be conducted onsite or even in the sewer systems (Zhang *et al.*, 2008). For instance, Salehin *et al.* (2019) evaluated different alternatives for reducing coagulants usage in urban water management. Specifically, Salehin *et al.* (2019) revealed that iron dosed to sewers is still available for hydrogen sulfide removal from biogas in the downstream WWTP. This study, which was performed under real-life conditions, confirmed the practical feasibility and effectiveness of the strategy through a year-long full-scale investigation. Regarding metal addition in AD systems, Lee *et al.* (2016) evaluated the effects of $FeCl_3$ addition on the operation of a staged anaerobic fluidized membrane bioreactor (SAF-MBR). Total sulfur removal increased to 87–95% with $FeCl_3$ addition. Sulfide removal increased to 90% with addition of $FeCl_3$ at the molar Fe^{3+}:S ratio of 0.54 and to 95% when the ratio was increased to 0.95. The effluent sulfide concentration also decreased to 0.3–0.6 $mg \cdot L^{-1}$ after the addition of $FeCl_3$. However, the addition of $FeCl_3$ negatively affected membrane fouling of anaerobic membrane bioreactors (AnMBR) due to the precipitation of inorganic foulants on the membrane surface. Other actions related to direct metal addition have been carried out for controlling H_2S production in AD systems. These actions reveal the potential of automated control of metal addition in order to minimize H_2S emissions. However, other factors such as phosphorus precipitation with metal and pH affecting H_2S dissociation must be also considered when developing control systems.

Nonetheless, not all metals are useful for potential automated control applications. For example, Akgul *et al.* (2017) assessed the effect of iron and aluminum-based coagulants on the formation of volatile sulfur compounds. Aluminum-based coagulants significantly increased volatile sulfur compound levels, while ferric chloride achieved up to 93% reduction of the biogas H_2S concentration.

The sulfide interaction with iron as regulator of the microbial community in AD systems is another factor to consider when designing automated control systems based on metal addition (Shakeri Yekta *et al.*, 2017). In this regard, control boundaries based on the S:Fe molar ratio might be defined since this ratio maintains a direct relation with methane production. An increase in the S:Fe molar ratio from 0.3 to 0.5 did not alter the methane production, while S:Fe molar ratios above 0.5–1.0 resulted in the accumulation of acetate in the digester and a decline in daily methane production.

9.4.3.2.2 Sulfide addition for selective metal precipitation

Besides the addition of metals for sulfide control, sulfide addition can also be used for selective metal precipitation. Different examples can be found in the literature using different types of ISE and pH sensors aiming to control chemical sulfide addition in a precipitation reactor. Table 9.1 shows some examples of control strategies applied for metal precipitation using chemical and biogenic sulfide.

Table 9.1 Examples of control strategies for metal precipitation from chemical and biogenic sulfide (extended from Cassidy *et al.*, 2015).

Controlled Variable	**Manipulated Variable**	**Objective**	**Reference**
Chemical sulfide concentration	Sulfide flow	Selective heavy metal precipitation (Cu-Zn and Pb-Zn)	Veeken *et al.* (2003)
Biogenic sulfide concentration	Sulfide flow	Zinc precipitation	König *et al.* (2006)
Biogenic sulfide concentration	Sulfide flow	Zinc precipitation	Esposito *et al.* (2006)
Chemical sulfide concentration	Sulfide flow	Copper and Zinc selective precipitation	Sampaio *et al.* (2009)
Chemical sulfide concentration	Sulfide flow	Zinc and Nickel selective precipitation	Sampaio *et al.* (2010)
Biogenic sulfide concentration	OLR	Sulfate reduction in a fluidized bed reactor	Villa-Gomez *et al.* (2014)

Veeken *et al.* (2003) proposed a process where a sulfide-selective ISE was used to control sulfide addition in order to remove heavy metals (Cd, Cu, Ni, Pb, and Zn). The heavy metals were removed to ppb levels at pH 6.0 by sulfide precipitation while maintaining the total sulfide concentration below 0.02 mg L^{-1}. By controlling the sulfide concentration at different levels, the metals in mixtures of Cu–Zn and Pb–Zn were selectively precipitated from solution, thus producing pure metal sulfide sludge that could be reused.

König *et al.* (2006) presented a dynamic model and a PI control strategy for the precipitation of ZnS. In this case, the sulfide concentration was also controlled using an ISE. The controller was able to handle sudden disturbances in the process conditions (pH, influent flow rate, or zinc and sulfide concentration). Esposito *et al.* (2006) also assessed the performance of a zinc sulfide precipitation process using a biogenic sulfide solution (the effluent of a sulfate-reducing bioreactor) as the sulfide source. A PI controller was used to control the pH and S^{2-} concentration at the desired level using a pH probe and a sulfide-selective ISE. Sampaio *et al.* (2009) evaluated the selective precipitation of Cu from Zn in a CSTR by using a sulfide-selective ISE. In this case, copper was continuously and selectively precipitated with Na_2S to concentrations below 0.3 ppb from water containing around 600 ppm of both Cu and Zn. Sampaio *et al.* (2010) studied the role of supersaturation within Zn–Ni sulfide selective precipitation. In this case, the selective removal of Zn with Na_2S from a mixture of Zn and Ni was studied in a CSTR operating at pH 5.

Villa-Gomez *et al.* (2014) evaluated progressive step changes in OLR to create sulfide responses for the design of a sulfide control in sulfate-reducing bioreactors. The sulfide was measured using a sulfide-selective ISE and the values obtained were used to calculate PID controller parameters. A rapid response and high sulfide increment were obtained through a stepwise increase in the influent COD concentration, while a stepwise decrease to the hydraulic retention time (HRT) resulted in a slower response with smaller sulfide increment. Regardless of the way that the OLR was decreased, the ISE response showed a time-varying behavior due to sulfide accumulation (HRT change) or utilization of substrate sources that were not accounted for (influent COD change). However, the ISE response was shown to be informative for application of substrate dosing in sulfate reducing bioreactors.

REFERENCES

Ahmed W. and Rodríguez J. (2018). Modelling sulfate reduction in anaerobic digestion: Complexity evaluation and parameter calibration. *Water Research*, **130**, 255–262.

Akgul D., Abbott T. and Eskicioglu C. (2017). Assessing iron and aluminum-based coagulants for odour and pathogen reductions in sludge digesters and enhanced digestate dewaterability. *Science of the Total Environment*, **598**, 881–888.

Arnell M., Astals S., Åmand L., Batstone D. J., Jensen P. D. and Jeppsson U. (2016). Modelling anaerobic co-digestion in Benchmark Simulation Model No. 2: Parameter estimation, substrate characterisation and plant-wide integration. *Water Research*, **98**, 138–146.

Angelidaki I., Treu L., Tsapekos P., Luo G., Campanaro S., Wenzel H. and Kougias P. G. (2018). Biogas upgrading and utilization: Current status and perspectives. *Biotechnology Advances*, **36**, 452–466.

Appels L., Baeyens J., Degrève J. and Dewil R. (2008). Principles and potential of the anaerobic digestion of waste-activated sludge. *Progress in Energy and Combustion Science*, **36**, 755–781.

Atta N. F., Galal A., Mark H. B., Yu T. and Bishop P. L. (1998). Conducting polymer ion sensor electrodes–III. Potentiometric sulfide ion selective electrode. *Talanta* **47**, 987–999.

Balasubramanian S. and Pugalenthi V. (2000). A comparative study of the determination of sulphide in tannery waste water by ion selective electrode (ISE) and iodimetry. *Water Research*, **34**, 4201–4206.

Bandekar, J., Sethna, R. and Kirshner, M. (1995). Quantitative determination of sulfur oxide species in white liquor bt FT-IR. *Applied Spectroscopy*, **49**(11), 1577–1582.

Barat R., Serralta J., Ruano M. V, Jiménez E., Ribes J., Seco A. and Ferrer J. (2013). Biological Nutrient Removal Model No. 2 (BNRM2): a general model for wastewater treatment plants. *Water Science and Technology*, **67**(7), 1481–1489.

Barrera E. L., Spanjers H., Dewulf J., Romero O. and Rosa E. (2013). The sulfur chain in biogas production from sulfate-rich liquid substrates: a review on dynamic modeling with vinasse as model substrate. *Journal of Chemical Technology and Biotechnology*, **88**, 1405–1420.

Barrera E. L., Spanjers H., Solon K., Amerlinck Y., Nopens I. and Dewulf J. (2015). Modeling the anaerobic digestion of cane-molasses vinasse: Extension of the Anaerobic Digestion Model No. 1 (ADM1) with sulfate reduction for a very high strength and sulfate rich wastewater. *Water Research*, **71**, 42–54.

Batstone D. J., Keller J., Angelidaki I., Kalyuzhnyi S. V., Pavlostathis S. G., Rozzi A., Sanders W. T. M., Siegrist H. and Vavilin V. A. (2002). The IWA Anaerobic Digestion Model No 1 (ADM1). *Water Science and Technology*, **45**(10), 65–73.

Batstone D. J., Keller J. and Steyer J. P. (2006). A review of ADM1 extensions, applications, and analysis: 2002–2005. *Water Science and Technology*, **54**(4), 1–10.

Batstone D. J., Puyol D., Flores-Alsina X. and Rodríguez J. (2015). Mathematical modelling of anaerobic digestion processes: applications and future needs. *Reviews in Environmental Science and Bio/Technology*, **14**, 595–613.

Bastin, G. and Dochain, D. (1990). On-line estimation and adaptive control of bioreactors, Elsevier Science Publishing, Amsterdam.

Berrocal M. J., Cruz A., Badr I. H. A. and Bachas L. G. (2000). Tripodal Ionophore with Sulfate Recognition Properties for Anion-Selective Electrodes. *Analytical Chemistry*, **72**, 5295–5299.

Camp C. E. and Sublette K. L. (1992). Control of a *Thiobacillus denitrificans* bioreactor using machine vision. *Biotechnology and Bioengineering*, **39**, 529–538.

Capson-Tojo G., Moscoviz R., Astals S., Robles Á. and Steyer J.-P. (2020). Unraveling the literature chaos around free ammonia inhibition in anaerobic digestion. *Renewable and Sustainable Energy Reviews*, **117**, 109487.

Cassidy J., Lubberding H. J., Esposito G., Keesman K. J. and Lens P. N. L. (2015). Automated biological sulphate reduction: A review on mathematical models, monitoring and bioprocess control. *FEMS Microbiology Reviews*, **39**, 823–853.

Celis C. A. (2012). Improvement of anaerobic digestion by using of microaerobic conditions. Department of Water Technology and Environmental Engineering, Ph.D., Institute of Chemical Technology in Prague, Prague, p. **181**.

Chen Y., Cheng J. J. and Creamer K. S. (2008). Inhibition of anaerobic digestion process: A review. *Bioresource Technology*, **99**, 4044–4064.

Chen Q., Wu W., Qi D., Ding Y. and Zhao Z. (2020). Review on microaeration-based anaerobic digestion: State of the art, challenges, and prospectives. *Science of the Total Environment*, **710**, 136388.

Cho J. H., Kim Y. W., Na K. J. and Jeon G. J. (2008). Wireless electronic nose system for real-time quantitative analysis of gas mixtures using micro-gas sensor array and neuro-fuzzy network. *Sensors Actuators B Chemical*, **134**, 104–111.

Cirne D. G., Van Der Zee F. P., Fernandez-Polanco M. and Fernandez-Polanco F. (2008). Control of sulphide during anaerobic treatment of S-containing wastewaters by adding limited amounts of oxygen or nitrate. *Reviews in Environmental Science and Biotechnology*, **7**, 93–105.

D'Acunto B., Esposito G., Frunzo L. and Pirozzi F. (2011). Dynamic modeling of sulfate reducing biofilms. *Computers & Mathematics with Applications*, **62**, 2601–2608.

Devinny J. S. and Ramesh J. (2005). A phenomenological review of biofilter models. *Chemical Engineering Journal*, **113**, 187–196.

Díaz I., Lopes A. C., Pérez S. I. and Fdz-Polanco M. (2011). Determination of the optimal rate for the microaerobic treatment of several H_2S concentrations in biogas from sludge digesters. *Water Science and Technology*, **64**(1), 233–238.

Dochain D. and Perrier M. (1993). Control Design for Nonlinear Wastewater Treatment Processes. *Water Science and Technology*, **28**(11–12), 283–293.

Driankov D., Hellendoorn H. and Reinfrank M. (1996). An Introduction to Fuzzy Control, 2nd edn. Springer-Verlag, Berlin, Heidelberg.

Durán F. (2013). Mathematical modelling of the anaerobic urban wastewater treatment including sulphate-reducing bacteria. Application to an anaerobic membrane bioreactor (Modelación matemática del tratamiento anaerobio de aguas residuales urbanas incluyendo las bacterias. Universitat Politècnica de València.

Durán F., Robles A., Seco A., Ferrer J., Ribes J. and Serralta J. (2017). Modelling the anaerobic treatment of urban wastewater: application to AnMBR technology. In: 15th IWA World Conference on Anaerobic Digestion (AD15), Beijing.

Esposito G., Veeken A., Weijma J. and Lens P. N.L. (2006). Use of biogenic sulfide for ZnS precipitation. *Separation and Purification Technology*, **51**, 31–39.

Estaben M., Polit M. and Steyer J. P. (1997). Fuzzy control for an anaerobic digester. *Control Engineering Practice*, **5**, 1303–1310.

Estrada-Vázquez C., Macarie H., Kato M. T., Rodríguez-Vázquez R., Esparza-García F. and Poggi-Varaldo H. M. (2003). The effect of the supplementation with a primary carbon source on the resistance to oxygen exposure of methanogenic sludge. *Water Science and Technology*, **48**(6), 119–124.

Fan L.-S., Leyva-Ramos R., Wisecarver K. D. and Zehner B. J. (1990). Diffusion of phenol through a biofilm grown on activated carbon particles in a draft-tube three-phase fluidized-bed bioreactor. *Biotechnology and Bioengineering*, **35**, 279–286.

Fdz.-Polanco M., Díaz I., Pérez S. I., Lopes A. C. and Fdz.-Polanco F. (2009). Hydrogen sulphide removal in the anaerobic digestion of sludge by micro-aerobic processes: pilot plant experience. *Water Science and Technology*, **60**(12), 3045–3050.

Fedorovich V., Lens P. and Kalyuzhnyi S. (2003). Extension of Enaerobic Digestion Model No. 1 with processes of sulfate reduction. *Applied Biochemistry and Biotechnology*, **109**, 33–45.

Feldman H., Flores-Alsina X., Ramin P., Kjellberg K., Jeppsson U., Batstone D. J. and Gernaey K. V (2019). Assessing the effects of intra-granule precipitation in a full-scale industrial anaerobic digester. *Water Science and Technology*, **79**(7), 1327–1337.

Fernández-Arévalo T., Lizarralde I., Fdz-Polanco F., Pérez-Elvira S. I., Garrido J. M., Puig S., Poch M., Grau P. and Ayesa E. (2017). Quantitative assessment of energy and resource recovery in wastewater treatment plants based on plant-wide simulations. *Water Research*, **118**, 272–288.

Fibbioli M., Berger M., Schmidtchen F. P. and Pretsch E. (2000). Polymeric Membrane Electrodes for Monohydrogen Phosphate and Sulfate. *Analytical Chemistry*, **72**, 156–160.

Fine G. F., Cavanagh L. M., Afonja A. and Binions R. (2010). Metal Oxide Semi-Conductor Gas Sensors in Environmental Monitoring. *Sensors* **10**, 5469–5502.

Flores-Alsina X., Solon K., Kazadi Mbamba C., Tait S., Gernaey K. V, Jeppsson U. and Batstone D. J. (2016). Modelling phosphorus (P), sulfur (S) and iron (Fe) interactions for dynamic simulations of anaerobic digestion processes. *Water Research*, **95**, 370–382.

Flores-Estrella R. A., Alcaraz-González V., García-Sandoval J. P. and González-Álvarez V. (2019). Robust output disturbance rejection control for anaerobic digestion processes. *Journal of Process Control*, **75**, 15–23.

Fomichev A. O. and Vavilin V. A. (1997). The reduced model of self-oscillating dynamics in an anaerobic system with sulfate-reduction. *Ecological Modelling*, **95**, 133–144.

Frunzo L., Esposito G., Pirozzi F. and Lens P. (2012). Dynamic mathematical modeling of sulfate reducing gas-lift reactors. *Process Biochemistry*, **47**, 2172–2181.

Funahashi K.-I. (1989). On the approximate realization of continuous mappings by neural networks. *Neural Networks*, **2**, 183–192.

Gaida D., Wolf C. and Bongards M. (2017). Feed control of anaerobic digestion processes for renewable energy production: A review. *Renewable and Sustainable Energy Reviews*, **68**, 869–875.

Ganjali M. R., Naji L., Poursaberi T., Taghizadeh M., Pirelahi H., Yousefi M., Yeganeh-Faal A. and Shamsipur M. (2002). Novel sulfate ion-selective polymeric membrane electrode based on a derivative of pyrilium perchlorate. *Talanta*, **58**, 359–366.

García-Calzada M., Marbán G. and Fuertes A. B. (1999). Potentiometric determination of sulphur in solid samples with a sulphide selective electrode. *Analytica Chimica Acta*, **380**, 39–45.

Gołębiewski P., Puciłowski B., Sommer F., Kubik S., Daniels M., Dehaen W., Sivasankaran U., Kumar K. G., Radecka H. and Radecki J. (2019). Electrochemical sensing of sulfate in aqueous solution with a cyclopeptide-dipyrromethene-Cu(II) or Co(II) complex attached to a gold electrode. *Sensors Actuators B Chemical*, **285**, 536–545.

Grootscholten T., Keesman K. and Lens P. (2008). Modelling and on-line estimation of zinc sulphide precipitation in a continuously stirred tank reactor. *Separation and Purification Technology*, **63**, 654–660.

Gupta A., Flora J. R.V., Sayles G. D. and Suidan M. T. (1994). Methanogenesis and sulfate reduction in chemostats—II. Model development and verification. *Water Research*, **28**, 795–803.

Guterman H., Ben-Yaakov S. and Abeliovich A. (1983). Determination of total dissolved sulfide in the pH range 7.5 to 11.5 by ion selective electrodes. *Analytical Chemistry*, **55**, 1731–1734.

Hao T., Xiang P., Mackey H. R., Chi K., Lu H., Chui H., van Loosdrecht M. C. M. and Chen G.-H. (2014). A review of biological sulfate conversions in wastewater treatment. *Water Research*, **65**, 1–21.

Harerimana C., Keffala C., Jupsin H. and Vasel J.-L. (2013). Development of a simple model for anaerobic digestion based on preliminary measurements of the bacterial sulphur activity in wastewater stabilization ponds. *Environmental Technology*, **34**, 2213–2220.

Hedrich S., Kermer R., Aubel T., Martin M., Schippers A., Johnson D. B. and Janneck E. (2018). Implementation of biological and chemical techniques to recover metals from copper-rich leach solutions. *Hydrometallurgy* **179**, 274–281.

Henze M., Gujer W., Mino T., Matsuo T., Wentzel M. C., Marais G. V. R. and Van Loosdrecht M. C. M. (1999). Activated sludge model No.2D, ASM2D. *Water Science and Technology*, **39**(1), 165–182.

Henze M., Gujer W., Mino T. and van Loosedrecht M. (2006). Activated Sludge Models ASM1, ASM2, ASM2d and ASM3. IWA Publishing.

Hornik K., Stinchcombe M. and White H. (1989). Multilayer feedforward networks are universal approximators. *Neural Networks*, **2**, 359–366.

Hughes M. N., Centelles M. N. and Moore K. P. (2009). Making and working with hydrogen sulfide: The chemistry and generation of hydrogen sulfide in vitro and its measurement in vivo: A review. *Free Radical Biology & Medicine*, **47**, 1346–1353.

Huisman J., Schouten G. and Schultz C. (2006). Biologically produced sulphide for purification of process streams, effluent treatment and recovery of metals in the metal and mining industry. *Hydrometallurgy*, **83**, 106–113.

Hussain A., Hasan A., Javid A. and Qazi J. I. (2016). Exploited application of sulfate-reducing bacteria for concomitant treatment of metallic and non-metallic wastes: a mini review. *3 Biotech*, **6**, 119.

Ito T., Yoshiguchi K., Ariesyady H. D. and Okabe S. (2012). Identification and quantification of key microbial trophic groups of methanogenic glucose degradation in an anaerobic digester sludge. *Bioresource Technology*, **123**, 599–607.

Janssen A. J.H., Sleyster R., van der Kaa C., Jochemsen A., Bontsema J. and Lettinga G. (1995). Biological sulphide oxidation in a fed-batch reactor. *Biotechnology and Bioengineering*, **47**, 327–333.

Janssen A. J.H., Meijer S., Bontsema J. and Lettinga G. (1998). Application of the redox potential for controling a sulfide oxidizing bioreactor. *Biotechnology and Bioengineering*, **60**, 147–155.

Jenicek P., Koubova J., Bindzar J. and Zabranska J. (2010). Advantages of anaerobic digestion of sludge in microaerobic conditions. *Water Science and Technology*, **62**(2), 427–434.

Jeppsson U. and Rosen C. (2006). Aspects on ADM1 Implementation within the BSM2 Framework.

Jimenez J., Latrille E., Harmand J., Robles A., Ferrer J., Gaida D., Wolf C., Mairet F., Bernard O., Alcaraz-Gonzalez V., Mendez-Acosta H., Zitomer D., Totzke D., Spanjers H., Jacobi F., Guwy A., Dinsdale R., Premier G., Mazhegrane S., Ruiz-Filippi G., Seco A., Ribeiro T., Pauss A. and Steyer J.-P. (2015). Instrumentation and control of anaerobic digestion processes: a review and some research challenges. *Reviews in Environmental Science andReviews in Environmental Science and Bio/Technology*, **14**, 615–648.

Jones B. D. and Ingle J. D. (2005). Evaluation of redox indicators for determining sulfate-reducing and dechlorinating conditions. *Water Research*, **39**, 4343–4354.

Kaksonen A. H. and Puhakka J. A. (2007). Sulfate Reduction Based Bioprocesses for the Treatment of Acid Mine Drainage and the Recovery of Metals. *Engineering in Life Sciences*, **7**, 541–564.

Kalyuzhnyi S., Fedorovich V., Lens P., Hulshoff Pol L. and Lettinga G. (1998). Mathematical modelling as a tool to study population dynamics between sulfate reducing and methanogenic bacteria. *Biodegradation*, **9**, 187–199.

Kazadi Mbamba C., Batstone D. J., Flores-Alsina X. and Tait S. (2015a). A generalised chemical precipitation modelling approach in wastewater treatment applied to calcite. *Water Research*, **68**, 342–353.

Kazadi Mbamba C., Tait S., Flores-Alsina X. and Batstone D. J. (2015b). A systematic study of multiple minerals precipitation modelling in wastewater treatment. *Water Research*, **85**, 359–370.

Kazadi Mbamba C. (2016). Dynamic modelling of Minerals Precipitation in Wastewater Treatment. The University of Queensland. https://doi.org/10.14264/uql.2016.350

Kazadi Mbamba C., Lindblom E., Flores-Alsina X., Tait S., Anderson S., Saagi R., Batstone D. J., Gernaey K. V. and Jeppsson U. (2019). Plant-wide model-based analysis of iron dosage strategies for chemical phosphorus removal in wastewater treatment systems. *Water Research*, **155**, 12–25.

Khanal S. K. and Huang J.-C. (2006). Online Oxygen Control for Sulfide Oxidation in Anaerobic Treatment of High-Sulfate Wastewater. *Water Environment Research*, **78**, 397–408.

Khoshnevisan B., Tsapekos P., Alfaro N., Díaz I., Fdz-Polanco M., Rafiee S. and Angelidaki I. (2017). A review on prospects and challenges of biological H_2S removal from biogas with focus on biotrickling filtration and microaerobic desulfurization. *Biofuel Research Journal*, **4**, 741–750.

Knobel A. N. and Lewis A. E. (2002). A mathematical model of a high sulphate wastewater anaerobic treatment system. *Water Research*, **36**, 257–265.

Kobayashi T., Li Y.-Y., Kubota K., Harada H., Maeda T. and Yu H.-Q. (2012). Characterization of sulfide-oxidizing microbial mats developed inside a full-scale anaerobic digester employing biological desulfurization. *Applied Microbiology and Biotechnology*, **93**, 847–857.

König J., Keesman K. J., Veeken A. and Lens P. N.L. (2006). Dynamic Modelling and Process Control of ZnS Precipitation. *Separation Science and Technology*, **41**, 1025–1042.

Koutsoukos P., Amjad Z., Tomson M. B. and Nancollas G. H. (1980). Crystallization of calcium phosphates. A constant composition study. *Journal of the American Chemical Society*, **102**, 1553–1557.

Kralj D., Brečević L. and Kontrec J. (1997). Vaterite growth and dissolution in aqueous solution III. Kinetics of transformation. *Journal of Crystal Growth*, **177**, 248–257.

Krayzelova L., Bartacek J., Kolesarova N. and Jenicek P. (2014). Microaeration for hydrogen sulfide removal in UASB reactor. *Bioresource Technology*, **172**, 297–302.

Krayzelova L., Bartacek J., Díaz I., Jeison D., Volcke E. I. P. and Jenicek P. (2015). Microaeration for hydrogen sulfide removal during anaerobic treatment: a review. *Reviews in Environmental Science and Bio/Technology*, **14**, 703–725.

Kurtz M. J. and Henson M. A. (1997). Input-output linearizing control of constrained nonlinear processes. *Journal of Process Control*, **7**, 3–17.

Lauwers J., Appels L., Thompson I. P., Degrève J., Van Impe J. F. and Dewil R. (2013). Mathematical modelling of anaerobic digestion of biomass and waste: Power and limitations. *Progress in Energy and Combustion Science*, **39**, 383–402.

Lee E., McCarty P. L., Kim J. and Bae J. (2016). Effects of $FeCl_3$ addition on the operation of a staged anaerobic fluidized membrane bioreactor (SAF-MBR). *Water Science and Technology*, **74**(1), 130–137.

Lens P. N. L. and Kuenen J. G. (2001). The biological sulfur cycle: Novel opportunities for environmental biotechnology. *Water Science and Technology*, **44**(8), 57–66.

Lewis A. E. (2010). Review of metal sulphide precipitation. *Hydrometallurgy*, **104**, 222–234.

Lewis A. E., Seckler M. M., Kramer H. and Van Rosmalen G. (2003). Industrial Crystallization in Practice from Process to Product, Cambridge University Press, UK.

Li Q. and Lancaster J. R. (2013). Chemical foundations of hydrogen sulfide biology. *Nitric Oxide*, **35**, 21–34.

Li Z.-Q., Liu G.-D., Duan L.-M., Shen G.-L. and Yu R.-Q. (1999). Sulfate-selective PVC membrane electrodes based on a derivative of imidazole as a neutral carrier. *Analytica Chimica Acta*, **382**, 165–170.

Li Y., Wang Y., Gao Y., Zhao H. and Zhou W. (2018). Seawater toilet flushing sewage treatment and nutrients recovery by marine bacterial-algal mutualistic system. *Chemosphere*, **195**, 70–79.

Liamleam W. and Annachhatre A. P. (2007). Electron donors for biological sulfate reduction. *Biotechnology Advances*, **25**, 452–463.

Liu Z., Maszenan A. M., Liu Y. and Ng W. J. (2015). A brief review on possible approaches towards controlling sulfate-reducing bacteria (SRB) in wastewater treatment systems. *Desalination and Water Treatment*, **53**, 2799–2807.

Lizarralde I., Fernández-Arévalo T., Brouckaert C., Vanrolleghem P., Ikumi D. S., Ekama G. A., Ayesa E. and Grau P. (2015). A new general methodology for incorporating physico-chemical transformations into multi-phase wastewater treatment process models. *Water Research*, **74**, 239–256.

Lizarralde I., Fernández-Arévalo T., Manas A., Ayesa E. and Grau P. (2019). Model-based opti mization of phosphorus management strategies in Sur WWTP, Madrid. *Water Research*, **153**, 39–52.

Lomako S. V., Astapovich R. I., Nozdrin-Plotnitskaya O. V., Pavlova T. E., Lei S., Nazarov V. A., Okaev E. B., Rakhman'ko E. M. and Egorov V. V. (2006). Sulfate-selective electrode and its application for sulfate determination in aqueous solutions. *Analytica Chimica Acta*, **562**, 216–222.

Lyew D. and Sheppard J. (2001). Use of conductivity to monitor the treatment of acid mine drainage by sulphate-reducing bacteria. *Water Research*, **35**, 2081–2086.

Majer V., Sedlbauer J. and Bergin G. (2008). Henry's law constant and related coefficients for aqueous hydrocarbons, CO_2 and H_2S over a wide range of temperature and pressure. *Fluid Phase Equilibria*, **272**, 65–74.

Martí N., Barat R., Seco A., Pastor L. and Bouzas A. (2017). Sludge management modeling to enhance P-recovery as struvite in wastewater treatment plants. *Journal of Environmental Management*, **196**, 340–346.

Mazloum Ardakani M., Akrami Z., Mansournia M. and Zare H. R. (2006). Sulfate-selective Electrode Based on a Complex of Copper. *Analytical Sciences*, **22**, 673–678.

Mazloum-Ardakani M., Manshadi A. D., Bagherzadeh M. and Kargar H. (2012). Impedimetric and Potentiometric Investigation of a Sulfate Anion-Selective Electrode: Experiment and Simulation. *Analytical Chemistry*, **84**, 2614–2621.

McDonagh C., Burke C. S. and MacCraith B. D. (2008). Optical Chemical Sensors. *Chemical Reviews*, **108**, 400–422.

McFarland M. J. and Jewell W. J. (1989). In situ control of sulfide emissions during the thermophilic (55°C) anaerobic digestion process. *Water Research*, **23**, 1571–1577.

Méndez-Acosta H. O., Palacios-Ruiz B., Alcaraz-González V., González-Álvarez V. and García-Sandoval J. P. (2010). A robust control scheme to improve the stability of anaerobic digestion processes. *Journal of Process Control*, **20**, 375–383.

Millero F. J. (1986). The thermodynamics and kinetics of the hydrogen sulfide system in natural waters. *Marine Chemistry*, **18**, 121–147.

Morigi M., Scavetta E., Berrettoni M., Giorgetti M. and Tonelli D. (2001). Sulfate-selective electrodes based on hydrotalcites. *Analytica Chimica Acta*, **439**, 265–272.

Musvoto E. V, Wentzel M. C., Loewenthal R. E. and Ekama G. A. (2000). Integrated chemical–physical processes modelling—I. Development of a kinetic-based model for mixed weak acid/base systems. *Water Research*, **34**, 1857–1867.

Muyzer G. and Stams A. J. M. (2008). The ecology and biotechnology of sulphate-reducing bacteria. *Nature Reviews Microbiology*, **6**, 441–454.

Nezamzadeh-Ejhieh A. and Afshari E. (2012). Modification of a PVC-membrane electrode by surfactant modified clinoptilolite zeolite towards potentiometric determination of sulfide. *Microporous and Mesoporous Materials*, **153**, 267–274.

Nezamzadeh-Ejhieh A. and Esmaeilian A. (2012). Application of surfactant modified zeolite carbon paste electrode (SMZ-CPE) towards potentiometric determination of sulfate. *Microporous and Mesoporous Materials*, **147**, 302–309.

Nghiem L. D., Manassa P., Dawson M. and Fitzgerald S. K. (2014). Oxidation reduction potential as a parameter to regulate micro-oxygen injection into anaerobic digester for reducing hydrogen sulphide concentration in biogas. *Bioresource Technology*, **173**, 443–447.

Nguyen D. and Khanal S. K. (2018). A little breath of fresh air into an anaerobic system: How microaeration facilitates anaerobic digestion process. *Biotechnology Advances*, **36**, 1971–1983.

Nguyen D., Gadhamshetty V., Nitayavardhana S. and Khanal S. K. (2015). Automatic process control in anaerobic digestion technology: A critical review. *Bioresource Technology*, **193**, 513–522.

Nguyen D., Wu Z., Shrestha S., Lee P. H., Raskin L. and Khanal S. K. (2019). Intermittent micro-aeration: New strategy to control volatile fatty acid accumulation in high organic loading anaerobic digestion. *Water Research*, **166**, 115080.

Nishizawa S., Bühlmann P., Xiao K. P. and Umezawa Y. (1998). Application of a bis-thiourea ionophore for an anion selective electrode with a remarkable sulfate selectivity. *Analytica Chimica Acta*, **358**, 35–44.

Noguera D. R., Brusseau G. A., Rittmann B. E. and Stahl D. A. (1998). A unified model describing the role of hydrogen in the growth of *Desulfovibrio vulgaris* under different environmental conditions. *Biotechnology and Bioengineering*, **59**(6), 732–746.

Noguera D. R., Pizarfo G., Stahl D. A. and Rittmann B. E. (1999). Simulation of multispecies biofilm development in three dimensions. *Water Science and Technology*, **39**(7), 123–130.

Olsson G., Carlsson B., Comas J., Copp J., Gernaey K. V., Ingildsen P., Jeppsson U., Kim C., Rieger L., Rodríguez-Roda I., Steyer J.-P., Takács I., Vanrolleghem P. A., Vargas A., Yuan Z. and Åmand L. (2014). Instrumentation, control and automation in wastewater – from London 1973 to Narbonne 2013. *Water Science and Technology*, **69**(7), 1373.

Overmeire A., Lens P. and Verstraete W. (1994). Mass transfer limitation of sulfate in methanogenic aggregates. *Biotechnology and Bioengineering*, **44**, 387–391.

Pandey S. K., Kim K.-H. and Tang K.-T. (2012). A review of sensor-based methods for monitoring hydrogen sulfide. *TRAC Trends in Analytical Chemistry*, **32**, 87–99.

Patón M., González-Cabaleiro R. and Rodríguez J. (2018). Activity corrections are required for accurate anaerobic digestion modelling. *Water Science and Technology*, **77**(8), 2057–2067.

Peces M., Astals S., Jensen P. D. and Clarke W. P. (2018). Deterministic mechanisms define the long-term anaerobic digestion microbiome and its functionality regardless of the initial microbial community. *Water Research*, **141**, 366–376.

Percheron, G., Bernet, N. and Moletta, R. (1996). A new method for the determination of dissolved sulfide in strongly colored anaerobically treated effluents. *Bioprocess Engineering*. **15**(6), 317–322.

Poinapen J. and Ekama G. (2010). Biological sulphate reduction with primary sewage sludge in an upflow anaerobic sludge bed reactor – Part 6: Development of a kinetic model for BSR. *Water SA*, **36**, 203–214.

Pokorna-Krayzelova L., Mampaey K. E., Vannecke T. P. W., Bartacek J., Jenicek P. and Volcke E. I. P. (2017). Model-based optimization of microaeration for biogas desulfurization in UASB reactors. *Biochemical Engineering Journal*, **125**, 171–179.

Premier G. C., Dinsdale R., Guwy A. J., Hawkes F. R., Hawkes D. L. and Wilcox S. J. (1999). A comparison of the ability of black box and neural network models of ARX structure to represent a fluidized bed anaerobic digestion process. *Water Research*, **33**, 1027–1037.

Puyol D., Batstone D. J., Hülsen T., Astals S., Peces M. and Krömer J. O. (2017). Resource recovery from wastewater by biological technologies: Opportunities, challenges, and prospects. *Frontiers in Microbiology*, **7**, 1–23.

Ramos I. and Fdz-Polanco M. (2014). Microaerobic control of biogas sulphide content during sewage sludge digestion by using biogas production and hydrogen sulphide concentration. *Chemical Engineering Journal*, **250**, 303–311.

Ramos I., Pérez R. and Fdz-Polanco M. (2014). The headspace of microaerobic reactors: Sulphide-oxidising population and the impact of cleaning on the efficiency of biogas desulphurisation. *Bioresource Technology*, **158**, 63–73.

Ramsing N. B., Kühl M. and Jørgensen B. B. (1993). Distribution of sulfate-reducing bacteria, O_2, and H_2S in photosynthetic biofilms determined by oligonucleotide probes and microelectrodes. *Applied and Environmental Microbiology*, **59**, 3840–3849.

Redondo R., Machado V. C., Baeza M., Lafuente J. and Gabriel D. (2008). On-line monitoring of gas-phase bioreactors for biogas treatment: hydrogen sulfide and sulfide analysis by automated flow systems. *Analytical and Bioanalytical Chemistry*, **391**, 789–798.

Renard P., Dochain D., Bastin G., Naveau H. and Nyns E.-J. (1988). Adaptive control of anaerobic digestion processes – a pilot-scale application. *Biotechnology and Bioengineering*, **31**, 287–294.

Robles Á., Ruano M. V., Charfi A., Lesage G., Heran M., Harmand J., Seco A., Steyer J.-P., Batstone D. J., Kim J. and Ferrer J. (2018). A review on anaerobic membrane bioreactors (AnMBRs) focused on modelling and control aspects. *Bioresource Technology*, **270**, 612–626.

Salehin S., Kulandaivelu J., Rebosura M., Khan W., Wong R., Jiang G., Smith P., McPhee P., Howard C., Sharma K., Keller J., Donose B. C., Yuan Z. and Pikaar I. (2019). Opportunities for reducing coagulants usage in urban water management: The Oxley Creek Sewage Collection and Treatment System as an example. *Water Research*, **165**, 114996.

Sampaio R. M. M., Timmers R. A., Xu Y., Keesman K. J. and Lens P. N. L. (2009). Selective precipitation of Cu from Zn in a pS controlled continuously stirred tank reactor. *Journal of Hazardous Materials*, **165**, 256–265.

Sampaio R. M. M., Timmers R. A., Kocks N., André V., Duarte M. T., van Hullebusch E. D., Farges F. and Lens P. N. L. (2010). Zn–Ni sulfide selective precipitation: The role of supersaturation. *Separation and Purification Technology*, **74**, 108–118.

Santegoeds C. M., Damgaard L. R., Hesselink G., Zopfi J., Lens P., Muyzer G. and de Beer D. (1999). Distribution of sulfate-reducing and methanogenic bacteria in anaerobic aggregates determined by microsensor and molecular analyses. *Applied and Environmental Microbiology*, **65**, 4618–4629.

Seco A., Ruano M. V, Ruiz-Martinez A., Robles A., Barat R., Serralta J. and Ferrer J. (2020). Plant-wide modelling in wastewater treatment: showcasing experiences using the Biological Nutrient Removal Model. *Water Science and Technology*, **81**(8), 1700–1714. https://doi.org/10.2166/wst.2020.056

Serrano A., Peces M., Astals S. and Villa-Gómez D. K. (2020). Batch assays for biological sulfate-reduction: a review towards a standardized protocol. *Critical Reviews in Environmental Science and Technology*, **50**, 1195–1223.

Shakeri Yekta S., Ziels R. M., Björn A., Skyllberg U., Ejlertsson J., Karlsson A., Svedlund M., Willén M. and Svensson B. H. (2017). Importance of sulfide interaction with iron as regulator of the microbial community in biogas reactors and its effect on methanogenesis, volatile fatty acids turnover, and syntrophic long-chain fatty acids degradation. *Journal of Bioscience and Bioengineering*, **123**, 597–605.

Shamsipur M., Yousefi M., Hosseini M., Ganjali M. R., Sharghi H. and Naeimi H. (2001). A Schiff Base Complex of Zn(II) as a Neutral Carrier for Highly Selective PVC Membrane Sensors for the Sulfate Ion. *Analytical Chemistry*, **73**, 2869–2874.

Shen C. F. and Guiot S. R. (1996). Long-term impact of dissolved O_2 on the activity of anaerobic granules. *Biotechnology and Bioengineering*, **49**, 611–620.

Solon K., Flores-Alsina X., Mbamba C. K., Volcke E. I. P., Tait S., Batstone D., Gernaey K. V. and Jeppsson U. (2015). Effects of ionic strength and ion pairing on (plant-wide) modelling of anaerobic digestion. *Water Research*, **70**, 235–245.

Solon K., Flores-Alsina X., Kazadi Mbamba C., Ikumi D., Volcke E. I.P., Vaneeckhaute C., Ekama G., Vanrolleghem P. A., Batstone D. J., Gernaey K. V. and Jeppsson U. (2017). Plant-wide modelling of phosphorus transformations in wastewater treatment systems: Impacts of control and operational strategies. *Water Research*, **113**, 97–110.

Stewart P. S. (1998). A review of experimental measurements of effective diffusive permeabilities and effective diffusion coefficients in biofilms. *Biotechnology and Bioengineering*, **59**, 261–272.

Steyer J.-P., Buffière P., Rolland D. and Moletta R. (1999). Advanced control of anaerobic digestion processes through disturbances monitoring. *Water Research*, **33**, 2059–2068.

Steyer J. P., Bernard O., Batstone D. J. and Angelidaki I. (2006). Lessons learnt from 15 years of ICA in anaerobic digesters. *Water Science and Technology*, **53**(4–5), 25–33.

Sun J., Dai X., Wang Q., Pan Y. and Ni B.-J. (2016). Modelling Methane Production and Sulfate Reduction in Anaerobic Granular Sludge Reactor with Ethanol as Electron Donor. *Scientific Reports*, **6**, 35312.

Tartakovsky B., Mehta P., Bourque J.-S. and Guiot S. R. (2011). Electrolysis-enhanced anaerobic digestion of wastewater. *Bioresource Technology*, **102**, 5685–5691.

Tigli O. and Zaghloul M. E. (2007). A Novel Saw Device in CMOS: Design, Modeling, and Fabrication. *Ieee Sensors Journal*, **7**, 219–227.

Tóth I., Solymosi P. and Szabó Z. (1988). Application of a sulphide-selective electrode in the absence of a pH-buffer. *Talanta* **35**, 783–788.

van Lier J. B., van der Zee F. P., Frijters C. T. M. J. and Ersahin M. E. (2015). Celebrating 40 years anaerobic sludge bed reactors for industrial wastewater treatment. *Reviews in Environmental Science and Bio/Technology*, **14**, 681–702.

Vavilin V. A., Vasiliev V. B., Ponomarev A. V. and Rytow S. V. (1994). Simulation model 'methane' as a tool for effective biogas production during anaerobic conversion of complex organic matter. *Bioresource Technology*, **48**, 1–8.

Vavilin V. A., Vasiliev V. B., Rytov S. V. and Ponomarev A. V. (1995). Modeling ammonia and hydrogen sulfide inhibition in anaerobic digestion. *Water Research*, **29**, 827–835.

Veeken A. H. M., de Vries S., van der Mark A. and Rulkens W. H. (2003). Selective Precipitation of Heavy Metals as Controlled by a Sulfide Selective Electrode. *Separation Science and Technology*, **38**, 1–19.

Villa-Gomez D. K. K., Cassidy J., Keesman K. J. J., Sampaio R. and Lens P. N. L. (2014). Sulfide response analysis for sulfide control using a pS electrode in sulfate reducing bioreactors. *Water Research*, **50**, 48–58.

Virji S., Fowler J. D., Baker C. O., Huang J., Kaner R. B. and Weiller B. H. (2005). Polyaniline nanofiber composites with metal salts: Chemical sensors for hydrogen sulfide. *Small* **1**, 624–627.

Wan Y., Zhang D. and Hou B. (2009). Monitoring microbial populations of sulfate-reducing bacteria using an impedimetric immunosensor based on agglutination assay. *Talanta* **80**, 218–223.

Wan Y., Zhang D. and Hou B. (2010a). Determination of sulphate-reducing bacteria based on vancomycin-functionalised magnetic nanoparticles using a modification-free quartz crystal microbalance. *Biosensors & Bioelectronics*, **25**, 1847–1850.

Wan Y., Zhang D. and Hou B. (2010b). Selective and specific detection of sulfate-reducing bacteria using potentiometric stripping analysis. *Talanta*, **82**, 1608–1611.

Wan Y., Zhang D., Wang Y., Qi P. and Hou B. (2011). Direct immobilisation of antibodies on a bioinspired architecture as a sensing platform. *Biosensors & Bioelectronics*, **26**, 2595–2600.

Ward A. J., Hobbs P. J., Holliman P. J. and Jones D. L. (2008). Optimisation of the anaerobic digestion of agricultural resources. *Bioresource Technology*, **99**, 7928–7940.

Wilcox S. J., Hawkes D. L., Hawkes F. R. and Guwy A. J. (1995). A neural network, based on bicarbonate monitoring, to control anaerobic digestion. *Water Research*, **29**, 1465–1470.

Wu D., Ekama G. A., Chui H.-K., Wang B., Cui Y.-X., Hao T.-W., van Loosdrecht M. C. M. and Chen G.-H. (2016). Large-scale demonstration of the sulfate reduction autotrophic denitrification nitrification integrated (SANI®) process in saline sewage treatment. *Water Research*, **100**, 496–507.

Wu D., Li L., Zhao X., Peng Y., Yang P. and Peng X. (2019). Anaerobic digestion: A review on process monitoring. *Renewable and Sustainable Energy Reviews*, **103**, 1–12.

Yang Y., Chen Q., Guo J. and Hu Z. (2015). Kinetics and methane gas yields of selected C1 to C5 organic acids in anaerobic digestion. *Water Research*, **87**, 112–118.

Zadeh L. A. (1965). Fuzzy sets. *Information and Control*, **8**, 338–353.

Zhang L., De Schryver P., De Gusseme B., De Muynck W., Boon N. and Verstraete W. (2008). Chemical and biological technologies for hydrogen sulfide emission control in sewer systems: A review. *Water Research*, **42**, 1–12.

Zancato, M., Pietrogrande, A. and Macca, C. (1990). Sequential determination of sulfur and chlorine in organic compounds by potentiometric titration of sulfate and chloride with an automatic titrator. *Annali di Chimica*, **80**(9–10), 445–452.

Part IV

Treatment of Gases Polluted by Sulfurous Compounds

Chapter 10

Measurement and assessment of odorous sulfur compounds

James E. Hayes, Hung V. Le, Ruth M. Fisher, Nhat Le-Minh and Richard M. Stuetz

10.1 INTRODUCTION

10.1.1 Odour of sulfur compounds

Hydrogen sulfide (H_2S) and other volatile sulfur compounds (VSCs) are found naturally and also as a by-product of many industries, in particular wastewater treatment, waste management and some food production industries (Bentley & Chasteen, 2004; Fisher *et al.*, 2018b; Rochat *et al.*, 2007). VSCs are probably the most impactful odorants due to their foul smell, low odour threshold, and relative abundance in natural and industrial processes. As a result, there are numerous methodologies to appropriately assess VSCs as an odour.

VSC monitoring and assessment is accomplished by a number of approaches that are partially determined by the desired outcomes. For instance, constant monitoring of known VSC emissions is possible through online sensors, such as "e-noses", which are useful for gross VSC emissions but are typically unable to provide detailed analysis. Comparatively, detailed results of VSC concentrations and speciation require a series of steps that evaluate the appropriateness of VSC capture options. From here gas chromatography (GC) separation, followed by subsequent analysis using various chemical detectors, is suitable depending on the VSCs under investigation. VSC sampling and monitoring are challenging due to their low odour thresholds and volatility. These qualities render sampling to be assessed for its capacity to appropriately store the VSCs.

DOI: 10.2166/9781789060966_0321

10.1.2 The impact of sulfur odour

It is challenging to estimate the impact of VSCs with regard to their contribution to odour pollution. VSCs have been so ubiquitous regarding odour impacts that it is only fairly recently that non-VSC odorants (typically volatile organic compounds, or VOCs) have been considered for monitoring purposes (Wang *et al.*, 2013). With regard to non-toxic but olfactory impact, communities regularly report individuals with considerable health concerns including dizziness, vomiting, headaches, and other indications of malaise due to malodour exposure (Neutra *et al.*, 1991; Seeber *et al.*, 2002; Shusterman, 2001). In turn, these illnesses and complaints have often led to expensive odour amelioration efforts, or even industry closures (Hayes *et al.*, 2014).

With respect to industry, odour sampling and analysis have steadily increased in sophistication and application as the demands of stakeholders, in particular local communities, have led to elevated numbers and severity of complaints (Hayes *et al.*, 2014; Lebrero *et al.*, 2011; Shusterman, 2001). While the analysis of H_2S is typically the most common assessment approach, it does not prioritise odour impact assessments due to the nature of VSC limitations in terms of sampling and

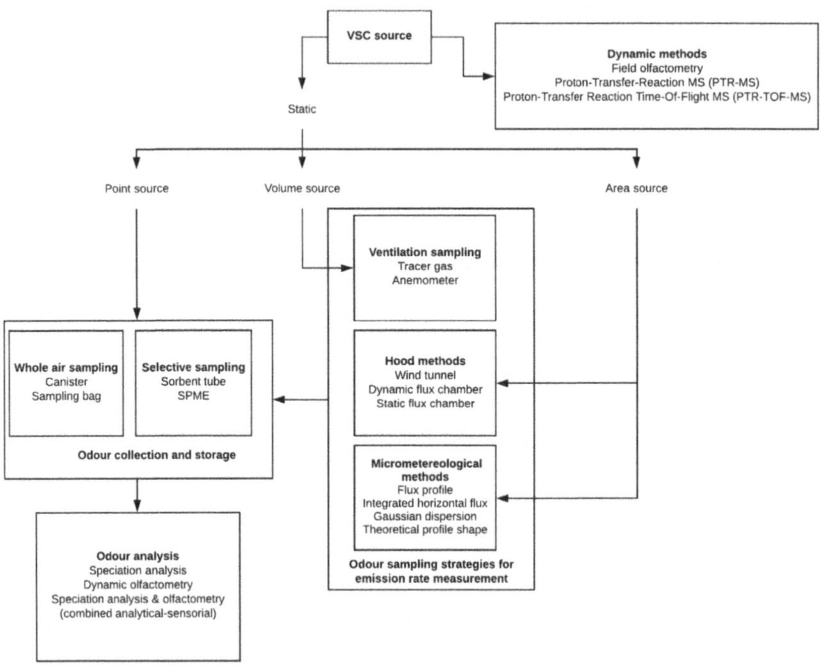

Figure 10.1 Pathways for VSC analysis (SPME: solid phase microextraction, MS: mass spectrometry).

measurement. This is because VSCs are unstable, are typically contained within complex sample matrices, have low concentrations, and possess a wide variation of concentrations between compounds (Kim *et al.*, 2006; Pandey & Kim, 2009; Wardnecki, 1998). Varying methodologies for odour sampling and analysis has been appraised elsewhere (Brattoli *et al.*, 2011; Hayes *et al.*, 2014; Laor *et al.*, 2014; Lebrero *et al.*, 2011; Muñoz *et al.*, 2010), so this precis is concerned with methods that can effectively encapsulate VSCs for analysis. The variables with regard to establishing effective VSC analysis include sample collection and storage, as well as methods of determining odour analysis type and emission rate estimation (Figure 10.1).

10.2 ESTABLISHING EFFECTIVE VSC EMISSION ANALYSIS

10.2.1 Introduction to VSC sample collection and storage

Characterisation of the odour profile in terms of its odour threshold, hedonics, and chemical speciation requires odour analysis. This begins with obtaining samples of the odour in question. VSC odour analysis begins with sampling procedures that are broadly divided between "static" and "dynamic" types. While it is possible to conduct some analysis of VSCs in the field, "dynamic" sampling methodologies are hampered by logistical issues ranging from transportation of human assessors of odour, the over-complicated environmental odour samples affecting analysis, and the lack of variable control compared to laboratory settings (Laor *et al.*, 2011; Laor *et al.*, 2014). Nevertheless, several products and techniques exist that are used for dynamic testing. These include products like the Nasal Ranger™, and techniques including membrane inlet mass spectrometry (MIMS) and proton-transfer-reaction mass spectrometry (PTR-MS) (Beckmann *et al.*, 2009; Blake *et al.*, 2009; Cesca *et al.*, 2007; Feilberg *et al.*, 2009; Nyord *et al.*, 2011).

Care must be taken in terms of sampling material selection, and suitability depends on VSC speciation, length of time the sample is stored, as well as the type of analysis desired (Fisher *et al.*, 2018b; Le *et al.*, 2013b). The four main methods of sample collection for VSCs are solid phase microextraction (SPME), sorbent tubes, canisters, and plastic bags (Figure 10.2). For odour sensory measurement (using dynamic olfactometry), sampling media are generally plastic bags made of materials that are inert, odour free, and can offer reasonable storage integrity and stability of odour samples (Capelli *et al.*, 2013; Gostelow *et al.*, 2003; Zarra *et al.*, 2012). For odour compound speciation analysis, other sampling media such as sorbent tubes, canisters and SPME can also be used with further requirements regarding artefact formation down the chain of chemical analysis (Cai *et al.*, 2006b; Kleeberg *et al.*, 2005). For all types of sample capture, VSCs require special considerations that may mean that preferred

Figure 10.2 Sample collection methods for VSCs include (a) SPME fibres, (b) sorbent tubes, (c) canisters, and (d) plastic bags.

methods for general odour sampling may not be appropriate. One crucial component is that VSCs degrade at different rates, often very quickly, and possibly into other VSCs.

10.2.2 Solid phase microextraction

Solid phase microextraction (SPME) is one of the most popular methods for odour collection. SPME has been widely used in beverage, food, and environmental assessments (Antoniou *et al.*, 2007; Cai *et al.*, 2006a; Campillo *et al.*, 2009; Hill & Smith, 2000; Lestremau *et al.*, 2004a; Lestremau *et al.*, 2004b; Salgado-Petinal *et al.*, 2005). In SPME, analytes from samples are extracted using an exposed fibre coated with an extracting phase (liquid or solid). This fibre is placed inside a protective needle and can be retractable. After sample extraction, the SPME fibre is transferred to the GC injection and thermally desorbed to release the analytes for speciation analysis.

The advantage of SPME is that it is compact, portable, quick, and cheap (Cai *et al.*, 2006b; Kabir & Kim, 2012). Additionally, SPME combines both sampling and pre-concentration into one phase with high sensitivity and low detection limits (Lestremau *et al.*, 2004a; Panighel & Flamini, 2014). SPME is simple to use and can easily be connected to instruments such as gas or liquid chromatography (Kleeberg *et al.*, 2005). Many studies have been conducted into the use of different SPME fibres for VSC sampling and analysis. The most common SPME fibres include Carboxen/polydimethylsiloxane (CAR/PDMS) (Antoniou *et al.*, 2007; Campillo *et al.*, 2009; Haberhauer-Troyer *et al.*, 1999; Hill & Smith, 2000; Lestremau *et al.*, 2004a), divinylbenzene/polydimethylsiloxane (DVB/PDMS) (Antoniou *et al.*, 2007; Cai *et al.*, 2006a; Hill & Smith, 2000; Lestremau *et al.*, 2004b), polydimethylsiloxane (PDMS) (Antoniou *et al.*, 2007; Hill & Smith, 2000; Salgado-Petinal *et al.*, 2005), polyacrylate (PA), Carboxen/divinylbenzene/polydimethylsiloxane (CAR/DVB/PDMS) (Antoniou *et al.*, 2007; Campillo *et al.*, 2009; Salgado-Petinal *et al.*, 2005), and Carbowax/divinylbenzene (CW/DVB) (Antoniou *et al.*, 2007;

Salgado-Petinal *et al.*, 2005). A potential SPME fibre with good application for sulfur analysis is graphene-supported zinc oxide (ZnO) as described by Zhang *et al.* (2012), however further research is required to optimise this new technology within the context of complex odours (Zhang *et al.*, 2012).

For VSC analysis, the CAR/PDMS fibre is the most common choice due its superior efficiency in extraction/desorption of many different VSCs (Haberhauer-Troyer *et al.*, 1999; Hill & Smith, 2000; Nielsen & Jonsson, 2002; Ras *et al.*, 2008). The second option is DVB/PDMS whereas other types of SPME fibres are not recommended due to the effectiveness of CAR/PDMS and DVB/PDMS (Antoniou *et al.*, 2007; Fang & Qian, 2005; Hill & Smith, 2000; Lestremau *et al.*, 2004a, b).

10.2.3 Sorbent tubes

Sorbent tubes are a popular and often legislative method of odour sampling (Koziel *et al.*, 2005; Perrault *et al.*, 2014; Woolfenden, 2010; Zhang *et al.*, 2010). In principle, the air sample is passed through and captured in a tube containing known amounts and compositions of appropriate sorbents for the target analytes. The capturing process can be either active or passive. In active sampling, a pump is used to pull/push the sample through the sorbent bed whereas in passive sampling, one end of the sorbent tube will be exposed to the sample for a certain period of time (Muñoz *et al.*, 2010). Passive sampling requires substantially more time than active sampling. After being captured, the analyte can be thermally desorbed and transferred to analytical devices.

A wide variety of sorbent tubes (single-bed and multiple-bed) have been tested for their ability to capture and store VSCs. Among the available sorbent tubes, the best VSC collection and storage results could be obtained with 2-bed sorbent tubes that combined Tenax TA and Unicarb (Mochalski *et al.*, 2009). Mochalski *et al.* (2009) studied a mixture of methanethiol (MeSH), ethanethiol (EtSH), dimethyl sulfide (DMS), H_2S, carbonyl sulfide (COS) and carbon disulfide (CS_2) and found that Tenax TA was a good sorption medium for MeSH and EtSH, whereas Unicarb was suitable for capturing COS, DMS and CS_2. In terms of storability, a study of nine VSCs and multiple odour VOCs showed no significant losses of the analytes over 3 days on Tenax TA – Unicarb sorbent tubes at 4°C (Mariné *et al.*, 2012).

10.2.4 Canisters

Canisters are another method of collecting odour samples, and have some application for VSCs (Król *et al.*, 2010; McGlenny *et al.*, 2012; Schweigkofler & Niesser, 1999; Trabue *et al.*, 2008). To minimise the loss of the analyte and any possible interference with the sample, the material for the canisters must be inert, i.e. passivated. This can be accomplished in a number of ways, the most popular perhaps using Summa technology, which involves the canisters having a layer of chromium and nickel oxides in order to reduce the number of surface active sites and reactivity (Król *et al.*, 2010; McGlenny *et al.*, 2012). However, Summa canisters have been observed to be unsuitable for storing VSCs; as a result other canister types such as fused-silica lined (FSL) stainless steel are preferred (Ochiai *et al.*, 2001; Rumsey *et al.*, 2014).

Among many types of canisters, high VSC recovery can be obtained with a SilcoCan canister (Brown *et al.*, 2015; Khan *et al.*, 2012; Sulyok *et al.*, 2001) as demonstrated by Khan *et al.* (2012), while comparing the storability of a mixture of VSCs in old and new generations of SilcoCan canisters. In the new generation of SilcoCan canisters, the recoveries of all the VSCs exceeded 97% after 3 days. After 7 days, more than 85% of H_2S and MeSH could still be recovered. Comparatively, in the old generation of SilcoCan canisters with thinner internal amorphous silicon coating, the recoveries of H_2S and MeSH were extremely poor with more than 50% losses within 1 day, whilst COS, DMS, and CS_2 remained stable even after 10 days (Khan *et al.*, 2012). Another canister type that provides very good recovery of VSCs in whole air samples are FSL canisters (Rumsey *et al.*, 2014). The study carried out by Rumsey *et al.* (2014) demonstrated that the most often used Summa canisters lost all traces of MeSH after 2 days, meanwhile FSL canisters retained 80% of MeSH over the same period.

10.2.5 Sample bags

Odour capture using sample bags is similar to using canisters in that they both provide whole air samples. Sample bags are useful for VSC collection as they capture the whole air sample matrix and are comparatively economical compared to other methods, although it should be noted that bags are disposed of after a single use (Fisher *et al.*, 2018a; Hansen *et al.*, 2011; Koziel *et al.*, 2005; Le *et al.*, 2015).

Storability of VSC samples in different types of plastic sampling bags has been thoroughly investigated (Brown *et al.*, 2015; Lau, 1989). To date, the most common and versatile bag material that produces reasonably good stability of VSC samples is Tedlar. Tedlar has been examined in multiple studies and indicates excellent storage for the most pertinent VSCs showing stability for EtSH and DMS for 14 days, with other VSCs recovered after approximately 12 days (Hansen *et al.*, 2011; Le *et al.*, 2015; Sulyok *et al.*, 2001). In a separate study, Lau (1989) concluded that Tedlar was the most effective sample bag medium for VSCs with the exception of SO_2 (Lau, 1989). Tedlar is not without its disadvantages, however: Jo *et al.*'s (2012) study suggests that VSC stability is affected by the size of the bag, which is typically 10 L and should not be decreased (Jo *et al.*, 2012).

Tedlar's dominance as VSC sample bag material is contested by other material types. Nalophan, a slightly less expensive material for sample bags, is less effective than Tedlar at VSC capture but offers comparable storability (Guillot *et al.*, 2012; Hansen *et al.*, 2011). It should also be noted that many studies comparing Tedlar to Nalophan have used thicker Tedlar bags. Further research using identical bag thicknesses are required (Guillot *et al.*, 2012; Le *et al.*, 2015). Mylar is another considered alternative with comparable benefits to Nalophan (Adams *et al.*, 2003; Le *et al.*, 2013b; Le *et al.*, 2015).

As a further alternative, bags made of interior fluorinated ethylene propylene (FEP) film and exterior aluminium foil (metallised FEP) are also a very good option for VSC storage (Zhu *et al.*, 2015b). Zhu *et al.* (2015b) investigated the storability of EtSH and some other VOCs in 1 L metallised FEP bags. These bags had less background contaminants emitted by the material than Tedlar bags (Zhu *et al.*, 2015b). Other studies by Kuster and Goldan (1987) on the use of FEP film only as the wall of emission flux chambers suggested that the material is good for handling H_2S, EtSH, dry SO_2, and marginally good for MeSH (Kuster & Goldan, 1987).

Two other bag types, namely Plastigas and Flexfoil balloon, were found to provide good storage stability for some of the VSCs. investigated the appropriateness of Plastigas and Flexfoil bags as well as Tedlar for the storage of VSC-containing breath samples (Winkel & Tangermann, 2008). After 3 days, Tedlar bags gave recoveries of 23% for H_2S, 80% for MeSH, and 88% for DMS. Over the same period, Plastigas produced much higher storage stability with recoveries of 80, 90, and 88% for H_2S, MeSH, and DMS, respectively. Stability was also good for these VSCs stored in foil balloons with recoveries of 70, 90, and 85% of H_2S, MeSH, and DMS, respectively. However, whilst foil balloons and Plastigas have somewhat similar retention rates, Plastigas is considerably more expensive (Winkel & Tangerman, 2008). While promising, these bag types require further investigation before they can replace Tedlar.

10.2.6 Comparative advantages and disadvantages of VSC sample collection and storage methods

The selection of the most suitable collection methods for ambient VSCs is a complicated process. Respective advantages and disadvantages of each sample capture type are provided in Table 10.1. Many aspects need to be considered before the most appropriate technique can be identified. Those aspects include measurement techniques (speciation analysis and olfactometry measurement), nature and characteristics of the odour source, collection materials, sampling duration and the number of samples, transportation, storage time and conditions (Zarra *et al.*, 2012).

For static VSC sampling, whole air collection using plastic bags or canisters is the preferred method if the odour source is not too far from the testing laboratory and cluttering is not a problem. However, when the distance between source and laboratory is great or transportation arrangements are difficult, collection methods using sorbent tubes, and sometimes small size plastic bags become a more realistic option.

Although SPME is compact and has some advantages, for most scenarios it is unsuitable for VSC sampling. This is because SPME extraction of VSCs is

Table 10.1 Relative advantages and disadvantages of VSC sample capture methodologies.

Methods	Advantages	Disadvantages	Reference
SPME	• Selectively capture sample depending on sorbents • Combines sampling and pre-concentration • Suitable for time-weighted sampling • Easy to connect to analytical equipment such as GC	• Bias adsorption between heavy and light VSCs • For accurate quantification, standard addition is recommended (although challenging with gas samples) • Extreme storage temperature required for acceptable stability of sorbed VSCs	Lestremau *et al.* (2003a) Haberhauer-Troyer *et al.* (1999) Lestremau *et al.* (2003b) Cai *et al.* (2006a) Antoniou *et al.* (2007) Kabir and Kim (2012) Murray (2001) Nielsen *et al.* (2002) Laor *et al.* (2008) Lopez *et al.* (2007)
Sorbent tubes	• Selectively captures sample depending on sorbents • Combines sampling and pre-concentration • Suitable for less volatile samples (polar and non-polar) • Suitable for time-weighted sampling • Simple cleaning procedures (thermal desorption of samples automatically cleans the tube) • Suitable for short term storage of VSCs • Easier and cheaper for transportation	Not suitable for measurements requiring whole air sample (e.g. dynamic olfactometry)	Woolfenden (1997) Black *et al.* (1978) Przyjazny (1985) Devai *et al.* (1996) Devai *et al.* (1997a) Devai *et al.* (1997b) Baltussen *et al.* (1999) Mariné *et al.* (2012) Mochalski *et al.* (2009)

Canister	• Whole air sampling • Suitable for highly volatile, non-polar compounds • Suitable for long time storage for VSCs provided the samples are dehumidified	• Not suitable for time-weighted sampling • Bulky, heavy, high transportation costs • Non-pressurised canister samples require addition of purified water and zero air before analysis • Difficult and time consuming cleaning procedure • Sample humidity affects recovery of H_2S and some thiols • Pre-concentration sometimes required for analysis	Woolfenden (1997) Khan *et al.* (2012) Rumsey *et al.* (2014) Ochiai *et al.* (2002) Trabue *et al.* (2008)
Plastic sample bags	• Capture whole air sample • Relatively cheap • Suitable for storage of VSCs up to 24 hours (big volume sample) or 6–8 hours (small volume sample)	• Not suitable for time-weighted sampling • Bulky, high transportation costs • Analytes can adsorb on or permeate through bags • Moisture can permeate into bags if exterior humidity higher than interior • Possible pre-concentration required for analysis	Woolfenden (1997) Lau (1989) Sulyok *et al.* (2001) Guillot *et al.* (2008) Winkel *et al.* (2008) Mochalski *et al.* (2009) Hansen *et al.* (2011) Le *et al.* (2013a, b) Le *et al.* (2015)

seriously affected by the sample matrix with regard to levels of oxygen, humidity, and alcohol on SPME extraction of VSCs and also due to the significant bias of extraction between light and heavy VSCs (Cai *et al.*, 2006b; Campillo *et al.*, 2009; Haberhauer-Troyer *et al.*, 1999; Kabir & Kim, 2012; Lestremau *et al.*, 2004b; Lopez *et al.*, 2007; Mochalski *et al.*, 2009; Murray, 2001; Nielsen & Jonsson, 2002). To correct this, several authors suggest standard "additions" to the SPME process to determine its VSC extraction efficiency, but this is time consuming and does not wholly restore the method's accuracy (Kabir & Kim, 2012; Mochalski *et al.*, 2009; Murray, 2001). Interestingly, this bias of extraction between light and heavy compounds by SPME appears to happen exclusively to VSCs (Laor *et al.*, 2008). Another factor that makes SPME unsuitable for VSC field sampling is the extreme conditions required for VSC stability, which includes storage of fibres at below 0°C (Haberhauer-Troyer *et al.*, 1999; Lestremau *et al.*, 2004a). In summary, while possible, SPME has serious drawbacks with regard to VSC sampling and other methods should be adopted.

In terms of logistical considerations and the relative merits of the method, sorbent tube collection of the VSC offers a clear advantage. Sorbent tubes are cheaper for transportation and offer reasonably good VSC stability under relatively simple storage conditions. The major drawback of sorbent tubes, however, is that the method is not suitable for olfactometry beyond gas-chromatography mass spectrometry/olfactometry (GC-MS/O), which does encapsulate Odour Units and has some drawbacks with regard to understanding the combined odour impact (Hayes *et al.*, 2014; Hayes *et al.*, 2017a). Furthermore, depending on the analytical process, VSCs in sorbent tubes may degrade while being extracted, which adds another level of interpretation in some instances (Sivret *et al.*, 2013). In the case where both speciation analysis and olfactometry measurement are required, the use of plastic sample bags is a viable alternative to sorbent tubes, and is cheaper and more efficient than canisters. However, the temporal storability of plastic sample bags is significantly lower than that of sorbent tubes.

10.3 ANALYTICAL AND SENSORIAL METHODOLOGIES FOR DETERMINING VSC SPECIATION

10.3.1 Gas chromatography

Whichever sampling strategy is adopted, chemical analysis of VSC odour samples is typically accomplished through the gas chromatography (GC) family of techniques (Pandey & Kim, 2009). Each methodology must be carefully considered, as no single method is effective at measuring every kind of VSC. GC analysis of VSCs can be affected by co-elution, so use of appropriate columns is critical (Sivret *et al.*, 2016). For quantification of VSCs, the most popular and effective methods include the use of cold traps on columns to ensure stability (Choi *et al.*, 2004; Fisher *et al.*, 2018b; Pandey & Kim, 2009).

From GC separation, targeted analysis of VSCs is most often accomplished through mass spectrometry (MS), mass spectrometry/olfactometry (MS/O), sulfur chemiluminescence detection (SCD), and flame ionisation detection (FID) with some studies also using flame photometric detection (FPD) (Muñoz *et al.*, 2010). The reason for this relative paucity of appropriate techniques is due to the extreme sensitivity required to detect minute concentrations of VSCs which are otherwise readily detected by the human nose (Hayes *et al.*, 2014; Lebrero *et al.*, 2011; Muñoz *et al.*, 2010). The use of FID for VSC detection has been criticised for its comparatively low sensitivity (Lebrero *et al.*, 2011; Muñoz *et al.*, 2010).

Of all potential techniques, MS is the most common, thanks to its comprehensive encapsulation of not just VSCs but other compounds as well (Lebrero *et al.*, 2011; Muñoz *et al.*, 2010). Care must be taken for appropriate column selection, and co-elution is a large risk given the relatively minute proportions in which VSCs can present themselves (Figure 10.3) (Barczak *et al.*, 2018; Gutierrez *et al.*, 2015; Niessen, 2001). MS also lends itself to be combined with olfactometry (MS/O), wherein the sample is split, and a user sniffs the sample in real time (Acree & van Ruth, 2003; d'Acampora *et al.*, 2008; Hayes *et al.*, 2017b; Zhang *et al.*, 2010). While there are some disadvantages to this technique, the outputs include being able to link specific odours to specific chemical compounds (Figure 10.3) (Hayes *et al.*, 2014). While further analysis has been attempted by food industries, environmental odour management has typically been interested in identifying malodour contributors (Debonneville *et al.*, 2002; Hayes *et al.*, 2017a).

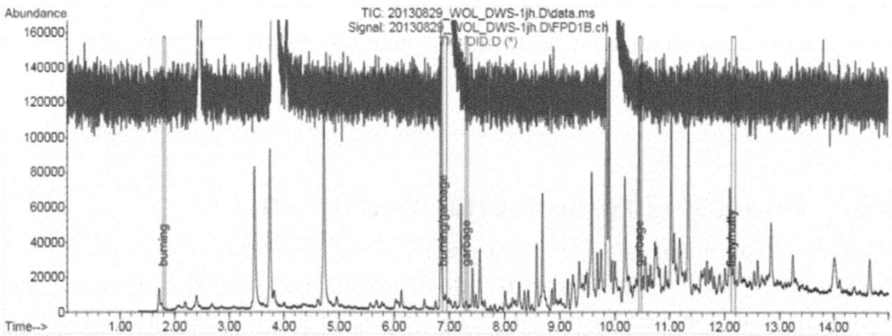

Figure 10.3 Example of a GC-MS/O output of an environmental odour sample, in this instance taken from dewatered solids at a wastewater treatment system. Red marks indicate where the user noted an odour with accompanying descriptor, while the black line indicates relative abundance of compounds. Note that the "burning" odour event at 1.9 minutes had a comparatively low compound abundance. This is characteristic of VSCs and other VSCs present may be "masked" by the co-elution of other compounds.

SCD is a relatively new technique that detects VSCs through their chemiluminescent species (Yan, 2002). It's limited in comparison to MS in that it is only capable of measuring VSCs and not other compounds, but its accuracy and ability to quantify VSCs means that it is relatively popular as a detection method (Choi *et al.*, 2004; Khan *et al.*, 2012; Pandey & Kim, 2009).

FPD is operates via the flame detection of sulfur and phosphorous compounds, and as such it offers excellent sensitivity for VSCs (Brody & Chaney, 1966; Sasamoto & Ochiai, 2010). It has been popular in food and beverages, and other non-industrial purposes due to its extreme sensitivity (Murata *et al.*, 2006; Ochiai & Sasamoto, 2011; Sasamoto & Ochiai, 2010; Zhu *et al.*, 2015a). However, its selectivity means that it is not as popular as MS, which can detect a range of odours beyond VSCs, and pure VSC analysis has predominantly implemented SCD-based methods (Lebrero *et al.*, 2011; Muñoz *et al.*, 2010). Nevertheless, it has seen some industrial application, especially when high sensitivity is required to investigate novel applications (Abbott & Eskicioglu, 2015; Dhar *et al.*, 2012; Ho *et al.*, 2008).

10.3.2 Sensors

Multiple sensors exist for the detection of H_2S, as well as so-called e-noses (Bootsma *et al.*, 2013; Hayes *et al.*, 2017a; Stuetz *et al.*, 1999). While some of these sensors are highly sensitive, it must also be recognised that they require regular, and sometimes delicate, upkeep. Comparatively, sensors that are placed in the field and monitored remotely over large periods of time may not be as accurate, but are useful for observing gross odour incursions (Guitiérrez *et al.*, 2014; Hayes *et al.*, 2014; Stuetz *et al.*, 2000). Special mention must be made of H_2S gold film analysers which have both field and plastic bag application and confer a respectable degree of accuracy (Pandey *et al.*, 2012). However, limited information is available on cross interference in the operation of H_2S-based monitors.

10.3.3 Olfactometry/sensorial measurement

Sensorial measurement is most often used to determine the relative strength of odours by determining the number of odour units (OUs), which is the number of dilutions from which 50% of panellists are unable to identify the odour compared to air (Hayes *et al.*, 2014; Lebrero *et al.*, 2011; Muñoz *et al.*, 2010). As numerous researchers have pointed out, the process is expensive, difficult to standardise, and in the case of VSCs there is some potential for hazardous exposure to the panellist (Bockreis & Steinberg, 2005; European Committee for Standardization, 2003; McGinley & McGinley, 2001). Nevertheless, the technique is invaluable from a legislative perspective, and the concept of OUs is taken for dispersion modelling and mapping (Hayes *et al.*, 2014).

Table 10.2 Common VSCs, their thresholds, and odour qualities. Note many qualities of VSCs are currently under researched.

Family	Compound	Chemical Formula	Abbreviation	Odour Threshold Value[1]	Odour Hedonic Quality[2]	Half-life of VSCs (Days)[3]
Thiols	Methanethiol	CH_3SH	MeSH	0.0003–38	Putrid/rotten cabbage	29
	Ethanethiol	C_2H_5SH	EtSH	0.046–21	Garlic/skunky	25
	2-Methyl-2-propanethiol	$(CH_3)_3CSH$		0.09–3.3	Skunky	
	1-Butanethiol	$CH_3(CH_2)_3SH$		1.5–3.0	Skunky	
Oxidised species	Sulfur dioxide	SO_2		870–1600	Burnt matches	0.02
Sulfides	Hydrogen sulfide	H_2S		0.21–270	Rotten eggs	10
	Dimethyl sulfide	$(CH3)_2S$	DMS		Cabbage	
	Ethyl methyl sulfide	$CH_3CH_2SCH_3$		62	Stinging	
	Dimethyl disulfide	CH_3SSCH_3	DMS	1.1–78	Garlic	
	Diethyl disulfide	$(C_2H_5)_2S_2$		0.3–23	Garlic	
	Dimethyl trisulfide	CH_3SSSCH_3	DMTS	0.06–7.5	Putrid	
	Carbon disulfide	CS_2		70–180	Sweet, similar to chloroform	58
	Carbonyl sulfide	COS	COS	250		100

[1]Compiled by van Gemert (2003).
[2]Descriptions from Fisher *et al.* (2017), and Ruth (1986).
[3]Values taken from Lau (1989).

10.3.4 Hedonic odour qualities of VSCs

As Sucker *et al.* (2008a) illustrates, odour hedonics, that is the descriptive qualities of a particular odour, are growing in importance with regard to odour legislation and the assessment of environmental impact (Sucker *et al.*, 2008a; Sucker *et al.*, 2008b; Sucker *et al.*, 2004). To that end, VSCs are unanimously considered among the most unpleasant odours by the majority of any given population (Le *et al.*, 2013a). VSCs differ with regard to their odour thresholds, and to a lesser extent their odour hedonics (Table 10.2). The challenges faced by industries are VSCs' near universal unpleasantness as well as some VSCs having extremely low odour thresholds, meaning that minute amounts are capable of producing high odour impacts (Fisher *et al.*, 2017; Ruth, 1986).

10.4 EMISSION RATE ESTIMATION, MAPPING AND DISPERSION MODELLING

10.4.1 Emission rate estimation

While analytical and sensorial methodologies evaluate the qualities of a VSC odour, the emission rate of an odour source is a crucial factor to consider in determining odour impact (Lebrero *et al.*, 2011; Standards Australia/New Zealand Standard, 2001; Standards Australia/Standards New Zealand, 2009). This is because emission rate estimations (alongside more in-depth air dispersion modelling and mapping) provide a means of understanding a VSC odour source's area of effect and determining some variables with regard to sampling methodologies. Depending on the geometrical configuration of the emission source, appropriate sampling equipment and strategies can be selected for emission rate estimation. Emission rates are important for odour mapping and dispersion modelling and, coupled with analytical/sensorial analysis, can provide a reasonable understanding of a source's environmental odour impact (Capelli *et al.*, 2013; Hobbs *et al.*, 2000). In general, odour source topography is categorised into three groups: point, area, and volume sources (Figure 10.1) (Bockreis & Steinberg, 2005; Capelli *et al.*, 2013; Gostelow *et al.*, 2003; Zarra *et al.*, 2012).

Point sources are controlled points where odour is emitted, usually through a stack or similar exhaust. The average emission flow rate in point sources can be easily obtained using anemometers or pitot tubes. Odour samples from point sources are generally extracted via a sampling probe inserted into the stack (Capelli *et al.*, 2013; Gostelow *et al.*, 2003; Zarra *et al.*, 2012).

Measurements of emission rates from area sources such as landfills and wastewater treatment plants are more complicated due to the large spatial and temporal variability in these emission rates. For solid or liquid areas with substantial spatial emission variation, and without the interference from upwind sources, micrometeorological estimations of emission rates from downwind odour measurements and wind velocity are preferred methods (Gostelow *et al.*, 2003;

Prata *et al.*, 2018). Emission rates from area sources can be estimated using different micrometeorological methods such as flux profile, integrated horizontal flux, theoretical profile shape, and Gaussian dispersion (Gostelow *et al.*, 2003). When the sites of area sources are complex with upwind interference or when detailed surveys of different emission sources are desired, hood sampling methods are recommended (Bockreis & Steinberg, 2005; Standards Australia/New Zealand Standard, 2001). For homogeneous sites with negligible wind effects, emissions from areas with active surfaces (with outward flow) can be sampled using static flux chambers, whereas sampling from passive surface areas (without outward flow) is generally achieved using dynamic flux chambers (Capelli *et al.*, 2013; Jiang, 2001; Zarra *et al.*, 2012). In cases where a wind effect is significantly affecting the environment, wind tunnels are the sampling equipment of choice for area sources (Capelli *et al.*, 2013; Gostelow *et al.*, 2003; Jiang, 2001; Zarra *et al.*, 2012). It should be noted that all of these sampling strategies are recommended for odours of all types, and in this regard VSCs have no additional considerations.

Volume sources are three-dimensional sources of odour, for example buildings and other structures. Emissions from volume sources can be intentional (for

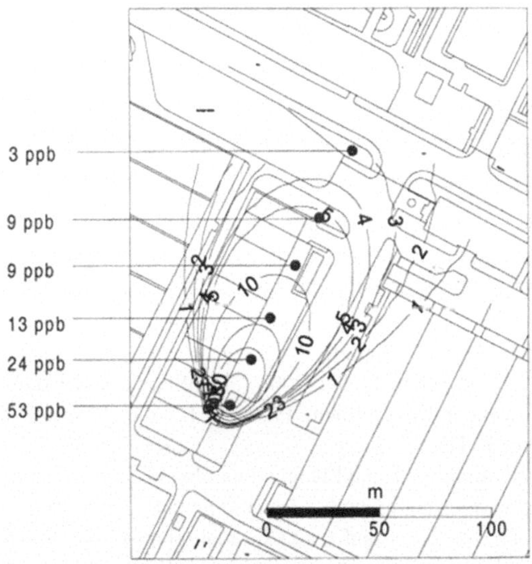

Figure 10.4 Example of an H_2S map to estimate the dispersion of H_2S. As the H_2S moves away from its source (in this instance at the 53 ppb marker), its concentration decreases. This does not tell the whole story, however, as during its time in an environment H_2S will degrade into varying compounds, some of which may have a lower threshold (and therefore stronger odour) than H_2S. Further research into H_2S degradation in the environment is essential for capturing this phenomenon (*Source*: Hobson & Yang, 2001).

example controlled forced ventilation) or unintentional (for example uncontrolled natural ventilation though windows or other inlets and outlets). For volume sources with controlled ventilation, anemometers can be used for emission rate estimation by measuring the flows at the inlets and outlets (Capelli *et al.*, 2013). For volume sources with natural ventilation, accurate emission rates can only be obtained by releasing tracer gases such as CO, CO_2 or SF_6 at a known rate and measuring the concentrations within the building (Capelli *et al.*, 2013; Gostelow *et al.*, 2003; Jiang, 2001; Zarra *et al.*, 2012). Additionally, tracer gas methods are also suitable for estimating emission rates from volume sources with forced ventilation (Gostelow *et al.*, 2003). Again, these are standard odour measuring qualities and as such VSCs do not require alternative strategies.

10.4.2 Mapping and dispersion modelling

Where the OU and hedonic quality are considered alongside the emission rate of a malodour source, mapping and dispersion modelling can be attempted (Figure 10.4). Other reviews have covered mapping and dispersion modelling effectively, and VSCs need no special consideration in this regard beyond the likelihood that they will degrade into other compounds in the environment, which further complicates analysis (Cavalini, 1994; Cesca *et al.*, 2007; Hobbs *et al.*, 2000; McIntyre, 2000; Sironi *et al.*, 2010).

REFERENCES

Abbott T. and Eskicioglu C. (2015). Effects of metal salt addition on odor and process stability during the anaerobic digestion of municipal waste sludge. *Waste Management* **46**, 449–458.

Acree T. E. and Van Ruth S. (2003). Gas Chromatography/Olfactometry. *CPFAC*, **10**, G1–8.

Adams G., Hargreaves R., Witherspoon J., Ong H., Burrowes P., Easter C., Dickey J., James F., Macpherson L., Porter R., Quigley C., Sloan A., Stone L., Card T., Corsi R., Hentz L., Balchunas B. and Kopchynski D. (2003). Identifying and Controlling Odor in the Municipal Wastewater Environment Phase I: Literature Search and Review. IWA publishing, London, UK.

Antoniou C. V., Koukouraki E. E. and Diamadopoulos E. (2007). Analysis of volatile and semivolatile organic compounds in municipal wastewater using headspace Solid-Phase Microextraction and Gas Chromatography. *Wat Environment Research*, **79**, 921–930.

Baltussen E., David F., Sandra P. and Cramers C. (1999). On the performance and inertness of different materials used for the enrichment of sulfur compounds from air and gaseous samples. *Journal of Chromatographic A*, **864**(2), 345–350.

Barczak R. J., Fisher R. M., Wang X. and Stuetz R. M. (2018). Variations of odorous VOCs detected by different assessors via gas chromatography coupled with mass spectrometry and olfactory detection port (ODP) system. *Water Science and Technology*, **77**(3), 759–765.

Beckmann K., Messinger J., Badger M. R., Wydrynski T. and Hillier W. (2009). On-line mass spectrometry: membrane inlet sampling. *Photosynthesis Research*, **102**, 511–522.

Beghi S. and Guillot J. M. (2008). Use of poly(ethylene terephtalate) film bag to sample and remove humidity from atmosphere containing volatile organic compounds. *Journal of Chromatography A*, **1183**(1–2), 1–5.

Bentley R. and Chasteen T. G. (2004). Environmental VOSCs – formation and degradation of dimethyl sulfide, methanethiol and related materials. *Chemosphere*, **55**, 291–317.

Black M. S., Herbst R. P. and Hitchcock D. R. (1978). Solid absorbent preconcentration and gas chromatographic analysis of sulfur gases. *Analytical Chemistry*, **50**(7), 848–851.

Blake R. S., Monks P. S. and Ellis A. M. (2009). Proton-Transfer Reaction Mass Spectrometry. *Chemical Reviews*, **109**, 861–896.

Bockreis A. and Steinberg I. (2005). Measurement of odour with focus on sampling techniques. *Waste Management*, **25**, 859–63.

Bootsma S., Bilsen I. and Milan B. (2013). E-noses for the detection of odor & safety relevant situations. 5th IWA Conference on Odours and Air Emissions, San Francisco, California, USA, International Water Association.

Brattoli M., De Gennaro G., De Pinto V., Loiotile A. D., Lovascio S. and Penza M. (2011). Odour detection methods: olfactometry and chemical sensors. *Sensors*, **11**, 5290–5322.

Brody S. S. and Chaney J. E. (1966). Flame Photometric Detector: the application of a specific detector for phosphorous and for sulfur compounds – sensitive to subnanogram quantities. *Journal of Chromatographic Science*, **4**, 42–46.

Brown A. S., Van Der Veen A. M. H., Arrhenius K., Murugan A., Culleton L. P., Ziel P. R. and Li J. (2015). Sampling of gaseous sulfur-containing compounds at low concentrations with a review of best-practice methods for biogas and natural gas applications. *TrAC Trends Analytical Chemistry*, **64**, 42–52.

Cai L., Gong S., Chen M. L. and Wu C. (2006a). Vinyl crown ether as a novel radical crosslinked sol-gel SPME fiber for determination of organophosphorus pesticides in food samples. *Analytica Chimica Acta*, **559**, 89–96.

Cai L., Koziel J. A., Lo Y. C. and Hoff S. J. (2006b). Characterization of volatile organic compounds and odorants associated with swine barn particulate matter using solid-phase microextraction and gas chromatography-mass spectrometry-olfactometry. *Journal of Chromatography A*, **1102**, 60–72.

Campillo N., Peñalver R., López-García I. and Hernández-Córdoba M. (2009). Headspace solid-phase microextraction for the determination of volatile organic sulphur and selenium compounds in beers, wines and spirits using gas chromatography and atomic emission detection. *Journal of Chromatography A*, **1216**, 6735–6740.

Capelli L., Sironi S., Del Rosso R. and Guillot J.-M. (2013). Measuring odours in the environment vs. dispersion modelling: a review. *Atmospheric Environment*, **79**, 731–743.

Cavalini P. (1994). Industrial odorants: the relationship between modeled exposure concentrations and annoyance. *Archives of Environmental Health*, **49**, 344–351.

Cesca J., Flanagan A. and Cunnington M. (2007). Case study: odour risk management at the WTP, one of Australia's largest & most unique WWTPS. 14th International Union of Air Pollution Prevention and Environmental Protection Associations (IUAPPA) World Congress 2007, Brisbane, QLD, Australia.

Choi Y. G., Ko D. H. and Chung T. H. (2004). Determination of nanogram level of VSCs using GC-SCD and an adsorption tube. *Water Science and Technology*, **49**(9), 329–334.

D'acampora Z. B., Dugo G. and Mondello L. (2008). Gas chromatography-olfactometry in food flavour analysis. *Journal of Chromatography A*, **1186**, 123–143.

Debonneville C., Orsier B., Flament I. and Chaintreau A. (2002). Improved hardware and software for quick chromatography-olfactometry using CHARM and GC-"SNIF" analysis. *Analytical Chemistry*, **74**, 2345–2351.

Devai, I. and DeLaune, R. D. (1996). Evaluation of various solid absorbents for sampling trace levels of methanethiol. *Organic Geochemistry*, **24**(8–9), 941–944.

Devai, I. and DeLaune R. D. (1997a). Field sampling of trace levels of hydrogen sulfide with the use of solid adsorbent preconcentration. *FACT*, **1**(3), 145–149.

Devai, I. and DeLaune R. D. (1997b). Trapping efficiency of various solid adsorbents for sampling and quantitative gas chromatographic analysis of carbonyl sulfide. *Analytical Letters*, **30**(1), 187–198.

Dhar B. R., Nakhla G. and Ray M. B. (2012). Techno-economic evaluation of ultrasound and thermal pretreatment for enhanced anaerobic digestion of municipal waste activated sludge. *Waste Management*, **32**, 542–549.

European Commitee for Standardization 2003. Air quality- Determination of odour concentration by dynamic olfactometry.

Fang Y. and Qian M. C. (2005). Sensitive quantification of sulfur compounds in wine by headspace solid-phase microextraction technique. *Journal of Chromatography A*, **1080**.

Feilberg A., Adamsen A. P. S., Lindholst S., Lyngbye M. and Schäfer A. (2009). Evaluation of biological air filters for livestock ventilation air by Membrane Inlet Mass Spectrometry. *Journal of Environmental Quality* **39**, 1085–1096.

Fisher R., Alvarez-Gaitan J. P., Hayes J. E., Vitanage D. and Stuetz R. (2017). Improving biosolids management through the use of odour wheels. Ozwater 2017, Sydney, Australia.

Fisher R. M., Barczak R. J. and Stuetz R. M. (2018a). Identification of odorant characters using GC-MS/O in biosolids emissions from aerobic and anaerobic stabilisation. *Water Science and Technology*, **2017**(3), 736–742.

Fisher R. M., Le-Minh N., Alvarez-Gaitan J. P., Moore S. J. and Stuetz R. M. (2018b). Emissions of volatile sulfur compounds (VSCs) throughout wastewater biosolids processing. *Science of the Total Environment*, **616–617**, 622–631.

Gostelow P., Longhurst P. J., Parsons S. and Stuetz R. M. (2003). Sampling Measurement for Odours. IWA Publishing.

Guillot J. M., Bilsen I., Both R., Hangartner M., Kost W. J., Kunz W., Nicolas J., Oxbol A., Secanella J., Van Belois H., Van Elst T., Van Harreveld T. and Milan B. (2012). The future European standard to determine odour in ambient air by using field inspection. *Water Science and Technology*, **66**(8), 1691–8.

Guitiérrez M. C., Chica A. F., Martín M. A. and Romain A. C. (2014). Compost pile monitoring using different approaches: GC-MS, e-nose and dynamic olfactometry. *Waste Biomass Valorization*, **5**, 469–479.

Gutierrez M. C., Martin M. A., Pagans E., Vera L., Garcia-Olmo J. and Chica A. F. (2015). Dynamic olfactometry and GC-TOFMS to monitor the efficiency of an industrial biofilter. *Science of the Total Environment*, **512–513**, 572–581.

Haberhauer-Troyer C., Rosenberg E. and Grasserbauer M. (1999). Evaluation of solid-phase microextraction for sampling of volatile organic sulfur compounds in air for subsequent gas chromatographic analysis with atomic emission detection. *Journal of Chromatography A*, **848**, 305–315.

Hansen M. J., Adamsen A. P. S., Feilberg A. and Jonassen K. E. N. (2011). Stability of odorants from pig production in sampling bags for olfactometry. *Journal of Environmental Quality*, **40**, 1096–1102.

Hayes J. E., Stevenson R. J. and Stuetz R. M. (2014). The impact of malodour on communites: a review of assessment techniques. *Science of the Total Environment*, **500**, 395–407.

Hayes J. E., Fisher R. M., Stevenson R. J., Mannebeck C. and Stuetz R. M. (2017a). Unrepresented community odour impact: improving engagement strategies. *Science of the Total Environment*, **609**, 1650–1658

Hayes J. E., Stuetz R. M. and Stevenson R. J. (2017b). A review of modern olfactometry, equipment and applications. In: Chemosensory Sensors and Systems: Evolutionary Significance, Biological Effects and New Insights, M. Brandt (ed.), Nova Science Publishers Inc.

Hill P. G. and Smith R. M. (2000). Determination of sulphur compounds in beer using headspace solid-phase microextraction and gas chromatographic analysis with pulsed flame photometric detection. *Journal of Chromatography A*, **872**(1–2).

Ho K.-L., Chung Y.-C., Lin Y.-H. and Tseng C.-P. (2008). Microbial populations analysis and field application of biofilter for the removal of volatile sulfur compounds from swine wastewater treatment system. *Journal of Hazardous Materials*, **152**, 580–588.

Hobbs S. E., Longhurst P. J., Sarkar U. and Sneath R. W. (2000). Comparison of dispersion models for assessing odour from municipal solid wastes. *Water Resources Management*, **18**, 420–428.

Hobson J. and Yang G. (2001). Odour mapping using H_2S measuremenets. In: Odours in Wastewater Treatment, R. M. Stuetz and F.-B. Frenchen (eds), IWA Publishing, London, UK, pp. 214–231.

Jiang J. (2001). What are the areas for further research?. *Water Science and Technology*, **44** (9), 17–25.

Jo S. H., Kim K. H., Shon Z. H. and Parker D. (2012). Identification of control parameters for the sulfur gas storability with bag sampling methods. *Analytica Chimica Acta*, **738**, 51–8.

Kabir E. and Kim K.-H. (2012). Use of solid phase microextraction (SPME) in the analysis of the reduced sulfur compounds (RSC) and its experimental limitations. *Microchemical Journal*, **103**, 42–48.

Khan M. A. H., Whelan M. E. and Rhew R. C. (2012). Analysis of low concentration reduced sulfur compounds (RSCs) in air: storage issues and measurement by gas chromatography with sulfur chemiluminescence detection. *Talanta*, **88**, 581–586.

Kim K.-H., Choi G.-H., Choi Y.-J., Song H.-N., Yang H. S. and Oh J. M. (2006). The effects of sampling materials selection in the collection of reduced sulfur compounds in air. *Talanta*, **68**, 1712–1719.

Kleeberg K. K., Liu Y., Jans M., Schlegelmilch M., Streese J. and Stegmann R. (2005). Development of a simple and sensitive method for the characterization of odorous waste gas emissions by means of solid-phase microextraction (SPME) and GC-MS/olfactometry. *Waste Management*, **25**, 872–9.

Koziel J. A., Spinhirne J. P., Lloyd J. D., Parker D. B., Wright D. W. and Kuhrt F. W. (2005). Evaluation of sample recovery of malodorous livestock gases from air sampling bags, solid-phase microextraction fibers, Tenax TA sorbent tubes, and sampling canisters. *Journal of the Air & Waste Management*, **55**, 1147–1157.

Król S., Zabiegała B. and Namieśnik J. (2010). Monitoring VOCs in atmospheric air II. Sample collection and preparation. *TrAC Trends Analytical Chemistry*, **29**, 1101–1112.

Kuster W. C. and Goldan P. D. (1987). Quantitation of the losses of gaseous sulfur compounds to enclosure walls. *Environmental Science & Technology*, **21**, 810–815.

Laor Y., Koziel J. A., Cai L. and Ravid U. (2008). Chemical-sensory characterization of dairy manure using headspace solid-phase microextraction and multidimensional gas chromatography mass spectrometry-olfactometry. *Journal of the Air & Waste Management Association*, **58**(9), 1187–1197.

Laor Y., Orenstein P., Ravid U., Baybikov R., Hanan A., Saadi I. and Abbou Y. (2011). A Screening Tool for Selection of Field Odor Assessors. *Journal of the Air & Waste Management*, **61**, 1353–1360.

Laor Y., Parker D. B. and Pagé T. (2014). Measurement, prediction, and monitoring of odors in the environment: a critical review. *Reviews in Chemical Engineering*, **30**, 139–166.

Lau Y. L. (1989). Measurement of sulphur gases in ambient air. *Environmental Monitoring and Assessment*, **13**, 69–74.

Le H., Sivret E. C., Parcsi G. and Stuetz R. M. (2013a). Stability of volatile sulfur compounds (VSCs) in sampling bags – impact of temperature. *Water Science and Technology*, **68**(8), 1880–7.

Le H., Sivret E. C., Parsci G. and Stuetz R. M. (2013b). Stability of volatile sulfer compounds (VSCs) in sample bags- impact of temperature. 5th IWA Conference on Odours and Air Emissions, San Francisco, California, USA, International Water Association.

Le H. V., Sivret E. C., Parsci G. and Stuetz R. M. (2015). Impact of storage conditions on the stability of volatile sulfur compounds in sampling bags. *Journal of Environmental Quality* **44**, 1523–1529.

Lebrero R., Bouchy L., Stuetz R. and Muñoz R. (2011). Odor assessment and management in wastewater treatment plants: a review. *Critical Reviews in Environmental Science and Technology*, **41**, 915–950.

Lestremau F., Desauziers V., Roux, J.-C. and Fanlo J.-L. (2003a). Development of a quantification method for the analysis of malodorous sulphur compounds in gaseous industrial effluents by solid-phase microextraction and gas chromatography-pulsed flame photometric detection. *Journal of Chromatography A*, **999**(1–2), 71–80.

Lestremau F., Andersson F. A. T., Desauziers V. and Fanlo J-L. (2003b). Evaluation of solid-phase microextraction for time-weighted average sampling of volatile sulfur compounds at ppb concentrations. *Analytical Chemistry*, **75**(11), 2626–2632.

Lestremau F., Andersson F. A. T. and Desauziers V. (2004a). Investigation of artefact formation during analysis of Volatile Sulphur Compounds using Solid Phase Microextraction (SPME). *Chromatographia*, **59**, 607–613.

Lestremau F., Desauziers V. and Fanlo J.-L. (2004b). Headspace SPME followed by GC-PFPD for the analysis of malodorous sulfur compounds in liquid industrial effluents. *Analytical and Bioanalytical Chemistry*, **378**, 190–196.

Lopez R., Lapena A. C., Cacho J. and Ferreira V. (2007). Quantitative determination of wine highly volatile sulfur compounds by using automated headspace solid-phase microextraction and gas chromatography-pulsed flame photometric detection. Critical study and optimization of a new procedure. *Journal of Chromatography A*, **1143**, 8–15.

Mariné S., Pedrouzo M., Marcé R. M., Fonseca I. and Borrull F. (2012). Comparison between sampling and analytical methods in characterization of pollutants in biogas. *Talanta*, **100**, 145–152.

Mcginley M. and Mcginley C. (2001). The New European Olfactometry Standard: implementation, experience, and perspectives. Air and Waste Management Association: 2001 Annual Conference Technical Program, Orlando, Florida.

Mcglenny W. A., Pleil J. D., Evans G. F., Oliver K. D., Holdren M. W. and Winberry W. T. (2012). Canister-Based Method for Monitoring Toxic VOCs in Ambient Air. *Journal of the Air & Waste Management*, **41**, 1308–1318.

Mcintyre A. (2000). Application of dispersion modelling to odour assessment: a pratical tool or a complex trap? *Water Science and Technology*, **41**(6), 81–88.

Mochalski P., Wzorek B., Śliwka I. and Amann A. (2009). Improved pre-concentration and detection methods for volatile sulphur breath constituents. *Journal of Chromatography B*, **877**, 1856–1866

Muñoz R., Sivret E., Parsci G., Lebrero R., Wang X., Suffet I. H. and Stuetz R. (2010). Monitoring techniques for odour abatement assessment. *Water Research*, **44**, 5129–5149.

Murata T., Rahardjo A., Fujiyama Y., Yamaga T., Hanada M., Yaegaki K. and Miyazaki H. (2006). Development of a compact and simple gas chromatography for oral malodor measurement. *Journal of Periodontology*, **77**, 1142–1147.

Murray R. A. (2001). Limitations to the use of solid-phase micoextraction for quantitation of mixtures of volatile organic sulfur compounds. *Analytical Chemistry*, **73**, 1646–1649.

Neutra R., Lipscomb J., Satin K. and Shusterman D. (1991). Hypotheses to explain the higher symptom rates observed around hazardous waste sites. *Environmental Health Perspectives*, **94**, 31–38.

Nielsen A. T. and Jonsson S. (2002). Quantification of volatile sulfur compounds in complex gaseous matrices by solid-phase microextraction. *Journal of Chromatography A*, **963**, 57–64.

Niessen W. M. A. (2001). Principles & instrumentation of gas chromatography-mass spectrometry. In: Current practice of Gas Chromatography-Mass Spectrometry, W. M. A. Niessen (ed.), Marcel Dekker Inc, New York, USA.

Nyord A. F. T., Hansen M. N. and Lindholst S. (2011). Chemical evaluation of odor reduction by soil injection of animal manure. *Journal of Environmental Quality*, **40**, 1674–1682.

Ochiai N., Tsuji A., Nakamura N., Daishima S. and Cardin D. B. (2002). Stabilities of 58 volatile organic compounds in fused-silica-lined and SUMMA polished canisters under various humidified conditions. *Journal of Environmental Monitoring*, **4**(6), 879–889.

Ochiai N. and Sasamoto K. (2011). Selectable one-dimensional or two-dimensional gas chromatography-olfactometry/mass spectrometry with preparative fracton collection for analysis of ultra-trace amounts or odor compounds. *Journal of Chromatography A*, **1218**, 3180–3185.

Ochiai N., Takino M., Daishima S. and Cardin D. B. (2001). Analysis of volatile sulphur compounds in breath by gas chromatography-mass spectrometry using a three-stage cryogenic trapping preconcentration system. *Journal of Chromatography B*, **762**, 67–75.

Pandey S. K. and Kim K.-H. (2009). A review of methods for the determination of reduced sulfur compounds (RSCs) in air. *Environmental Science & Technology*, **43**, 3020–3029.

Pandey S. K., Kim K. H. and Tang K.-T. (2012). A review of sensor-based methods for monitoring hydrogen sulfide. *TrAC Trend Analytical Chemistry*, **32**, 87–99.

Panighel A. and Flamini R. (2014). Applications of solid-phase microextraction and gas chromatography/mass spectrometry (SPME-GC/MS) in the study of grape and wine volatile compounds. *Molecules*, **19**, 21291–21309.

Perrault K., Stuart B. and Forbes B. (2014). A longitudinal study of decomposition odour in soil using sorbent tubes and solid phase microextraction. *Chromatography* **1**, 120–140.

Prata A. A., Lucernoni F., Santos J., Capelli L., Sironi S., Le-Minh N. and Stuetz R. M. (2018). Mass transfer inside a flux hood for the sampling of gaseous emission from liquid surface – experimental assessment and emission rate rescaling. *Atmospheric Environment*, **179**, 227–238.

Przyjazny, A. (1985). Preconcentration of volatile organosulphur compounds from the atmosphere on selected porous polymers. *Journal of Chromatography A*, **333**(1), 327–336.

Ras M. R., Marcé R. M. and Borrull F. (2008). Solid-phase microextraction–gas chromatography to determine volatile organic sulfur compounds in the air at sewage treatment plants. *Talanta*, **77**, 774–778.

Rochat S., Laumer J. Y. and Chaintreau A. (2007). Analysis of sulfur compounds from the in-oven roast beef aroma by comprehensive two-dimensional gas chromatography. *Journal of Chromatography A*, **1147**, 85–94.

Rumsey I. C., Aneja V. P. and Lonneman W. A. (2014). Characterizing reduced sulfur compounds emissions from a swine concentrated animal feeding operation. *Atmospheric Environment*, **94**, 458–466.

Ruth J. H. (1986). Odor thresholds and irritation levels of several chemicals substances: a review. *American Industrial Hygiene Association Journal*, **47**, 142–151.

Salgado-Petinal C., Alzaga R., García-Jares C., Llompart M. and Bayona J. M. (2005). Low part per trillion determination of reactive alkanethiols in wastewater by in situ Derivatization-Solid-Phase Microextraction followed by GC-MS. *Analytical Chemistry*, **77**, 6012–6018.

Sasamoto K. and Ochiai N. (2010). Selectable one-dimensional or two-dimensional gas chromatography-mass spectrometry with simultaneous olfactometry or element-specific detection. *Journal of Chromatography A*, **1217**, 2903–2910.

Schweigkofler M. and Niesser R. (1999). Determination of siloxanes and VOC in landfill gas and sewage gas by canister sampling and GC-MS/AES analysis. *Environmental Science & Technology*, **33**, 3680–3685.

Seeber A., Van Thriel C., Haumann K., Kiesswetter E., Blaszkewicz M. and Golka K. (2002). Psychological reactions related to chemosensory irritation. *International Archives of Occupational and Environmental Health*, **75**, 314–25.

Shusterman D. (2001). Odor-associated health complaints: competing explanatory models. *Chemical Senses*, **26**, 338–343.

Sironi S., Capelli L., Céntola P., Del Rosso R. and Pierucci S. (2010). Odour impact assessment by means of dynamic olfactometry, dispersion modelling and social participation. *Atmospheric Environment*, **44**, 354–360.

Sivret E. C., Wang B., Kenny S., Bustamante H. and Stuetz R. M. (2013). Beyond H_2S: chemical characterisation of sewer odour emissions. *Water* **40**, 58–61.

Sivret E. C., Wang B., Parcsi G. and Stuetz R. M. (2016). Prioritisation of odorants emitted from sewers using odour activity values. *Water Research*, **88**, 308–321.

Standards Australia/New Zealand Standard. (2001). Australian/New Zealand Standard: Stationary Source Emissions Part 4: Area Source Sampling; Flux Chamber Technique. Standards Australia.

Standards Australia/Standards New Zealand. (2009). AS/NZ 4323.4:2009 stationary source emissions method 4: area source sampling- flux chamber technique. In: Standards, and Australia, E. (ed.), Standards Australia/Standards New Zealand. Sydney, Australia.

Stuetz R. M., Fenner R. A. and Engin G. (1999). Assessment of odours from sewage treatment works by an electronic nose: H_2S analysis and olfactometry. *Water Research*, **33**, 453–461.

Stuetz R. M., Fenner R. A., Hall S. J., Stratful I. and Loke D. (2000). Monitoring of wastewater odours using an electronic nose. *Water Science and Technology*, **41**(6), 41–47.

Sucker K., Müller F., Bischoff M. and Winneke G. (2004). Assessment of frequency, intensity and hedonic tone of environmental odours in the field: a comparison of trained and untrained residents. *VDI Berichte*, **1850**, 219–288.

Sucker K., Both R., Bischoff M., Guski R., Kramer U. and Winneke G. (2008a). Odor frequency and odor annoyance Part II: dose-response associations and their modification by hedonic tone. *International Archives of Occupational and Environmental Health*, **81**, 683–94.

Sucker K., Both R., Bischoff M., Guski R. and Winneke G. (2008b). Odor frequency and odor annoyance. Part I: assessment of frequency, intensity and hedonic tone of environmental odors in the field. *International Archives of Occupational and Environmental Health*, **81**, 671–82.

Sulyok M., Haberhauer-Troyer C., Rosenberg E. and Grasserbauer M. (2001). Investigation of the storage stability of selected volatile sulfur compounds in different sampling conditions. *Journal of Chromatography A*, **917**, 367–374.

Trabue S. L., Scoggin K. D., Li H., Burns R. and Xin H. (2008). Field sampling method for quantifying odorants in humid environments. *Environmental Science & Technology*, **42**, 3745–3750.

Van Germert, L. J. (2003). "Odour Thresholds". Compilations of odour threshold values in air, water and other media. Oliemans, Punter & Partners BV, Utrecht.

Wang B., Sivret E. C., Parcsi G., Le N. M., Kenny S., Bustamante H. and Stuetz R. M. (2013). Is H_2S a suitable process indicator for odour abatement performance of sewer odours? *Water Science and Technology*, **69**(1), 92–98.

Wardnecki W. (1998). Problems with the determination of environmental sulphur compounds by gas chromatography. *Journal of Chromatography A*, **793**, 1–19.

Winkel E. G. and Tangerman A. (2008). Appropriate sample bags and syringes for preserving breath samples in breath odor research: a technical note. *Journal of Breath Research*, **017011**.

Woolfenden E. (1997). Monitoring VOCs in air using sorbent tubes followed by thermal desorption-capillary GC analysis: summary of data and practical guidelines. *Journal of the Air & Waste Management*, **47**(1), 20–36.

Woolfenden E. (2010). Sorbent-based sampling methods for volatile and semi-volatile compounds in air. Part 2. Sorbent selection and other aspects of optimizing air monitoring methods. *Journal of Chromatography A*, **1217**, 2685–2694.

Yan X. (2002). Sulfur and nitrogen chemiluminscence detection in gas chromatographic analysis. *Journal of Chromatography A*, **976**, 3–10.

Zarra T., Giuliani S., Naddeo V. and Belgiorno V. (2012). Control of odour emission in wastewater treatment plants by direct and undirected measurement of odour emission capacity. *Water Science and Technology*, **66**(8), 1627–1633.

Zhang S., Cai L., Koziel J. A., Hoff S. J., Schmidt D. R., Clanton C. J., Jacobson L. D., Parker D. B. and Heber A. J. (2010). Field air sampling and simultaneous chemical and sensory analysis of livestock odorants with sorbent tubes and GC–MS/olfactometry. *Sensors and Actuators B-Chemical*, **146**, 427–432.

Zhang S., Du Z. and Li G. (2012). Graphene-supported zinc oxide solid-phase microextraction coating with enhanced selectivity and sensitivity for the determination of sulfur volatiles in *Allim* species. *Journal of Chromatography A*, **1260**, 1–8.

Zhu J., Chen F., Wang L., Niu Y., Yu D., Shu C., Chen H., Wang H. and Ziao Z. (2015a). Comparison of aroma-active volatiles in Oolong tea infusions using GC-Olfactometry, GC-FPD, and GC-MS. *Journal of Agricultural and Food Chemistry*, **63**, 7499–7510.

Zhu W., Koziel J. A., Cai L., Wright D. and Kuhrt F. (2015b). Testing odorants recovery from a novel metallized fluorinated ethylene propylene gas sampling bag. *Journal of the Air & Waste Management Association*, **65**, 1434–1445.

Chapter 11

Biological H_2S removal from gases

Markéta Andreides, Lucie Pokorná-Krayzelová, Jan Bartáček and Pavel Jeníček

11.1 INTRODUCTION

Hydrogen sulfide (H_2S) is an odorous and poisonous gas present in biogas (the main product of anaerobic digestion), in sour gas (i.e. natural gas with a high content of H_2S) or in general in the gases coming as a concomitant with crude oil extraction. H_2S can also be produced abiotically, e.g. through the reaction of anhydrite and hydrocarbons as reported in deep carbonate gas reservoirs (Worden & Smalley, 1996). In environmental engineering applications, H_2S is most often generated during anaerobic fermentation of substrates containing sulfur compounds. Sulfate-reducing bacteria (SRB) with a respiratory metabolism use sulfate as an electron acceptor while producing sulfide, which is toxic both in its liquid as well as in its gaseous form.

Gaseous hydrogen sulfide causes the corrosion of concrete and steel digesters (see Chapter 5) and other equipment such as pipes, burners and combined heat and power (CHP) units, which increases the operational costs since the engine oil is acidified and its replacement interval is significantly shortened. While burning, the acid rain precursors SO_2 and SO_3 are emitted to the environment. Last but not the least, gaseous hydrogen sulfide is extremely toxic and causes health risks for the operators of fermenters and sewer maintainers. The olfactory nerve responding to the smell of hydrogen sulfide is paralyzed after a few inhalations of 100–150 ppm,

DOI: 10.2166/9781789060966_0345

which can lead to death (Zenz *et al.*, 1994). H_2S concentrations over 100 ppm are common in sewers (Wang *et al.*, 2020). In the biogas, the H_2S concentration normally ranges from 50–3000 ppm and in some cases it can reach well over 10,000 ppm (Krayzelova *et al.*, 2015). In the liquid form, sulfide can inhibit methanogenesis even at very low concentration, i.e. 50 mg L^{-1} (see chapter 6).

It is important to remove hydrogen sulfide from biogas before its utilization. For biogas combustion, the concentration of hydrogen sulfide should not exceed 100 ppm (Goss, 2017), while for biomethane production, the concentration should be extremely low, in most countries less than 5 ppm (Awe *et al.*, 2017). Alongside anaerobic digesters, all the above-mentioned problems related to the presence of hydrogen sulfide can also appear in the sewer systems. The corrosion of concrete decreases the lifetime of sewer systems (Chapter 5), while gaseous hydrogen sulfide severely threatens workers maintaining sewer systems.

There are many technologies for hydrogen sulfide removal available on the market, based on biochemical or physicochemical processes. This chapter focuses on the techniques utilizing biochemical oxidation of hydrogen sulfide to harmless sulfur products.

11.2 FUNDAMENTALS OF BIOLOGICALLY DRIVEN PROCESSES FOR H_2S REMOVAL FROM GASES

11.2.1 Mechanisms of the oxidation of reduced sulfur compounds

The oxidation of hydrogen sulfide can be either abiotic or biotic. Chemical oxidation of hydrogen sulfide occurs spontaneously in parallel to biological oxidation. This phenomenon can become an advantage for biological oxidation since it decreases the concentration of hydrogen sulfide and enhances the removal process (Pokorna & Zabranska, 2015). During chemical oxidation, partially oxidized products such as thiosulfate are generated (González-Sánchez & Revah, 2009):

$$2\,HS^- + O_2 \rightarrow S_2O_3^{2-} + H_2O \quad \Delta G^\circ = -387.35\,\text{kJ/mol of reaction} \qquad (11.1)$$

Thiosulfate may be (1) bio-oxidized to tetrathionate (Vishniac, 1952) and further metabolized to sulfate, (2) disproportionated biologically to sulfide and sulfate or (3) oxidized by sulfur-oxidizing bacteria (SOB) (Equation 11.2), since the energetic yield from this reaction is highly advantageous (Chen & Morris, 1972).

$$S_2O_3^{2-} + H_2O + 2\,O_2 \rightarrow 2\,SO_4^{2-} + 2\,H^+ \quad \Delta G^\circ = -818.3\,\text{kJ/mol of reaction} \qquad (11.2)$$

In nature, biological oxidation of H_2S is preferred over the chemical one (Yavuz *et al.*, 2007) and is performed by phototrophic or chemotrophic SOB (Pokorna & Zabranska, 2015). In environmental technologies, the chemolithotrophic species are more important, since the delivery of light for phototrophic metabolism limits process kinetics. Oxygen (or sometimes nitrate) serves as an electron acceptor in the metabolism of SOB. The concentration of dissolved oxygen affects the thermodynamics of the reaction and hence the final oxidation product as shown in Equations 11.3–11.5 (Tang *et al.*, 2009):

$$HS^- + 0.5\,O_2 \rightarrow S^0 + OH^- \qquad \Delta G^\circ = -209.4\,\text{kJ/mol of reaction} \tag{11.3}$$

$$S^0 + 1.5\,O_2 + H_2O \rightarrow SO_4{}^{2-} + 2\,H^+ \qquad \Delta G^\circ = -587.1\,\text{kJ/mol of reaction} \tag{11.4}$$

$$HS^- + 2\,O_2 \rightarrow SO_4{}^{2-} + H^+ \qquad \Delta G^\circ = -798.2\,\text{kJ/mol of reaction} \tag{11.5}$$

The amount of available electron acceptor to donor (i.e. the O_2/S^{2-} or $NO_x{}^-/S^{2-}$) molar ratio is the most important factor in biological sulfide oxidation. In addition, the pH value, temperature, ionic strength and the presence of inhibitors or catalysts also play an important role (Chen & Morris, 1972; Janssen *et al.*, 1998; Nielsen *et al.*, 2004).

In practice, it is difficult to keep the O_2/S^{2-} ratio stable, since the concentration of hydrogen sulfide in real systems fluctuates (Janssen *et al.*, 1998). The theoretical molar ratio of O_2/S^{2-} needed for sulfide oxidation to elemental sulfur is 0.5 (Equation 11.3), however, in reality, this is usually not sufficient since oxygen is also used for different oxidative reactions (Pokorna, 2017).

When nitrate or nitrite acts as an electron acceptor, the process is called autotrophic denitrification and is often used for nitrate (or nitrite) removal from ground and wastewater as well (Chapter 6). Elemental sulfur, sulfide and thiosulfate can serve as electron donors and the intermediate oxidized sulfur species are similar to those in oxidation by oxygen. The oxidation of sulfur is a two-step process, where sulfide is oxidized by nitrate to elemental sulfur in the first step, and elemental sulfur is further oxidized to sulfate in the second step. However, the nitrate removal rate is lower in the second step (An *et al.*, 2010). If thiosulfate is the electron donor, the energy yield is high, and the required amount of nitrate is the same as for sulfide oxidation (Sørensen *et al.*, 1980). For SOB, thiosulfate is energetically the most advantageous electron donor contrasting with the least advantageous elemental sulfur (Pokorna & Zabranska, 2015).

The final products of autotrophic denitrification strongly depend on the N/S molar ratio. Theoretically, an N/S ratio equal to 1 or lower results in the lowest energy yield and sulfide is incompletely oxidized to elemental sulfur. The complete oxidation to sulfate (and reduction of nitrate to nitrogen gas) stably occurs when the N/S ratio reaches 1.6 (Gommers *et al.*, 1988; Krishnakumar & Manilal, 1999; Reyes *et al.*, 2004). When the nitrate concentration is low, it could be dosed externally, but the dosing must be strictly controlled in practice to minimize operational costs (Khanongnuch *et al.*, 2019). The final product of autotrophic denitrification varies, as shown in Equations 11.6–11.17:

$$S^{2-} + 0.4\,NO_3^- + 2.4\,H^+ \rightarrow S^0 + 0.2\,N_2 + 1.2\,H_2O$$
$$\Delta G^\circ = -191.0\,\text{kJ/mol of reaction} \quad (11.6)$$

$$S^0 + 1.2\,NO_3^- + 0.4\,H_2O \rightarrow SO_4^{2-} + 0.6\,N_2 + 0.8\,H^+$$
$$\Delta G^\circ = -547.6\,\text{kJ/mol of reaction} \quad (11.7)$$

$$S^{2-} + 1.6\,NO_3^- + 1.6\,H^+ \rightarrow SO_4^{2-} + 0.8\,N_2 + 0.8\,H_2O$$
$$\Delta G^\circ = -743.9\,\text{kJ/mol of reaction} \quad (11.8)$$

$$S^{2-} + 4\,NO_3^- \rightarrow SO_4^{2-} + 4\,NO_2^-$$
$$\Delta G^\circ = -501.4\,\text{kJ/mol of reaction} \quad (11.9)$$

$$S^{2-} + NO_3^- + 2\,H^+ \rightarrow S^0 + NO_2^- + H_2O$$
$$\Delta G^\circ = -130.4\,\text{kJ/mol of reaction} \quad (11.10)$$

$$5\,HS^- + 8\,NO_3^- + 3H^+ \rightarrow 5\,SO_4^{2-} + 4\,N_2 + 4\,H_2O$$
$$\Delta G^\circ = -3848\,\text{kJ/mol of reaction} \quad (11.11)$$

$$5\,HS^- + 5\,NO_3^- + 5H^+ \rightarrow 2.5\,S^0 + 2.5\,SO_4^{2-} + 2.5\,N_2 + 5\,H_2O$$
$$\Delta G^\circ = -2564\,\text{kJ/mol of reaction} \quad (11.12)$$

$$5\,HS^- + 2\,NO_3^- + 7\,H^+ \rightarrow 5\,S^0 + N_2 + 6\,H_2O$$
$$\Delta G^\circ = -1264\,\text{kJ/mol of reaction} \quad (11.13)$$

$$S^{2-} + 0.67\,NO_2^- + 2.67\,H^+ \rightarrow S^0 + 0.33\,N_2 + 1.33\,H_2O$$
$$\Delta G^\circ = -240.3\,\text{kJ/mol of reaction} \quad (11.14)$$

$$S^{2-} + 2.67\,NO_2^- + 2.67\,H^+ \rightarrow SO_4^{2-} + 1.33\,N_2 + 1.33\,H_2O$$
$$\Delta G^\circ = -920.3\,\text{kJ/mol of reaction} \quad (11.15)$$

$$S^0 + 3\,NO_3^- + H_2O \rightarrow SO_4^{2-} + 3\,NO_2^- + 2\,H^+$$
$$\Delta G^\circ = -191.0\,\text{kJ/mol of reaction} \quad (11.16)$$

$$S^0 + 2\,NO_2^- \rightarrow SO_4^{2-} + N_2$$
$$\Delta G^\circ = -191.0\,\text{kJ/mol of reaction} \quad (11.17)$$

Sources: Equations 11.6–11.10 (Cardoso et al., 2006), Equations 11.11–11.13 (Cai et al., 2008), Equations 11.14–11.15 (Le Borgne & Baquerizo, 2019), Equations 11.16–11.17 (An et al., 2010).

11.2.2 Microorganisms involved in H_2S removal

As mentioned above, the main organisms responsible for hydrogen sulfide removal are SOB. SOB can use energy from light (phototrophic) or from chemical reactions (chemotrophic) (Tang *et al.*, 2009). SOB are autotrophic organisms, which means that the energy yielded is later used for the utilization of CO_2 in the Calvin cycle for production of organic compounds.

11.2.2.1 Phototrophic SOB

Phototrophic SOB use hydrogen sulfide as an electron donor, so there is no oxygen emission during their photosynthesis (Pokorna & Zabranska, 2015). The best-known bacterial groups are green sulfur bacteria and purple sulfur bacteria (Tang *et al.*, 2009). Green sulfur bacteria use an extraordinary CO_2 utilization method, which is the reverse tricarboxylic acid (TCA) cycle (Madigan *et al.*, 2014) and, originally, they were assumed to be the only organisms using this pathway. Later, other groups of bacteria using the reverse TCA cycle such as some Proteobacteria or microaerophilic bacteria from the phylum Aquificae were described as well (Madigan *et al.*, 2014). Compared to other phototrophic bacteria, green sulfur bacteria can function at very low light intensity and the photosynthesis can be performed under anaerobic conditions (Tang *et al.*, 2009). H_2S oxidation is then performed according to the following equation (Madigan *et al.*, 2014):

$$2\,CO_2 + H_2S + 2\,H_2O \xrightarrow{\text{light}} 2\,(CH_2O) + H_2SO_4$$
$$\Delta G^\circ = -191.0\,\text{kJ/mol of reaction} \tag{11.18}$$

Phototrophic SOB have an important role in natural ecosystems, but are not popular in environmental technologies, since their growth rate is very slow compared to chemotrophs and the intensive light source required increases both investment and operational costs (Oh *et al.*, 1998).

11.2.2.2 Chemolithotrophic SOB

Chemolithotrophic SOB, also known as colorless sulfur bacteria, are gram negative and can be present in nature in various environments even in extreme conditions, i.e. superthermophylic, acidophilic and halophilic (Kelly & Wood, 2000; Syed *et al.*, 2006). Regarding the oxygen tolerance, SOB can be strictly aerobic (*Acidithiobacillus thioxidans, Thiobacillus thioparus* and *T. novellus*), facultatively anaerobic (*Thiobacillus denitrificans* and *T. ferrooxidans*) and microaerophilic (*Thiothrix nivea* and *T. micro-aerophila*) (Syed *et al.*, 2006).

Besides oxygen, chemolithotrophic SOB can use oxidized nitrogen species (NO_3^- and NO_2^-) as an electron acceptor while reducing it to nitrite, dinitrogen gas (N_2) or even to ammonia through dissimilatory nitrate reduction to ammonia (DNRA). The final products mainly depend on the S/N ratio in the feed (Dolejs *et al.*, 2015). They use many reduced sulfur species as electron donors (S^{2-}, S^0, $S_2O_3^{2-}$, $S_4O_6^{2-}$ and SO_3^{2-}) and their oxidation end products are mainly elemental sulfur and sulfate (Tang *et al.*, 2009).

Denitrifying SOB are very attractive for technological applications because of their high affinity to sulfide, low nutrient requirements, and high oxidation rates. The most cited genera are *Thiobacillus sp., Beggiatoa sp.* and *Thiothrix sp.* (Janssen *et al.*, 1998) and the most cited species are *Thiobacillus denitrificans, Thiomicrospira denitrificans, Thiobacillus versutus, Thiosphaera pantotropha* and *Pseudomonas denitrificans (Tang et al., 2009).*

A specific case within the *Thiobacillus* genera is *T. ferrooxidans.* This acidophilic bacterium extracts energy from the oxidation of Fe^{2+} to Fe^{3+} (Equation 11.19). Hydrogen sulfide then reacts with ferric ion while producing elemental sulfur according to Equation 11.20 (Giro *et al.*, 2006):

$$4\,Fe^{2+} + O_2 + 4\,H^+ \rightarrow 4\,Fe^{3+} + 2\,H_2O \qquad (11.19)$$

$$H_2S + Fe_2(SO_4)_3 \rightarrow S^0 + 2\,FeSO_4 + H_2SO_4 \qquad (11.20)$$

11.3 TECHNOLOGIES FOR H_2S REMOVAL FROM GASES

11.3.1 Physicochemical methods

11.3.1.1 In-situ precipitation

In-situ precipitation can be used to control H_2S concentrations in the biogas by the reaction of Fe^{2+} or Fe^{3+} (may be supplied as $FeCl_2$, $FeCl_3$, or $FeSO_4$) with hydrogen sulfide to generate insoluble FeS directly in the anaerobic fermenter (Petersson & Wellinger, 2009). This technology can remove high concentrations of H_2S. However, it is not possible to decrease the H_2S concentration in the gas cost-effectively below 100–150 ppmv (Persson *et al.*, 2006). The high operating costs associated mainly with the purchase of chemical reagents represent the main disadvantage of this otherwise simple H_2S control method (Munoz *et al.*, 2015).

11.3.1.2 Chemical and physical adsorption

Chemical adsorption of H_2S on Fe_2O_3, $Fe(OH)_3$ or ZnO has become a popular technology because of its simplicity, high efficiency, and fast oxidation kinetics (Petersson & Wellinger, 2009). These reagents are often immobilized on wood chips to increase the superficial area of the adsorbent (Persson *et al.*, 2006). H_2S removal can also be carried out using adsorption on non-impregnated or impregnated activated carbons (Munoz *et al.*, 2015). However, the high cost associated with the regeneration and replacement of the adsorbent material limits the application of adsorption mainly to small- and medium-scale digesters (Abatzoglou & Boivin, 2009).

11.3.1.3 Chemical and physical absorption

The absorption of H_2S from biogas can be carried out purely physically by using either water or organic solvents or chemically by using aqueous chemical solutions with a conversion of hydrogen sulfide to elemental sulfur or metal sulfides (Wellinger & Lindberg, 1999). Both water and organic solvent scrubbing are suitable for the removal of low concentrations of H_2S. The addition of chemicals such as NaOH, arsenic, vanadium or the most common one iron – in the form of $FeCl_2$ or $Fe(OH)_3$ – can create high H_2S concentration gradients between the gaseous and aqueous phase, thus reducing the liquid to biogas ratio required for efficient H_2S mass transfer (Abatzoglou & Boivin, 2009). In iron-based oxidation, ferric oxide is used as an oxidant contained in neutral or weak alkaline solutions. Ferric oxide oxidizes hydrogen sulfide while being reduced to Fe^{2+}. The catalyst solution is then regenerated and Fe^{2+} is oxidized to Fe^{3+} by air (Xiao *et al.*, 2017).

11.3.1.4 Membrane separation

Membrane separation processes are based on the selective permeability of certain membranes for H_2S and the corresponding retention of CH_4 on the other side of the membrane. The membrane serves as a barrier between the phases that allows the transport of special species from one phase to the other without dispersing (Gabelman & Hwang, 1999; Lu *et al.*, 2010). Membrane separation is used to overcome problems such as foaming, unloading, flooding, and corrosion typically observed in the conventional absorption processes (Mansourizadeh *et al.*, 2010; Rajabzadeh *et al.*, 2009). It is usually connected with the treatment of natural gas. The separation efficiency of H_2S is strongly dependent on the material of the membrane. In the last few years, many new materials (i.e. silicone, CH_4-type zeolite and polydimethylsiloxane) for membranes were developed (Maghsoudi & Soltanieh, 2014; Pokorna-Krayzelova *et al.*, 2018a; Tilahun *et al.*, 2018) and their prices decreased rapidly.

Recently, a high silica – CH_4-type zeolite membrane was used for the simultaneous removal of CO_2 and H_2S (Maghsoudi & Soltanieh, 2014). H_2S removal is based on the physical adsorption of H_2S in the zeolite framework and depends on the Al/Si ratio in the material forming the membrane (Maghsoudi & Soltanieh, 2014). The residual concentrations of H_2S in the treated gas were too high for direct ventilation into the atmosphere. The authors recommended the use of an additional sulfur recovery unit after the membrane, which, however, made the process too complex for biomethane production.

The use of ionic liquid membranes, as a promising technology for H_2S removal, has already been tested at lab-scale (Dolejš *et al.*, 2014; Zhang *et al.*, 2017). However, further research is needed to better understand the removal mechanisms and to increase their reliability.

11.3.2 Biological methods

The physicochemical methods are burdened with disadvantages such as high investment cost, high operational costs, the necessity of additional chemicals, and the production of more waste which is neither feasible nor environmentally friendly (Buisman *et al.*, 1989). A significant effort has therefore been devoted to the research of biological methods (Khoshnevisan *et al.*, 2017).

Biological treatment methods are based on the ability of naturally occurring SOB to remove hydrogen sulfide from biogas. These methods are good alternatives to physicochemical treatment since they have the potential to overcome some or all of the disadvantages of physicochemical processes (Elias *et al.*, 2002; Kim *et al.*, 2002; Oyarzún *et al.*, 2003; Sercu *et al.*, 2005). Biological methods based on the biochemical oxidation of sulfide to sulfate, thiosulfate and elemental sulfur involve lower operational costs with lower or no need for chemical addition (Buisman *et al.*, 1989; Syed *et al.*, 2006).

Biological H_2S treatment (or its combination with physicochemical processes) can occur in many types of reactors such as photobioreactors (Bahr *et al.*, 2014; Kobayashi *et al.*, 1983), biotrickling filters (Gabriel & Deshusses, 2003; Soreanu *et al.*, 2008), biofilters (Chung *et al.*, 1996; Elias *et al.*, 2002) or bioscrubbers (Mesa *et al.*, 2002; Nishimura & Yoda, 1997) with oxygen or nitrate and nitrite used as electron acceptors. All these reactors may be constructed as gas fed-batch (Basu *et al.*, 1996; Janssen *et al.*, 1995) or continuous flow (Buisman *et al.*, 1990; Henshaw & Bewtra, 1998) reactors.

11.3.2.1 Photobioreactors and microalgae-based H_2S removal

Usually, photobioreactors contain either green or purple sulfur bacteria performing desulfurization. An advantageous strategy can be a combination of bacteria and algae. Algal-bacterial symbiosis in photobioreactors can support the simultaneous removal of H_2S and CO_2 in a single process (Bahr *et al.*, 2014). Thus, the O_2 supplied by microalgal photosynthesis during CO_2 biofixation is used by SOBs to convert H_2S to either sulfate or elemental sulfur. The highest cost for this type of desulfurization comes from the light supply, which can be natural light in transparent reactors (i.e. glass) or artificial white light or IR with wavelength around 700–900 nm (Kim & Chang, 1991). Tubular transparent reactors impenetrable for oxygen were operated and tested with a H_2S removal efficiency of 95% (Kobayashi *et al.*, 1983) and 92–95% (Henshaw & Zhu, 2001). The combination of anoxic denitrifying organisms during the night and phototrophic organisms during the day was also tested for hydrogen sulfide removal (Syed *et al.*, 2006). Photosynthetic biogas upgrading in a semi-industrial scale algae-bacterial photobioreactor was published by Rodero *et al.* (2019). There, biogas was upgraded to biomethane by operating an adsorption column. The waste CO_2 was led to a pond with algae, that use CO_2 as a carbon source and produce oxygen for SOB to oxidize hydrogen sulfide to elemental sulfur (Figure 11.1). Most of the studies evaluating H_2S removal in photobioreactors reported complete removal regardless of the use of stand-alone photobioreactors

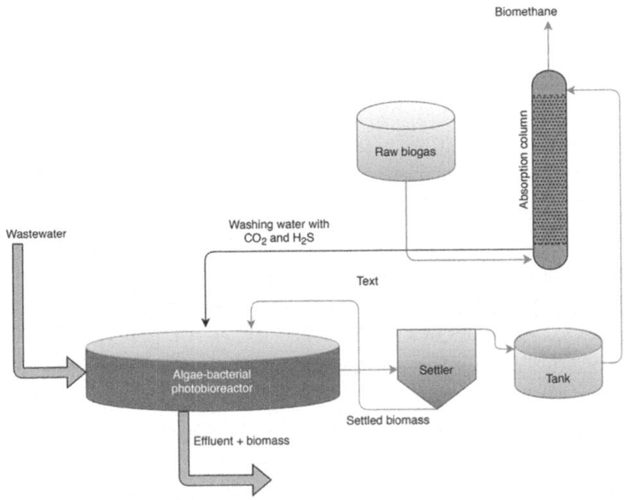

Figure 11.1 The scheme of an algae-bacterial photobioreactor (after Rodero *et al.*, 2019).

with *in-situ* biogas sparging or two-stage with an absorption column (Bahr *et al.*, 2014; Mann *et al.*, 2009).

11.3.2.2 Biofilters, biotrickling filters, and bioscrubbers

In biofilters, biotrickling filters, and bioscrubbers, two main processes take place: firstly, the pollutants are transferred to a liquid medium or biofilm, and secondly, the pollutants are degraded by the microorganisms that are present in the liquid medium or in the biofilm.

In biofilters, the biological packing material covered by biofilm is periodically sprinkled with water or nutrient-rich solutions and no recycling of the used water or buffer is present. This technology is already used at full-scale for treating biogas from many industrial fields (Omri *et al.*, 2011) and the removal efficiency is usually more than 99% (Aita *et al.*, 2016). On the other hand, the long-term operation of a biofilter can lead to serious problems with clogging by sulfur. Also, sulfuric acid can be produced during H_2S removal. In the lowest zones of the biofilter, where the H_2S concentration is the highest, the pH can decrease below 1. This condition leads to the inhibition of the bacteria and forced replacement of the packing material (Barbusinski *et al.*, 2017; Ben Jaber *et al.*, 2017; Jaber *et al.*, 2014; Mudliar *et al.*, 2010; Rabbani *et al.*, 2016).

Biotrickling filters are more sophisticated than biofilters. Buffer solutions with nutrients are continuously spread and recycled. Biotrickling filters have been successfully operated at laboratory- and full-scale using both O_2 (supplied via aeration) and NO_3^- as electron acceptors for the treatment of H_2S concentrations ranging from 500 to 10,000 ppmv (Munoz *et al.*, 2015). SOB are immobilized in the bed material such as open pore polyurethane foam, polypropylene HD Q-PAC®, coconut husk with cylindrical plastic pieces (Chaiprapat *et al.*, 2015) or glass fibers reinforced plastic filled by polypropylene Pall rings (Rodriguez *et al.*, 2014). This technology is a very promising method for treating gas with very high H_2S concentration, i.e. more than 12,000 ppm (Fortuny *et al.*, 2008). The SOB involved are both chemotrophic and phototrophic. However, the chemotrophic SOB prevail, since their growth is faster and tolerance to sulfides higher (Tang *et al.*, 2009). The acidic filtration is more cost-effective compared to the alkaline one since sulfuric acid can be produced if the oxidation of sulfide is complete (González-Sánchez & Revah, 2007). In general, H_2S biofiltration exhibits surprisingly high robustness and lower operating costs than physicochemical technologies (Fortuny *et al.*, 2011). The main limitations of this technology are packing media clogging and higher operational costs derived from the increase in pressure drop and the need for the regeneration or replacement of packing media caused by the accumulation of elemental sulfur (Montebello *et al.*, 2014; Rodriguez *et al.*, 2014).

Bioscrubbers are two-stage technologies with separated gas adsorption (first step) and degradation (second step). Absorption can be achieved in many ways such as bubble columns, spray towers, or packed columns (Nishimura & Yoda, 1997; Syed *et al.*, 2006). The absorbed gas is later degraded in the bioreactor, where microorganisms are dispersed in the liquid phase. The main advantage of bioscrubbers is their ability to treat higher H_2S concentrations in the inlet as well as to cope with severe fluctuations (Potivichayanon *et al.*, 2006). Bioscrubbers are great options when space is limited. The removal efficiency is usually in the range of 90–99% (Nishimura & Yoda, 1997). Moreover, the liquid medium can be recycled to achieve higher removal efficiencies and lower operational costs (Hansen & Rindel, 2000).

11.3.2.3 Main commercially available technologies

The commercial use of biological methods currently outweighs the use of traditional physicochemical methods, thanks to the significant difference in operational costs and due to the savings in chemicals and energy. However, most of the offered technologies combine the physicochemical separation of H_2S with the biological oxidation step.

One of the most used commercial desulfurization units is THIOPAQ® (Thiopaq, 2020) and Sulfothane™ (Sulfothane, 2020). Both units are based on the combination of absorption in a scrubber with biological oxidation in a separate bioreactor (Hulshoff Pol *et al.*, 2001). In these two-step units, hydrogen sulfide from biogas is captured in washing water (pH 8–9) by the reaction with hydroxide ions. In the second step, the washing water with sulfide enters a bioreactor, where sulfide is oxidized by autotrophic colorless sulfur bacteria and the washing water is regenerated. The high hydrogen sulfide removal rate (>99.5%) is achieved by the large difference between sulfide concentration in the washing water at the entrance of the scrubber and in the biogas. The upgraded biogas constantly has a very low H_2S concentration and can be used for combustion as well as for biogas upgrading to biomethane. THIOPAQ® can be used for gas flow rates up to 2500 Nm^3/h, while standard Sulfothane™ units are designed for 100–1500 Nm^3 of biogas. Both methods produce pure elemental sulfur, which may decrease the operational costs.

Other biotechnologies for biogas desulfurization are also available, such as the biotrickling filter systems SulfurexBF® (DMT Environmental Technology, The Netherlands) and BiogasCleaner® (BioGasclean, Denmark) (Dumont, 2015; Lin *et al.*, 2018). On a commercial basis, there are also desulfurization units based on microaeration, the full-scale experience is described in the following section on microaeration (Jeníček *et al.*, 2017).

11.4 MODIFICATIONS OF ANAEROBIC FERMENTERS TO DECREASE THE H_2S CONCENTRATION IN BIOGAS

All of the above mentioned biological methods are based on *ex-situ* purification of biogas with high H_2S concentrations generated in anaerobic fermenters (or gases of other origins). This section concentrates on the *in-situ* minimization of sulfide and hydrogen sulfide production.

11.4.1 Two-stage anaerobic digestion

In single-stage anaerobic digestion, all steps of anaerobic degradation occur in the same reactor whereas, in two-stage anaerobic treatment, hydrolysis and acidogenesis take place in the first reactor, while acetogenesis and methanogenesis occur in the second one. The pH maintained in each reactor and the presence of different substrates result in the growth of different microorganisms in each reactor. The pH range 4.5–6.5 in the first reactor provides favorable conditions for hydrolytic and acidogenic bacteria, while the pH in the second reactor (6.5–7.0) provides favorable conditions for acetogenic and methanogenic bacteria (Budiastuti *et al.*, 2018).

According to Yun *et al.* (2016), the role of the first (sulfidogenic)-stage digester is not only acidogenesis but also sulfidogenesis (sulfate reduction to H_2S), which would minimize the input of H_2S-generating substrates in the subsequent second (methanogenic)-stage digester. The H_2S content in the biogas produced in the conventional single-stage digester, used as a control, reached 1650 ($\pm$25) ppmv, while biogas produced in the second (methanogenic)-stage digester had H_2S concentrations of 200 ($\pm$15) ppmv.

Dai *et al.* (2017) operated a two-stage fermentation with the first stage maintained at pH 10. The high pH influenced iron present in the extracellular polymers (EPS) of the biomass. This EPS of the sludge became more unconsolidated and unstable and Fe in the exchangeable fraction of the EPS was more likely to precipitate with negative bivalent sulfur ions (S^{2-}) originating from the reduction of sulfate. This method was not as efficient as the previous one. Also, the addition of chemicals increases the amount of sludge, which is undesirable while operating wastewater treatment plants (WWTPs).

11.4.2 Microaeration

Microaeration is a widely used biological desulfurization method often applied in full-scale reactors. It is defined as the supply of a small amount of air or oxygen into the anaerobic fermenter to support the growth of sulfide-oxidizing bacteria. These bacteria form a biofilm on the wall (or other surfaces) of the fermenter's headspace and on the gas-liquid interface (Figure 11.2) and oxidize hydrogen sulfide to harmless elemental sulfur (Ramos *et al.*, 2014a). This sulfur forms in the bacterial cytoplasm or outside the bacterial cells. As it detaches from the walls, the S^0 rich biofilm falls into the liquid and is discharged from the fermenter with the effluent.

When referring to microaeration, the amount of oxygen supplied is crucial. According to Krayzelova *et al.* (2015), the term microaeration refers to the maximum dose of 1 L O_2 per 1 L of feed. However, the concentration of oxygen referred to as 'microaerobic' largely varies across the literature (Krayzelova *et al.*, 2015).

The dosing of oxygen or air to remove sulfide does not require any additional space, chemicals, or high investments, which is the main advantage of this method for already existing anaerobic fermenters. Compared to other technologies, microaeration is a highly effective, simple, and inexpensive method (Syed *et al.*, 2006). From the technological point of view, the ideal end-product of oxidation is

Figure 11.2 (a) Formation of massive sulfur-rich biofilms on the surface of a mixer shaft (arrow 1) and on the membrane ceiling of the fermenter near to the location of the air inlet (arrow 2). (b) Sulfur biofilms on a fermenter's concrete ceiling and at the gas-liquid interface, *Courtesy of Jindrich Prochazka*.

elemental sulfur that can be recovered and potentially reused (Krayzelova *et al.*, 2015). On the other hand, the formation of elemental sulfur can lead to the clogging of pipes conveying biogas to a gasholder. Therefore, the amount of elemental sulfur created in the fermenter must be controlled and manually removed if necessary (Krayzelova *et al.*, 2015).

The actual end-product of microaeration is affected by the O_2/S^{2-} ratio. Theoretically, 0.5 mol O_2/mol S^{2-} is necessary for the oxidation of sulfide to elemental sulfur (Equation 11.3) (Alcántara *et al.*, 2004). However, at specific conditions, the optimum ratio has been as high as 0.7 (Janssen *et al.*, 1995). The differences in optimum ratio reported by various authors can be explained by the influence of other parameters such as sulfide concentration (Klok *et al.*, 2013) or pH value (Munz *et al.*, 2009). If the fermenter is overloaded with oxygen or oxygen is consumed too slowly, chemical sulfide oxidation plays a significant role during microaeration, and thiosulfate appears in the system (Díaz *et al.*, 2011a; van den Ende & van Gemerden, 1993).

11.4.2.1 Technological and physical factors influencing microaeration

11.4.2.1.1 Oxygen supply

Oxygen is usually dosed in the form of air either into the reactor's headspace or into the liquid phase. When dosed into the headspace, less oxygen is used for the oxidation of organic matter in the liquid. This results in a lower air dose requirement and therefore, the content of residual nitrogen in the biogas is also lower (Díaz *et al.*, 2010, 2011b). Also, the anaerobic conditions in the liquid phase are not disturbed and the consumption of air by SOB in the headspace is quick.

Dosing air into the liquid phase can have additional advantages. Partial oxidation of hardly degradable substances can increase the methane yield (Xu *et al.*, 2014). The sulfide concentration in the liquid can decrease about 20–30% (Krayzelova *et al.*, 2014), which enhances methanogenesis by lowering the concentration of toxic compounds in the liquid phase (Krayzelova *et al.*, 2015). Previously, methanogens were assumed to be strictly anaerobic; however, recent research shows that in a mixed culture, methanogens are either able to tolerate a limited amount of oxygen or can be protected by facultative anaerobic bacteria. This protection was observed both in suspended as well as granular sludge (Estrada-Vazquez *et al.*, 2003; Kato *et al.*, 1993; Shen & Guiot, 1996).

The choice of the dosing point is partially dependent on the configuration of the fermenter (1 stage or 2 stage reactors) and the mixing strategy (mixing with biogas recirculation, sludge recirculation, or by propellers). The most frequent combinations are shown in Figure 11.3.

Better control over the localization of sulfur depositions can be gained by using membranes to supply air or oxygen. Membranes can be placed either into the liquid

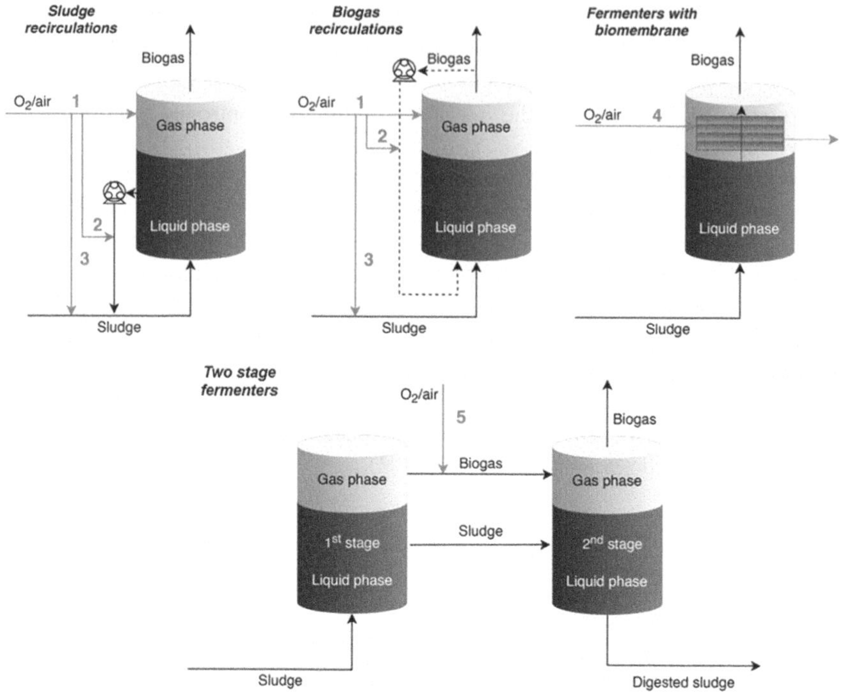

Figure 11.3 The most common strategies of O_2/air dosing for microaeration as H_2S control method in anaerobic digestion: (1) Directly into the headspace, (2) into the sludge recirculation stream, (3) into the biogas recirculation, (4) directly into the liquid, (5) into the headspace through the biomembrane or (6) into the biogas stream between the 1st and 2nd stage of a two-stage fermenter.

phase (Camiloti *et al.*, 2014) or into the headspace (Pokorna-Krayzelova *et al.*, 2017a). Membrane microaeration in the headspace (Figure 11.4) has been already tested in both lab- and pilot-scale reactors (Pokorna-Krayzelova *et al.*, 2017a) and showed an efficiency of more than 99%.

Alternatively, oxygen can also be supplied by the electrolysis of water, which produces oxygen directly in the fermenter. Simultaneously emitted hydrogen can significantly enhance methanogenesis by supporting hydrogenotrophic methanogens (Tartakovsky *et al.*, 2011). However, this technique is so far only at the stage of lab-based research (Osset-Álvarez *et al.*, 2019; Wang *et al.*, 2019).

11.4.2.1.2 Reactor configuration

Oxygen/air can be either (1) directly supplied into the anaerobic fermenter (application *in-situ*), or (2) supplied into a separate unit treating biogas (application *ex-situ*). *In-situ* microaeration with oxygen/air supplied directly inside

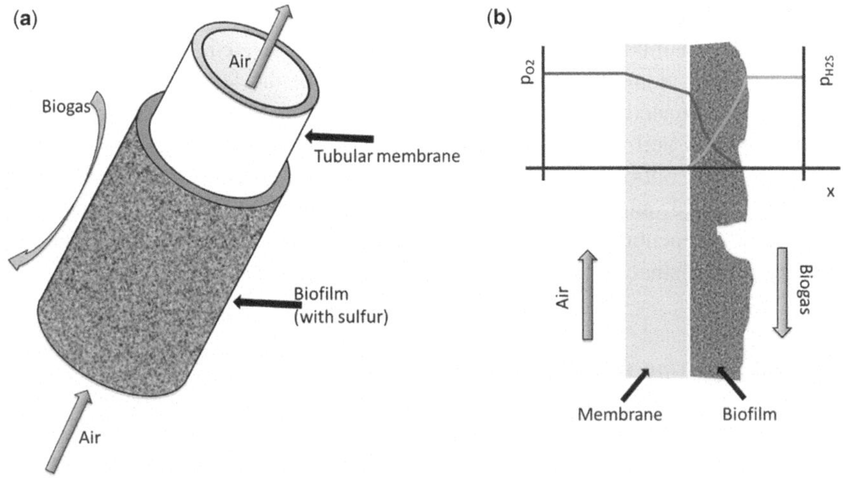

Figure 11.4 (a) Microaeration through a tubular membrane placed in the headspace of an anaerobic digester and (b) Oxygen and hydrogen sulfide profiles in the microaerobic biofilm covering the membrane used for microaeration.

the reactor has already been successfully applied in various reactor types such as completely mixed digesters (Díaz *et al.*, 2011b; Fdz-Polanco *et al.*, 2009; Jenicek *et al.*, 2014; Kobayashi *et al.*, 2012; Nghiem *et al.*, 2014; Schneider *et al.*, 2002 and others), continuously stirred tank reactors (Duangmanee *et al.*, 2007), anaerobic sludge blanket reactors (Krayzelova *et al.*, 2014; Zhou *et al.*, 2007), expanded granular sludge bed reactors (Díaz *et al.*, 2010; Rodriguez *et al.*, 2012), and fluidized bed reactors (van der Zee *et al.*, 2007). An overview of anaerobic reactors where microaeration has been reported can be found in the review written by Krayzelova *et al.* (2015).

At full scale, microaeration is most often used in anaerobic digesters treating sewage sludge or in biogas plants fed by energy crops, municipal waste, and/or agricultural wastes. The technique can successfully remove up to 34,000 ppmv of H_2S with efficiencies higher than 97% (Díaz *et al.*, 2010; Fdz-Polanco *et al.*, 2009; Jenicek *et al.*, 2014, 2008, 2010; Ramos *et al.*, 2014c). Currently, microaeration is not applied for the treatment of industrial wastewaters, even though at lab-scale it has been used for the treatment of brewery wastewater, or wastewater from sugar and paper industries that commonly present high sulfur loads (Krayzelova *et al.*, 2014). When microaeration is performed directly in high-rate reactors, such as upflow anaerobic sludge blanket (UASB), expanded granular sludge bed (EGSB) or internal circulation (IC) reactors, the main danger comes from the possible clogging of the three-phase separators, which are crucial components of these reactors. More research at pilot- and full-scale is needed to estimate the probability of this complication and the potential mitigation actions.

In the *ex-situ* configuration, a microaerobic unit (or compartment) is added to the process, thus maintaining the anaerobic digester unaerated, which avoids the accumulation of elemental sulfur in the headspace of the fermenter. Biogas with H_2S can be treated separately in a sulfide oxidation unit (SOU) where microaeration is performed with an efficiency of up to 99% (Khanal & Huang, 2006). Another option is conveying raw biogas into the SOU inoculated by anaerobic sludge and with microaerobic conditions in the headspace. The precipitated elemental sulfur can be easily removed from the SOU without affecting performance of the fermenter (Ramos *et al.*, 2013).

11.4.2.1.2.1 Mixing method. While dosing air into the liquid, the efficiency of microaeration can be enhanced by proper mixing. Sulfide oxidation is then improved thanks to the more intensive contact with oxygen. Mixing by biogas recirculation can analogically help to improve the contact between oxygen and the liquid phase. This type of mixing diminishes the difference between air dosing into the headspace and into the liquid phase, because of similar gas/liquid distribution of oxygen. On the other hand, the oxygen consumption in both cases increases due to the reactions with organic compounds (Díaz *et al.*, 2011b; Fdz-Polanco *et al.*, 2009). In general, increased mixing intensity will increase the gas-liquid mass transfer coefficient. Sulfide is then easily transferred to the gas phase where microaeration usually takes place (Krayzelova *et al.*, 2015).

11.4.2.1.2.2 Location of sulfide oxidation and sulfur accumulation. As mentioned above, microaeration can take place both in the liquid phase and in the headspace based on the type of sludge (granules or suspension) and on the reactor configuration. Most authors reported the growth of SOB exclusively on the walls of the headspace (Díaz *et al.*, 2011b; Kobayashi *et al.*, 2012; Ramos *et al.*, 2014b). The headspace volume is probably the crucial parameter for proper microaeration. Ramos *et al.* (2014a) achieved complete hydrogen sulfide removal when the headspace of a 100 L reactor had a volume of 25 L and almost no removal when the headspace volume was decreased to close to zero.

The oxidation of hydrogen sulfide in the headspace leads to the accumulation of elemental sulfur present in the biofilm. These yellowish deposits can clog pipes and cause leakages of biogas due to the resulting overpressure (Jeníček *et al.*, 2017). On the other hand, many authors observed very little or no accumulation (Díaz *et al.*, 2011a; Krayzelova *et al.*, 2014). In Jeníček *et al.* (2017), no clogging of pipes was observed in any of seven compared full-scale units. It could be caused by the dosing of air into the biogas recirculation stream where sulfur is to a high extent produced in the liquid phase and removed from the digester with the digestate. When air is dosed into the headspace, clogging might be observed more frequently (Ramos *et al.*, 2012).

11.4.2.1.2.3 Liquid surface area. Based on common knowledge, there is an assumption, that the liquid surface area can be an important parameter for correct microaeration design. Even though the effect of the liquid surface area was not sufficiently explored, Mulbry *et al.* (2017) tested the effect of the liquid surface area by microaeration in barrels placed vertically and horizontally. In the horizontal position, the liquid surface area increased from 0.099 m^2 to 0.150 m^2. The microaeration rate was approximately two times higher in the horizontal position compared to the vertical one. The authors suggested an additional positive effect of a larger biofilm area, which could also play a role in the faster sulfide removal, but no further evaluation has been carried out yet.

11.4.2.1.2.4 Oxygen flow rate and biogas residence time in headspace. There is no consensus on the exact concentration of oxygen which leads to efficient desulfurization. Krayzelova *et al.* (2015) presented an overview of anaerobic reactors, where the use of microaeration has been reported. Based on this overview, oxygen concentrations between 0.3% and 3% can lead to the successful desulfurization of biogas from the digestion of sewage sludge, agricultural materials, and manure.

In practice, oxygen supply is usually controlled based on the residual oxygen concentration in the biogas, which must be low enough not to adversely affect the process of anaerobic digestion. However, the COD/S ratio of the sulfate-rich industrial wastewater (brewery, paper and sugar industry) is usually very low and a high amount of oxygen per volume of biogas is required.

11.4.2.1.2.5 Influence of pH. Typically, anaerobic digestion occurs at neutral pH around 7 to achieve the ideal conditions for methanogens. At this pH, sulfur appears in the liquid phase dominantly as HS^- and H_2S (Migdisov *et al.*, 2002), see the $H_2S - H_2O$ distribution diagram in Pitter (1999). With decreasing pH, the concentration of H_2S increases. In addition, the distribution of H_2S between the gas and liquid phase occurs (Henry's constant $= c(g)/c(l) = 0.5$ (dimensionless)). This is very important since H_2S removal only takes place in the gas phase when dosing air into the headspace of the reactor. When the air is dosed into the headspace, it is thus necessary to increase the contact with the liquid phase containing dissolved H_2S removal or decrease the pH to favor the stripping of H_2S (Krayzelova *et al.*, 2015). On the other hand, the operation of anaerobic digestion at low pH could be tricky, since methanogens are very sensitive to pH changes (O'Flaherty *et al.*, 1998).

11.4.2.2 Positive and adverse effects of microaeration

11.4.2.2.1 Positive effects of microaeration

Dosing of oxygen into an anaerobic reactor has, alongside hydrogen sulfide removal, additional advantages such as the enhancement of hydrolysis, better recovery from shock loadings or acid shocks, better sludge quality, and the possibility of elemental sulfur reuse (Krayzelova *et al.*, 2015).

Indeed, hydrolysis is the crucial step of anaerobic digestion and it is often the rate-limiting step of the methane production from complex substrates such as manure or sewage sludge (Myint *et al.*, 2007). Thus, enhancing hydrolysis can improve the whole anaerobic digestion (Botheju & Bakke, 2011). Zhu *et al.* (2009) showed that sufficient microaeration could enhance the hydrolysis of easily biodegradable organics, promote acidogenesis, and avoid the accumulation of lactic acid. Moreover, microaerobic assays presented a shorter lag-phase than anaerobic assays in the study conducted by Díaz *et al.* (2011b).

Microaeration can shorten the recovery time after overloading the system or after a strong pH decrease. Oxygen seems to be a good tool for dealing with overloading when the bacteria are partially inhibited (Jenicek *et al.*, 2010). Tale *et al.* (2015) carried out additional inoculation of anaerobic fermenters by oxygen-tolerant methanogens to achieve faster recovery after the system was overloaded. After a shock-load of sucrose, the pH in the completely mixed methanogenic reactors recovered more quickly under microaeration conditions (Zitomer & Shrout, 1998). Aeration may also prevent pH decreases in highly loaded systems since volatile acids are oxidized and carbon dioxide and hydrogen are stripped out. Microaeration also appears to improve the quality of the digested sludge in terms of lower foaming potential, better dewaterability (Jenicek *et al.*, 2011a, b, 2014), and better degradability of volatile suspended solids (Jeníček *et al.*, 2017).

Last, but not least, the elemental sulfur produced can be used in bioleaching processes (Tichý *et al.*, 1994; Chapter 15) or for autotrophic sulfur-oxidizing denitrification (Krayzelova *et al.*, 2014; Chapter 13). The biologically produced sulfur is more bioavailable and more suitable for microorganisms compared to chemically produced sulfur since the density of biologically produced sulfur is lower and the particles have a hydrophilic surface (Janssen *et al.*, 2009; Kleinjan *et al.*, 2003).

11.4.2.2.2 Negative effects of microaeration

Some drawbacks of microaeration have already been mentioned, such as oxygen toxicity to methanogens, dilution of biogas by nitrogen from dosed air, and the deposition of elemental sulfur in biogas pipes. It has also been mentioned that oxygen dosing into the liquid phase can enhance hydrolysis by partial oxidation of the hardly degradable organic substrate. On the other hand, in some cases, lower methane production was observed (Khanal & Huang, 2003b; Kobayashi *et al.*, 2012; Ramos & Fdz-Polanco, 2013; Rodriguez *et al.*, 2012). This was probably due to organic matter oxidation as the result of incorrect control of air dosing. Oxygen overdosing leads to the imbalance of the system, which can cause the inhibition of methanogens as well as excessive oxidation of organic matter.

In general, the mixing of methane with oxygen can create an explosive mixture. The explosive range is 85–95% of oxygen in methane. In the case of microaeration, the oxygen content is mostly between 0.3–3% which excludes this potential risk. Therefore, the risk of methane emissions to the atmosphere is much higher and must be considered while operating anaerobic fermenters (Krayzelova *et al.*, 2015).

11.4.2.3 Full-scale applications of microaeration

The ability of microaeration to successfully remove hydrogen sulfide from biogas has already been demonstrated in many full-scale digesters treating agricultural wastes (Schneider *et al.*, 2002), cow manure (Kobayashi *et al.*, 2012) and sewage sludge (Díaz *et al.*, 2015; Jenicek *et al.*, 2010, 2017). An extensive study of 12 years' (2003–2015) operation of microaeration in seven large WWTPs in Central Europe was published by Jeníček *et al.* (2017). Microaeration can be used in small as well as large WWTPs (the smallest and the largest WWTPs had a population equivalent 19,000 and 350,000, respectively) with a removal efficiency over 90% in most cases. In all WWTPs, air was dosed into the liquid and no clogging of the pipes was observed. The dilution of biogas by nitrogen was negligible, thus the switch to pure oxygen was not considered. The microaeration had a relatively long start-up period (3 to 12 weeks) dedicated to the air dose optimization. However, this could be shortened by reliable predictive process control.

11.4.2.4 Model-based optimization and process control

11.4.2.4.1 Model-based optimization

Although microaeration is used in full-scale digesters, advanced process control is still not developed. Especially when the reactor is mixed intermittently or the substrate is dosed irregularly, the hydrogen sulfide concentration in the biogas fluctuates considerably. A predictive control system should thus be designed to avoid these fluctuations.

In general, biochemical processes are complex systems of many biochemical reactions, which must be well understood for proper operation. For this purpose, mathematical modeling is a great tool, which allows the study each process separately to determine the importance of individual reactions. For modeling of anaerobic digestion, the generally accepted Anaerobic Digestion Model no.1 (ADM1) (Batstone *et al.*, 2002) is used. In Pokorna-Krayzelova *et al.* (2017b), ADM1 was extended by sulfur reactions (sulfate reduction and sulfide oxidation) in anaerobic/microaerobic fermenters. The resulting model ADM1-S/O was later extended by chemical oxidation of hydrogen sulfide (Pokorna-Krayzelova *et al.*, 2018b).

11.4.2.4.2 Process control

A good control strategy is necessary for the proper operation of each process. In the case of microaeration, no exact control strategy has been developed yet. For the development of a predictive process control system, many parameters that can influence microaeration such as biofilm area, mixing intensity, and others still have to be properly described.

In general, the oxygen dose can be estimated by the online measurement of the hydrogen sulfide concentration in the headspace. Ramos & Fdz-Polanco (2014) used a proportional-integration-derivation (PID) controller for this purpose and found the resulting removal efficiency of hydrogen sulfide was more than 99%, while the residual concentration of oxygen in the biogas was around 0.05%. On the other hand, the incurred delay in response can lead to fluctuations of the hydrogen sulfide concentration in the resulting biogas. Another parameter tested was biogas flow. When the sulfur load was stable, 3.5 to 5 L of oxygen per m^3 of biogas was required to remove 99% of hydrogen sulfide at the initial H_2S concentration of 0.01% v/v (Ramos & Fdz-Polanco, 2014).

In some papers, oxidation-reduction potential (ORP) is used as a control parameter (Chang *et al.*, 2014; Khanal & Huang, 2003a; Nghiem *et al.*, 2014). The concentration of residual oxygen is maintained at a value which does not affect anaerobic conditions. The standard ORP in microaerobic conditions reported by Nguyen and Khanal (2018) ranges from 0 to −300 mV with reference to the standard hydrogen electrode. The bottleneck of using ORP is its sensitivity to many parameters such as temperature, pH and ionic strength. The solution could be the use of an automatic monitoring control system such as supervisory control and data acquisition system (SCADA) (Nguyen & Khanal, 2018).

11.5 REMOVAL OF ORGANIC SULFUR COMPOUNDS FROM BIOGAS

11.5.1 Emissions of organic sulfur compounds

Alongside H_2S, volatile organic sulfur compounds (VSCs) are also present in biogas (Fisher *et al.*, 2018). Regarding real wastewater treatment plants, VSC concentrations are at least one order of magnitude lower than H_2S concentrations. In a full-scale study performed in the Czech Republic, the concentration of H_2S and organic sulfur (S_{org}) was monitored in the biogas produced from 15 municipal wastewater treatment plants with the anaerobic digestion of sludge. The range of H_2S concentrations was 26–274 mg/m^3, whereas the range of S_{org} concentrations was only 0.2–2.9 mg/m^3 (Jenicek, unpublished data).

The most common VSCs are methanethiol, ethanethiol, dimethyl sulfide and dimethyl disulfide (Calbry-Muzyka *et al.*, 2019). Emissions of VSCs usually relate to the anaerobic treatment of sulfur-rich substrates such as sulfolipids or amino acids (methionine, cysteine, and their derivates) (Andersson *et al.*, 2004; Visscher, 1996). H_2S and VSCs are usually present together in biogas. For example, when methoxyaromatic compounds are present in the substrate, the resulting biogas contains both H_2S and methanethiol as precursors for dimethyl sulfide production (Kiene, 1996). Note that while produced in anaerobic digestion, VSCs can be simultaneously consumed by SRB and methanogens as well (Visscher, 1996).

The harmful effects of VSCs are similar to those of H_2S since they can cause both technical and serious health issues (Andersson *et al.*, 2004) and contribute to foul odors (Chapter 10). In recent years, the upgrading of biogas to biomethane became popular (Rönsch *et al.*, 2016). To classify gas resulting from anaerobic digestion as a biomethane, deep cleaning and removal of undesirable trace substances such as VSCs are required to achieve the desired quality (Calbry-Muzyka *et al.*, 2019).

11.5.2 VSC removal technologies

Since the H_2S and VSCs are usually present together in biogas, it is reasonable to remove these pollutants together. Many technologies for VSC degradation by advanced oxidation processes have been published (Kamal *et al.*, 2016; Shayegan *et al.*, 2018; Son, 2017), but the high investment and operational costs of these technologies make them disadvantageous. Traditional absorption can also be used. In Calbry-Muzyka *et al.* (2019), a two-stage absorption column with various commercial absorbents (SulfaTrap) was tested. In the first stage, robust H_2S removal took place, and the trace VSCs were eliminated in the second stage. In the second stage, siloxanes were removed as well. It was shown, that activated carbon-based absorbents were the most efficient sorbents for VSC removal.

Biological methods such as biofilters/biotrickling filters are adaptable and reliable methods for the removal of both H_2S and VSCs (Jaber *et al.*, 2014). On the other hand, one step removal can be difficult due to the occasional inhibitory effects of elevated H_2S concentrations on VSC oxidation (Sun *et al.*, 2019). Two-step removal processes can deal with both pollutants with high efficiency (Calbry-Muzyka *et al.*, 2019; Sun *et al.*, 2019). Sun *et al.* (2019) operated a two-stage biotrickling filter with polypropylene ring carriers. In the first stage (acid biotrickling filter), H_2S was captured and oxidized, whereas in the second stage (neutral biotrickling filter), the residual H_2S and VSCs were removed. Further research is needed to establish the process, but the removal efficiency of more than 86% (excluding dimethyl disulfide – 62%) and 92% for VSCs and H_2S, respectively, shows this technology is promising.

REFERENCES

Abatzoglou N. and Boivin S. (2009). A review of biogas purification processes. *Biofuels, Bioproducts and Biorefining*, **3**(1), 42–71.

Aita B. C., Mayer F. D., Muratt D. T., Brondani M., Pujol S. B., Denardi L. B., Hoffmann R. and da Silveira D. D. (2016). Biofiltration of H_2S-rich biogas using Acidithiobacillus thiooxidans. *Clean Technologies and Environmental Policy*, **18**(3), 689–703.

Alcántara S., Velasco A., Muñoz A., Cid J., Revah S. and Razo-Flores E. (2004). Hydrogen sulfide oxidation by a microbial consortium in a recirculation reactor system: Sulfur formation under oxygen limitation and removal of phenols. *Environmental Science & Technology*, **38**(3), 918–923.

An S., Tang K. and Nemati M. (2010). Simultaneous biodesulphurization and denitrification using an oil reservoir microbial culture: Effects of sulphide loading rate and sulphide to nitrate loading ratio. *Water Research*, **44**(5), 1531–1541.

Andersson F. A. T., Karlsson A., Svensson B. H. and Ejlertsson J. (2004). Occurrence and Abatement of Volatile Sulfur Compounds during Biogas Production. *Journal of the Air & Waste Management Association*, **54**(7), 855–861.

Awe O. W., Zhao Y., Nzihou A., Minh D. P. and Lyczko N. (2017). A Review of Biogas Utilisation, Purification and Upgrading Technologies. *Waste and Biomass Valorization*, **8**(2), 267–283.

Bahr M., Díaz I., Dominguez A., González Sánchez A. and Muñoz R. (2014). Microalgal-Biotechnology As a Platform for an Integral Biogas Upgrading and Nutrient Removal from Anaerobic Effluents. *Environmental Science & Technology*, **48**(1), 573–581.

Barbusinski K., Kalemba K., Kasperczyk D., Urbaniec K. and Kozik V. (2017). Biological methods for odor treatment – A review. *Journal of Cleaner Production*, **152**, 223–241.

Basu R., Clausen E. C. and Gaddy J. L. (1996). Biological conversion of hydrogen sulfide into elemental sulfur. *Environmental Progress and Sustainable Energy*, **15**(4), 234–238.

Batstone D. J., Keller J., Angelidaki I., Kalyuzhnyi S. V., Pavlostathis S. G., Rozzi A., Sanders W. T. M., Siegrist H. and Vavilin V. A. (2002). The IWA anaerobic digestion model no. 1 (ADM1). *Water Science and Technology*, **45**(10), 65–73.

Ben Jaber M., Couvert A., Amrane A., Le Cloirec P. and Dumont E. (2017). Removal of hydrogen sulfide in air using cellular concrete waste: Biotic and abiotic filtrations. *Chemical Engineering Journal*, **319**, 268–278.

Botheju D. and Bakke R. (2011). Oxygen effects in anaerobic digestion: A review. *The Open Waste Management Journal*, **4**, 1–19.

Budiastuti H., Ghozali M., Wicaksono H. K. and Hadiansyah R. (2018). Two Stage Anaerobic Reactor Design and Treatment To Produce Biogas From Mixed Liquor of Vegetable Waste. *Journal of Physics: Conference Series*, **953**, 012041.

Buisman C., Post R., Ijspeert P., Geraats G. and Lettinga G. (1989). Biotechnological process for sulphide removal with sulphur reclamation. *Acta Biotechnologica*, **9**(3), 255–267.

Buisman C. J. N., Geraats B. G., Ijspeert P. and Lettinga G. (1990). Optimization of sulphur production in a biotechnological sulphide-removing reactor. *Biotechnology and Bioengineering*, **35**(1), 50–56.

Cai J., Zheng P. and Mahmood Q. (2008). Effect of sulfide to nitrate ratios on the simultaneous anaerobic sulfide and nitrate removal. *Bioresource Technology*, **99**(13), 5520–5527.

Calbry-Muzyka A. S., Gantenbein A., Schneebeli J., Frei A., Knorpp A. J., Schildhauer T. J. and Biollaz S. M. A. (2019). Deep removal of sulfur and trace organic compounds from biogas to protect a catalytic methanation reactor. *Chemical Engineering Journal*, **360**, 577–590.

Camiloti P. R., Mockaitis G., Domingues Rodrigues J. A., Rissato Zamariolli Damianovic M. H., Foresti E. and Zaiat M. (2014). Innovative anaerobic bioreactor with fixed-structured bed (ABFSB) for simultaneous sulfate reduction and organic matter removal. **89**(7), 1044–1050.

Cardoso R. B., Sierra-Alvarez R., Rowlette P., Flores E. R., Gomez J. and Field J. A. (2006). Sulfide oxidation under chemolithoautotrophic denitrifying conditions. *Biotechnology and Bioengineering*, **95**(6), 1148–1157.

Chaiprapat S., Charnnok B., Kantachote D. and Sung S. (2015). Bio-desulfurization of biogas using acidic biotrickling filter with dissolved oxygen in step feed recirculation. *Bioresource Technology*, **179**, 429–435.

Chang Y.-J., Chang Y.-T., Hung C.-H., Lee J.-W., Liao H.-M. and Chou H.-L. (2014). Microbial community analysis of anaerobic bio-corrosion in different ORP profiles. *International Biodeterioration & Biodegradation*, **95**, 93–101.

Chen K. Y. and Morris J. C. (1972). Kinetics of oxidation of aqueous sulfide by oxygen. *Environmental Science & Technology*, **6**(6), 529–537.

Chung Y. C., Huang C. and Tseng C. p. (1996). Reduction of H_2S/NH_3 production from pig feces by controlling environmental conditions. *Journal of Environmental Science and Health. Part A: Environmental Science and Engineering and Toxicology*, **31**(1), 139–155.

Dai X., Hu C., Zhang D. and Chen Y. (2017). A new method for the simultaneous enhancement of methane yield and reduction of hydrogen sulfide production in the anaerobic digestion of waste activated sludge. *Bioresource Technology*, **243**, 914–921.

Díaz I., Lopes A. C., Pérez S. I. and Fdz-Polanco M. (2010). Performance evaluation of oxygen, air and nitrate for the microaerobic removal of hydrogen sulphide in biogas from sludge digestion. *Bioresource Technology*, **101**(20), 7724–7730.

Díaz I., Lopes A. C., Perez S. I. and Fdz-Polanco M. (2011a). Determination of the optimal rate for the microaerobic treatment of several H_2S concentrations in biogas from sludge digesters. *Water Science and Technology*, **64**(1), 233–8.

Díaz I., Pérez S. I., Ferrero E. M. and Fdz-Polanco M. (2011b). Effect of oxygen dosing point and mixing on the microaerobic removal of hydrogen sulphide in sludge digesters. *Bioresource Technology*, **102**(4), 3768–3775.

Díaz I., Ramos I. and Fdz-Polanco M. (2015). Economic analysis of microaerobic removal of H_2S from biogas in full-scale sludge digesters. *Bioresource Technology*, **192**, 280–286.

Dolejš P., Poštulka V., Sedláková Z., Jandová V., Vejražka J., Esposito E., Jansen J. C. and Izák P. (2014). Simultaneous hydrogen sulphide and carbon dioxide removal from biogas by water–swollen reverse osmosis membrane. *Separation and Purification Technology*, **131**(0), 108–116.

Dolejs P., Paclík L., Maca J., Pokorna D., Zabranska J. and Bartacek J. (2015). Effect of S/N ratio on sulfide removal by autotrophic denitrification. *Applied Microbiology and Biotechnology*, **99**(5), 2383–2392.

Duangmanee T., Kumar S. and Sung S. (2007). Micro-aeration for sulfide removal in anaerobic treatment of high-solid wastewater: a pilot-scale study. *Proceedings of the Water Environment Federation*, **2007**(16), 2748–2760.

Dumont E. (2015). H_2S removal from biogas using bioreactors: a review. *Nternational Journal of Energy and Environment*, **6**(5), 479–498.

Elias A., Barona A., Arreguy A., Rios J., Aranguiz I. and Peñas J. (2002). Evaluation of a packing material for the biodegradation of H_2S and product analysis. *Process Biochemistry*, **37**(8), 813–820.

Estrada-Vazquez C., Macarie H., Kato M. T., Rodriguez-Vazquez R., Esparza-Garcia F. and Poggi-Varaldo H. M. (2003). The effect of the supplementation with a primary carbon source on the resistance to oxygen exposure of methanogenic sludge. *Water Science and Technology*, **48**(6), 119–24.

Fdz-Polanco M., Diaz I., Perez S. I., Lopes A. C. and Fdz-Polanco F. (2009). Hydrogen sulphide removal in the anaerobic digestion of sludge by micro-aerobic processes: pilot plant experience. *Water Science and Technology*, **60**(12), 3045–50.

Fisher R. M., Le-Minh N., Alvarez-Gaitan J. P., Moore S. J. and Stuetz R. M. (2018). Emissions of volatile sulfur compounds (VSCs) throughout wastewater biosolids processing. *Science of The Total Environment*, **616–617**, 622–631.

Fortuny M., Baeza J. A., Gamisans X., Casas C., Lafuente J., Deshusses M. A. and Gabriel D. (2008). Biological sweetening of energy gases mimics in biotrickling filters. *Chemosphere*, **71**(1), 10–17.

Fortuny M., Gamisans X., Deshusses M. A., Lafuente J., Casas C. and Gabriel D. (2011). Operational aspects of the desulfurization process of energy gases mimics in biotrickling filters. *Water Research*, **45**(17), 5665–5674.

Gabelman A. and Hwang S.-T. (1999). Hollow fiber membrane contactors. *Journal of Membrane Science*, **159**(1), 61–106.

Gabriel D. and Deshusses M. A. (2003). Retrofitting existing chemical scrubbers to biotrickling filters for H_2S emission control, **100**(11), 6308–6312.

Giro M. E. A., Garcia O. and Zaiat M. (2006). Immobilized cells of Acidithiobacillus ferrooxidans in PVC strands and sulfite removal in a pilot-scale bioreactor. *Biochemical Engineering Journal*, **28**(2), 201–207.

Gommers P. J. F., van Schie B. J., van Dijken J. P. and Kuenen J. G. (1988). Biochemical limits to microbial growth yields: An analysis of mixed substrate utilization. *Biotechnology and Bioengineering*, **32**(1), 86–94.

González-Sánchez A. and Revah S. (2007). The effect of chemical oxidation on the biological sulfide oxidation by an alkaliphilic sulfoxidizing bacterial consortium. *Enzyme and Microbial Technology*, **40**(2), 292–298.

González-Sánchez A. and Revah S. (2009). Biological sulfide removal under alkaline and aerobic conditions in a packed recycling reactor. *Water Science and Technology*, **59**(7), 1415–1421.

Goss M. T. (2017). Internal combustion engines. Water Environment Federation, Alexandria, Virginia, USA.

Hansen N. G. and Rindel K. (2000). Bioscrubbing, an effective and economic solution to odour control at wastewater treatment plants. *Water Science and Technology*, **41**(6), 155–164.

Henshaw P. F. and Bewtra B. (1998). Biological conversion of sulfide to elemental sulfur. *Indian Journal of Engineering & Material Sciences*, **5**, 202–210.

Henshaw P. F. and Zhu W. (2001). Biological conversion of hydrogen sulphide to elemental sulphur in a fixed-film continuous flow photo-reactor. *Water Research*, **35**(15), 3605–3610.

Hulshoff Pol L. W., Lens P. N. L., Weijma J. and Stams A. J. M. (2001). New developments in reactor and process technology for sulfate reduction. *Water Science and Technology*, **44**, 67–76.

Jaber M. B., Anet B., Amrane A., Couriol C., Lendormi T., Cloirec P. L., Cogny G. and Fillières R. (2014). Impact of nutrients supply and pH changes on the elimination of hydrogen sulfide, dimethyl disulfide and ethanethiol by biofiltration. *Chemical Engineering Journal*, **258**, 420–426.

Janssen A. J. H., Sleyster R., Van der Kaa C., Jochemsen A., Bontsema J. and Lettinga G. (1995). Biological sulphide oxidation in a fed-batch reactor. *Biotechnology and Bioengineering*, **47**(3), 327–333.

Janssen A. J. H., Meijer S., Bontsema J. and Lettinga G. (1998). Application of the redox potential for controling a sulfide oxidizing bioreactor. *Biotechnology and Bioengineering*, **60**(2), 147–155.

Janssen A. J., Lens P. N., Stams A. J., Plugge C. M., Sorokin D. Y., Muyzer G., Dijkman H., Van Zessen E., Luimes P. and Buisman C. J. (2009). Application of bacteria involved in the biological sulfur cycle for paper mill effluent purification. *Science of The Total Environment*, **407**(4), 1333–43.

Jenicek P., Celis C. A., Koubova J. and Pokorna D. (2011a). Comparison of microbial activity in anaerobic and microaerobic digesters. *Water Science and Technology*, **63**(10), 2244–2249.

Jenicek P., Celis C. A., Koubova J. and Ruzickova I. (2011b). Change of the digested sludge quality at microaerobic digestion. *Journal of Residuals Science & Technology*, **8**, 39–44.

Jenicek P., Celis C. A., Krayzelova L., Anferova N. and Pokorna D. (2014). Improving products of anaerobic sludge digestion by microaeration. *Water Science and Technology*, **69**(4), 803–9.

Jenicek P., Keclik F., Maca J. and Bindzar J. (2008). Use of microaerobic conditions for the improvement of anaerobic digestion of solid wastes. *Water Science and Technology*, **58**, 1491–1496.

Jenicek P., Koubova J., Bindzar J. and Zabranska J. (2010). Advantages of anaerobic digestion of sludge in microaerobic conditions. *Water Science and Technology*, **62**(2), 427–434.

Jeníček P., Horejš J., Pokorná-Krayzelová L., Bindzar J. and Bartáček J. (2017). Simple biogas desulfurization by microaeration – Full scale experience. *Anaerobe*, **46**, 41–45.

Kamal M. S., Razzak S. A. and Hossain M. M. (2016). Catalytic oxidation of volatile organic compounds (VOCs) – A review. *Atmospheric Environment*, **140**, 117–134.

Kato M. T., Field J. A. and Lettinga G. (1993). High tolerance of methanogens in granular sludge to oxygen. *Biotechnology and Bioengineering*, **42**(11), 1360–6.

Kelly D. P. and Wood A. P. (2000). Reclassification of some species of Thiobacillus to the newly designated genera Acidithiobacillus gen. nov., Halothiobacillus gen. nov. and Thermithiobacillus gen. nov. *International Journal of Systematic and Evolutionary Microbiology*, **50**(2), 511–516.

Khanal S. K. and Huang J.-C. (2003a). ORP-based oxygenation for sulfide control in anaerobic treatment of high-sulfate wastewater. *Water Research*, **37**(9), 2053–2062.

Khanal S. K. and Huang J. C. (2003b). Anaerobic treatment of high sulfate wastewater with oxygenation to control sulfide toxicity. *Journal of Environmental Engineering*, **129**(12), 1104–1111.

Khanal S. K. and Huang J. C. (2006). Online oxygen control for sulfide oxidation in anaerobic treatment of high-sulfate wastewater. *Water Environment Research*, **78**(4), 397–408.

Khanongnuch R., Di Capua F., Lakaniemi A.-M., Raj E. and Lens P. N. L. (2019). Long-term performance evaluation of an anoxic sulfur oxidizing moving bed biofilm reactor under nitrate limited conditions. *Environmental Science: Water Research & Technology.*

Khoshnevisan B., Tsapekos P., Alfaro N., Díaz I., Fdz-Polanco M., Rafiee S. and Angelidaki I. (2017). A review on prospects and challenges of biological H2S removal from biogas with focus on biotrickling filtration and microaerobic desulfurization. *Biofuel Research Journal*, **4**(4), 741–750.

Kiene R. P. (1996). Microbiological controls on dimethylsulfide emissions from wetlands and the ocean. In: Microbiology of Atmospheric Trace Gases, J. Colin Murrell, Donovan P. Kelly (eds), Springer, Berlin, pp. 205–225.

Kim B. W. and Chang H. N. (1991). Removal of hydrogen sulfide by Chlorobium thiosulfatophilum in immobilized-cell and sulfur-settling free-cell recycle reactors. *Biotechnology Progress*, **7**(6), 495–500.

Kim H., Kim Y. J., Chung J. S. and Xie Q. (2002). Long-Term Operation of a Biofilter for Simultaneous Removal of H_2S and NH_3. *Journal of the Air & Waste Management Association*, **52**(12), 1389–1398.

Kleinjan W., Keizer A. and Janssen A. H. (2003). Biologically produced sulfur. In: Elemental Sulfur and Sulfur-Rich Compounds I, R. Steudel (ed.), Springer Berlin Heidelberg, Berlin, Vol. **230**, pp. 167–188.

Klok J. B. M., de Graaff M., van den Bosch P. L. F., Boelee N. C., Keesman K. J. and Janssen A. J. H. (2013). A physiologically based kinetic model for bacterial sulfide oxidation. *Water Research*, **47**(2), 483–492.

Kobayashi H. A., Stenstrom M. and Mah R. A. (1983). Use of photosynthetic bacteria for hydrogen sulfide removal from anaerobic waste treatment effluent. *Water Research*, **17**(5), 579–587.

Kobayashi T., Li Y. Y., Kubota K., Harada H., Maeda T. and Yu H. Q. (2012). Characterization of sulfide-oxidizing microbial mats developed inside a full-scale anaerobic digester employing biological desulfurization. *Applied Microbiology and Biotechnoology*, **93**(2), 847–57.

Krayzelova L., Bartacek J., Kolesarova N. and Jenicek P. (2014). Microaeration for hydrogen sulfide removal in UASB reactor. *Bioresource Technology*, **172**(0), 297–302.

Krayzelova L., Bartacek J., Díaz I., Jeison D., Volcke E. P. and Jenicek P. (2015). Microaeration for hydrogen sulfide removal during anaerobic treatment: a review. *Reviews in Environmental Science and Bio/Technology*, **14**(4), 703–725.

Krishnakumar B. and Manilal V. B. (1999). Bacterial oxidation of sulphide under denitrifying conditions. *Biotechnology Letters*, **21**(5), 437–440.

Le Borgne S. and Baquerizo G. (2019). Microbial Ecology of Biofiltration Units Used for the Desulfurization of Biogas. *ChemEngineering*, **3**, 72.

Lin S., Mackey H. R., Hao T., Guo G., van Loosdrecht M. C. M. and Chen G. (2018). Biological sulfur oxidation in wastewater treatment: A review of emerging opportunities. *Water Research*, **143**, 399–415.

Lu Y., Xinhai Y., Shan-Tung T., Yan J. and Dahlquist E. (2010). Wetting of polypropylene hollow fiber membrane contactors. *Journal of Membrane Science*, **362**(1–2), 444–452.

Madigan M. T., Martinko J. M., Bender K. S., Buckley D. H. and Stahl D. A. (2014). Brock Biology of Microorganisms. Pearson, Glenview (Illinois).

Maghsoudi H. and Soltanieh M. (2014). Simultaneous separation of H_2S and CO_2 from CH4 by a high silica CHA-type zeolite membrane. *Journal of Membrane Science*, **470**, 159–165.

Mann G., Schlegel M., Schumann R. and Sakalauskas A. (2009). Biogas-conditioning with microalgae. *Agronomy Research*, **7**(1), 33–38.

Mansourizadeh A., Ismail A. F. and Matsuura T. (2010). Effect of operating conditions on the physical and chemical CO_2 absorption through the PVDF hollow fiber membrane contactor. *Journal of Membrane Science*, **353**(1), 192–200.

Mesa M., Macias M., Cantero D. J. C. and quarterly b. e. (2002). Biological iron oxidation by Acidithiobacillus ferrooxidans in a packed-bed bioreactor. **16**(2), 69–74.

Migdisov A. A., Williams-Jones A. E., Lakshtanov L. Z. and Alekhin Y. V. (2002). Estimates of the second dissociation constant of H_2S from the surface sulfidation of crystalline sulfur. *Geochimica et Cosmochimica Acta*, **66**(10), 1713–1725.

Montebello A. M., Mora M., Lopez L. R., Bezerra T., Gamisans X., Lafuente J., Baeza M. and Gabriel D. (2014). Aerobic desulfurization of biogas by acidic biotrickling filtration in a randomly packed reactor. (1873–3336 (Electronic)). *Journal of Hazardous Materials*, **280**, 200–208.

Mudliar S., Giri B., Padoley K., Satpute D., Dixit R., Bhatt P., Pandey R., Juwarkar A. and Vaidya A. (2010). Bioreactors for treatment of VOCs and odours – A review. *Journal of Environmental Management*, **91**(5), 1039–1054.

Mulbry W., Selmer K. and Lansing S. (2017). Effect of liquid surface area on hydrogen sulfide oxidation during micro-aeration in dairy manure digesters. *PLOS ONE*, **12**(10), e0185738.

Munoz R., Meier L., Diaz I. and Jeison D. (2015). A review on the state-of-the-art of physical/chemical and biological technologies for biogas upgrading. *Reviews in Environmental Science and Bio-Technology*, **14**(4), 727–759.

Munz G., Gori R., Mori G. and Lubello C. (2009). Monitoring biological sulphide oxidation processes using combined respirometric and titrimetric techniques. *Chemosphere*, **76**(5), 644–50.

Myint M., Nirmalakhandan N. and Speece R. E. (2007). Anaerobic fermentation of cattle manure: Modeling of hydrolysis and acidogenesis. *Water Research*, **41**(2), 323–332.

Nghiem L. D., Manassa P., Dawson M. and Fitzgerald S. K. (2014). Oxidation reduction potential as a parameter to regulate micro-oxygen injection into anaerobic digester for reducing hydrogen sulphide concentration in biogas. *Bioresource Technology*, **173**, 443–447.

Nguyen D. and Khanal S. K. (2018). A little breath of fresh air into an anaerobic system: How microaeration facilitates anaerobic digestion process. *Biotechnol Adv*, **36**(7), 1971–1983.

Nielsen A. H., Vollertsen J. and Hvitved-Jacobsen T. (2004). Chemical sulfide oxidation of wastewater: Effects of pH and temperature. *Water Science and Technology*, **50**(4), 185–92.

Nishimura S. and Yoda M. (1997). Removal of hydrogen sulfide from an anaerobic biogas using a bio-scrubber. *Water Science and Technology*, **36**(6), 349–356.

O'Flaherty V., Mahony T., O'Kennedy R. and Colleran E. (1998). Effect of pH on growth kinetics and sulphide toxicity thresholds of a range of methanogenic, syntrophic and sulphate-reducing bacteria. *Process Biochemistry*, **33**(5), 555–569.

Oh K. J., Kim D. and Lee I.-H. (1998). Development of effective hydrogen sulphide removing equipment using Thiobacillus sp.IW. *Environmental Pollution*, **99**(1), 87–92.

Omri I., Bouallagui H., Aouidi F., Godon J.-J. and Hamdi M. (2011). H_2S gas biological removal efficiency and bacterial community diversity in biofilter treating wastewater odor. *Bioresource Technology*, **102**(22), 10202–10209.

Osset-Álvarez M., Rovira-Alsina L., Pous N., Blasco-Gómez R., Colprim J., Balaguer M. D. and Puig S. (2019). Niches for bioelectrochemical systems on the recovery of water, carbon and nitrogen in wastewater treatment plants. *Biomass and Bioenergy*, **130**.

Oyarzún P., Arancibia F., Canales C. and Aroca G. E. (2003). Biofiltration of high concentration of hydrogen sulphide using *Thiobacillus thioparus*. *Process Biochemistry*, **39**(2), 165–170.

Persson M., Jonsson O. and Wellinger A. (2006). Task 37 – Biogas upgrading to vehicle fuel standards and grid injection. *IEA Bioenergy*, **37**, 20–21.

Petersson A. and Wellinger A. (2009). Biogas upgrading technologies – Developments and innovations. http://typo3.dena.de/fileadmin/biogas/Downloads/Studien/IEA-BiogasUpgradingTechnologies2009.pdf (accessed 9 March 2015).

Pitter P. (1999). Hydrochemie VŠCHT Praha 1999.

Pokorna-Krayzelova L., Bartacek J., Vejmelkova D., Alvarez A. A., Slukova P., Prochazka J., Volcke E. I. P. and Jenicek P. (2017a). The use of a silicone-based biomembrane for microaerobic H_2S removal from biogas. *Separation and Purification Technology*, **189**, 145–152.

Pokorna-Krayzelova L., Mampaey K. E., Vannecke T. P. W., Bartacek J., Jenicek P. and Volcke E. I. P. (2017b). Model-based optimization of microaeration for biogas desulfurization in UASB reactors. *Biochemical Engineering Journal*, **125**, 171–179.

Pokorna-Krayzelova L., Bartacek J., Theuri S. N., Segura Gonzalez C. A., Prochazka J., Volcke E. I. P. and Jenicek P. (2018a). Microaeration through a biomembrane for biogas desulfurization: lab-scale and pilot-scale experiences. *Environmental Science: Water Research & Technology*, **4**(8), 1190–1200.

Pokorna-Krayzelova L., Vejmelková D., Selan L., Jenicek P., Volcke E. I. P. and Bartacek J. (2018b). Final products and kinetics of biochemical and chemical sulfide oxidation under microaerobic conditions. *Water Science and Technology*, **78**(9), 1916–1924.

Pokorna L. (2017). Microaeration for biogas desulfurization – experimental and simulation study of various reactor types. In: Department of Water Technology and Environmental Engineering; Department of Biosystems Engineering, P. Jeníček and E. Volcke (eds), Vol. Ph.D., University of Chemistry and Technology Prague; Universiteit Gent, Prague, pp. 198.

Pokorna D. and Zabranska J. (2015). Sulfur-oxidizing bacteria in environmental technology. *Biotechnology Advances*, **33**(6), 1246–1259.

Potivichayanon S., Pokethitiyook P. and Kruatrachue M. (2006). Hydrogen sulfide removal by a novel fixed-film bioscrubber system. *Process Biochemistry*, **41**(3), 708–715.

Rabbani K. A., Charles W., Kayaalp A., Cord-Ruwisch R. and Ho G. (2016). Pilot-scale biofilter for the simultaneous removal of hydrogen sulphide and ammonia at a wastewater treatment plant. *Biochemical Engineering Journal*, **107**, 1–10.

Rajabzadeh S., Yoshimoto S., Teramoto M., Al-Marzouqi M. and Matsuyama H. (2009). CO_2 absorption by using PVDF hollow fiber membrane contactors with various membrane structures. *Separation and Purification Technology*, **69**(2), 210–220.

Ramos I. and Fdz-Polanco M. (2013). The potential of oxygen to improve the stability of anaerobic reactors during unbalanced conditions: Results from a pilot-scale digester treating sewage sludge. *Bioresource Technology*, **140**, 80–85.

Ramos I. and Fdz-Polanco M. (2014). Microaerobic control of biogas sulphide content during sewage sludge digestion by using biogas production and hydrogen sulphide concentration. *Chemical Engineering Journal*, **250**, 303–311.

Ramos I., Diaz I. and Fdz-Polanco M. (2012). The role of the headspace in hydrogen sulfide removal during microaerobic digestion of sludge. *Water Science and Technology*, **66**(10), 2258–64.

Ramos I., Pérez R. and Fdz-Polanco M. (2013). Microaerobic desulphurisation unit: A new biological system for the removal of H_2S from biogas. *Bioresource Technology*, **142**, 633–640.

Ramos I., Peña M. and Fdz-Polanco M. (2014a). Where does the removal of H_2S from biogas occur in microaerobic reactors? *Bioresource Technology*, **166**, 151–157.

Ramos I., Pérez R. and Fdz-Polanco M. (2014b). The headspace of microaerobic reactors: Sulphide-oxidising population and the impact of cleaning on the efficiency of biogas desulphurisation. *Bioresource Technology*, **158**, 63–73.

Ramos I., Pérez R., Reinoso M., Torio R. and Fdz-Polanco M. (2014c). Microaerobic digestion of sewage sludge on an industrial-pilot scale: The efficiency of biogas desulphurisation under different configurations and the impact of O_2 on the microbial communities. *Bioresource Technology*, **164**, 338–346.

Reyes J., Razo-Flores E. and Gomez J. (2004). Simultaneous biological removal of nitrogen, carbon and sulfur by denitrification. *Water research*, **38**, 3313–21.

Rodero M. d. R., Lebrero R., Serrano E., Lara E., Arbib Z., García-Encina P. A. and Muñoz R. (2019). Technology validation of photosynthetic biogas upgrading in a semi-industrial scale algal-bacterial photobioreactor. *Bioresource Technology*, **279**, 43–49.

Rodriguez E., Lopes A., Fdz-Polanco M., Stams A. J. and Garcia-Encina P. A. (2012). Molecular analysis of the biomass of a fluidized bed reactor treating synthetic vinasse at anaerobic and micro-aerobic conditions. *Applied Microbiology and Biotechnology*, **93**(5), 2181–91.

Rodriguez G., Dorado A. D., Fortuny M., Gabriel D. and Gamisans X. (2014). Biotrickling filters for biogas sweetening: Oxygen transfer improvement for a reliable operation. *Process Safety and Environmental Protection*, **92**(3), 261–268.

Rönsch S., Schneider J., Matthischke S., Schlüter M., Götz M., Lefebvre J., Prabhakaran P. and Bajohr S. (2016). Review on methanation – From fundamentals to current projects. *Fuel*, **166**, 276–296.

Sercu B., Núñez D., Van Langenhove H., Aroca G. and Verstraete W. (2005). Operational and microbiological aspects of a bioaugmented two-stage biotrickling filter removing hydrogen sulfide and dimethyl sulfide. *Biotechnology and Bioengineering*, **90**(2), 259–269.

Shayegan Z., Lee C.-S. and Haghighat F. (2018). TiO_2 photocatalyst for removal of volatile organic compounds in gas phase – A review. *Chemical Engineering Journal*, **334**, 2408–2439.

Shen C. F. and Guiot S. R. (1996). Long-term impact of dissolved O_2 on the activity of anaerobic granules. *Biotechnology and Bioengineering*, **49**(6), 611–620.

Schneider R., Quicker P., Anzer T., Prechtl S. and Faulstich M. (2002). Grundlegende Untersuchungen zur effektiven, kostengunstigen Entfernung von Schwefelwasserstoff aus Biogas.

Son Y.-S. (2017). Decomposition of VOCs and odorous compounds by radiolysis: A critical review. *Chemical Engineering Journal*, **316**, 609–622.

Soreanu G., Béland M., Falletta P., Edmonson K. and Seto P. (2008). Laboratory pilot scale study for H_2S removal from biogas in an anoxic biotrickling filter. *Water Science and Technology*, **57**(2), 201–207.

Sørensen J., Tiedje J. M. and Firestone R. B. (1980). Inhibition by sulfide of nitric and nitrous oxide reduction by denitrifying Pseudomonas fluorescens. *Applied and Environmental Microbiology*, **39**(1), 105.

Sulfothane (2020). http://www.biothane.com/biothane/ressources/files/1/33134,Leaflet_Sulfothane_LR.pdf).

Sun S., Jia T., Chen K., Peng Y. and Zhang L. (2019). Simultaneous removal of hydrogen sulfide and volatile organic sulfur compounds in off-gas mixture from a wastewater treatment plant using a two-stage bio-trickling filter system. *Frontiers of Environmental Science & Engineering*, **13**(4), 60.

Syed M., Soreanu G., Falletta P. and Béland M. (2006). Removal of hydrogen sulfide from gas streams using biological processes – A review. *Canadian Biosystems Engineering*, **48**, 2.1–2.14.

Tale V. P., Maki J. S. and Zitomer D. H. (2015). Bioaugmentation of overloaded anaerobic digesters restores function and archaeal community. *Water Research*, **70**(0), 138–147.

Tang K., Baskaran V. and Nemati M. (2009). Bacteria of the sulphur cycle: An overview of microbiology, biokinetics and their role in petroleum and mining industries. *Biochemical Engineering Journal*, **44**(1), 73–94.

Tartakovsky B., Mehta P., Bourque J. S. and Guiot S. R. (2011). Electrolysis-enhanced anaerobic digestion of wastewater. *Bioresource Technology*, **102**(10), 5685–5691.

Thiopaq (2020). https://en.paques.nl/mediadepot/297229d56452/WEB5800000249 brochureThiopaq.pdf.

Tichý R., Janssen A., Grotenhuis J. T. C., Lettinga G. and Rulkens W. H. (1994). Possibilities for using biologically-produced sulphur for cultivation of *Thiobacilli* with respect to bioleaching processes. *Bioresource Technology*, **48**(3), 221–227.

Tilahun E., Sahinkaya E. and Çalli B. (2018). A hybrid membrane gas absorption and bio-oxidation process for the removal of hydrogen sulfide from biogas. *International Biodeterioration & Biodegradation*, **127**, 69–76.

van den Ende F. P. and van Gemerden H. (1993). Sulfide oxidation under oxygen limitation by a *Thiobacillus thioparus* isolated from a marine microbial mat. *FEMS Microbiology Ecology*, **13**(1), 69–77.

van der Zee F. P., Villaverde S., García P. A. and Fdz.-Polanco F. (2007). Sulfide removal by moderate oxygenation of anaerobic sludge environments. *Bioresource Technology*, **98**(3), 518–524.

Vishniac W. (1952). The metabolism of Thiobacillus thioparus. I. The oxidation of thiosulfate. *Journal of bacteriology*, **64**(3), 363–373.

Visscher P. T. (1996). Microbial turn-over of volatile sulfur compounds. In: Microbiology of Atmospheric Trace Gases, J. Colin Murrell and Donovan P. Kelly (eds), Springer, Berlin, pp. 227–242.

Wang Y., Lin H. and Hu B. (2019). Electrochemical removal of hydrogen sulfide from swine manure. *Chemical Engineering Journal*, **356**, 210–218.

Wang T., Wu K., Kan L. and Wu M. (2020). Current understanding on microbiologically induced corrosion of concrete in sewer structures: a review of the evaluation methods and mitigation measures. *Construction and Building Materials*, **247**.

Wellinger A. and Lindberg A. (1999). Biogas upgrading and utilization. Task 24: energy from biological conversion of organic wastes. IEA Bioenergy, 3–20.

Worden R. H. and Smalley P. C. (1996). H2S-producing reactions in deep carbonate gas reservoirs: Khuff Formation, Abu Dhabi. *Chemical Geology*, **133**(1), 157–171.

Xiao C., Ma Y., Ji D. and Zang L. (2017). Review of desulfurization process for biogas purification. *IOP Conference Series: Earth and Environmental Science*, **100**, 012177.

Xu S., Selvam A. and Wong J. W. C. (2014). Optimization of micro-aeration intensity in acidogenic reactor of a two-phase anaerobic digester treating food waste. *Waste Management*, **34**(2), 363–369.

Yavuz B., Türker M. and Engin G. Ö. (2007). Autotrophic Removal of Sulphide from Industrial Wastewaters Using Oxygen and Nitrate as Electron Acceptors. *Environmental Engineering Science*, **24**(4), 457–470.

Yun Y. M., Sung S., Shin H. S., Han J. I., Kim H. W. and Kim D. H. (2016). Producing desulfurized biogas through removal of sulfate in the first-stage of a two-stage anaerobic digestion. (1097–0290 (Electronic)), *Biotechnology and Bioengineering*, **114**, 970–979.

Zenz C., Dickerson O. B. and Horvath E. P. (1994). Occupational medicine. Mosby, St. Louis.

Zhang X., Tu Z., Li H., Huang K., Hu X., Wu Y. and MacFarlane D. R. (2017). Selective separation of H_2S and CO_2 from CH_4 by supported ionic liquid membranes. *Journal of Membrane Science*, **543**, 282–287.

Zhou W., Imai T., Ukita M., Li F. and Yuasa A. (2007). Effect of limited aeration on the anaerobic treatment of evaporator condensate from a sulfite pulp mill. *Chemosphere*, **66**(5), 924–929.

Zhu M., Lü F., Hao L.-P., He P.-J. and Shao L.-M. (2009). Regulating the hydrolysis of organic wastes by micro-aeration and effluent recirculation. *Waste Management*, **29**(7), 2042–2050.

Zitomer D. H. and Shrout J. D. (1998). Feasibility and benefits of methanogenesis under oxygen-limited conditions. *Waste Management*, **18**(2), 107–116.

Part V

Sulfur Cycle Conversions for Advanced Wastewater Treatment and Resource Recovery

Chapter 12

The role of sulfur radicals in advanced oxidation processes

Aizhong Ding, Wenjuan Jia, Lirong Cheng, Dayang Wang, Lei Zheng and Shurong Zhang

12.1 AN OVERVIEW OF SULFUR RADICALS IN ADVANCED OXIDATION

12.1.1 Introduction to sulfur radicals

Sulfur radicals are involved in a variety of biological and non-biological reactions, found as intermediates in these processes (Glass, 1999). Sulfur compounds play important roles in the building of cellular structures and life activities, while sulfur radicals act as mediators in bio-electron transfer, and in bio-redox processes in living cells. In the field of inorganic synthesis, the feature of reaction media containing sulfur radicals promotes the high conductivity of sulfur-containing materials (Abedinzadeh, 2001; Hu & Long, 2015).

In recent years, advanced oxidation technology has been applied to generate in situ free radicals for oxidizing refractory pollutants (Bolton *et al.*, 2001). These free radicals are mainly hydroxyl radicals (OH•), sulfate radicals ($SO_4{\bullet}^-$) and chlorine radicals (Cl•), of which $SO_4{\bullet}^-$ is a typical representative of sulfur radicals in advanced oxidation technology (Anipsitakis & Dionysiou, 2004a, b; Deng & Ezyske, 2011; Hui *et al.*, 2009; Tizaoui *et al.*, 2011). Many emerging pollutants, such as pharmaceuticals and personal care products (PPCPs) and endocrine disruptors (ECDs) have been found present in surface water and

 The chapter is from the book *Environmental Technologies to Treat Sulfur Pollution: Principles and Engineering, 2nd Edition*, Piet N.L. Lens (Ed.).
DOI: 10.2166/9781789060966_0379

groundwater (Michael-Kordatou *et al.*, 2015). These pollutants are complex in structure and non-biodegradable, which leads to difficulties in effective removal with traditional sewage treatment technology (Wang *et al.*, 2014). However, advanced oxidation technologies have become some of the most promising technologies to remove these refractory pollutants from water with their highly oxidizing and strong mineralization ability. As one of these technologies, sulfur radicals, mainly $SO_4{\bullet}^-$, have been widely used in groundwater remediation and wastewater treatment in recent decades (Ji *et al.*, 2015; Liu *et al.*, 2014; Wang *et al.*, 2017).

12.1.2 Characteristics of $SO_4{\bullet}^-$

The $SO_4{\bullet}^-$ radicals can exist in solutions in the range of pH=2~8. In alkaline conditions, $SO_4{\bullet}^-$ will react with OH^- in the solution and form hydroxyl radicals (•OH) and SO_4^{2-}. If there are other inorganic ions present in the solutions, such as Cl^- and CO_3^{2-}, $SO_4{\bullet}^-$ will react with these inorganic ions as well (Grebel *et al.*, 2010; Li *et al.*, 2017; Yang *et al.*, 2014). The half-life of $SO_4{\bullet}^-$ is 30–40 s, which is much larger than that of •OH, 25 s.

Compared with other free radicals commonly used in advanced oxidation technologies, $SO_4{\bullet}^-$ has a higher oxidation capacity (2.5–3.1 V) than the hydroxyl radical •OH (2.8 V) (Ghanbari & Moradi, 2017). $SO_4{\bullet}^-$ can oxidize many kinds of refractory pollutants, especially those pollutants with aromatic ring and olefin structures (Wright *et al.*, 2001).

The $SO_4{\bullet}^-$ radical can be produced in a variety of ways, such as radiation, photolysis, pyrolysis, and transition metal catalysis of peroxymonosulfate (PMS) or persulfate (PS), whose chemical formula are $S_2O_8^{2-}$ and HSO_5^-, respectively. The precursor oxides, PS or PMS, are cheap and easy to store and transport (Tsao & Wilmarth, 2002; Zhang *et al.*, 2015). Therefore, many researchers have an interest in $SO_4{\bullet}^-$ radicals for advanced wastewater treatment.

12.1.3 Detection of $SO_4{\bullet}^-$

There are two methods used to measure $SO_4{\bullet}^-$ in solution. The first method is that a scavenger of $SO_4{\bullet}^-$ is added in solution to inhibit the reactivity of $SO_4{\bullet}^-$, then the difference in the target pollutant removal efficiency between the scavenged and the control is measured from which the content of $SO_4{\bullet}^-$ can be calculated (Ghanbari & Moradi, 2017; Yang *et al.*, 2011). Common scavengers of $SO_4{\bullet}^-$ are phenol, methanol (MeOH), and tert-butyl alcohol (t-BuOH). Table 12.1 shows their reaction rate constants with $SO_4{\bullet}^-$.

The second method to detect the $SO_4{\bullet}^-$ radical is the use of 5, 5-dimethyl pyridine N-oxide (DMPO) to capture $SO_4{\bullet}^-$ in the solution (Ghanbari & Moradi, 2017). The electron spin resonance (ESR) spectrum is then adopted to detect the signal of DMPO-$SO_4{\bullet}^-$ in solution. Under different pH conditions, it is possible to convert the $SO_4{\bullet}^-$ and •OH radicals in the solution into each other.

Table 12.1 Scavengers of $SO_4{\bullet}^-$ (Yang *et al.*, 2011).

Scavenger	Reaction Rate Constant
Phenol	$8.8 \times 10^9\ M^{(-1)}S^{(-1)}$
MeOH	$6 \times 10^7–7.7 \times 10^7\ M^{(-1)}S^{(-1)}$
t-BuOH	$4.5 \times 10^5–9.1 \times 10^5\ M^{(-1)}S^{(-1)}$

The content of •OH in solution is usually much higher than that of $SO_4{\bullet}^-$ even in the systems of PS or PMS activation in alkaline conditions. When measuring the content of $SO_4{\bullet}^-$ in the solution, the •OH in the solution should be removed according to excess tert-butanol (Ghanbari & Moradi, 2017).

12.1.4 The application of $SO_4{\bullet}^-$ in advanced oxidation processes

It has been known since the 1950s that the $SO_4{\bullet}^-$ radicals can be generated using PS activation (Moortgat *et al.*, 1994). It has been discovered recently that $SO_4{\bullet}^-$ can also be generated by the activation of PMS (Matta *et al.*, 2011). Subsequently, the application of PS and PMS to generate $SO_4{\bullet}^-$ radicals in advanced oxidation processes has been more widely investigated (Li *et al.*, 2015; Waclawek *et al.*, 2017). The recent publications related to PS and PMS research are shown in Figure 12.1.

Most research focuses on the activation methods and target pollutants that react with $SO_4{\bullet}^-$. The activation methods with PS or PMS are usually by energy activation, transition metal activation, and heterogeneous activation (Ghauch & Tuqan, 2012; Li *et al.*, 2015; Liang & Bruell, 2008; Waldemer *et al.*, 2007). Energy activation includes thermal activation and UV activation, while transition metal activation includes iron activation, cobalt activation or silver activation

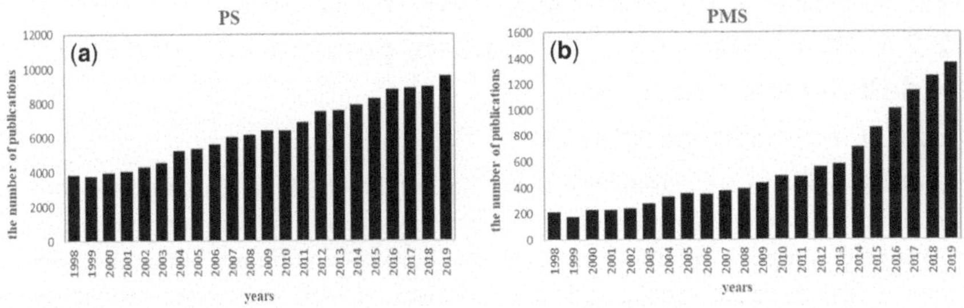

Figure 12.1 Annual number of publications concerning (a) PS and (b) PMS citations. *Source*: Web of science (data as of January, 2020).

(Anipsitakis & Dionysiou, 2004a, b; Hu and Long, 2015; Yan *et al*., 2011). The disadvantage of transition metals as the activators is that the metals can be new pollutants to the treatment system. The activation effectiveness is most influenced by the pH of the solution. Heterogeneous activators are usually non-metallic materials packed with transition metal ions to overcome the disadvantages of transition metal ions as activators only. Thus, heterogeneous activators of different new materials and loading means have recently received more attention (Li *et al*., 2015).

The oxidation targets of the $SO_4^{\bullet -}$ radical range from specific pollutants in solutions to actual contaminated water bodies. The specific pollutants investigated include various emerging pollutants, such as antibiotics, and sterol hormones, while the research focuses on the removal efficiency and removal mechanism of these specific pollutants. The actual contaminated water bodies studied widely are landfill leachates, of which the contaminants include dissolved organic matter (DOM) and ammonia nitrogen (Deng & Ezyske, 2011; Wang *et al*., 2017).

One of the applications of $SO_4^{\bullet -}$ is the in situ chemical oxidation (ISCO) technology for soil and groundwater remediation, which can clean the sources of pollution much faster than other in situ remediation technologies. ISCO is a technology that transports strong oxidants to the pollution zone to oxidize the target pollutants into carbon dioxide, water, and inorganic acids (Michael-Kordatou *et al*., 2015). The oxidants used include potassium permanganate ($KMnO_4$), ozone (O_3), hydrogen peroxide (H_2O_2), and PS. Among these oxidants, PS is more stable than H_2O_2 in groundwater, and oxides more pollutants than $KMnO_4$ and O_3 do. In ISCO technology, the most commonly used activation methods of PS are thermal activation and iron activation (Zhang *et al*., 2015).

With the successful application of $SO_4^{\bullet -}$ in groundwater remediation, its application in advanced wastewater treatment has also received extensive attention as the $SO_4^{\bullet -}$ radical has a high oxidation capacity and strong selectivity to target pollutants (Ye *et al*., 2017).

12.2 GENERATION OF THE $SO_4^{\bullet -}$ RADICAL

12.2.1 Overview of $SO_4^{\bullet -}$ radicals generation in advanced oxidation processes

12.2.1.1 Precursor substances

Generally, oxidants are classified into peroxy bonds (O—O) and non-peroxy bonds containing substances (Waclawek *et al*., 2017). Oxidants containing peroxy bonds can degrade target contaminants with the peroxy-containing bonds broken to produce free radicals, such as the oxidant H_2O_2. PMS or PS are derivatives of hydrogen peroxide in which hydrogen atoms are substituted by sulfonyl groups in a single or symmetrical pattern. For this reason, $SO_4^{\bullet -}$ can be generated by

Table 12.2 Properties of the two precursors to generate $SO_4{\bullet}^-$ radicals (Waclawek *et al.*, 2017).

Oxidant	Bond Energy of Peroxy Bond (kJ/mol)	Solubility in Water at 25°C (g/l)	Average Stable Time in Groundwater	Price (USD/kg)	Price (USD/mol)
PS	92	730	>Five months	0.74	0.18
PMS	377	298	About one day	2.20	1.36

means of PMS or PS excited by energy or catalyzed by transition metals. Table 12.2 shows the properties of PMS and PS (Waclawek *et al.*, 2017).

Compared with PMS, PS has greater water-solubility and stability, has a lower price, and breaks bonds with energy excitation more easily. Therefore, it is preferentially applied by engineers for the removal of pollutants in practical projects (Waclawek *et al.*, 2017).

12.2.1.2 PS and PMS activation

Ultraviolet (UV), thermal and transition metals are commonly used to activate PS or PMS to produce $SO_4{\bullet}^-$ in advanced oxidation processes (Ghauch & Tuqan, 2012; Liang *et al.*, 2008; Rachel *et al.*, 2007). Metal catalysis oxidation technology can be divided into homogeneous and heterogeneous catalysis. The homogeneous system usually shows better treatment effects under acidic conditions because it tends to deactivate metals with precipitation in the alkaline environment, while the heterogeneous system can perform well in a wide pH range (Ahmed *et al.*, 2010; Hui *et al.*, 2009).

Iron (II) is the most popular transition metal used to activate PS or PMS for its cost and availability. However, it needs to be used in acidic conditions, otherwise it will become complexed (Antoniou *et al.*, 2010). In addition, excess iron (II) reacts with $SO_4{\bullet}^-$ and remains in the treated solution when activating PS (Rastogi *et al.*, 2009; Yan *et al.* (2011)). In the heterogeneous catalysis process, iron exists in the form of zero value iron loaded on the surface of heterogeneous materials and can play a more important role with heterogeneous catalysis processes (Liang *et al.*, 2004). In this way, the leakage of iron is small (Xiong *et al.*, 2014).

12.2.2 Homogeneous activation of PS

12.2.2.1 Energy activation

The generation of $SO_4{\bullet}^-$ from activation of PS or PMS involves the process through which peroxy-containing bonds in PS or PMS get broken. The process requires sufficient external energy input for rapid and effective production of $SO_4{\bullet}^-$.

Hence, the energy excitation method has become one of the main excitation methods in advanced oxidation technology. The energy applied is heating, UV, ultrasonic, or radiation proven to be effective in activating PMS or PS to produce $SO_4\bullet^-$ for degrading target contaminants (Ghanbari & Moradi, 2017). Equations 12.1 and 12.2 show the pathways of $SO_4\bullet^-$ generation given as:

$$S_2O_8{}^{2-} \xrightarrow{\text{Energy}} 2SO_4\bullet^- \tag{12.1}$$

$$HSO_5{}^- \xrightarrow{\text{Energy}} SO_4\bullet^- + \cdot OH \tag{12.2}$$

The different peroxy-bonds containing oxidants are different in bond energy and bond length, while the energy demand is different for breaking the bonds. The bond lengths of PS and PMS are 1.497 Å and 1.460 Å, while the bond energies are 92 and 377 kJ/mol, respectively (Anipsitakis & Dionysiou, 2004a, b). Therefore, different activation methods for PS or PMS show different effectiveness in generating $SO_4\bullet^-$.

Thermal activation is an effective method to activate PS or PMS to generate $SO_4\bullet^-$. Wang *et al.* (2018) investigated the effectiveness of PS and PMS (1–20 mM) activated at different heating temperatures (40, 60, and 80°C), respectively, to treat secondary effluent for 3 h (Figure 12.2). The content of dissolved organic carbon (DOC) in the effluent treated decreased significantly with the increase in temperature. The highest DOC removal efficiency of 90% occurred in the PS system at a temperature of 80°C. The experimental results show that temperature plays a significant role in the process of $SO_4\bullet^-$ generation.

It is clear that the degradation trends of DOC in both PMS and PS systems showed a similar behavior, however, the PMS system showed a better degradation rate than the PS system at 40°C (Figure 12.2). This might be due to the difference in activation energy among PS and PMS. The peroxy-bonds of PMS are easily activated at lower temperatures, while at higher temperature, the removal effect of DOC in the PS system is slightly higher than in the PMS

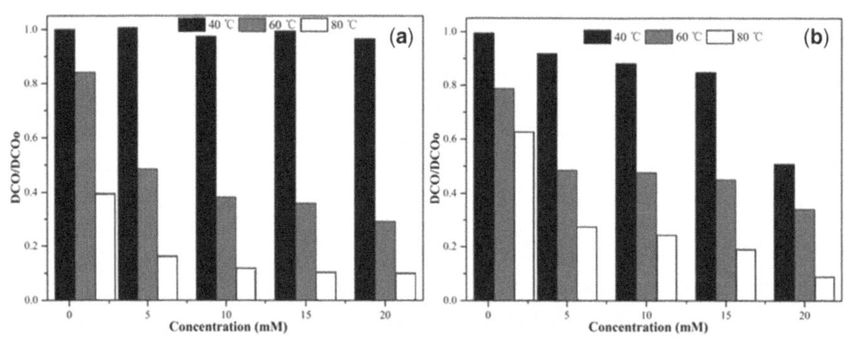

Figure 12.2 A degradation of DOC by (a) PS and (b) PMS at different temperatures (Wang *et al.*, 2018).

system. This is because PS has a lower bond energy, one PS molecule can generate two $SO_4^{\bullet-}$ radicals, whereas one PMS molecule can produce only one $SO_4^{\bullet-}$ radical with one •OH radical. Compared with •OH, $SO_4^{\bullet-}$ plays a more important role in the degradation of DOC in water (Li *et al.*, 2016; Wang *et al.*, 2015).

To investigate the impact of reaction time and temperature on DOC removal with $SO_4^{\bullet-}$, Wang *et al.* (2018) carried out an experiment with PS (20 mM) and PMS (20 mM) (Figure 12.3). The experiment showed that DOC was hard to remove during the first 2 h and thereafter removal increased at 40°C in the PS system, but the overall removal efficiency was as low as 20%. The removal of DOC in the other two PS reaction systems (60 and 80°C) increased significantly during the first 2 h and then increased slowly, finally reaching constant removal. The PMS system displayed a high DOC removal efficiency in the initial 2 h at different temperatures (40, 60, and 80°C).

UV is another method to activate PS and PMS to generate $SO_4^{\bullet-}$. Previous research indicated that PS is more photosensitive than PMS (Zhang *et al.*, 2015). Therefore, Wang *et al.* (2018) carried out an experiment to investigate the activation of PS using UV (254 nm) to treat secondary effluent. Two UV lights with a power of 20 W (UV intensity 0.15 mJ/cm^2) and 40 W (UV intensity 0.25 mJ/cm^2) were used in the experiment, and the PS concentration ranged from 1 to 20 mM (Figure 12.4).

Figure 12.4 shows that the degradation of DOC increased with PS concentration at different UV power. When the concentration of PS reached 10 mM, the DOC degradation efficiency reached 80%. By comparing with the heating activation system (Figure 12.3), a PS concentration of 20 mM was used to investigate the relationship of DOC removal and reaction time. The experimental results showed that the DOC removal efficiency and reaction time were positively correlated for the two UV powers applied.

The activation using UV can lead to a high $SO_4^{\bullet-}$ yield, however the $SO_4^{\bullet-}$ yield cannot significantly increase with the increase in UV intensity. Researchers showed that the higher UV intensity could not lead to a higher removal efficiency for

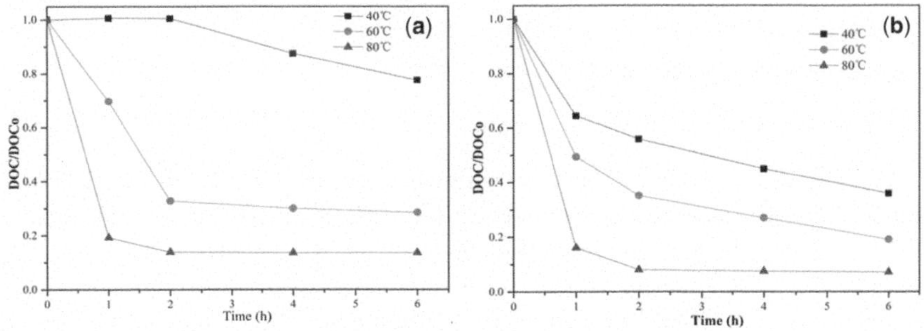

Figure 12.3 Optimal reaction degradation of DOC in (a) PS and (b) PMS systems at different temperatures as a function of time (Wang *et al.*, 2018).

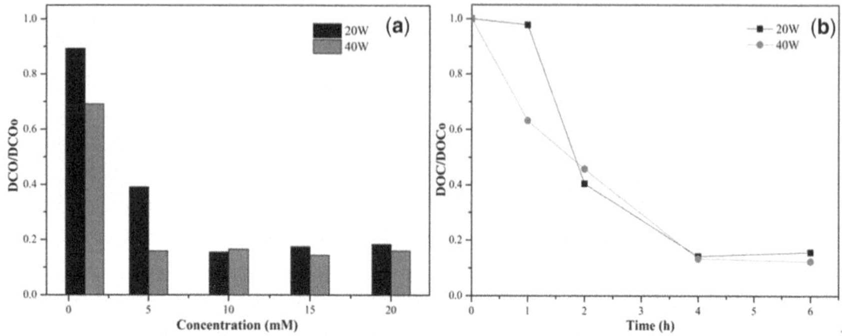

Figure 12.4 UV excited PS (a) and DOC removal (b) in the $SO_4 \bullet^-$ radical based system to treat secondary effluent (Wang *et al.*, 2018).

degrading natural organic matter in reservoir water using H_2O_2 activated with UV (Goslan *et al.*, 2006; Michael-Kordatou *et al.*, 2015).

12.2.2.2 Transition metals activation

In the Fenton system, ferrous ions are used as catalysts to catalyze H_2O_2 to produce •OH. PS and PMS, as derivatives of H_2O_2, can be activated using transition metals (Waclawek *et al.*, 2017). Anipsitakis and Dionysiou (2004a, b) used nine transition metals Ag(I), Ce(III), Co(II), Fe(II), Fe(III), Mn(II), Ni(II), Ru(III), and V(III) to catalyze PMS, PS, and H_2O_2 to degrade 2, 4-dichlorophenol (2, 4-dcp). Co(II) is suitable for catalyzing PMS to produce $SO_4 \bullet^-$ to degrade organic matter, Ag(I) is suitable for PS as well as Fe(II), and Fe(III) for H_2O_2. Equations 12.3–12.8 show transition metal catalysis of PS and PMS:

$$HSO_5^- + M^{n+} \rightarrow M^{(n+1)+} + SO_4 \bullet^- + OH^- \quad (12.3)$$

$$HSO_5^- + M^{n+} \rightarrow M^{(n+1)+} + SO_4^{2-} + \cdot OH \quad (12.4)$$

$$HSO_5^- + M^{(n+1)+} \rightarrow M^{n+} + SO_5 \bullet^- + H^+ \quad (12.5)$$

$$S_2O_8^{2-} + M^{n+} \rightarrow M^{(n+1)+} + SO_4 \bullet^- + SO_4^{2-} \quad (12.6)$$

$$SO_4 \bullet^- + M^{n+} \rightarrow M^{(n+1)+} + SO_4^{2-} \quad (12.7)$$

$$S_2O_8^{2-} + 2M^{n+} \rightarrow 2M^{(n+1)+} + 2SO_4^{2-} \quad (12.8)$$

Wang *et al.* (2018) carried out an experiment to compare Ag(I)/PS, Co(II)/PMS, and ZVI (zero valent iron)/PS systems for degrading effluent organic matter (EfOM) with 3 h reaction time. Figure 12.5 shows that each system could degrade the DOC present in the secondary effluent, however, they behaved differently. The Ag(I)/PS system exhibited the highest DOC removal efficiency, followed by the ZVI/PS system and then the Co(II)/PMS system. The Ag(I)/PS system could reach 90% DOC removal, while the Co(II)/PMS system reached

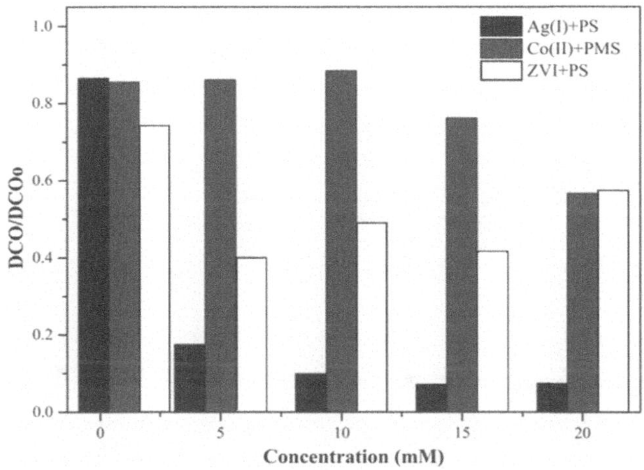

Figure 12.5 Effects of different catalysts on PS or PMS to degrade DOC present in EfOM (Wang *et al.*, 2018).

only 50%. The results indicated that Co(II) catalyzed PMS might not be suitable for treating secondary effluent.

In a ZVI/PS system, DOC removal and the PS concentration were not correlated, which makes it difficult to predict the removal efficiency of the treatment system. The reason for this might be that the Fe(II) was the active component for activating PS, it can react with $SO_4\bullet^-$ as well at a faster reaction rate than that of PS catalyzed by Fe (II). The number of Fe ions increased as the ZVI concentration increased (Alshamsi & Thomson, 2013). Equations 12.9 to 12.14 show the reactions:

$$Fe^0 + S_2O_8^{2-} \rightarrow +2SO_4^{2-} \tag{12.9}$$

$$2Fe^0 + O_2 + 2H_2O \rightarrow 4Fe^{2+} + 4OH^- \tag{12.10}$$

$$Fe^0 + 2H_2O \rightarrow Fe^{2+} + 2OH^- + H_2 \tag{12.11}$$

$$Fe^{2+} + S_2O_8^{2-} \rightarrow Fe^{3+} + SO_4^{2-} + SO_4\bullet^- \tag{12.12}$$

$$Fe^{2+} + S_2O_8^{2-} \rightarrow Fe^{3+} + SO_4^{2-} + SO_4\bullet^- \tag{12.13}$$

$$Fe^{2+} + SO_4\bullet^- \rightarrow Fe^{3+} + SO_4^{2-} \tag{12.14}$$

On removal efficient (Alshamsi and Thomson 2013; Wei *et al.*, 2016) revealed that the $SO_4\bullet^-$ radical can combine with a precursor substance or itself which leads to free radicals consumption. Therefore, a high dose precursor substance might have a negative effect on the removal efficiency. Equations 12.15 and 12.16 show the reactions:

$$S_2O_8^{2-} + SO_4\bullet^- \rightarrow S_2O_8\bullet^- + SO_4^{2-} \tag{12.15}$$

$$SO_4\bullet^- + SO_4\bullet^- \rightarrow S_2O_8^{2-} \tag{12.16}$$

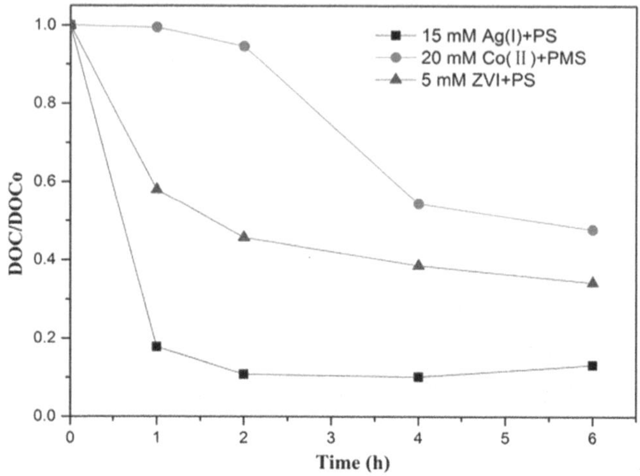

Figure 12.6 DOC degradation efficiency changes with time (Wang *et al.*, 2018).

The experiment of Wang *et al.* (2018) aimed to determine the relationship between DOC removal and reaction time. Figure 12.6 shows that DOC removal increased with reaction time. The Ag(I)/PS system had the highest removal efficiency: it could degrade DOC by more than 80% within an hour. The removal of DOC in the ZVI/PS system increased significantly in the first two hours and finally reached 70%. The Co(II)/PMS system had the lowest removal rate for DOC, it was less than 10% in the first two hours, but finally reached 50% at the end of the reaction.

12.2.3 Heterogeneous system for activating PS

The heterogeneous system for activating PS is mainly categorized into two methods. One method is to load non-metallic porous materials with transitional metals or metal oxides, such as activated carbon (Guan *et al.*, 2015), zeolite (Cerjan-Stefanovic *et al.*, 1996; Vaca *et al.*, 2001), and bentonite (Sheng *et al.*, 2014). Non-metallic materials often have a larger specific surface area, while the target pollutants and $SO_4^{\bullet-}$ radical can be adsorbed simultaneously on the surface, which can improve the efficiency of the oxidation reaction. However, one concern is that the metal-containing catalyst might lead to secondary pollution due to leakage of metal ions to the solution.

Therefore, metal-free catalytic materials such as carbon nanotubes and graphene have been developed to catalyze PS (Liang *et al.*, 2018). Figure 12.7 shows the schematic diagram of the second method of activation of PS by heterogeneous systems using biochar for the treatment of landfill leachate (Wang *et al.*, 2017). Wang *et al.* (2017) used biochar made from activated sludge. Because there are

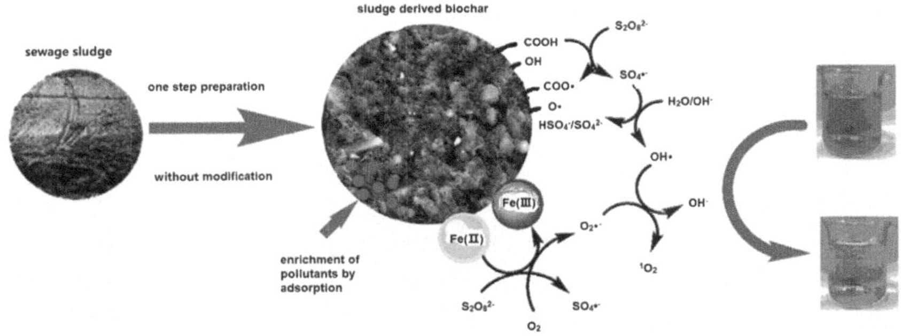

Figure 12.7 Diagram of activation of persulfate with biochar (adapted from Wang *et al.*, 2017).

many Fe ions and organic functional groups on the surface of the biochar, these effectively activated PS, producing $SO_4{\bullet}^-$ and other active substances that effectively treated landfill leachate.

12.3 MECHANISMS OF $SO_4{\bullet}^-$ OXIDATION

12.3.1 Comparison of sulfate radicals with other oxidizing substances

In advanced oxidation processes of wastewater treatment, the commonly used oxidants are $SO_4{\bullet}^-$, •OH, Cl• and O_3. Table 12.3 compares these oxidizing substances. The reaction mechanism of $SO_4{\bullet}^-$ with organic matter is that it mainly acts on unsaturated carbon-carbon bonds of unsaturated aromatic ring and olefin structures of organic pollutants by means of a single electron transfer (SET) and addition of unsaturated bonds (Wright *et al.*, 2001). This is because the SET reaction typically exhibits a lower energy barrier than radical addition. The rate constant through SET with $SO_4{\bullet}^-$ is higher. For most electron-rich aromatic components, the SET reaction from ring structure to $SO_4{\bullet}^-$ is dominant (Varanasi *et al.*, 2018). $SO_4{\bullet}^-$ radicals have a high mineralizing ability for organic pollutants without small molecules (e.g. organic acids) generated in the oxidation reaction, while they have a low reactivity with aliphatic acids and alcohols (Varanasi *et al.*, 2018).

HO• radicals react preferentially with high molecular weight protein-like compounds and soluble microbial products, and degrade these biomolecular substances at a higher rate than that for humic acid and low molecular weight components (González *et al.*, 2013; Wei *et al.*, 2013). The degradation of small molecule substances using HO• is low and is accompanied with the generation of low molecular weight compounds like

Table 12.3 Comparison of reaction properties of $SO_4^{\bullet -}$ with other common oxidants.

Oxidant	Redox Potential	Oxidized Object	Oxidation Method	Access	Reference
$SO_4^{\bullet -}$	2.5–3.1 V	Aromatic ring and olefin structures	Single electron transfer addition of unsaturated bond	Activation of PS/PMS	(Varanasi *et al.*, 2018; Wright *et al.*, 2001)
•OH	2.8 V	High molecular weight protein-like compounds	addition of unsaturated bond abstraction hydrogen reaction	Fenton	(González *et al.*, 2013; Wei *et al.*, 2013)
Cl•	1.36 V	Compounds containing N/S	Single electron transfer abstraction hydrogen reaction	UV/HClO	(Varanasi *et al.*, 2018)
O_3	2.07 V	High molecular weight hydrophobic compounds	Electrophilic	Electrolysis of oxygen	(Beltran, 2004; Gong *et al.*, 2008)

biopolymers and humic acids. The main oxidation pathways of HO• are the unsaturated C-C bonds they add and hydrogen abstraction of C-H bonds in aliphatic compounds.

Cl• can attack the O-H bonds of oxygen-containing organic compounds through SET and abstract hydrogen from C-H bonds. Cl• has a higher reaction rate with compounds containing N or S as it can attack N-containing functional groups (such as NOx and NHx) or S-containing functional groups (such as sulfides) in organic compounds through SET reactions.

O_3 preferentially reacts with high molecular weight hydrophobic compounds (humic acid, fulvic acid, and protein-like compounds), resulting in a significant reduction in their aromaticity (Audenaert *et al.*, 2013; Filloux *et al.*, 2012). The reaction order of O_3 with organic substances in solution depends on their hydrophily: hydrophobicity, transition hydrophilicity, and hydrophilicity (Gong *et al.*, 2008). This is because O_3 is an electrophilic reagent, it is more sensitive to aromatic compounds and has a higher reaction rate with aromatic compounds (Beltran, 2004). The use of O_3 for refractory wastewater treatment can improve the biodegradability of the organics presented in the water. However, the removal of DOC from the water is still relatively low (Linlin *et al.*, 2011; Rosario-Ortiz, 2008), and the removal efficiency depends on the solution pH (Tizaoui *et al.*, 2011). Table 12.3 shows the comparison of the oxidizing substances.

12.3.2 Removal mechanism of organic pollutants by $SO_4^{\bullet-}$ radicals

12.3.2.1 Removal mechanism of single organic pollutant by $SO_4^{\bullet-}$ radicals

Many researchers showed that $SO_4^{\bullet-}$ can degrade different types of refractory organic pollutants with a high removal rate (Anipsitakis & Dionysiou, 2004a, b; Huang *et al.*, 2005; Liang *et al.*, 2003). Liang *et al.* (2003) found that PS excited by heating (40–60°C) could rapidly degrade trichloroethane (TCA) with a removal efficiency of nearly 60%. Huang *et al.* (2005) investigated the degradation of 59 volatile organic compounds (VOCs) using a heating PS system and showed that each VOC was degraded at a different rate and 37 of the 59 were completely degraded. As for the photolysis method, PS or PMS combined with ultraviolet light (UV) were used to produce $SO_4^{\bullet-}$ radicals. Anipsitakis and Dionysiou (2004a, b) used UV to activate PS and PMS for 2, 4-dichlorophenol (2, 4-DCP) degradation and showed that the removal efficiencies of the UV_{254} /PS system and the UV_{254}/PMS system were 82 and 74%, respectively. Research was carried out using ultrasound as the energy activation method (Adewuyi & Owusu, 2006; Li *et al.*, 2013). These researchers used ultrasonic excited PS for degradation of trichloroethane (TCA) and trichloroethylene (TCE) and found that 50 ppm was completely degraded in 100 min. Other researchers investigated the ultrasonic/PS or PMS system for refractory organic pollutants degradation, with successful degradation of pollutants such as the drug amoxicillin (Gayathri *et al.*, 2010; Su *et al.*, 2012), textile dyes, and the solvent stabilizer 1, 4 - dioxane (Li & Zhu, 2016).

Transition metal catalysis and metal catalysis systems for degradation of different kinds of refractory organic compounds have been studied. Some researchers use the combination of energy excitation and metal catalysis to activate PMS or PS with a fairly good degradation efficiency (Hu & Long, 2015; Li *et al.*, 2016).

12.3.2.2 Change of organic compounds in complex systems

Most of the current research focuses on the degradation rate and mechanism of specific refractory organic matter in PMS or PS systems excited by different methods with few studies on actual complex waters. Deng and Ezyske (2011) used heat excited PS to treat garbage leachate, and found that it could not only effectively remove organic pollutants, but also ammonia nitrogen as well, while the Fenton system only removed organic pollutants. Joo (2014) studied the effect of nanoscale zero valent iron (nZVI) on PS activation when treating reverse osmosis concentrate (ROC), and the results showed that it could effectively remove dissolved organic carbon (DOC). These studies indicate that $SO_4^{\bullet-}$ radicals have a great potential for treatment of complex actual wastewater.

In addition, the removal effect and mechanism of organic pollutants in secondary effluent from a sewage treatment plant were investigated by Wang (2018). Figure 12.8 shows the fluorescence spectrum characteristics of the secondary effluent treated by a UV/PS system over 180 min. Figure 12.8 shows that the $SO_4{\bullet}^-$ radical quickly removes organic substances with fluorescence characteristics. When the reaction time reaches 120 min, all organic substances with fluorescence characteristics in the water are almost completely removed, then the water sample shows no further fluorescence response.

Figure 12.9 shows the results from experiments on changes of fluorescence spectra of EfOM in secondary effluent from a sewage treatment plant treated by O_3. Compared with O_3 reactions, $SO_4{\bullet}^-$ is still slower in degrading organic materials contributing to the fluorescence response which indicates that the $SO_4{\bullet}^-$ radical is less sensitive to fluorescence than O_3. This is also related to the characteristics of reactions with different technologies. Gas agitation and mixing induced by the O_3 bubbles can make better and faster contact with dissolved pollutants and degrade them. In the $SO_4{\bullet}^-$ radical, treatment first needs metal catalysis or energy to break the precursor material for producing $SO_4{\bullet}^-$ radicals. Therefore, the $SO_4{\bullet}^-$ radical technology changes the fluorescence characteristics at a slightly slower rate.

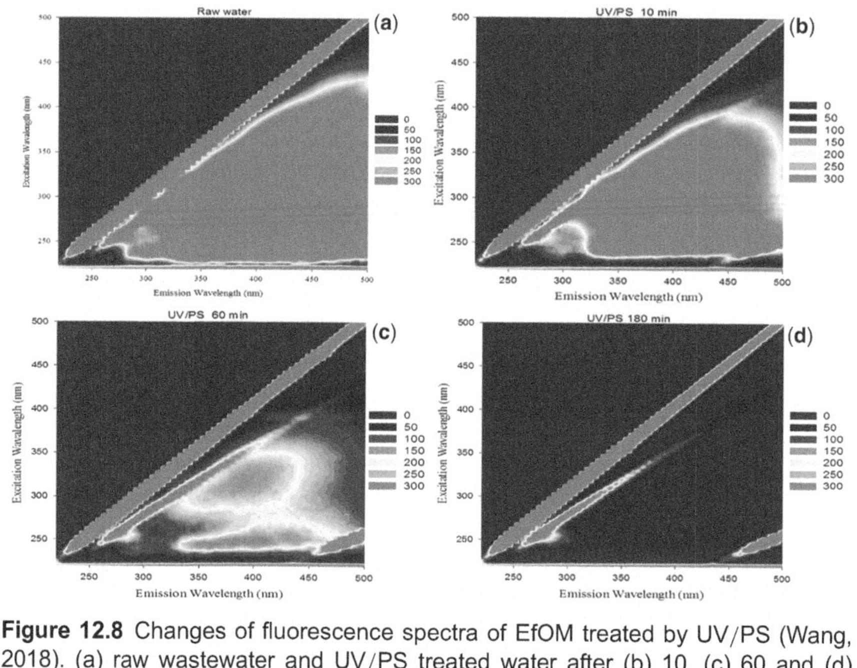

Figure 12.8 Changes of fluorescence spectra of EfOM treated by UV/PS (Wang, 2018). (a) raw wastewater and UV/PS treated water after (b) 10, (c) 60 and (d) 180 min.

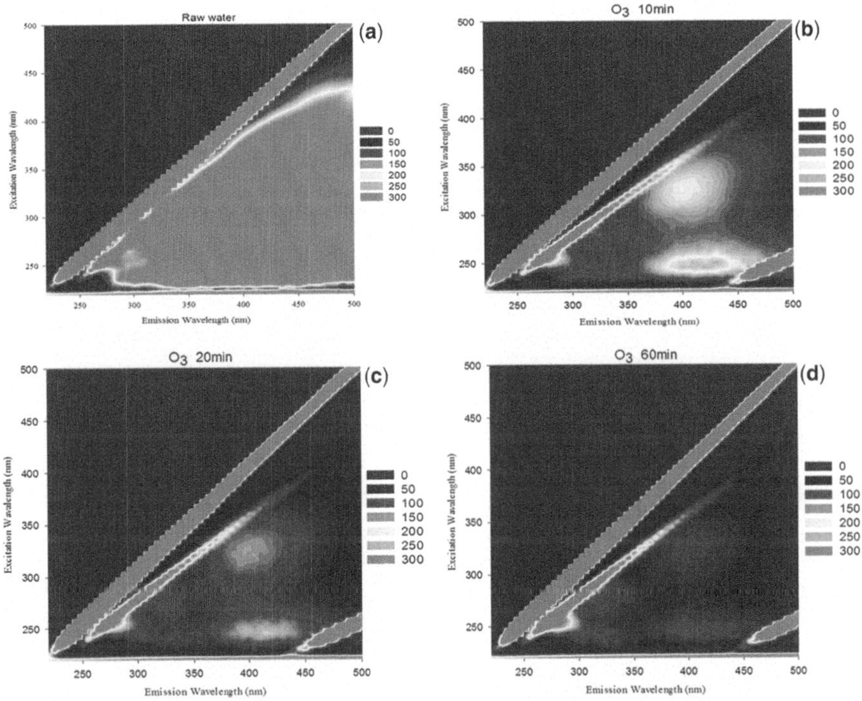

Figure 12.9 Changes of fluorescence spectra of EfOM treated by O_3 (Wang, 2018). (a) raw wastewater and O_3 treated water after (b) 10, (c) 20 and (d) 60 min.

The removal trend of organic matter in the whole emission wavelength by $SO_4\bullet^-$ radicals was basically the same, indicating that the $SO_4\bullet^-$ radical has a higher overall selectivity for the degradation of organic pollutants. Figure 12.10 shows the variation characteristics of the molecular weight distribution of secondary effluent treated by UV/PS technology. Regarding the change of substance composition during the treatment process with $SO_4\bullet^-$ radicals, Figure 12.10 shows that the content of large molecular substances decreased with treatment time, while the content of small molecular substances increased accordingly, and new lower molecular substances appeared in the system. This indicates that new degradation by-products were generated in the $SO_4\bullet^-$ radical oxidation process.

12.3.3 Effect of coexisting inorganic ions

The existence of some inorganic ions like Cl^-, CO_3^{2-} and HCO_3^- in the solution can react with $SO_4\bullet^-$ produced, the by-products of other radicals such as $Cl\bullet$ ($Cl_2\bullet^-$), $CO_3\bullet^-$ and $HCO_3\bullet$ thus affect the removal rate and efficiency of target pollutants in the system (Grebel *et al.*, 2010; Li *et al.*, 2017; Yang *et al.*, 2014). The coexisting halogen and carbonate ions react with the $SO_4\bullet^-$ radicals

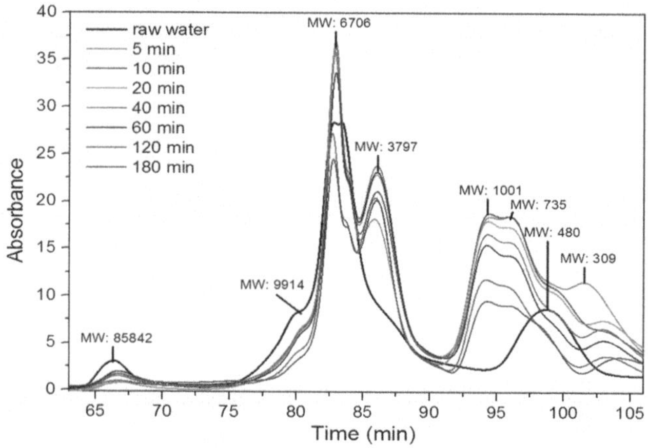

Figure 12.10 Molecular weight distribution of EfOM under UV/PS treatment for different exposure times (Wang, 2018).

produced in the system, as shown in the following equations:

$$Cl^- + SO_4\bullet^- \rightarrow Cl^\bullet + SO_4^{2-} \quad (12.17)$$

$$Cl^- + Cl^\bullet \rightarrow Cl_2\bullet^- \quad (12.18)$$

$$CO_3^{2-} + SO_4\bullet^- \rightarrow CO_3\bullet^- + SO_4^{2-} \quad (12.19)$$

$$HCO_3^- + SO_4\bullet^- \rightarrow HCO_3^\bullet + SO_4^{2-} \quad (12.20)$$

which will change the removal efficiency of target pollutants in the system (Tan *et al.*, 2013; Kong *et al.*, 2016)

12.3.4 Synergetic removal of multiple pollutants

As in the actual sewage, refractory organic matter exists together with the common pollutants such as phosphorus and ammonia nitrogen. Traditional treatment is usually applied to remove nitrogen and phosphorus from domestic sewage, however, under some circumstances, it is difficult to meet the discharge requirements (Guaya *et al.*, 2015). Therefore, advanced treatment is required to remove multiple pollutants simultaneously. Jia (2020) showed that the heterogeneous catalyst system based on $SO_4\bullet^-$ can simultaneously remove organic pollutants, phosphorus, and ammonia nitrogen from wastewater.

To investigate the synergetic removal of multiple pollutants from actual secondary effluent, a system was designed to activate activated carbon loaded ZVI to treat sewage collected from WWTPs in Beijing (China). The parameters of PS dosage, catalytic materials and pH value were investigated to reveal their impact on co-removal of multiple pollutants (Figure 12.11). With the increase in PS concentration (Figure 12.11(a)), the removal efficiency of total organic carbon

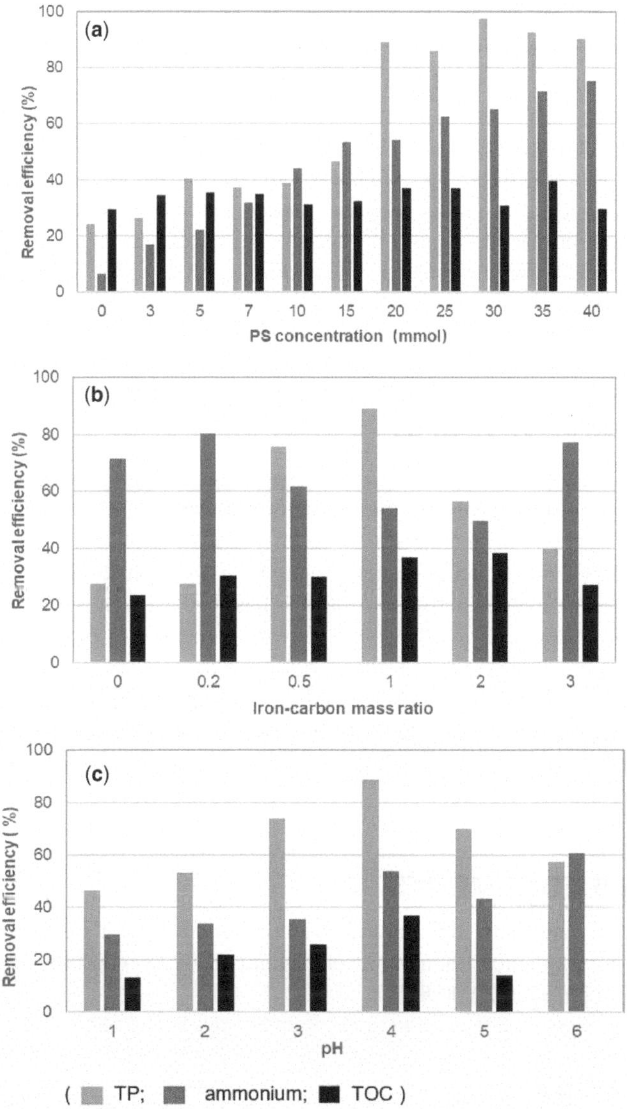

Figure 12.11 Influence of (a) PS concentration, (b) iron-carbon mass ratio and (c) pH on pollutant removal effect (Jia, 2020).

(TOC), total phosphorus (TP) and ammonium increased. The total removal of the three pollutants was the highest when the mass ratio of Fe/C was 1. When the Fe/C ratio was higher, excessive iron ions could eliminate $SO_4^{\bullet -}$ in the system, resulting in interference of the removal efficiency. When the Fe/C ratio was lower, the activation ability of PS was not sufficient, resulting in the insufficient yield of $SO_4^{\bullet -}$ radicals. Therefore, the amount of precursor oxidant and metal ions should be appropriate for a better co-removal of multiple pollutants.

The pH value showed its impact on the removal of the pollutants because the pH affects the form of pollutants, the type of free radicals, and the adsorption effect of non-metallic catalytic materials. The pH value affects the NH_4^+/NH_3 ratio in the solution, the ratio of $SO_4^{\bullet -}$ and •OH, as well as the adsorption effect of activated carbon on ammonia nitrogen, organic matter and phosphorus. In conclusion, simultaneous removal of organic pollutants, ammonia nitrogen, and phosphorus can be achieved by optimizing the parameters in the heterogeneous PS activation system.

12.4 RESEARCH PROSPECT OF SULFUR RADICALS IN ADVANCED OXIDATION

At present, the research on sulfur radicals in an advanced oxidation process mainly concerns $SO_4^{\bullet -}$, specifically on the production means of $SO_4^{\bullet -}$ radicals and the removal efficiency and mechanism of specific organic pollutants. The activation methods widely investigated are energy activation and transition metal activation. Recently, heterogeneous activation of PS has been given more attention to overcome the disadvantages of energy activation and transition metal activation. The heterogeneous activation of PS is believed to be a likely future prospect. There are few studies on the removal of pollutants from actual waters using $SO_4^{\bullet -}$ and application of the ISCO technology. Research related to engineering will lead to a better understanding of the mechanism and extended use of the technology.

ACKNOWLEDGEMENTS

This work was funded by the Natural Science Foundation of Beijing (Z170004), National Natural Science Foundation of China (41672227) and the National Key R&D Project (2018YFC1800905).

REFERENCES

Abedinzadeh Z. (2001). Sulfur-centered reactive intermediates derived from the oxidation of sulfur compounds of biological interest. *Canadian Journal of Physiology and Pharmacology*, **79**(2), 166–170.

Adewuyi Y. G. and Owusu S. O. (2006). Ultrasound-induced aqueous removal of nitric oxide from flue gases: effects of sulfur dioxide, chloride, and chemical oxidant. *Journal of Physical Chemistry A*, **110**(38), 11098–11107.

Ahmed S., Rasul M. G., Martens W. N., Brown R. and Hashib M. A. (2010). Heterogeneous photocatalytic degradation of phenols in wastewater: a review on current status and developments. *Desalination*, **261**(1–2), 3–18.

Alshamsi M. A. and Thomson N. R. (2013). Treatment of organic compounds by activated Persulfate using Nanoscale Zerovalent iron. *Industrial and Engineering Chemistry Research*, **52**(38), 13564–13571.

Anipsitakis G. P. and Dionysiou D. D. (2004a). Radicals generation by the interaction of transition metals with common oxidants. *Environmental Science and Technology*, **38**(13), 3705–3712.

Anipsitakis G. P. and Dionysiou D. D. (2004b). Transition metal/UV-based advanced oxidation technologies for water decontamination. *Applied Catalysis B: Environmental*, **54**(3), 155–163.

Antoniou M.G., de la Cruz A.A. and Dionysiou D.D. (2010). Degradation of microcystin-LR using sulfate radicals generated through photolysis, thermolysis and e−transfer mechanisms. *Applied Catalysis B: Environmental*, **96**(3–4), 290–298.

Audenaert W. T. M., Vandierendonck D., Van Hulle S. W. H. and Nopens I. (2013). Comparison of ozone and HO · induced conversion of effluent organic matter (EfOM) using ozonation and UV/H_2O_2 treatment. *Water Research*, **47**(7), 2387–2398.

Beltran F. J. (2004). Ozone Reaction Kinetics for Water and Wastewater Systems. CRC Press, Boca Raton, America, p. 384.

Bolton J. R., Bircher K. G., Tumas W. and Tolman C. A. (2001). Figures-of-merit for the technical development and application of advanced oxidation processes. *Journal of Advanced Oxidation Technologies*, **1**(1), 13–17.

Cerjan-Stefanovic S., Curkovic L. and Tugomir F. (1996). Metal ion exchange by natural zeolites. *Croatica Chemica Acta*, **6**(31), 1379–1382.

Deng Y. and Ezyske C. M. (2011). Sulfate radicals-advanced oxidation process (SR-AOP) for simultaneous removal of refractory organic contaminants and ammonia in landfill leachate. *Water Research*, **45**(18), 6189–94.

Filloux E., Gallard H. and Croue J. P. (2012). Identification of effluent organic matter fractions responsible for low-pressure membrane fouling. *Water Research*, **46**(17), 5531–40

Gayathri P., Dorathi R. P. J. and Palanivelu K. (2010). Sonochemical degradation of textile dyes in aqueous solution using sulphate radicals activated by immobilized cobalt ions. *Ultrasonics Sonochemistry*, **17**(3), 566–571.

Ghanbari F. and Moradi M. (2017). Application of peroxymonosulfate and its activation methods for degradation of environmental organic pollutants: Review. *Chemical Engineering Journal*, **310**, 41–62.

Ghauch A. and Tuqan A. M. (2012). Oxidation of bisoprolol in heated persulfate/H_2O_2 systems: Kinetics and products. *Chemical Engineering Journal*, **183**, 162–171.

Glass R. S. (1999). Sulfur radical cations. *Topics in Current Chemistry*, **205**, 1–87.

Gong J., Liu Y. and Sun X. (2008). O_3 and UV/O_3 oxidation of organic constituents of biotreated municipal wastewater. *Water Research*, **42**(4–5), 1238–1244.

Gonzalez O., Justo A., Bacardit J., Ferrero E., Malfeito J. J. and Sans C. (2013). Characterization and fate of effluent organic matter treated with UV/H_2O_2 and ozonation. *Chemical Engineering Journal*, **226**(24),402–408.

Goslan E. H., Gurses F., Banks J. and Parsons S. A. (2006). An investigation into reservoir NOM reduction by UV photolysis and advanced oxidation processes. *Chemosphere*, **65** (7), 1113–9.

Grebel J. E., Pignatello J. J. and Mitch W. A. (2010). Effect of halide ions and carbonates on organic contaminant degradation by hydroxyl radical-based advanced oxidation processes in saline waters. *Environmental Science and Technology*, **44**(17), 6822–6828.

Guan X., Sun Y., Qin H., Li J., Lo I., He D. and Dong H. (2015). The limitations of applying zero-valent iron technology in contaminants sequestration and the corresponding countermeasures: The development in zero-valent iron technology in the last two decades (1994–2014). *Water Research*, **75**, 224–248.

Guaya D., Valderrama C., Farran A., Armijos C. and Cortina J. L. (2015). Simultaneous phosphate and ammonium removal from aqueous solution by a hydrated aluminum oxide modified natural zeolite. *Chemical Engineering Journal*, **271**, 204–213.

Hu P. and Long M. (2015). Cobalt-catalyzed sulfate radical-based advanced oxidation: a review on heterogeneous catalysts and applications. *Applied Catalysis B: Environmental*, **181**, 103–117.

Huang K. C., Zhao Z., Hoag G. E., Dahmani A. and Block P. A. (2005). Degradation of volatile organic compounds with thermally activated persulfate oxidation. *Chemosphere*, **61**(4), 0–560.

Hui Z., Hao F. and Zhang D. (2009). Degradation of C.I. Acid Orange 7 by ultrasound enhanced heterogeneous Fenton-like process. *Journal of Hazardous Materials*, **172** (2), 654–660.

Ji Y., Dong C., Kong D., Lu J. and Zhou Q. (2015). Heat-activated persulfate oxidation of atrazine: Implications for remediation of groundwater contaminated by herbicides. *Chemical Engineering Journal*, **263**, 45–54.

Jia W. (2020). Study on Removal of Secondary Effluent Pollutants by Activated Persulfate with Activated Carbon Supported by Iron. M.A.Thesis, Beijing Normal University, Beijing.

Joo S. H. (2014). Advanced treatment of reverse osmosis concentrate by integrated activated carbon and iron-activated persulfate oxidation. *Water, Air, and Soil Pollution*, **225**(8), 2076, DOI: 10.1007/s11270–014-2076-9.

Kong X., Jiang J., Ma J., Yang Y., Liu W. and Liu Y. (2016). Degradation of atrazine by UV/chlorine: Efficiency, influencing factors, and products. *Water Research*, **90**, 15–23.

Li B. and Zhu J. (2016). Simultaneous degradation of 1,1,1–trichloroethane and solvent stabilizer 1,4-dioxane by a sono-activated persulfate process. *Chemical Engineering Journal*, **284**, 750–763.

Li B., Li L., Lin K., Zhang W., Lu S. and Luo Q. (2013). Removal of 1,1,1–trichloroethane from aqueous solution by a sono-activated persulfate process. *Ultrasonics Sonochemistry*, **20**(3), 855–863.

Li Z., Yang Q., Zhong Y., Li X., Zhou L., Li X. and Zeng G. (2015). Granular activated carbon supported iron as heterogeneous persulfate catalyst for the pretreatment of mature landfill leachate. *RSC Advances*, **6**, 987–994.

Li Y., Liu L., Liu L., Liu Y., Zhang H. and Han H. (2016). Efficient Oxidation of Phenol by Persulfate Using Manganite as a Catalyst. *Journal of Molecular Catalysis A: Chemical*, **411**, 264–271.

Li W., Orozco R., Camargos N. and Liu H. (2017). Mechanisms on the impacts of alkalinity, pH, and chloride on persulfate-based groundwater remediation. *Environmental Science and Technology*, **51**(7), 3948–3959.

Liang C. and Bruell C. J. (2008). Thermally activated persulfate oxidation of trichloroethylene: experimental investigation of reaction orders. *Industrial and Engineering Chemistry Research*, **47**(9), 2912–2918.

Liang C. J., Bruell C. J., Marley M. C. and Sperry K. L. (2003). Thermally activated persulfate oxidation of trichloroethylene (TCE) and 1,1,1–trichloroethane (TCA) in aqueous systems and soil slurries. *Soil and Sediment Contamination*, **12**(2), 207–228.

Liang C., Bruell C. J., Marley M. C. and Sperry K. L. (2004). Persulfate oxidation for in situ remediation of TCE. II. Activated by chelated ferrous ion. *Chemosphere*, **55**(9), 1225–1233.

Liang L., Zhou M., Yang W. and Jiang L. (2018). Enhanced activation of persulfate by carbohydrate-derived carbon cryogels for effective removal of organic pollutants. *Chemical Engineering Journal*, **352**, 673–681.

Linlin W., Xuan Z. and Meng Z. (2011). Removal of dissolved organic matter in municipal effluent with ozonation, slow sand filtration and nanofiltration as high quality pre-treatment option for artificial groundwater recharge. *Chemosphere*, **83**(5), 693–9.

Liu H., Bruton T., Doyle F. and Sedlak D. (2014). In Situ Chemical Oxidation of Contaminated Groundwater by Persulfate: Decom position by Fe(III)- and Mn (IV)-Containing Oxides and Aquifer Materials. *Environmental Science and Technology*, **48**(17), 10330–10336.

Matta R., Tlili S., Chiron S. and Barbati Stephane. (2011). Removal of carbamazepine from urban wastewater by sulfate radical oxidation. *Environmental Chemistry Letters*, **9**(3), 347–353.

Michael-Kordatou I., Michael C., Duan X., He X., Dionysiou D. D., Mills M. A. and Fatta-Kassinos D. (2015). Dissolved effluent organic matter: characteristics and potential implications in wastewater treatment and reuse applications. *Water Research*, **77**, 213–248.

Moortgat G. K., Barnes A. J., Bras G. L. and Sodeau J. R. (1994). Low-Temperature Chemistry of the Atmosphere. Springer, Berlin, Heidelberg.

Rachel H. W., Paul G. T., Richard L. J. and James T. N. (2007). Oxidation of Chlorinate Ethenes by heat-activated persulfate Kinetics and products. *Environmental Science and Technology*, **41**, 1010–1015.

Rastogi A., Al-Abed S. R. and Dionysiou D. D. (2009). Effect of inorganic, synthetic and naturally occurring chelating agents on Fe(II) mediated advanced oxidation of chlorophenols. *Water Research*, **43**(3), 684–694.

Rosario-Ortiz F. L. (2008). Quantitative Correlation of Absolute Hydroxyl Radicals Rate Constants with Non-Isolated Effluent Organic Matter Bulk Properties in Water. *Environmental Science and Technology*, **42**(16), 5924–30.

Sheng G., Shao X., Li Y., Li J., Dong H., Cheng W., Gao X. and Huang Y. (2014). Enhanced removal of uranium(VI) by nanoscale zerovalent iron supported on Na–bentonite and an investigation of mechanism. *The Journal of Physical Chemistry A*, **118**(16), 2952–2958.

Su S., Guo W., Yi C., Leng Y. and Ma Z. (2012). Degradation of amoxicillin in aqueous solution using sulphate radicals under ultrasound irradiation. *Ultrasonics Sonochemistry*, **19**(3): 469–74.

Tan C., Gao N., Deng Y., Zhang Y., Sui M., Deng J. and Zhou S. (2013). Degradation of antipyrine by UV, UV/H_2O_2 and UV/PS. *Journal of Hazardous Materials*, **260**, 1008–1016.

Tizaoui C., Grima N. and Hilal N. (2011). Degradation of the antimicrobial triclocarban (TCC) with ozone. *Chemical Engineering and Processing Process Intensification*, **50** (7), 637–643.

Tsao M. S. and Wilmarth W. K. (2002). The aqueous chemistry of inorganic free radicals. i. the mechanism of the photolytic decomposition of aqueous persulfate ion and evidence regarding the sulfate hydroxyl radical interconversion equilibrium. *Journal of Physical Chemistry*, **63**(3), 346–353.

Vaca M. M., Lopez Callejas R., Gehr R., Jimenez Cisneros B. E. and Alvarez P. J. (2001). Heavy metal removal with mexican clinoptilolite: multi-component ionic exchange. *Water Research*, **35**(2), 373–378.

Varanasi L., Coscarelli E., Khaksari M., Mazzoleni L. R. and Minakata D. (2018). Transformations of dissolved organic matter induced by uv photolysis, hydroxyl radicals, chlorine radicals, and sulfate radicals in aqueous-phase UV-based advanced oxidation processes. *Water Research*, **135**, 22–30.

Waclawek S., Lutze H. V., Grubel K., Padil V. V. T. and Dionysiou D. D. (2017). Chemistry of persulfates in water and wastewater treatment: a review. *Chemical Engineering Journal*, **330**, 44–62.

Waldemer R. H., Tratnyek P. G., Johnson R. L. and Nurmi J. T. (2007). Oxidation of chlorinated ethenes by heat-activated persulfate: kinetics and products. *Environmental Science and Technology*, **41**(3), 1010–1015.

Wang D. (2018). The study of based on sulfate radical on removal of organic matter from secondary effluents and ammonia. M.A.Thesis, Beijing Normal University, Beijing.

Wang S., Zhang X., Wang Z. W., Li X. and Ma J. (2014). In-depth characterization of secondary effluent from a municipal wastewater treatment plant located in northern china for advanced treatment. *Water Science & Technology*, **69**(7), 1482.

Wang Y., Sun H., Ang H. M., Tade Moses O. and Wang S. (2015). The 3d-hierarchically structured MnO_2 for catalytic oxidation of phenol solutions by activation of peroxymono sulfate: structure dependence and mechanism. *Applied Catalysis B: Environmental*, **164**, 159–167.

Wang J., Liao Z., Ifthikar J., Shi L. and Chen Z. (2017). Treatment of refractory contaminants by sludge-derived biochar/persulfate system via both adsorption and advanced oxidation process. *Chemosphere*, **185**, 754–763.

Wang D., Cheng L., Wang M., Zhang X., Xue D., Zhuo W., Zheng L. and Ding A. (2018). The performance of a sulfate-radical mediated advanced oxidation process in the degradation of organic matter from secondary effluents. *Environmental science: Water Research & Technology*, **4**, 773–782.

Wei L., Nanaboina V., Zhou Q. and Korshin G. V. (2013). Changes of excitation/emission matrixes of wastewater caused by Fenton- and Fenton-like treatment and their associations with the generation of hydroxyl radicals, oxidation of effluent organic matter and degradation of trace-level organic pollutants. *Journal of Hazardous Materials*, **244–245**, 698–708.

Wei X., Gao N., Li C., Deng Y., Zhou S. and Li L. (2016). Zero-valent iron (ZVI) activation of persulfate (PS) for oxidation of bentazon in water. *Chemical Engineering Journal*, **285**, 660–670.

Wright J. S., Johnson E. R. and Dilabio G. A. (2001). Predicting the activity of phenolic antioxidants: theoretical method, analysis of substituent effects, and application to major families of antioxidants. *Journal of the American Chemical Society*, **123**(6), 1173–1183.

Xiong X., Sun B., Zhang J., Gao N., Shen J., Li J. and Guan X. (2014). Activating persulfate by Fe^0 coupling with weak magnetic field: performance and mechanism. *Water Research*, **62**, 53–62.

Yan J., Lei M., Zhu L., Anjum M. N., Zou J. and Tang H. (2011). Degradation of sulfamonomethoxine with Fe_3O_4 magnetic nanoparticles as heterogeneous activator of persulfate. *Journal of Hazardous Materials*, **186**(2–3), 1398–1404.

Yang S., Yang X., Shao X., Niu R. and Wang L. (2011). Activated carbon catalyzed persulfate oxidation of azo dye acid orange 7 at ambient temperature. *Journal of Hazardous Materials*, **186**(1), 659–666.

Yang Y., Pignatello J. J., Ma J. and Mitch W. A. (2014). Comparison of halide impacts on the efficiency of contaminant degradation by sulfate and hydroxyl radical-based advanced oxidation processes (AOPs). *Environmental Science and Technology*, **48**(4), 2344–2351.

Ye T., Wei Z., Spinney R., Tang C. J., Luo S., Xiao R. and Dionysiou D. D. (2017). Chemical structure-based predictive model for the oxidation of trace organic contaminants by sulfate radical. *Water Research*, **116**, 106–115.

Zhang B., Zhang Y., Teng Y. and Fan M. (2015). Sulfate radicals and its application in decontamination technologies. *Crit Rev Environ Sci Technol*, **45**, 1756–1800.

Chapter 13

Interactions of the sulfur and nitrogen cycles: microbiology and process technology

Philippe M. Chazal, Maria F. Carboni and Piet N. L. Lens

13.1 INTRODUCTION

Microorganisms play a key role in the biogeochemical cycling of numerous elements. This is particularly the case for the elements of living biomass such as carbon, hydrogen, oxygen, nitrogen, sulfur and phosphorus. When studying these cycles, the interactions between the different elements are of utmost importance. This chapter describes the interactions between the sulfur and nitrogen cycles, with particular attention paid to processes allowing nitrogen removal from polluted environments using sulfur cycle conversions.

13.1.1 The nitrogen cycle

The nitrogen cycle largely depends on microbial activities (Figure 13.1). Gaseous N_2 from the air is reduced in soil or water to ammonium by bacterial nitrogen fixation – step (1) in Figure 13.1. These bacteria possess the *Nif* genes that code for a specialized enzyme, nitrogenase, which catalyzes this N_2 reduction under well-defined conditions. In soil, it is present in free-living microorganisms such as *Azotobacter sp.* and *Klebsiella sp.* or in symbiotic bacteria such as *Rhizobium sp.* growing in association with plants. Nitrogen fixation also occurs in aquatic habitats by Cyanobacteria. Ammonium thus formed is incorporated into organic

DOI: 10.2166/9781789060966_0403

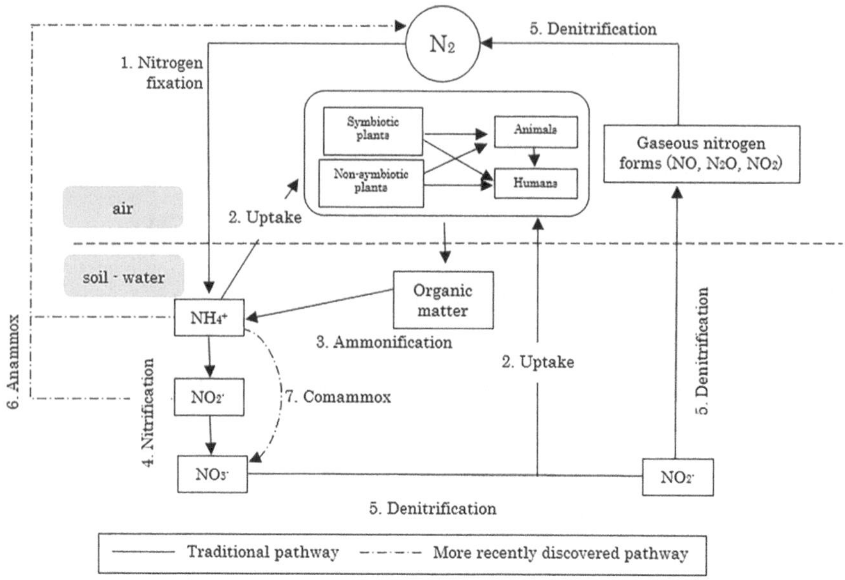

Figure 13.1 Schematic representation of the nitrogen cycle. See text for details on the different steps.

matter (e.g. proteins and nucleic acids) by plants, animals and humans – step (2). Organic matter released by these eukaryotes in the environment is degraded by other bacteria such as members of the Myxobacteriales. During the decay, ammonium is released via ammonification – step (3) in Figure 13.1.

Ammonium is then oxidized by nitrifying bacteria in the nitrification phase – step (4). In a first step, *Nitrosomonas sp.* and related genera oxidize ammonia to nitrous acid (HNO_2). In a second step, *Nitrobacter sp.* and related genera oxidize nitrites to nitrates. Nitrates are the most common form of nitrogenous compounds used by plants. When present in excess to the plants' requirements, nitrates can be reduced to gaseous forms of nitrogen (mainly N_2, but also NO and N_2O) by several denitrifying bacteria such as *Pseudomonas sp.*, *Bacillus sp.*, *Thiobacillus sp.* and *Spirillum sp.* – step (5) in Figure 13.1.

The nitrogen cycle (Figure 13.1) is a synoptic overview and has been extensively and periodically reviewed during previous years. Recent data suggest that this cycle is much more complex (Stein & Klotz, 2016; Hu & He, 2017) because of newly isolated bacterial species, numerous microbial interactions in microbial communities and interferences between (well-known as well as newly discovered) metabolic pathways.

For instance, methanotrophs (Costa *et al.*, 2019) and even thermophilic bacteria (Pel *et al.*, 1997) are also involved in nitrification or denitrification. Nitrifiers can communicate cell to cell because of the release of pheromone-like molecules such as homoserine derivatives (Mellbye *et al.*, 2017). This would allow nitrifying bacteria to form biofilms that could help them to survive better when the environmental conditions are unfavorable, for instance in the case of substrate starvation (Mellbye *et al.*, 2017). Nitrites can be reduced to dinitrogen gas in the presence of ammonia via hydroxylamine and hydrazine in the so-called anaerobic ammonium oxidation (Anammox – step (6) in Figure 13.1) process (Kartal *et al.*, 2011; Ma *et al.*, 2016; Pereira *et al.*, 2017). Wan *et al.* (2016) reported that ammonia can be oxidized to hydroxylamine under the catalysis of ammonia monooxygenase and is then oxidized to nitrite catalyzed by hydroxylamine oxidoreductase. The oxidation of NH_2OH provides the energy and electrons for the oxidation of ammonia.

A complete review of these new aspects of the nitrogen cycle has been presented by Kuypers *et al.* (2018). The most recent discovery of the cycle concerns the nitrification step that conventionally is divided into two stages: from ammonia to nitrite and from nitrite to nitrate (Stein & Klotz, 2016). So far, these have always been distinguished as separate groups: ammonia oxidizing bacteria/archaea and nitrite oxidizing bacteria, respectively. However, Van Kessel *et al.* (2015) and Daims *et al.* (2015) showed that complete oxidation of ammonia to nitrate by one organism (complete ammonia oxidation; comammox – step 7) is feasible.

13.1.2 Anthropic alteration of the nitrogen cycle and its environmental consequences

Nitrogen compounds which are biologically available enter into the terrestrial biosphere at elevated rates compared to the preindustrial level, due to human activities such as use of fertilizer and fossil fuel consumption (Galloway *et al.*, 2004). Significant amounts of mobilized N enter the water cycle. Groundwater and surface water are transported through freshwater river systems to coastal marine systems, causing a worldwide environmental issue: eutrophication (Posadas *et al.*, 2013). Additionally, nitrates can be reduced to nitrites in the human gastrointestinal tract. Nitrites can thus combine with amino acids to release nitrosamines, which are carcinogenic (Della Rocca *et al.*, 2006). Nitrites also combine with hemoglobin in blood to form methemoglobin, which is subsequently no longer able to carry oxygen through the body (Aslan & Cakici, 2007). This can lead to serious forms of anemia, which can even be lethal. Therefore, the 2000 European Union Directive on drinking water parameters reported that nitrate concentrations must not exceed 50 mg L^{-1}, whereas for nitrite and ammonium values are 0.5 and 0.3 mg L^{-1}, respectively.

Northern European countries such as Belgium, The Netherlands, Germany and Denmark are severely affected by nitrate contamination of surface and groundwater. This is also the case for numerous French areas where animal culture and breeding are carried out at intensive levels, such as Bretagne,

Nord-Pas-de-Calais, Beauce and Brie. The magnitude of this problem has been gradually increasing over the past decades.

13.1.3 Nitrate pollution abatement techniques

Preventive procedures have been developed to limit nitrate contamination in the environment. Nevertheless, this prevention policy is restricted to nitrates that are released nowadays. One must take into account that the nitrates polluting surface or groundwaters now were released in soils 15–20 years ago (Mariotti, 1992). This is why the development of remedial procedures to treat pollution is also necessary.

Denitrification is the only nitrogen transformation that removes nitrogen from ecosystems, essentially irreversibly, and it roughly balances the amount of nitrogen fixed by microorganisms. It converts nitrate to dinitrogen in the following sequence:

$$NO_3^- \rightarrow NO_2^- \rightarrow NO \rightarrow N_2O \rightarrow N_2 \tag{13.1}$$

It is a four sequential step process catalyzed by specific enzymes: nitrate reductase, nitrite reductase, nitric oxide reductase and nitrous oxide reductase (Wang & Chu, 2016).

Both physico-chemical and biological procedures have been developed for nitrate removal. Physico-chemical procedures (denitritation) are mainly based on ion exchange resins (Sibony, 1979), osmosis and electrodialysis (Richard & Leprince 1982; Philipot & Sibony 1986). These procedures are very effective for nitrate removal and produce drinking water that meets legal standards. However, they also present disadvantages. They induce high production costs and all methods accumulate nitrates, e.g. saline eluates of ion exchange resins. These concentrated waste brines require further treatment or disposal (Zhou *et al.*, 2009).

Biological procedures are nowadays regularly used for nitrate removal even at industrial scale. They are mostly based on the use of denitrifying heterotrophic bacteria. Denitrifiers are generally (facultative) aerobes, that can use alternative electron acceptors such as nitrate or nitrite in the absence of oxygen. In anaerobiosis, but in the presence of nitrates, oxidative metabolism can be achieved (Figure 13.2). The electrons coming from the oxidation of the organic substrate are transferred to nitrate which is thus reduced to N_2 (Figure 13.2) via different intermediate forms as shown in Reaction 13.1.

This process is carried out by heterotrophic bacteria and requires the presence of organic carbon (Ashok & Hait, 2015). The latter could already be present in the wastewater or may need to be added because it is absent or present in low concentrations, e.g. in groundwater, leachate of old landfills or industrial effluents. Therefore, carbon sources like methanol, ethanol, acetic acid, glycerol or sugar are commonly added to the heterotrophic denitrification process

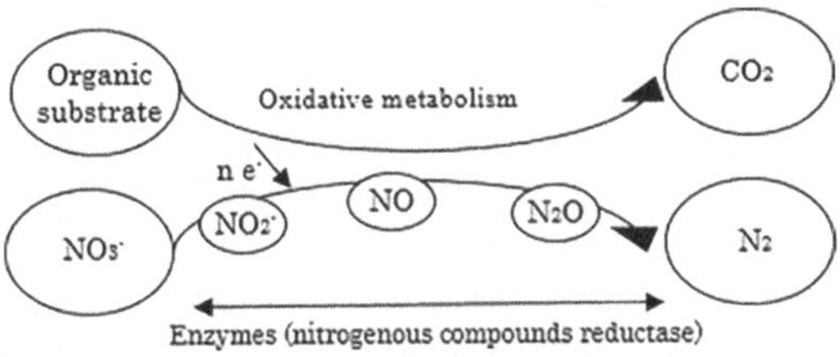

Figure 13.2 Scheme of nitrate reduction in anaerobiosis by heterotrophs coupled to the oxidation of organic substrate.

(Mohseni-Bandpi *et al.*, 2013). The complete degradation of those compounds releases carbon dioxide.

For drinking water production, several reactor configurations have been developed to perform complete denitrification, e.g. Nitrazur (Richard & Partos, 1986), Biodenit (Philipot & Patte, 1982), Denipor (Roennefhart, 1986) and Nitraflux (Guerber *et al.*, 1998). The crude water is recirculated in a reactor containing solid material (Biolite, Biodagene, sand or pouzzolane). Organic carbon sources are added at the appropriate concentration to provide enough carbon for bacterial growth. During start-up, the denitrifying bacteria colonize the filter. When enough biomass has colonized the support material (10^{11}–10^{12} g biomass per g of material), the reactor is fed with the nitrate-rich water in a continuous way. In the first part of the filter, dissolved oxygen is consumed without nitrate removal. When oxygen is depleted, nitrate reduction takes place in the anaerobic part of the filter. Optimized heterotrophic denitrification plants can achieve groundwater denitrification at volumetric denitrification rates of 1–1.5 kg NO_3–N m^{-3} d^{-1} (Richard & Partos, 1986; Ravarini *et al.*, 1988; Guerber *et al.*, 1998).

The addition of organic carbon is a cost that needs to be added to the overall process costs. Its complete mineralization does not always happen, potentially resulting in secondary organic pollution at the end of the treatment. Moreover, some wastewaters have a final disinfection step, mostly achieved with chlorination. The combination of chlorine and organics can lead to the formation of trihalomethanes (THMs) and haloacetic acids (HAAs) that are extremely toxic compounds (Quintiliani *et al.*, 2018; Di Capua *et al.*, 2019). Therefore, denitrification carried out by autotrophic bacteria that do not need the presence of organic carbon has been widely studied in recent decades.

13.2 MICROBIAL ASPECTS OF AUTOTROPHIC DENITRIFICATION PROCESSES

13.2.1 Sulfur-oxidizing bacteria

The sulfur and nitrogen cycles interact via sulfur-oxidizing bacteria. These interactions are encountered during denitrification of wastewaters containing sulfurous compounds, drinking water production as well as during natural denitrification in aquifers or in sediments located in well-defined geographical areas. Sulfur-oxidizing bacteria belong to a very heterogeneous bacterial group of which the members share the ability to use reduced sulfur compounds as an energy source (Robertson & Kuenen, 1992). This group was named in 1890 by Winogradsky the "colorless sulfur bacteria", because these bacteria do not contain pigments. They are able to oxidize reduced sulfur forms (Equations 13.2 to 13.5):

$$S^{2-} \rightarrow S^{\circ} + 2e^- \tag{13.2}$$

$$S^{\circ} + 4\,H_2O \rightarrow SO_4^{\ 2-} + 8\,H^+ + 6\,e^- \tag{13.3}$$

$$S_2O_3^{\ 2-} + 5\,H_2O \rightarrow 2\,SO_4^{\ 2-} + 10\,H^+ + 8\,e^- \tag{13.4}$$

$$S_4O_6^{\ 2-} + 10\,H_2O \rightarrow 4\,SO_4^{\ 2-} + 20\,H^+ + 14\,e^- \tag{13.5}$$

These oxidative reactions are coupled to the reduction of O_2 or NO_3^- (Equations 13.6 and 13.7):

$$O_2 + 4\,e^- + 4\,H^+ \rightarrow 2\,H_2O \tag{13.6}$$

$$2\,NO_3^- + 12\,H^+ + 10\,e^- \rightarrow 3\,N_2 + 6\,H_2O \tag{13.7}$$

Equation 13.7 integrates the reduction of nitrates to nitrites and the reduction of nitrites to N_2 via NO and N_2O.

When both oxygen and nitrates are present, the reduction of oxygen occurs preferentially because of the more favorable Gibb's free energy (ΔG^0) (McKenney *et al.*, 2001). In anaerobic conditions, nitrate reduction can occur in the presence of denitrifying bacteria. The colorless sulfur bacteria have versatility in their growth characteristics, but are mainly aerobic. However, *Thiobacillus denitrificans* is able to reduce nitrate under anaerobic conditions. This has resulted in its taxonomic subdivisions (Kuenen *et al.*, 1992; Chapter 3).

13.2.2 *Thiobacillus denitrificans*, a denitrifying "colorless sulfur bacterium"

Taylor and Hoare (1971) developed a selective growth medium for *T. denitrificans* which is still in use today even if under a slightly modified composition (Koenig & Liu, 1996; Trouvé *et al.*, 1998a). The most accurate way to determine genera and species is the sequencing of 16S RNA, but traditional dichotomic keys remain available, although they are less precise (Kuenen, 1989; Kelly & Harrison, 1989). The principle of this determination key is described in Box 13.1. Its main morphological and growth characteristics are given in Table 13.1. Tables 13.2 and 13.3 summarize, respectively, the main enzymatic activities and growth rates of *T. denitrificans*.

BOX 13.1 DETERMINATION KEY FOR *THIOBACILLUS DENITRIFICANS*

A1 The microorganism is a prokaryote that uses H_2S, S^{2-}, polysulfides, S°, $S_2O_3^{2-}$, (poly)thionates and SO_3^{2-} as energy source. The main oxidation by-products are S° and SO_4^{2-} even if polysulfides, $S_2O_3^{2-}$, SO_3^{2-} or polythionates can be produced.
A2 This microorganism does not accumulate intracellularly S°, even when grown in the presence of O^2 and S^{2-}.
A3 The optimum growth temperature is below 55°C. The microorganism growth is strictly chemolithotrophic.
A4 Rod type cellular morphology, motile or non-motile.

→ Genus *Thiobacillus* (Kuenen, 1989)

B1 Neutrophilic microorganism (optimum growth pH between 6 and 8, but potentially initiated at pH 3).
B2 Strictly autotrophic and chemolithotrophic microorganism. Grows on media with $S_2O_3^{2-}$ as the sole energy source. Unable to use glucose as sole energy source. Fitted out with ubiquinone Q-8.
B3 Facultative anaerobe, produces N_2 from nitrate, nitrite and N_2O reduction in the absence of O_2. GC% content in DNA varies from 63 to 68%.

→ *Thiobacillus denitrificans* (Kelly & Harrison, 1989)

Several physico-chemical parameters influence denitrification by *T. denitrificans* using thiosulfate as the electron donor (Claus & Kutzner, 1985). These authors determined optimum pH values between 7.5 and 8. The optimum temperature was located between 25 and 35°C, and no denitrification was noticed below 7–8°C. Denitrification inhibition by nitrite started at approximately

Table 13.1 Morphological and growth characteristics of *T. denitrificans* (After: Kuenen, 1989; Kelly & Harrison, 1989)

Characteristic	Description
Shape	Gram negative rod 0.5 × 1–3 μm
Motility	Motile with polar flagellum
Aspect of colonies on agar slants containing $S_2O_3^{2-}$ and NO_3^-	Opalescent
Reduced sulfur compound that supports growth	S^{2-}, S^0, $S_2O_3^{2-}$, $S_4O_6^{2-}$
Denitrification in the presence of	S^{2-}, S^0, $S_2O_3^{2-}$
Mean growth rate	0.06–0.11 h^{-1}

Table 13.2 Enzymatic activities of *T. denitrificans* (for the complete thiosulfate metabolism, see Figure 13.3)

Type of Metabolism	Enzymes (Purified or Simply Evidenced)	Reference
Carbon assimilation (acetate can also be assimilated but its contribution to the total cellular amount never exceeds 10%)	Benson-Calvin enzymes Uncomplete tricarboxylic acids cycle	Baalsrud and Baalsrud (1954) Taylor and Hoare (1971) Kelly (1989)
Nitrate reduction	Nitrate reductase, nitrite reductase, nitrous oxide reductase	Adams *et al.* (1971) Le Gall *et al.* (1979) Hole *et al.* (1996)
Sulfur oxidation	Adenosine phosphosulfate reductase, sulfite reductase	Kelly (1982) Taylor (1989) Schedel and Trüper (1980) Sawhney and Nicolas (1977)

100 mg L^{-1} and 100% inhibition occurred at 2 g L^{-1}. Sulfates and chlorides also inhibited denitrification, respectively, at 4 and 22 g L^{-1}. The electron donor and acceptor inhibited denitrification as well at high concentrations: in case of both nitrate and thiosulfate, inhibition started at 12 g L^{-1}.

Some contradictory results have been published in the literature. For instance, according to Baalsrud and Baalsrud (1954) *T. denitrificans* is unable to use

Table 13.3 Growth yields of *T. denitrificans*

	Growth Yield	Reference
Carbon assimilation	0.38 fixed CO_2 mole/mole $S_2O_3^{2-}$	Hoor (1976, 1981)
Nitrate reduction	7.95 g dry mass/mole NO_3^- 8.55 g dry mass/mole NO_3^-	Justin and Kelly (1978a, b) and Claus and Kutzner (1985)
Thiosulfate oxidation	11.37 g dry mass/mole NO_3^- 9.52 g dry mass/mole NO_3^-	Justin and Kelly (1978a, b) and Claus and Kutzner (1985)

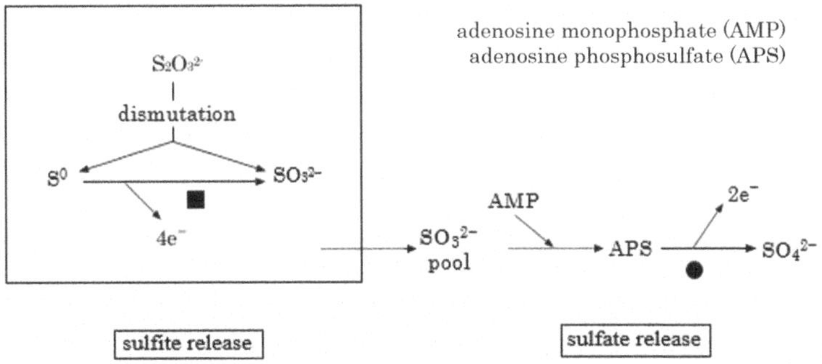

Figure 13.3 Schematic representation of the thiosulfate oxidation pathway by *T. denitrificans*. ■, Sulfite reductase; ●, APS reductase.

nitrates as a cellular nitrogen source. In contrast, the growth of some strains of this bacterium has been demonstrated in the absence of ammonium, but in the presence of nitrate as the sole nitrogen source (Baldensperger & Garcia, 1975; Claus & Kutzner, 1985). This suggests that, at least in the case of such strains, the nitrate metabolism not only implies their denitrification to N_2 (the nitrates being the electron acceptors), but also nitrate reduction to ammonia. The NH_4^+ thus produced is incorporated into cellular molecules, for instance amino acids, proteins, nucleotides and nucleic acids.

Trouvé *et al.* (1998a) isolated, using the procedure of Taylor and Hoare (1971), strains of *T. denitrificans* from the environment that denitrified in non-standard physico-chemical conditions compared with the results published by Claus and Kutzner (1985). In the case of strain 96.C2 isolated by Trouvé *et al.* (1998a) a 100% denitrification efficiency was obtained between pH 5.5 and 8, and this efficiency was still 83% at pH 5. The pH range thus appears to be much wider than the one determined by Claus and Kutzner (1985). For temperature effects, the nitrate removal rate (determined as the amount of nitrate removed per litre and per hour) was optimum at 10°C and was only half at 5°C. These results significantly differ from those published by Claus and Kutzner (1985) who found an optimum temperature of 25–35°C.

13.2.3 Other sulfur-oxidizing and nitrate-utilizing microorganisms

Although less cited in the technological literature, other microorganisms present in different ecological habitats, such as sediments or microbial mats, are also able to oxidize reduced sulfur forms along with nitrate reduction. Below, some of these bacterial species are listed although this list is not exclusive.

Two other *Thiobacillus* species have been reported to exhibit nitrate reduction coupled to sulfide oxidation: *T. versutus* and *T. thyasiris* (Kuenen *et al.*, 1992).

The GC% content (percentage of GC base pairs in DNA) of *T. thyasiris* is 52%, this percentage is between 65–68% for the species *T. versutus,* which is close to the 63–68% determined for *T. denitrificans*.

Thiomicrospira denitrificans (GC%: 36) shares the same properties (Kuenen *et al.*, 1992). This bacterium has a spiral shaped morphology that avoids confusion with the rod-shaped *Thiobacillus sp*.

Among the other sulfur bacteria, the gliding filamentous Beggiatoa sp. are often found to grow at the surface of sediments (freshwater or sulfide-rich marine water). When growing in such environmental conditions, they develop conspicuous white mats in which sulfide oxidation is coupled to nitrate reduction to N_2 (Sweerts *et al.*, 1990).

Thiosphaera panthotropha is able to denitrify using reduced sulfur compounds as the electron donor (Kuenen *et al.*, 1992). This mixotrophic bacterium possesses a complex metabolism. Denitrification is also obtained with hydrogen or a wide range of organic compounds (except those containing one carbon) as electron donors. Evidenced as a heterotrophic nitrifier, *T. panthotropha* can also denitrify aerobically (at an oxygen concentration of 3.8 mg L^{-1}) and heterotrophically.

Watsuntorn *et al.* (2017) isolated the new strain *Paracoccus versutus* MAL 1HM19 (GC%: 67.6) from the Mae Um Long Luang hot spring (Thailand). It simultaneously converts H_2S and NO_3^- to SO_4^{2-} and N_2 at 25°C (Watsuntorn *et al.*, 2020a).

Thiomargarita namibiensis, a 0.75 mm long bacterium was isolated in Africa and in South America (Schulz *et al.*, 1999). This "giant" bacterium is of interest because of its large size compared with that of "usual" bacteria (about 1 μm) but also because of its denitrifying capacity in the presence of reduced sulfur compounds. This bacterium is found in an environment containing high hydrogen sulfide levels which are toxic for most animal life. Another peculiarity of this bacterium is that it accumulates nitrates in vacuoles that are consumed during starvation periods (Schulz *et al.*, 1999). Microbiologists are awaiting new developments in the physiology of this microorganism for potential applications.

13.2.4 Denitrification in natural environments and at oxygen interfaces

Literature provides few data regarding the quantitative contribution of *T. denitrificans* and sulfur-oxidizing bacteria to natural denitrification phenomena. Firstly, its presence has not always been correlated with natural denitrification. Because the nitrate concentration in the environment is generally low, *T. denitrificans* is only present in low numbers. Moreover, it is strongly adsorbed to the rocky matrix where its nutrients are located. Secondly, it appears difficult to correlate the presence of this type of microorganism in a given habitat to its quantitative contribution to the total denitrification.

Jørgensen *et al.* (2009) reported that denitrification occurs in many Danish areas with intensive agriculture containing pyrite according to Reaction 13.8 (Bosch *et al.*, 2012):

$$FeS_2 + 3\,NO_3^- + 2\,H_2O \rightarrow Fe(OH)_3 + 1.5\,N_2 + 2\,SO_4^{2-} + H^+ \qquad (13.8)$$

Between 65 and 80% of nitrate reduction was due to anoxic oxidation of pyrite, resulting in a release of sulfate. Zhang *et al.* (2009a) reported similar behavior in Dutch sandy aquifers in which the nitrate concentration was up to 8 mM due to intense fertilizer application. Total nitrate reduction was accompanied by a build-up of sulfate, ferrous iron, and (pyrite-associated) trace metals.

Golterman (1991) noticed denitrifying activity in sediments from the Camargue area (France) and this activity increased after the addition of FeS. Garcia-Gil and Golterman (1993) demonstrated that the denitrification kinetics in these sediments after FeS addition were of the Michaelis-Menten type. Furthermore, they observed that denitrification was accompanied by sulfate release according to the stoichiometry of autotrophic denitrification.

Sweerts *et al.* (1990) reported data implying mixotrophic *Beggiatoa sp.* play a role in natural denitrification at the surface of freshwater sediments. Using microsensors, Sweerts *et al.* (1990) demonstrated that neither oxygen nor nitrate penetrated the sediment below 600 μm from the surface of the *Beggiatoa sp.* mat. They also demonstrated that anaerobic sulfide oxidation was coupled to nitrate reduction in the mat. This type of interaction occurs at oxygen interfaces present in microbial mats or, generally speaking, biofilms. Unraveling of the exact microbial composition of the denitrifying population present in both natural and man-made ecosystems warrants further research using specific analytical techniques, e.g. microelectrodes for nitrate and ammonium in combination with ribosomal RNA probes (Schramm *et al.*, 1996).

13.3 TECHNOLOGICAL ASPECTS OF AUTOTROPHIC DENITRIFICATION PROCESSES

13.3.1 Process parameters

When planning to use *T. denitrificans* or other sulfur-oxidizing bacteria in a technological process, it is a prerequisite to pay special attention to some parameters that affect the growth of these bacteria: (i) sulfide concentration (when it is the sulfur source), (ii) oxygen concentration, (iii) temperature, (iv) sterile conditions and (v) presence of organic carbon.

Among the reduced sulfur forms, sulfide exhibits an inhibitory effect on the growth of *T. denitrificans*. The process should thus be operated under sulfide concentrations below the inhibitory levels. The latter has to be determined for each bacterial strain used in a specific application. Sublette *et al.* (1998) reported that an increase in tolerance can be obtained by repeated exposure to increasing sulfide concentrations (in the Na_2S form). The *T. denitrificans* wild type strain was inhibited by sulfide concentrations between 0.1–0.2 mM, whereas the resistant F strain could resist up to 2.5 mM (over 10 times more) sulfide.

Oxygen and nitrates are both electron acceptors for the oxidation of reduced sulfur compounds (see Equations 13.5 and 13.6). In the presence of oxygen, sulfur oxidation is preferentially coupled with oxygen reduction. Sublette *et al.* (1998) reported that denitrification using sulfide as an electron donor cannot occur until O_2 levels are below 1 mg L^{-1}.

The temperature under which *T. denitrificans* usually denitrifies and oxidizes sulfur is in the mesophilic range (20–30°C). However, it is possible to select *T. denitrificans* strains able to denitrify and oxidize reduced sulfur compounds at temperatures around 5°C (Trouvé *et al.*, 1998a).

Sterile conditions can be obtained in the laboratory, but the maintenance of such conditions at industrial level is illusory. Even if the influent organic matter level is poor, the heterotrophic counts can be up to 10^8 colony forming units (CFU) mL^{-1} (Sublette *et al.*, 1998). The autotrophic growth of *T. denitrificans* may indeed release organic carbon as well as cell lysis material. With high organic matter concentrations (wastewater) one cannot predict the influence of heterotrophs on the contribution of *T. denitrificans* to denitrification. This should be considered separately for each type of influent. One way to proceed could be to compare the theoretical and experimental ratio of the [reacted oxidized products]/[reacted reduced substrates] for sulfur and the [remaining oxidized forms]/[initial oxidized forms] for nitrate. Nevertheless, further research is needed to control and minimize heterotrophic growth in a *T. denitrificans* bioreactor.

13.3.2 Process design – drinking water

Table 13.4 summarizes the characteristics of some of the main autotrophic denitrification processes for drinking water production using *T. denitrificans*.

13.3.2.1 Fixed-bed reactors containing elemental sulfur

Blecon *et al.* (1983) studied a fixed-bed reactor with upflow feeding. In the reactor, dolomite was used as bacterial growth support and also as a carbon and alkalinity source. In alkaline waters, the dolomite content should not exceed 10%. Higher dolomite concentrations reduced the denitrification efficiency. Several works followed this preliminary study. A mathematical model of the procedure was proposed in continuous flow and taking into account the variations of volumetric or hydraulic loading rates (Le Cloirec *et al.*, 1985). A 100% denitrification efficiency was reached for a volumetric nitrate loading rate of 0.18 kg NO_3^-–N/S° m^3 per day. The biomass production amounted 0.23 g dry mass per gram nitrate removed in a reactor containing half elemental sulfur and half maërl. The nitrate removal kinetics were determined as first-order reactions. These

Table 13.4 Characteristics of some of the main autotrophic denitrification processes using *Thiobacillus denitrificans* for drinking water production

Nature of the Influent	Reactor Type and Sulfur Source	Efficiency	Reference
Tap water enriched with nitrate loading rate of 0.18 kg NO_3^--N/S°-m^3.day	Fixed-bed reactor with elemental sulfur and dolomite (1:1 ratio)	100% denitrification efficiency was obtained. The authors determined that the reactions were first order kinetics.	Le Cloirec *et al.* (1985)
Groundwater enriched with nitrate (15.8 mg NO_3^--N L^{-1})	Slow sulfur/limestone filtration (1:1 ratio). Filter operating at a rate of 25 m h^{-1}	Denitrification efficiency: 65%. The nitrate content was reduced from 15.8 to less than 5.6 mg NO_3^--N L^{-1} (the European Guide Level). No backwashing required for a period of time of over 9 months because of slow filtration.	Schippers *et al.* (1987)
Crude water (chemically well defined) containing about 11 mg NO_3^--N L^{-1}	The first reactor is filled with iron wool and the second contains biolite (fixed bed type) – sulfide driven autotrophic denitrification	The process allows the production of drinking water with nitrate levels close to the European Guide Level. Little maintenance is required.	Montiel and Welté (1994)
Tap water enriched with nitrate volumetric loading rate of 1.5 kg NO_3^--N/m^3. day	Fixed-bed reactor with a mixture of pouzzolane and Neutralg (3:1 ratio). Thiosulfate as the sole energy source. Pilot scale experiments at low temperature (10°C)	100% denitrification efficiency	Trouvé and Chazal (1999)

(*Continued*)

Table 13.4 Characteristics of some of the main autotrophic denitrification processes using *Thiobacillus denitrificans* for drinking water production (*Continued*).

Nature of the Influent	Reactor Type and Sulfur Source	Efficiency	Reference
Synthetic nitrogen contaminated drinking water with 20–50 mg NO_3^--N L^{-1}	Elemental sulfur membrane bioreactor temperature controlled at 32–34°C	96% denitrification efficiency was reached after 7 day of operation that became 99% at the 17th day of operation. Membrane fouling was very low when the flux was 20 LMH ($L/m^2/h$)	Sahinkaya *et al.* (2015)
Synthetic nitrogen contaminated drinking water with 250 mg NO_3^--N L^{-1}	Inoculation by denitrifying culture from a SBR[1] performing simultaneous nitrification denitrification – sulfur-oxidizing membrane bioreactors with biogenic and chemical elemental sulfur	The SMBR[2] showed complete denitrification up to 286 ($\pm$34) mg N-NO_3^- L^{-1} d^{-1}, 1.7 times higher than that observed in the SMBRchem 65 $\pm$ 27 mg N-NO_3^- L^{-1} d^{-1}	Ucar *et al.* (2020)
Synthetic nitrogen contaminated drinking water with 30 mg NO_3^--N L^{-1}	Bioeletrochemical – sulfur autotrophic denitrification reactor	Denitrification efficiency between 90% and 100%	Wang and Qu (2003)

[1]Sequencing batch reactor.
[2]Sulfur-fed membrane bioreactor.

authors also determined the experimental values of the kinetic constants k_1 and k_2 of the denitrification process, according to Equation 13.9:

$$NO_3^- \xrightarrow{K_1} NO_2^- \xrightarrow{K_2} N_2 \tag{13.9}$$

They obtained values for k_1 and k_2 amounting to 3.15×10^{-3}/mg h^{-1} and 2.75×10^{-1}/mg h^{-1}, respectively.

During the same period of time a research program was developed in The Netherlands to adapt this process to small-size drinking water production units. The purpose of this project was to elaborate a cheap procedure that did not require excessive maintenance. Schippers *et al.* (1987) developed a sulfur/limestone process (ratio 1:1) functioning at a capacity of 35 $m^3\ h^{-1}$. The nitrate concentration range in the influent and effluent was, respectively, 14.7–15.8 and 0–5.6 mg NO_3^-–N L^{-1}.

13.3.2.2 Addition of elemental sulfur

Sahinkaya *et al.* (2015) studied a sulfur based autotrophic denitrification system for drinking water using a membrane bioreactor (MBR). They used flat sheet microfiltration polyethersulfone membranes (0.45 μm pore size) mounted into plexiglas modules. The reactor operated at 32–34°C and the synthetic contaminated water had a nitrate concentration of 25–50 mg L^{-1}. Powder sulfur (chemical) with a particle size <150 μm was added once a week in order to have a sulfur/nitrate ratio (w/w) between 2 and 3. They achieved complete denitrification with a nitrate loading rate up to 0.24 g $L^{-1}\ d^{-1}$). Low membrane fouling occurred when the flux was 20 L $m^{-2}\ h^{-1}$).

A comparison between the use of biogenic and chemical sulfur as an electron donor using the same type of membrane was performed by Ucar *et al.* (2020). They inoculated the two reactors with a mixed denitrifying culture that performed simultaneous nitrification and denitrification at 30°C. The synthetic feed solution consisted of tap water with a nitrate concentration of 250 mg L^{-1}, while the S^0 concentration was always kept at 750 mg L^{-1}. They obtained complete denitrification in the MBR with biogenic S^0, while for the MBR with chemical S^0, nitrate was never completely removed.

Wang and Qu (2003) coupled a bioeletrochemical and sulfur autotrophic denitrification system to treat a synthetic nitrate contaminated drinking water containing 30 mg-N L^{-1}. As shown in Figure 13.4 the sulfur denitrification was carried out in the bottom part, while the bioeletrochemical system was in the top part and it used the protons released in the sulfur denitrification allowing hydrogenotrophic denitrification; in this way no alkalinity source was required. The nitrate removal efficiency was between 90% and 100% depending on the hydraulic retention time (HRT) (1.9–5 h) without nitrite accumulation and a maximum sulfate concentration in the effluent of 170 mg L^{-1}.

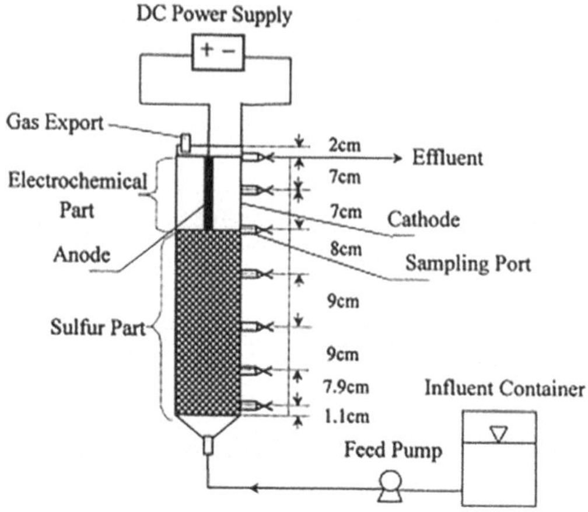

Figure 13.4 Bioeletrochemical set-up for sulfur autotrophic denitrification system (After: Wang & Qu, 2003).

13.3.2.3 Addition of soluble sulfur sources

Montiel and Welté (1994) developed a laboratory scale biological denitrification process, based on the existence of synergetic relations between chemolithotrophic bacteria able to coexist in fixed beds. These microorganisms originated from the natural microbiota of the influent. Iron, sulfur and nitrogen cycling was successively combined in this process. The pilot consisted of two main modules: the first module was loaded with iron wool and the second was filled with biolite (Figure 13.5). In the first module metal iron of the iron wool was oxidized to ferrous iron by contact with incoming water (which also contained nitrates). Thus, the E_h value and the dissolved oxygen concentration decreased. This phenomenon was immediately followed by the reduction of incoming sulfates to sulfides by sulfate-reducing bacteria. This led to the formation of ferrous sulfide, which was then reoxidized to sulfates, accompanied by the reduction of NO_3^- into N_2 by denitrifying thiobacilli. In the second module, iron bacteria oxidized ferrous iron into insoluble ferric iron (FeOOH) that remained in the bed. The influent nitrate concentration was decreased from 43 to 22 mg L^{-1} in the effluent. The sulfate concentration (11 mg L^{-1}) was not affected. The residence time (under optimization) was 12 h. According to the authors this process is suitable for the treatment of water for small denitrification plants owing to its low costs.

Entrapment of both *T. denitrificans* and nitrifying bacteria in alginate beads has also been investigated as a process allowing full nitrification/denitrification (Lewandowski *et al.*, 1987). In addition, $CaCO_3$ was incorporated, thus

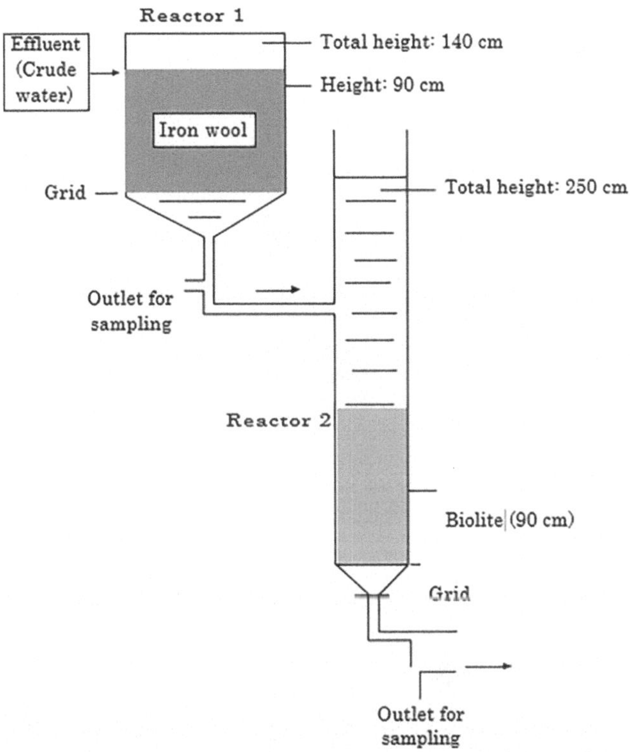

Figure 13.5 Scheme of the pilot reactor for autotrophic denitrification developed by Montiel and Welté (1994).

increasing the bead stability (after Ca^{2+} release) and providing the carbon source for *T. denitrificans*. This process was developed to remove any soluble form of nitrogen and obtained an efficiency of 4.8 mg N removed per litre and per hour. The beads were stable and active for 3 months in a continuous flow system.

Lee and Sublette (1990) evidenced that *T. denitrificans* can also remove gaseous nitrogen compounds, such as NO in the presence of thiosulfate and carbon dioxide. They demonstrated that NO can act as an electron acceptor for this bacterium. Up to 96% of NO removal was observed with an average $NO/S_2O_3^{2-}$ ratio of 4:1.

Trouvé *et al.* (1998a) reported some *T. denitrificans* strains achieved good denitrification at low temperature (5°C) and at acid (pH 5) or basic (pH 8) pH in batch conditions. Attempts made in the presence of ferrous sulfide from microbial origin (released by sulfate-reducing bacteria) did not allow as efficient denitrification as that obtained in the presence of ferrous sulfide from chemical origin (Trouvé *et al.*, 1998a).

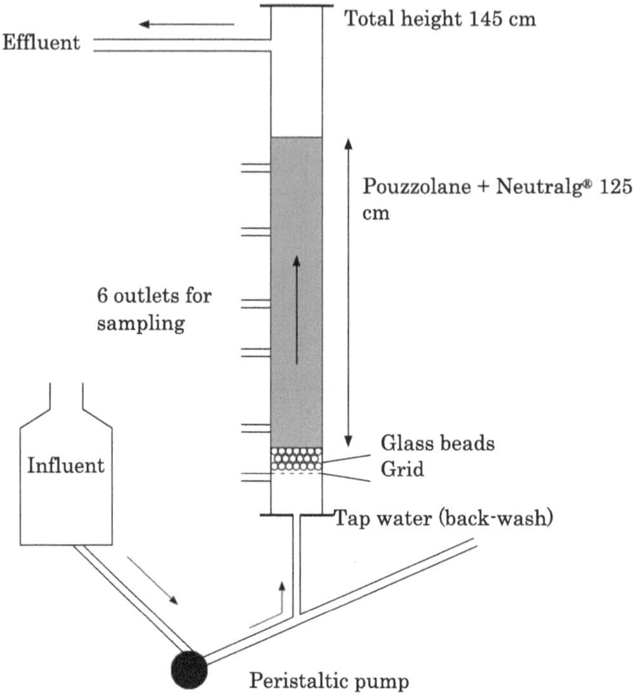

Figure 13.6 Scheme of the fixed bed pilot reactor for autotrophic denitrification used by Trouvé and Chazal, (1999).

This last experiment was done at pilot scale under continuous flow (Trouvé *et al.*, 1998b). In the presence of ferrous sulfide from chemical origin, a 100% denitrification efficiency was obtained at a residence time of 1 hour allowing the removal of a volumetric nitrate loading rate of 0.32 kg N-NO_3^- per cubic metre and per day. Furthermore, using the same type of strains, Trouvé and Chazal (1999) attempted denitrification of crude water to which only thiosulfates and nitrates were added at low temperature (10°C). This temperature value seems to be a limit for classical heterotrophic denitrification processes. In the presence of *T. denitrificans* 96.C2, a pilot scale fixed-bed reactor (containing a mixture of pouzzolane and Neutralg) was operated in continuous flow (Figure 13.6). A 100% denitrification efficiency was observed at a 30 minutes residence time and a volumetric nitrate loading rate of 1.5 kg N-NO_3^- per m^3 and per day.

The experimental [sulfate]/[thiosulfate], [thiosulfate]/[nitrate] and [sulfate]/[nitrate] ratios were respectively 2, 0.845 and 1.689, close to the theoretical ones (Driscoll & Bisogni, 1978). Figure 13.7 shows the evolution of the concentrations of nitrogenous and sulfurous compounds as a function of the column height in this reactor.

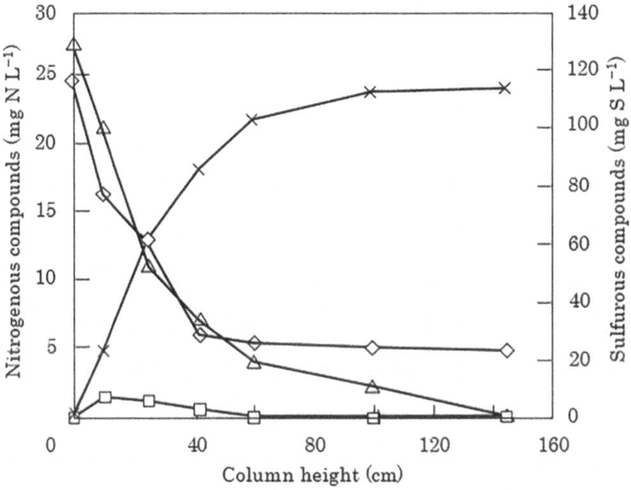

Figure 13.7 Autotrophic denitrification by *T. denitrificans* at 10°C. Evolution of the concentrations of NO_3^--N (Δ), NO_2^--N (■), $S_2O_3^{2-}$-S (◊) and SO_4^{2-}-S (×) as a function of the reactor height, at a nitrate loading rate of 1.5 kg NO_3^--N m^{-3} d^{-1} and a HRT of 30 min (After: Trouvé & Chazal, 1999).

13.3.3 Process design – wastewater treatment

The coupling of the sulfur and nitrogen cycles has been used for the treatment of different wastewaters, either with sulfide or nitrate removal as the objective. It appears difficult to establish comparisons between all the experimental processes, because of differences concerning the objective, and the versatility of the operating conditions. In general (with the exception of a few cases) a reduced sulfur source has to be added when nitrate removal is the main purpose. This can be a solid (filter beds containing elemental sulfur) or soluble (e.g. sulfide or thiosulfate) sulfur source. In contrast, nitrates have to be added when sulfide removal is the objective (Khanongnuch *et al.*, 2019c). In the latter case, one must consider that there is no net sulfur removal because sulfide is converted into elemental sulfur or sulfate.

Below, some technological approaches for nitrate removal from wastewater are presented. Table 13.5 summarizes the characteristics of some of the main processes developed in this field.

13.3.3.1 Filter beds containing solid sulfur sources

Kong *et al.* (2016) performed a comparative investigation on two biofilters containing elemental sulfur or pyrite for the treatment of domestic wastewater. In both columns, they had an aerobic phase to remove the total organic carbon (TOC), NH_4^+ and for the phosphorous accumulation and an anoxic phase

Table 13.5 Characteristics of some processes using *Thiobacillus denitrificans* for the autotrophic denitrification of wastewater

Nature of the Influent	Reactor Type and Sulfur Source	Efficiency	Reference
Synthetic wastewater with 250 mg L^{-1} NO_3^--N	*T. denitrificans* MP. Sulfur-limestone column reactor (22 ± 2°C) and a fermenter (30 ± 1°C).	Denitrification efficiency: 97.5% after 14.3 h in the column reactor. SO_4^{2-} concentrations increased from 190 mg L^{-1} in the influent to 1800 mg L^{-1} in the effluent.	Gu *et al.* (2004)
Secondary effluent from WWTP	Enriched cultures of *T. denitrificans* from WWTP anaerobic sludge. Pyrrhotite autotrophic denitrification biofilter.	Denitrification efficiency: 99% at HRT of 1.2 h. N concentration decreased from 13.81 (±1.52) mg L^{-1} in the WW to 0.05 (±0.01) mg L^{-1} in the effluent.	Yang *et al.* (2017)
Synthetic wastewater	Culture from tidal flats. Sulfur packed bed reactor (35°C)	The maximum nitrate removal rates were from 3.9 to 4.5 kg NO_3^--N m^{-3} d^{-1} depending on the nitrate concentration in the influent (175–500 mg $L^{-1)}$	Park *et al.* (2002)
Synthetic wastewater	Fluidized-bed reactor with thiosulfate as sulfur source	100% denitrification efficiency was achieved and 99% $S_2O_3^{2-}$ removal efficiency was obtained with N/S ratio of 0.5, 20°C, HRT of 5 h and influent pH of 6.9 (±0.1).	Khanongnuch *et al.* (2018)
Synthetic wastewater	Anoxic sulfur-oxidizing moving-bed biofilm reactor treating thiosulfate and nitrate contaminated wastewater	Complete nitrate removal was obtained 0.5, 0.3 and 0.1 N/S ratios while thiosulfate removal efficiencies were 98, 82.3 and 37.7% respectively.	Khanongnuch *et al.* (2019b)

Synthetic wastewater	*T. denitrificans* DSM12475. Thiosulfate fluidized-bed reactor (20 ± 2°C–30 ± 0.2°C)	Complete denitrification and no nitrite accumulation with nitrate loading rate of 600 mg L^{-1} h^{-1} and HRT of 10 min. Decrease of HRT to 5 min resulted in 50% nitrate removal efficiency.	Zou *et al.* (2016)
Synthetic wastewater	Continuous flow sulfide mixed reactor inoculated with a mixed culture containing *Thiobacillus denitrificans* (30°C)	Complete denitrification and sulfide removal were achieved at a nitrate loading rate of 7.96 mM N L^{-1} d^{-1} and a N/S ratio of 0.89.	Manconi *et al.* (2007)
Municipal wastewater type influent	Ananox process: an elaborate multichamber design. *T. denitrificans* oxidizes in an anoxic chamber, sulfides released in an anaerobic chamber. The sulfide rich effluent is mixed with nitrates coming from an oxic nitrifying chamber.	Complete denitrification. Denitrification rates varied between 3.6 and 2.84 mg NO_3^--N g VSS^{-1} h^{-1} according to the anoxic chamber height.	Xiushan *et al.* (1993)
Ion exchange saline eluates	Elemental sulfur in fixed-bed reactor at 20°C. Inoculation by enrichment of *T. denitrificans* in thiosulfate synthetic medium	Potential application for saline influents: 90% denitrification observed for NaCl concentration up to 30 g L^{-1}, for a volumetric loading rate of 1 kg NO_3^- m^{-3} d^{-1}	Sant'Anna *et al.* (1996)

with S^0/FeS_2 for the denitrification step. The inoculum was an activated sludge collected from a wastewater reuse plant. They used, respectively, in the sulfur column and in the pyrite column 150.86 g of sulfur with 274.29 g of crushed oyster shells and 171.43 g of pyrite with 35.71 g of crushed oyster shells. The sulfur column reached a higher nitrate removal efficiency than the pyrite column (89.74% vs 80.72%), while the pyrite column produced less sulfate (27.54 mg L^{-1} vs 41.55 mg L^{-1}) and was able to better maintain the pH (7.13 vs 6.31). The pyrite column, due to these advantages, seems to be better in practical use but anoxic condition and absence of organic carbon need to be guaranteed.

Yang *et al.* (2017) used pyrrothite ($Fe_{1-x}S$, $0 < x < 0.125$) as the electron donor and carrier material in a biofilter to perform autotrophic denitrification and phosphate removal. They treated real secondary wastewater (coming from a previous conventional anoxic/oxic activated sludge process) with a total oxidized nitrogen concentration of 13.81 (±1.52) mg L^{-1}. The inoculum source was an anaerobic sludge from a municipal wastewater treatment plant (WWTP) enriched using the method described by Li *et al.* (2016). They tried different hydraulic retention times (HRTs) and the best HRT – 1.2 h – enabled an effluent N concentration of 0.05 (±0.01) mg L^{-1} to be achieved. *Thiobacillus* was the relatively most dominant genus (87% relative abundance) and was able to oxidize sulfide to SO_4^{2-} and then, the latter was reduced to polysulfides (S_n^{2-}) due to the activity of sulfate-reducing bacteria. This work shows the potential use of pyrrothite as a sulfur source for autotrophic denitrification.

The production of $CaSO_4$ might cause problems during practical applications, because of extra (chemical) sludge production and clogging of pipes. Kuai and Verstraete (1999) compared autotrophic (using an elemental sulfur lime bed) to heterotrophic denitrification for the treatment of wastewaters from small-scale treatment systems (fewer than 10,000 inhabitant equivalents). The sulfur lime bed removed 75% of the total oxidizable nitrogen at a loading rate of 0.2 total oxidizable nitrogen L^{-1} d^{-1} and a hydraulic retention time of 3 h. The filter bed also retained some chemical oxygen demand (COD) (about 10%) and reduced fecal indicator organisms by about 1 log unit. The major drawback of the sulfur lime filter was its poor phosphorus removal efficiency (Kuai & Verstraete, 1999).

Sant'Anna *et al.* (1996) studied how to valorize exchange ion resin eluates using a fixed-bed reactor containing elemental sulfur (Figure 13.8). NaCl is abundant in this type of eluate (25 g L^{-1} as mean value). They examined the influence of this salt on the denitrification efficiency. The modeling procedure they developed allowed the determination of an inhibition constant (K_i) of 41 g NaCl L^{-1}. This opens perspectives for future developments of this technology, as the NaCl concentration in these eluates is usually far below this K_i. For a 30 g L^{-1} NaCl in the eluates, a 90% denitrification efficiency was obtained at 20°C at a nitrate volumetric loading rate of 1 kg NO_3^-/S° m^3 per day.

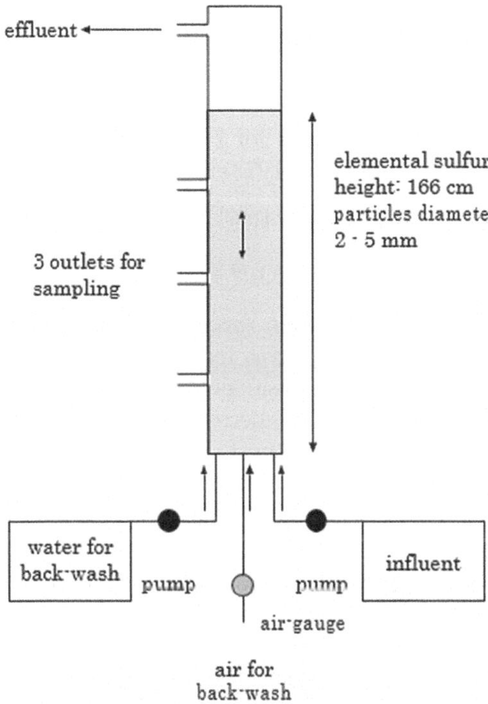

Figure 13.8 Pilot reactor scheme for autotrophic denitrification of ion exchange resin eluates, using an elemental sulfur filter bed (After: Sant'Anna *et al.*, 1996).

13.3.3.2 Elemental sulfur addition to wastewater

Kostrytsia *et al.* (2018a) used biogenic sulfur, a by-product of gas desulfurization, as a novel electron donor to treat nitrate contaminated wastewater in a moving bed bioreactor (MBBR). They have shown that this process has 10 times faster kinetics compared to autotrophic denitrification with chemically synthesized elemental sulfur (Kostrytsia *et al.*, 2018b). Moreover, using a biomass enriched on nitrite they were able to degrade up to 240 mg L^{-1} NO_2^-–N with a denitritation activity of 339.5 mg NO_2^-–N g per volatile suspended solid (VSS) per day. The low nitrite accumulation allowed a specific denitrification activity of 223.0 mg NO_3^-–N g VSS^{-1} d^{-1}. In this study an activated sludge collected from a denitrifying tank of a municipal wastewater treatment plant was used as inoculum and was enriched to perform denitrification and denitritation separately. After the enrichment, *Thiobacillus* was the most abundant genus and after that, *Moheibacter* and *Thermomonas* were predominant for, respectively, denitrification and denitritation (Kostrytsia *et al.*, 2018b).

Koenig and Liu (2004) studied an autotrophic denitrification process of high salinity wastewater coming from a flue gas desulfurization plant using elemental sulfur as an electron donor. They used as inoculum a sediment from the tidal flats of marshes in Hong Kong, as a potential source of autotrophic sulfur-oxidizing bacteria. They enriched these cultures for *T. denitrificans* to follow the process modeled by Batchelor and Lawrence (1978) using Equation 13.10:

$$1.10\,S^0 + NO_3^- + 0.40\,CO_2 + 0.08\,HCO_3^- + 0.08\,NH_4^+ \rightarrow$$
$$1.10\,SO_4^{2-} + 0.05\,N_2 + 0.08\,C_5H_7O_2N + 1.28\,H^+ \quad (13.10)$$

They found, with a series of batch tests, specific denitrification rates of 6 to 8 mg NO_3^- g VSS^{-1} h^{-1}. The best result was found with a pH between 7.0 and 8.0 and temperature of 30°C. When the osmotic pressure exceeded 19 atm (70% of seawater) the denitrification activity decreased and when it reached a salt concentration (independent of the type of salt ions) equal to seawater, approximately 70% of the activity was reached (Koenig & Liu, 2004).

Batchelor and Lawrence (1978) further emphasized an additional advantage of autotrophic denitrification: it would be more stable under transient loading conditions. They also showed that the maximum nitrate removal rate is a linear function of the ratio between the sulfur (S) and biomass (X) concentration over the S/X range 45–194 (expressed as mg S/mg organic N). The stoichiometry for autotrophic denitrification was relatively constant over the temperature range of 12–30°C and for the S/X range previously cited. In general, biomass production is low in an autotrophic denitrification reactor, compared with a heterotrophic reactor.

13.3.3.3 Addition of soluble sulfur sources to the wastewater

Bisogni and Driscoll (1977) studied and demonstrated the efficiency of $S_2O_3^{2-}$ and S^{2-} as soluble sulfur energy sources for denitrification of synthetic wastewaters. They modeled the system by Equations 13.11 (thiosulfates) and 13.12 (sulfides) where $C_5H_7O_2N$ is the formula of cellular biomass:

$$0.844\,S_2O_3^{2-} + NO_3^- + 0.0347\,CO_2 + 0.0865\,HCO_3^- + 0.0865\,NH_4^+$$
$$+ 0.434\,H_2O \rightarrow 1.689\,SO_4^{2-} + 0.5\,N_2 + 0.0865\,C_5H_7O_2N + 0.697\,H^+ \quad (13.11)$$

$$0.421\,S^{2-} + 0.421\,HS^- + NO_3^- + 0.346\,CO_2 + 0.0865\,HCO_3^- + 0.0865\,NH_4^+$$
$$\rightarrow 0.842\,SO_4^{2-} + 0.5\,N_2 + 0.0865\,C_5H_7O_2N + 0.697\,H^+ + 0.413\,H_2O \quad (13.12)$$

The same authors confirmed the efficiency of S^{2-} addition compared to S° in an upflow reactor for wastewater treatment (Driscoll & Bisogni, 1978). They suggest the use of limestone as an alkalinity source. The alkalinity consumption was 3.74 mg $CaCO_3$ per milligram of NO_3^--N reduced. Furthermore, Driscoll and Bisogni (1978) underlined the importance of the support surface. They suggested that the larger the surface of the support material compared to its volume, the more important the biofilm surface was. Thus, at a high surface ratio, the same denitrification efficiency can be achieved for lower residence times, and thus at higher volumetric loading rates.

Denitrification of mining wastewater enriched with nitrates was studied by Bouffard *et al.* (1998). Those effluents can be contaminated with high levels of nitrates (up to 1 g L^{-1}) owing to the use of nitrate-based explosives and nitric acid leaching for metal recovery. The anaerobic closed-cycle system is based on the co-culture of *T. denitrificans* and sulfate-reducing bacteria. The latter bacteria generate "*in situ*" sulfide, which is then used by *T. denitrificans* for denitrification. This type of bacterial co-culture can make the denitrification cycle a sustainable process (Yang *et al.*, 2017).

Di Capua *et al.* (2016) compared thiosulfate, chemically synthesized and biogenic elemental sulfur in batch bioassays with different S/N ratios at different temperatures. Thiosulfate had the highest denitrification rate (52.5 mg N-NO_3^- · L^{-1} d^{-1}), and was 10 times faster in nitrate removal compared to S^0. $S_2O_3^{2-}$ concentrations exceeding 2.2 g S–$S_2O_3^{2-}$/L were inhibitory for *T. denitrificans*, but beneficial for mixotrophic denitrification. Treatment of synthetic mine water by thiosulfate-driven denitrification was further tested in a fluidized-bed reactor (FBR) under acidic conditions, up to pH 4.75 (Di Capua *et al.*, 2017a), at psychrophilic temperatures (from 20 to 3°C) (Di Capua *et al.*, 2017b) and under the effect of increasing nickel concentrations (5–200 mg Ni L^{-1}) (Di Capua *et al.*, 2017c). A 99% denitrification efficiency was achieved at feed and effluent pH of 5.75 and 5.30, respectively. Below this pH value the denitrification activity dropped and they added a carbonation unit feeding CO_2 to the FBR that enabled complete denitrification even at pH as low as 4.75. A 100% denitrification efficiency was obtained at temperatures as low as 3°C for 98 days at an HRT and nitrate loading rate (NLR) of 1 h and 3.3 kgN-NO_3^- m^{-3} d^{-1}, respectively.

When using natural inocula from various origins, wastewaters were also successfully treated by autotrophic denitrification. Different enrichment methods for denitrification using thiosulfate (Pu *et al.*, 2015) or elemental sulfur (Hashimoto *et al.*, 1987) from such inocula have been described. *T. denitrificans* has also been employed in an industrial plant for the removal of organic matter and nitrogen from municipal wastewater, in the Ananox® process (Xiushan *et al.*, 1993). Denitrification is integrated in a complex treatment system involving different processes such as methanogenesis, sulfate reduction, sulfide oxidation, nitrification and denitrification.

13.3.4 Process design – waste gas treatment

Several physical-chemical processes are used for sulfide removal from waste gases (see Lagas, 2000) but they are very expensive if compared with the possibility of combining them with nitrate removal in a sulfide driven autotrophic denitrification system. In the last decade, these processes have been tested for the removal of H_2S from biogas (Dolejs *et al.*, 2015; Bayrakdar *et al.*, 2016; Khanongnuch *et al.*, 2019a).

The S/N ratio is a key parameter of the process, for this reason Dolejs *et al.* (2015) focused their study on testing several S/N ratios, ranging from 0.3 to 2.4 mol/mol, in a completely stirred tank reactor. They showed that complete desulfurization was achieved at a ratio equal to 0.85 mol/mol and the autotrophic denitrification efficiency was up to 80%, while nitrogen reduction to ammonia occurred and subsequently, elemental sulfur accumulated when the S/N ratio was higher than 1.3 mol/mol. Khanongnuch *et al.* (2019a) used an anoxic biotrickling filter (BTF) at feed N/S ratios of 1.2–1.7 mol/mol. A maximum H_2S elimination capacity of 19.2 g S m^{-3} h^{-1} (>99% efficiency) was achieved at NO_3^- loading rates between 2.9 and 12.9 NO_3^--N m^{-3} h^{-1} (N/S ratios = 1.2–1.7 mol mol^{-1}. Adding acetate at a C/N ratio of 0.2 stimulated the growth and activity of heterotrophic denitrifiers in the BTF, but biomass accumulation in the filter bed occurred causing a reduction of the HRT and subsequently to insufficient nitrate for oxidizing the H_2S load. Moreover, Khanongnuch *et al.* (2019b) evaluated the performance of a BTF testing different liquid flow rates, wet and dry operation and sulfide shock loads. They showed that wet–dry bed operations at 2–2 h and 24–24 h reduced the H_2S removal efficiency to 60–80%, while wet–dry operation in a cycle of 1–1 h and 12–12 h did not have much effect. When a H_2S shock load was applied by increasing the gas flow rate from 60 to 200 L h^{-1} and the H_2S concentration in the influent (from 112 (±15) to 947 (±151) ppmv), the BTF continued to show a good desulfurization efficiency (>93%).

Watsuntorn *et al.* (2019) tested three different strains closely related to *Paracoccus sp.* (>99%) separately and mixed for H_2S oxidation and NO_3^- reduction. The utilization of the mixed strains gave better results in terms of both hydrogen sulfide and nitrate removal than the single strain incubations. The best result in terms of H_2S oxidation was obtained with the three strains together. Watsuntorn *et al.* (2017) and Watsuntorn *et al.* (2020a, b) further tested the strain MAL 1HM19, closely related to *Paracoccus* sp., in a batch system and anoxic BTF to treat a H_2S rich biogas. In the batch test this strain was able to remove up to 800 ppmv H_2S within 10 hours, while in the BTF the maximum elimination capacity reached was 113.5 (±6.4) g S m^{-3} h^{-1} with a 97% removal efficiency during a H_2S shock load of ~4000 ppmv.

13.3.5 Optimization of the fixed-bed reactor design

Some problems can affect autotrophic denitrification by *T. denitrificans* in fixed-bed reactors, treating wastewater or drinking water, some of them may have the same origin:

- dysfunctioning releasing excess nitrate and nitrite concentrations in the effluent
- biomass accumulation – requires optimization of the back wash frequency
- gaseous oversaturation due to the release of N_2 or other gaseous forms of nitrogen (NO_x)
- hydraulic short-circuits
- production of unwanted sulfurous compounds

Nitrate release in the effluent can occur due to intensive bacterial proliferation (Kruithof *et al.*, 1988): excess biomass limits mass transfer of nitrate inside the biofilm (Di Capua *et al.*, 2015). The other consequence of this phenomenon is that excessive N_2 release leads to gaseous oversaturation and bubble formation. This is the origin of hydraulic short circuits in the filter beds that significantly modify the residence time, and thus affect the denitrification efficiency.

Nitrite release has been explained differently according to different authors. Van der Hoek *et al.* (1992) noticed that the presence of nitrite was also accompanied by excess nitrate release. According to Haider *et al.* (1988), the presence of nitrites in the effluent is a function of the fixed bed height for a given volumetric loading rate. For Driscoll and Bisogni (1978) nitrites appeared when the sulfur source was limiting in the growth medium, or in other words, if the ratio between electron donor and acceptor was unbalanced, as noticed by Blecon *et al.* (1983). The decrease in residence time and the rapid increase of the loading rate have also been implied (Sikora & Keeney, 1976; Le Cloirec & Martin, 1988), because k_1 (kinetic constant for nitrate reduction to nitrite) is superior to k_2 (kinetic constant for nitrite reduction to N_2). According to Lazarova *et al.* (1994) nitrite release is independent of (i) operational parameters (O_2, pH, and temperature), (ii) the hydraulic/volumetric loading rates, and (iii) the biofilm thickness. Lazarova *et al.* (1994) postulated that nitrite release is governed by (i) the type of the bacterium used in the filter and (ii) the density and the homogeneity of the biofilm developed by this bacterium. It also depends on the ratio between proteins and polysaccharides within the biofilm (Lazarova *et al.*, 1994).

Sulfides may also be released by autotrophic denitrification processes utilizing *T. denitrificans* using S° or $S_2O_3^{2-}$. The complete removal of nitrates in a limited part of the biofilter creates locally anaerobic areas in which high sulfate concentrations can accumulate. In these anaerobic microniches, sulfate or sulfur reducing bacteria such as *Desulforomonas acetoxidans* can grow using organic matter from cell lysis (Van der Hoek *et al.*, 1992). Sulfide can inhibit denitrification by *T. denitrificans* (Xiushan *et al.*, 1993). Sulfide may also induce clogging or increase effluent turbidity when further oxidized to S° (Hijnen *et al.*, 1988).

The preliminary gas removal (using a vacuum system) from the influent just before injection in the reactor has been proposed by Schippers *et al.* (1987) and Soares *et al.* (1988). O_2 removal prevents aerobic bacterial growth as well as the aerobic oxidation of S° (which results in sulfate formation), thus lowering the effluent sulfate concentration. Furthermore, the risk of gaseous oversaturation inside the reactor is greatly decreased, thus minimizing bubble and gas bag accumulation and related short circuits.

Control of biomass growth can be achieved by different means. According to Blecon *et al.* (1983), the plants were given a back wash every 10 days. This procedure lowered the biomass concentration, thus recreating the initial hydraulic characteristics which guaranteed good operating conditions. Koch and Siegrist (1997) noticed that mainly total suspended solids (TSS) retention and accumulation of biomass and N_2 contributed to the development of a headloss. N_2 solubility is a function of temperature and hydrostatic pressure. Its solubility decreases when the temperature is lowered or when the headloss increases. The gas bags have to be removed, and the required removal frequency increases all along the operational running time. This frequency can be determined by headloss measurements. Washing procedures mainly remove opportunistic biomass and TSS deposits. The biofilm, mainly comprising of denitrifying microorganisms, remains in the reactor and is thus made more performant by the back washing (Boller *et al.*, 1997).

Numerous authors prevented clogging and short-circuits by using alternative reactor concepts. The latter ones have been developed to treat larger volumetric loading rates, while avoiding nutrient diffusion problems independently of the biofilm thickness. The following reactor designs can be cited: biofilm airlift suspension reactors (Picioreanu *et al.*, 1997; Van Benthum *et al.*, 1997), membrane reactors (Kimura *et al.*, 2002), and upflow anaerobic sludge blanket reactors (Tarre *et al.*, 1995; Rouse *et al.*, 1999) with both microaerophilic and anaerobic areas (Guerrero & Zaiat, 2018).

When using fixed beds, the nature of the support material inside the reactor influences the process. According to Boehler and Haldenwang (1992), this material has to be porous and wrinkled. The more porous the material is, the greater the biofilm surface becomes, which can develop outside and inside the support. Furthermore, the same bacterial concentration dispatched on a greater surface induces thinner, higher-performing biofilms (Lazarova *et al.*, 1994). The material roughness has to be calibrated to allow good bacterial adhesion. It also influences the residual fixed bacterial concentration after a washing procedure (Hanus & Bernard 1988). Nevertheless, a lot of work remains to be done before fixed-bed processes can be applied at the industrial scale, although this is of great interest from a technological point of view.

13.4 NITROGEN CONVERSIONS IN ANAEROBIC REACTORS TREATING SULFATE-RICH WASTEWATER

13.4.1 Effect of sulfide on aerobic post-treatment

The versatility of bioconversions in the sulfur cycle also allows the integration of nitrogen and sulfur removal from wastewater. Nitrogen elimination has been combined with sulfur reclamation from concentrated photographical wastewater (Rooden *et al.*, 1995). Thiosulfate is converted into sulfide in an upflow anaerobic sludge bed (UASB) reactor, and S° is produced in a subsequent reactor. In the aerobic post-treatment, nitrogen and residual COD are removed by combined nitrification/denitrification.

In this respect, the sulfide present in the effluent of the anaerobic stage can be toxic to nitrifiers (Hooper & Terry, 1973) and chemolithotrophic sulfide oxidizers (Cardoso *et al.*, 2006) and can thus upset nitrification in the aerobic post-treatment. On the other hand, sulfide can contribute to nitrate removal, when its reoxidation to S° or sulfate by *T. denitrificans* can be scaled up. Garuti *et al.* (1992) reported that sulfide oxidation in the denitrification reactor reduced the need for external carbon source (see 13.3.3).

13.4.2 Combined denitrification and methanogenesis

In anaerobic process technology, direct integration of denitrification and methanogenesis in a single reactor has been suggested as an alternative method to the classical concept of anaerobic organic carbon removal followed by nitrification/denitrification for nitrogen removal (Akunna *et al.*, 1992; Hendriksen & Ahring, 1996). In addition, the combination of denitrification and methanogenesis has also been considered in drinking water production; both immobilized mixed cultures and bioreactors have been investigated to remove excess carbon source, mostly methanol, from denitrified effluents by methanogenesis (Lin & Chen, 1995).

13.4.3 Ammonification

In addition to denitrification (nitrate reduction to nitrogen gas), also nitrate reduction to ammonium (ammonification) can occur in anaerobic environments. Nitrite is an intermediate of both pathways. Ammonification can be expected in sulfidogenic reactors, as (i) pure cultures of sulfate-reducing bacteria convert nitrate to ammonium (Widdel 1988), (ii) anaerobic inocula contain large populations of fermentatives and obligate anaerobes, i.e. ammonium formers, compared with true denitrifiers (Kasper *et al.*, 1981) and (iii) certain carbon sources, e.g. glycerol and glucose, favor ammonification (Akunna *et al.*, 1993). With acetate as the carbon source, nitrate is totally converted into nitrogen gas via denitrification (Akunna *et al.*, 1993; Guynot *et al.*, 1998; Lens *et al.*, 1999). This is very beneficial, as nitrate reduction to ammonium is of no interest from a wastewater treatment point of view.

13.4.4 Combined denitrification and sulfidogenesis

In high-rate sulfidogenic reactors which are optimized for sulfate reduction, acetate removal is the rate-limiting step (Hulshoff Pol *et al.*, 1998). One way to optimize the acetate removal of these reactor systems is to increase the activity of facultative (an) aerobic bacteria as acetate scavengers. This can be achieved by dosing minor amounts of alternative electron acceptors, e.g. oxygen or nitrate, into the reactor system (Lens *et al.*, 1998; Chapter 6). Nitrate addition is an attractive option, as nitrate is much more soluble in water (several grams per litre) than oxygen and acetate is an excellent carbon source for denitrifying bacteria (Ganaye *et al.*, 1996).

In methanogenic (Percheron *et al.*, 1998) and sulfidogenic (Lens *et al.*, 1999) granular sludge, sulfide disappears upon the addition of nitrate. Percheron *et al.* (1998) showed that denitrifiers can use sulfide as the electron donor. In sulfidogenic granular sludge, however, sulfate-reducing bacteria (SRB) can also be responsible for the combined sulfide and nitrate removal. Indeed, SRB are able to completely reoxidize sulfide to sulfate in the presence of either oxygen or nitrate as electron acceptors (Dannenberg *et al.*, 1992). The latter phenomenon has been reported for pure cultures of *Desulfovibrio desulfuricans* and *Desulfobulbus propionicus* (Dannenberg *et al.*, 1992) as well as for anaerobic fresh-water sediments (Brunet & Garcia-Gil, 1996). However, the complete reoxidation of sulfide to sulfate is coupled to the ammonification of nitrate or nitrite, which contrasts the observed constant ammonium concentration (Lens *et al.*, 1999). The constant ammonium concentration also indicated the absence of the anammox process, i.e. the concomitant removal of nitrate/nitrite and ammonium (Van de Graaf *et al.*, 1996) in the studied sulfidogenic granular sludge. However, a possible turnover of nitrogen compounds in the system which cannot be detected by the nitrogen determinations applied cannot be excluded. Further research using for example ^{15}N-labeled substrates (Van de Graaf *et al.*, 1996) is needed for the elucidation of the exact prevailing nitrogen conversions in methanogenic and sulfidogenic granular sludge.

13.4.5 Role of sulfur compounds in anaerobic ammonium oxidation

In the last two decades, new interactions between the nitrogen and sulfur cycles have been discovered while studying anaerobic ammonium oxidation. A new process for simultaneous anaerobic ammonium oxidation and sulfate reduction was observed for the first time by Fdz-Polanco *et al.* (2001) – during anaerobic treatment of vinasse – who noticed that 50% of the nitrogen entering the reactor as total Kjeldahl nitrogen (TKN) was found as N_2 in the gas phase, and at the same time, 20% of the SO_4^{2-} originally present in the influent appears as S^{2-} in the effluent or H_2S in the gas, with a sulfur removal efficiency of 80%. Fdz-Polanco *et al.* (2001) proposed the following as a possible pathway:

$$SO_4^{2-} + 2\,NH_4^+ \rightarrow S + N_2 + 4\,H_2O \quad \Delta G = -47.8\,\text{kJ/mol} \tag{13.13}$$

This reaction could be obtained as a combination of the reactions involving nitrite formation in the anammox process:

$$3\,SO_4^{2-} + 4\,NH_4^+ \rightarrow 3\,S^{2-} + 4\,NO_2^- + 4\,H_2O + 8\,H^+ \tag{13.14}$$

$$3\,S^{2-} + 2\,NO_2^- + 8\,H^+ \rightarrow 3\,S + N_2 + 4\,H_2O \tag{13.15}$$

$$2\,NO_2^- + 2\,NH_4^+ \rightarrow N_2 + 4\,H_2O \tag{13.16}$$

There is no evidence yet that anammox bacteria possess the ability to oxidize ammonium with sulfate as an electron donor (Wang *et al.*, 2017; Bi *et al.*, 2020). Other studies confirmed, nevertheless, the pathway described in Equation 13.13 (Yang *et al.*, 2009; Rios-Del Toro *et al.*, 2018) even at low temperatures (14–15°C) (Wu *et al.*, 2020). Zhang *et al.* (2009b) observed that sulfate driven anaerobic ammonium oxidation occurs with acclimated anaerobic digested sludge in the absence of organic matter, but does not tend to occur spontaneously due to its low ΔG value. They showed that the reaction is enhanced by high sulfate and ammonia concentrations (NH_4^+-N and SO_4^{2-}-S concentrations were in a range between 84–270 mg L^{-1} and 450–740 mg L^{-1}, respectively) and low oxidation-reduction potential (ORP) (below −100 mV).

REFERENCES

Adams M. E., Warnes G. M. and Nicolas D. J. D. (1971). A sulfite dependent nitrate reductase from *Thiobacillus denitrificans*. *Biochimica et Biophysica Acta*, **235**, 398–406.

Akunna J. C., Bizeau C. and Moletta R. (1992). Denitrification in anaerobic digesters: possibilities and influence of wastewater COD/NOx-N ratio. *Environmental Technology*, **13**, 825–836.

Akunna J. C., Bizeau C. and Moletta R. (1993). Nitrate and nitrite reductions with anaerobic sludge using various carbon sources: glucose, glycerol, acetic acid, lactic acid and methanol. *Water Research*, **27**, 1303–1312.

Ashok V. and Hait S. (2015). Remediation of nitrate-contaminated water by solid-phase denitrification process—a review. *Environmental Science and Pollution Research*, **22**(11), 8075–8093. https://doi.org/10.1007/s11356-015-4334-9

Aslan S. and Cakici H. (2007). Biological denitrification of drinking water in a slow sand filter. *Journal of Hazardous Materials*, **148**(1–2), 253–258. https://doi.org/10.1016/j.jhazmat.2007.02.012

Baalsrud K. and Baalsrud K. S. (1954). Studies on *Thiobacillus denitrificans*. *Archiv für Mikrobiologie*, **20**, 34–62.

Baldensperger J. and Garcia J. L. (1975). Reduction of oxidized inorganic nitrogen compounds by a new strain of *Thiobacillus denitrificans*. *Archives of Microbiology*, **103**, 31–36.

Batchelor B. and Lawrence A. W. (1978). Autotrophic denitrification using elemental sulfur. *Journal Water Pollution Control Federation*, **50**(8), 1986–2001.

Bayrakdar A., Tilahun E. and Calli B. (2016). Biogas desulfurization using autotrophic denitrification process. *Applied Microbiology and Biotechnology*, **100**(2), 939–948. https://doi.org/10.1007/s00253-015-7017-z

Bi Z., Wanyan D., Li X. and Huang Y. (2020). Biological conversion pathways of sulfate reduction ammonium oxidation in anammox consortia. *Frontiers of Environmental Science & Engineering*, **14**, 1–11.

Bisogni J. J. and Driscoll C. T. (1977). Denitrification using thiosulfate and sulfide. *Journal of Environmental Engineering Division, American Society of Civil Engineers*, **103**, 593–604.

Blecon G., Gillet M., Martin G. and Philipot J. M. (1983). Autotrophic denitrification process by *Thiobacillus denitrificans* on sulfur-Maërl. *Revue Française des Sciences de l'Eau*, **2**, 267–279.

Boehler E. and Haldenwang L. (1992). The NEBIO tube reactor process for nitrate removal from drinking water. *Water Supply*, **10**(3), 71–77.

Boller M., Tschui M. and Gujer W. (1997). Effects of transient nutrient concentrations in tertiary biofilm reactors. *Water Science and Technology*, **36**(1), 101–109.

Bosch J., Lee K. Y., Jordan G., Kim K. W. and Meckenstock R. U. (2012). Anaerobic, nitrate-dependent oxidation of pyrite nanoparticles by *Thiobacillus denitrificans*. *Environmental Science and Technology*, **46**(4), 2095–2101. https://doi.org/10.1021/es2022329

Bouffard S. C., Hackl R. P. and Sobolewski A. (1998). Novel nitrate/sulphate closed-cycle biological treatment of mining effluents. European Conference on new advances in biological nitrogen anf phosphorus removal for municipal or industrial wastewaters Narbonne, France, October 12–14, 279–282.

Brunet R. C. and Garcia-Gil L. J. (1996). Sulfide-induced dissimilatory nitrate reduction to ammonia in anaerobic freshwater sediments. *FEMS Microbiology Ecology*, **21**, 131–138.

Cardoso R. B., Sierra-Alvarez R., Rowlette P., Flores E. R., Gómez J. and Field J. A. (2006). Sulfide oxidation under chemolithoautotrophic denitrifying conditions. *Biotechnology and Bioengineering*, **95**(6), 1148–1157. https://doi.org/10.1002/bit.21084

Claus G. and Kutzner H. J. (1985). Physiology and kinetics of autotrophic denitrification by *Thiobacillus denitrificans*. *Applied Microbiology and Biotechnology*, **22**, 283–288.

Costa R. B., Okada D. Y., Delforno T. P. and Foresti E. (2019). International Biodeterioration & Biodegradation Methane-oxidizing archaea, aerobic methanotrophs and nitrifiers coexist with methane as the sole carbon source. *International Biodeterioration and Biodegradation*, **138**, 57–62. https://doi.org/10.1016/j.ibiod.2019.01.005

Daims H., Lebedeva E., Pjevac P., Han P., Herbold C., Albertsen M., Jehmlich N., Palatinszky M., Vierheilig J., Bulaev A., Kirkegaards R. H., Von Bergen M., Rattei T., Bendinger B., Nielsen P. H. and Wagner M. (2015). Complete nitrification by Nitrospira bacteria. *Nature*, **528**, 504–509. https://doi.org/10.1038/nature16461

Dannenberg S., Kroder M., Dilling W. and Cypionka H. (1992). Oxidation of H_2, organic compounds and inorganic sulfur compounds coupled to reduction of O_2 or nitrate by sulfate-reducing bacteria. *Archives of Microbiology*, **158**, 93–99.

Della Rocca C., Belgiorno V. and Meriç S. (2006). An heterotrophic/autotrophic denitrification (HAD) approach for nitrate removal from drinking water. *Process Biochemistry*, **41**(5), 1022–1028. https://doi.org/10.1016/j.procbio.2005.11.002

Di Capua F., Papirio S., Lens P. N. L. and Esposito G. (2015). Chemolithotrophic denitrification in biofilm reactors. *Chemical Engineering Journal*, **280**, 643–657. https://doi.org/10.1016/j.cej.2015.05.131

Di Capua F., Ahoranta S. H., Papirio S., Lens P. N. L. and Esposito G. (2016). Impacts of sulfur source and temperature on sulfur-driven denitrification by pure and mixed cultures of *Thiobacillus*. *Process Biochemistry*, **51**(10), 1576–1584. https://doi.org/10.1016/j.procbio.2016.06.010

Di Capua F., Lakaniemi A. M., Puhakka J. A., Lens P. N. L. and Esposito G. (2017a). High-rate thiosulfate-driven denitrification at pH lower than 5 in fluidized-bed reactor. *Chemical Engineering Journal*, **310**, 282–291. https://doi.org/10.1016/j.cej.2016.10.117

Di Capua F., Milone I., Lakaniemi A. M., Lens P. N. L. and Esposito G. (2017b). High-rate autotrophic denitrification in a fluidized-bed reactor at psychrophilic temperatures. *Chemical Engineering Journal*, **313**, 591–598. https://doi.org/10.1016/j.cej.2016.12.106

Di Capua F., Milone I., Lakaniemi A. M., van Hullebusch E. D., Lens P. N. L. and Esposito G. (2017c). Effects of different nickel species on autotrophic denitrification driven by thiosulfate in batch tests and a fluidized-bed reactor. *Bioresource Technology*, **238**, 534–541. https://doi.org/10.1016/j.biortech.2017.04.082

Di Capua F., Pirozzi F., Lens P. N. L. and Esposito G. (2019). Electron donors for autotrophic denitrification. *Chemical Engineering Journal*, **362**(3), 922–937. https://doi.org/10.1016/j.cej.2019.01.069

Dolejs P., Paclík L., Maca J., Pokorna D., Zabranska J. and Bartacek J. (2015). Effect of S/N ratio on sulfide removal by autotrophic denitrification. *Applied Microbiology and Biotechnology*, **99**(5), 2383–2392. https://doi.org/10.1007/s00253-014-6140-6

Driscoll C. T. and Bisogni J. J. (1978). The use of sulfur and sulfide in packed bed reactors for autotrophic denitrification. *Journal of Water Pollution Control Federation*, **50**, 569–577.

Fdz-Polanco F., Fdz-Polanco M., Fernandez N., Urena M., Garcia P. and Villaverde S. (2001). New process for simultaneous removal of nitrogen and sulphur under anaerobic conditions. *Water Research*, **35**, 1111–1114.

Galloway J. N., Dentener F. J., Capone D. G., Boyer E. W., Howarth R. W., Seitzinger S. P., Asner G. P., Cleveland C. C., Green P. A., Holland E. A., Karl D. M., Michaels A. F., Porter J. H., Townsend A. R. and Vöosmarty C. J. (2004). Nitrogen cycles: past, present, and future. *Biogeochemistry*, **70**, 153–226. https://doi.org/10.1007/s10533-004-0370-0

Ganaye V., Fass S., Urbain V., Manem J. and Block J. C. (1996). Biodegradation of volatile fatty acids by three species of nitrate-reducing bacteria. *Environmental Technology*, **17**, 1145–1149.

Garcia-Gil L. J. and Golterman H. L. (1993). Kinetics of FeS-mediated denitrification in sediments from the Camargue. *FEMS Microbiology Ecology*, **13**, 85–92.

Garuti G., Dohanyos M. and Tilche A. (1992). Anaerobic-aerobic combined process for the treatment of sewage with nutrient removal: the ananox process. *Water Science and Technology*, **25**, 383–394.

Golterman H. L. (1991). Influence of FeS on denitrification in shallow waters. *Verhandlungen des Internationalen Verein Limnologie*, **24**, 3025–3028.

Gu J. D., Qiu W., Koenig A. and Fan Y. (2004). Removal of high NO_3^-concentrations in saline water through autotrophic denitrification by the bacterium *Thiobacillus denitrificans* strain MP. *Water Science and Technology*, **49**, 105–112.

Guerber F., Gerbier C. and Chazal P. M. (1998). Le Nitraflux, un nouveau procédé pour l'élimination des ions nitrate des eaux destinées à la potabilisation. Techniques, Sciences et Méthodes, L'eau, **4**, 107–114.

Guerrero R. B. S. and Zaiat M. (2018). Wastewater post-treatment for simultaneous ammonium removal and elemental sulfur recovery using a novel horizontal mixed aerobic-anoxic fixed-bed reactor configuration. *Journal of Environmental Management*, **215**, 358–365. https://doi.org/10.1016/j.jenvman.2018.03.074

Guynot M. E., Toribio A., Quevedo M. and Muxi L. (1998). Microflora of dissimilative nitrate reduction in a denitrifying reactor. *Applied Microbiology and Biotechnology*, **50**, 396–400.

Haider N., Morvan J., Le Cloirec P. and Martin G. (1988). Denitrification in a reactor packed with iron pyrite: laboratory studies. *Environmental Technology Letters*, **9**(5), 411–420.

Hanus F. and Bernard C. (1988). Dénitrification des eaux potables dans un réacteur biologique à élimination continue de la biomasse en excès: un procédé inédit. T.S.M. 243–246.

Hashimoto S., Furukawa K. and Shioyama M. (1987). Autotrophic denitrification using elemental sulfur. *Journal of Fermentation Technology*, **65**(6), 683–692.

Hendriksen H. V. and Ahring B. K. (1996). Integrated removal of nitrate and carbon in an upflow anaerobic sludge blanket (UASB) reactor: operating performance. *Water Research*, **30**, 1451–1458.

Hijnen W. A. M., Koning D., Kruithof J. C. and Van der Kooij D. (1988). The effect of bacteriological nitrate removal on the concentration of bacterial biomass and easily assimilable organic carbon compounds in ground water. *Water Supply*, **6**, 265–273.

Hole U. H., Vollack K. U., Zumft W. G., Eisenman E., Siddiqui R. O., Friedrich B. and Kroneck P. M. H. (1996). Characterization of the membranous denitrification enzimes nitrite reductase (cytochrome cd1) and copper containing nitrous oxide reductase from *Thiobacillus denitrificans*. *Archives of Microbiology*, **165**(1), 55–61.

Hooper A. B. and Terry K. R. (1973). Specific inhibitors of ammonia oxidation in Nitrosomonas. *Journal of Bacteriology*, **115**, 480–485.

Hoor A. T. (1976). Energetic aspect of the metabolism of reduced compounds in *Thiobacillus denitrificans*. *Antonie van Leewenhoek*, **42**, 483–492.

Hoor A. T. (1981). Cell yields and bioenergetics of *Thiomicrospira denitrificans* compared with *Thiobacillus denitrificans*. *Antonie van Leewenhoek*, **47**, 231–243.

Hu H. W. and He J. Z. (2017). Comammox—a newly discovered nitrification process in the terrestrial nitrogen cycle. *Journal of Soils and Sediments*, **17**(12), 2709–2717. https://doi.org/10.1007/s11368-017-1851-9

Hulshoff Pol L., Lens P., Stams A. J. M., Lettinga G. (1998). Anaerobic treatment of sulfate-rich wastewaters. *Biodegradation*, **9**, 213–224.

Jørgensen C. J., Jacobsen O. S., Elberling B. and Aamand J. (2009). Microbial oxidation of pyrite coupled to nitrate reduction in anoxic groundwater sediment. *Environmental Science and Technology*, **43**(13), 4851–4857. https://doi.org/10.1021/es803417s

Justin P. and Kelly D. P. (1978a). Growth kinetics of *Thiobacillus denitrificans* in anaerobic and aerobic chemostat culture. *Journal of General Microbiology*, **107**, 123–130.

Justin P. and Kelly D. P. (1978b). Metabolic changes in *Thiobacillus denitrificans* accompanying the transition from aerobic to anaerobic growth in continuous chemostat culture. *Journal of General Microbiology*, **107**, 131–137.

Kartal B., Maalcke W., De Almeida N., Ciprus I., Gloerich J., Geerts W., Op den Camp H. J. M., Harhangi H. R., Janssen-Megens E. M., Francoijs K.-J., Stunnenberg H. G., Keltjens J. T., Jetten M. S. M. and Strous M. (2011). Molecular mechanism of anaerobic ammonium oxidation. *Nature*, **479**, 127–130. https://doi.org/10.1038/nature10453

Kasper H. F., Tiedje J. M. and Firestone R. B. (1981). Denitrification and dissimilatory nitrate reduction to ammonium in digested sludge. *Canadian Journal of Microbiology*, **27**, 878–885.

Kelly D. P. (1982). Biochemistry of the chemolithotrophic oxidation of inorganic sulphur. *Philosophical Transactions of the Royal Society of London*, *B***298**, 499–528.

Kelly D. P. (1989). Physiology and biochemistry of unicellular sulfur bacteria. In: Autotrophic bacteria, H. G. Schlegel and B. Bowien (eds), Springer-Verlag, Science Tech Publisher, pp. 193–217.

Kelly D. P. and Harrison A. P. (1989). The genus *Thiobacillus*. In: Bergey's manual of systematic bacteriology, J. T. Staley, M. P. Bryant, N. Pfennig and J. D. Holt (eds), Vol. 3, pp. 1842–1858.

Khanongnuch R., Di Capua F., Lakaniemi A. M., Rene E. R. and Lens P. N. L. (2018). Effect of N/S ratio on anoxic thiosulfate oxidation in a fluidized bed reactor: Experimental and artificial neural network model analysis. *Process Biochemistry*, **68**, 171–181. https://doi.org/10.1016/j.procbio.2018.02.018

Khanongnuch R., Di Capua F., Lakaniemi A. M., Rene E. R. and Lens P. N. L. (2019a). H_2S removal and microbial community composition in an anoxic biotrickling filter under autotrophic and mixotrophic conditions. *Journal of Hazardous Materials*, **367**, 397–406. https://doi.org/10.1016/j.jhazmat.2018.12.062

Khanongnuch R., Di Capua F., Lakaniemi A. M., Rene E. R. and Lens P. N. L. (2019b). Long-term performance evaluation of an anoxic sulfur oxidizing moving bed biofilm reactor under nitrate limited conditions. *Environmental Science: Water Research and Technology*, **5**(6), 1072–1081. https://doi.org/10.1039/c9ew00220 k

Khanongnuch R., Di Capua F., Lakaniemi A. M., Rene E. R. and Lens P. N. L. (2019c). Transient–state operation of an anoxic biotrickling filter for H_2S removal. *Journal of Hazardous Materials*, **377**(3), 42–51. https://doi.org/10.1016/j.jhazmat.2019.05.043

Kimura K., Nakamura M. and Watanabe Y. (2002). Nitrate removal by a combination of elemental sulfur-based denitrification and membrane filtration. *Water Research*, **36**(7), 1758–1766. https://doi.org/10.1016/S0043-1354(01)00376-1

Koch G. and Siegrist H. (1997). Denitrification with methanol in tertiary filtration at wastewater treatment plant Zürich-Werdhölzli. *Water Science and Technology*, **36**(1), 165–172.

Koenig A. and Liu L. H. (1996). Autotrophic denitrification of landfill leachate using elemental sulphur. *Water Science and Technology*, **34**(5-6), 469–476.

Koenig A. and Liu L. (2004). Autotrophic denitrification of high-salinity wastewater using elemental sulfur: batch tests. *Water Environment Research*, **76**(1), 37–46. https://doi.org/10.2175/106143004×141564

Kong Z., Li L., Feng C., Dong S. and Chen N. (2016). Comparative investigation on integrated vertical-flow biofilters applying sulfur-based and pyrite-based autotrophic denitrification for domestic wastewater treatment. *Bioresource Technology*, **211**, 125–135.

Kostrytsia A., Papirio S., Frunzo L., Mattei M. R., Porca E., Collins G., Lens N. L. P. and Esposito G. (2018a). Elemental sulfur-based autotrophic denitrification and denitritation: microbially catalyzed sulfur hydrolysis and nitrogen conversions. *Journal of Environmental Management*, **211**, 313–322. https://doi.org/10.1016/j.jenvman.2018.01.064

Kostrytsia A., Papirio S., Morrison L., Ijaz U. Z., Collins G., Lens P. N. L. and Esposito G. (2018b). Biokinetics of microbial consortia using biogenic sulfur as a novel electron donor for sustainable denitrification. *Bioresource Technology*, **270**, 359–367. https://doi.org/10.1016/j.biortech.2018.09.044

Kruithof J. C., Van Bennekom C. A., Dierx H. A. L., Hijnen W. A. M., Van Paassen J. A. M. and Schippers J. C. (1988). Nitrate removal from groundwater by sulphur/limestone filtration. *Water Supply*, **6**, 207–217.

Kuai L. and Verstraete W. (1999). Autotrophic denitrification with elemental sulfur in small-scale wastewater treatment facilities. *Environmental Technology*, **20**, 201–209.

Kuenen J. G. (1989). Colorless sulfur bacteria. In: Bergey's manual of systematic bacteriology, J. T. Staley, M. P. Bryant, N. Pfennig and J. D. Holt (eds), Vol. 3, Springer, New York, NY, pp. 1834–1837.

Kuenen J. G., Robertson L. A. and Tuovinen O. H. (1992). The genera *Thiobacillus*, *Thiomicrospira* and *Thiosphaera*. In: The Prokaryotes, A. Balows, H. G. Trüper, M. Dworkin, W. Harder and K. H. Schlefer (eds), Springer-Verlag, New York, Vol. 3, pp. 2638–2657.

Kuypers M. M. M., Marchant H. K. and Kartal B. (2018). The microbial nitrogen-cycling network. *Nature Reviews Microbiology*, **16**(5), 263–276. https://doi.org/10.1038/nrmicro.2018.9

Lagas J. A. (2000). Survey of H_2S and SO_2 removal processes. In: Environmental technologies to treat sulfur pollution: principles and engineering, P. N. L. Lens and L. Hulshoff Pol (eds), IWA Publishing, London, pp. 237–264.

Lazarova V., Capdeville B. and Nikolov L. (1994). Influence of seeding conditions on nitrite accumulation in a denitrifying fluidized bed reactor. *Water Research*, **28**(5), 1189–1197.

Le Cloirec P. and Martin G. (1988). Simulation of the biological denitrification in a sulfur-calcium carbonate reactor. *Environmental Technology Letters*, **9**, 207–218.

Le Cloirec P., Martin G., Benbarka B. and Leroux A. Y. (1985). Un modèle mathématique de la dénitrification sur des filtres soufre-carbonate de calcium. *Chemical Engineering Journal*, **31**, B9–B18.

Le Gall J., Payne W. J., Morgan T. V. and Der Vartanian D. (1979). On the purification of nitrite reductase from *Thiobacillus denitrificans* and its reaction with nitrite under reducing conditions. *Biochemical and Biophysical Research Communications*, **87**(2), 355–362.

Lee K. H. and Sublette K. L. (1990). Reduction of nitric oxide to elemental nitrogen by *Thiobacillus denitrificans*. *Applied Biochemistry and Biotechnology*, **24/25**, 441–445.

Lens P. N. L., Visser A., Janssen A. J. H., Hulshoff Pol L. W. and Lettinga G. (1998). Biotechnological treatment of sulfate rich wastewaters. *Critical Reviews in Environmental Science and Technology*, **28**, 41–88.

Lens P., Sipma J., Hulshof Pol L. and Lettinga G. (1999). Effect of staging and nitrate addition on sulfidogenic acetate removal. *Water Research*, **34**, 31–42.

Lewandowski Z., Bakke R. and Characklis W. G. (1987). Nitrification and autotrophic denitrification in calcium alginate beads. *Water Science and Technology*, **19**(1/2), 175–182.

Li R., Morrison L., Collins G., Li A. and Zhan X. (2016). Simultaneous nitrate and phosphate removal from wastewater lacking organic matter through microbial oxidation of pyrrhotite coupled to nitrate reduction. *Water Research*, **96**, 32–41. https://doi.org/10.1016/j.watres.2016.03.034

Lin Y.-F. and Chen K.-C. (1995). Denitrification and methanogenesis in a co-immobilized mixed culture system. *Water Research*, **29**, 35–43.

Ma B., Wang S., Cao S., Miao Y., Jia F., Du R. and Peng Y. (2016). Biological nitrogen removal from sewage via anammox: recent advances. *Bioresource Technology*, **200**, 981–990. https://doi.org/10.1016/j.biortech.2015.10.074

Manconi I., Carucci A. and Lens P. (2007). Combined removal of sulfur compounds and nitrate by autotrophic denitrification in bioaugmented activated sludge system. *Biotechnology and Bioengineering*, **98**(3), 551–560. https://doi.org/10.1002/bit.21383

Mariotti A. (1992). La dénitrification in situ dans les eaux souterraines, processus naturels ou provoqués. *Les dossiers de la cellule Environnement*, **4**, 41–62.

McKenney D. J., Drury C. F. and Wang S. W. (2001). Effects of oxygen on denitrification inhibition, repression, and derepression in soil columns. *Soil Science Society of America Journal*, **65**(1), 126–132. https://doi.org/10.2136/sssaj2001.651126x

Mellbye B. L., Spieck E., Bottomley P. J. and Sayavedra-Soto L. A. (2017). Acyl-Homoserine lactone production in nitrifying bacteria of the genera. *Applied and Environmental Microbiology*, **83**(22), 1–13.

Mohseni-Bandpi A., Elliott D. J. and Zazouli M. A. (2013). Biological nitrate removal processes from drinking water supply – a review. *Journal of Environmental Health Science and Engineering*, **11**(1), 1–11. https://doi.org/10.1186/2052-336X-11-35

Montiel A. and Welté B. (1994). Nouveau procédé de dénitrification des eaux par le fer métallique. 11ème congrès JIE, Poitiers 1994.

Park J. H., Shin H. S., Lee I. S. and Bae J. H. (2002). Denitrification of high NO_3^--N containing wastewater using elemental sulfur; nitrogen loading rate and N_2O production. *Environmental Technology (United Kingdom)*, **23**(1), 53–65. https://doi.org/10.1080/09593332508618431

Pel R., Olhenduis R., Brand W., Vols A., Gottschal J. C. and Zwart K. B. (1997). Stable isotope analysis of a combiend nitrification-denitrification sustained by thermophilic methanotrophs under low-oxygen conditions. *Applied and Environmental Microbiology*, **49**, 1131–1141.

Percheron G., Michaud S., Bernet N. and Moletta R. (1998). Nitrate and nitrite reduction of a sulfide-rich environment. *Journal of Chemical Technology & Biotechnology*, **72**, 213–220.

Pereira A. D., Cabezas A., Etchebehere C., Chernicharo C. A. de L. and de Araújo J. C. (2017). Microbial communities in anammox reactors: a review. *Environmental Technology Reviews*, **6**(1), 74–93. https://doi.org/10.1080/21622515.2017.1304457

Philipot J. M. and Patte A. (1982). La dénitrification des eaux par voie biologique. *Techniques, Sciences et Méthodes, L'eau*, **77**(4), 165–172.

Philipot J. M. and Sibony J. (1986). L'élimination des nitrates dans les eaux potables. 16e Cong. Inter. & Expo. Distr. D'Eau, Rome, 11–15.

Picioreanu C., Van Loosdrecht M. C. M. and Heijnen J. J. (1997). Modelling the effect of oxygen concentration on nitrite accumulation in a biofilm airlift suspension reactor. *Water Science and Technology*, **36**(1), 147–156.

Posadas E., García-Encina P. A., Soltau A., Domínguez A., Díaz I. and Muñoz R. (2013). Carbon and nutrient removal from centrates and domestic wastewater using algal-bacterial biofilm bioreactors. *Bioresource Technology*, **139**, 50–58. https://doi.org/10.1016/j.biortech.2013.04.008

Pu J., Feng C., Liu Y., Li R., Kong Z., Chen N., Tong S., Hao C. and Liu Y. (2015). Pyrite-based autotrophic denitrification for remediation of nitrate contaminated groundwater. *Bioresource Technology*, **173**, 117–123. https://doi.org/10.1016/j.biortech.2014.09.092

Quintiliani C., Di Cristo C. and Leopardi A. (2018). Vulnerability assessment to trihalomethane exposure in water distribution systems. *Water (Switzerland)*, **10**(7), 1–15. https://doi.org/10.3390/w10070912

Ravarini P., Coutelle J. and Damez F. (1988). L'usine de Dennemont. Une unité de dénitrification-nitrification à grande échelle. *Techniques, Sciences et Méthodes, L'eau*, **83**(4), 235–240.

Richard Y. and Leprince A. (1982). Pollution par les nitrates: traitements disponibles. *Trib. Cebedeau*, **458**(35), 21–23.

Richard Y. and Partos J. (1986). Elimination biologique des nitrates en vue de la production d'eau potable. *Techniques, Sciences et Méthodes, L'eau*, **81**(3), 141–146.

Rios-Del Toro E. E., Valenzuela E. I., López-Lozano N. E., Cortés-Martínez M. G., Sánchez-Rodríguez M. A., Calvario-Martínez O., Sanchez-Carrillo S. and Cervantes F. J. (2018). Anaerobic ammonium oxidation linked to sulfate and ferric iron reduction fuels nitrogen loss in marine sediments. *Biodegradation*, **29**(5), 429–442. https://doi.org/10.1007/s10532-018-9839-8

Robertson L. A. and Kuenen J. G. (1992). The colourless sulfur bacteria. In: The prokaryotes, A. Balows, H. G. Trüper, M. Dworkin, W. Harder and K. H. Schlefer (eds), Springer-Verlag, New York, Vol. 3, pp. 385–413.

Roennefhart K. W. (1986). Nitrate elimination with heterotrophic aquatic microorganisms in fixed bed reactors with buoyant carriers. *Aqua*, **5**, 283–285.

Rooden A., Bosma W. W., Buisman C. N. L. and De Munck P. R. (1995). The treatment of photographic waste streams. *Med. Fac. Landbouww. Univ. Gent.*, **60**, 2729–2734.

Rouse J. D., Sumida K., Kida K. and Furukawa K. (1999). Maintainability of denitrifying granular sludge in soft to marginally hard waters in an upflow sludge-blanket reactor. *Environmental Technology*, **20**, 219–225.

Sahinkaya E., Yurtsever A., Aktaş Ö, Ucar D. and Wang Z. (2015). Sulfur-based autotrophic denitrification of drinking water using a membrane bioreactor. *Chemical Engineering Journal*, **268**, 180–186. https://doi.org/10.1016/j.cej.2015.01.045

Sant'Anna F. P. S., Martin G. and Taha S. (1996). Effect of sodium chloride on the denitrification of ion exchange resin eluates by a sulphur-oxidizing chemoautotrophic bacterium. *Revue des Sciences de l'Eau*, **3**, 333–350.

Sawhney V. and Nicolas D. J. D. (1977) Sulphite-and NADH-dependent nitrate reductase from *Thiobacillus denitrificans*. *Journal of General Microbiology*, **100**, 49–58.

Schedel M. and Trüper H. G. (1980). Anaerobic oxidation of thiosulfate and elemental sulfur in *Thiobacillus denitrificans*. *Archives of Microbiology*, **124**, 205–210.

Schippers J. C., Kruithof J. C., Mulder F. G. and Van Lieshout J. W. (1987). Removal of nitrate by slow sulphur/limestone filtration. *Aqua*, **5**, 274–280.

Schramm A., Larsen L. H., Revsbech N. P., Ramsing N. B., Amann R. and Schleifer K.-H. (1996). Structure and function of a nitrifying biofilm determined by in situ hybridisation and the use of microelectrodes. *Applied and Environmental Microbiology*, **62**, 4641–4647.

Schulz H. N., Brinkhoff T., Ferdelman T. G., Hernández Mariné M., Teske A. and Jørgensen B. B. (1999). Dense populations of a giant sulfur bacterium in namibian shelf sediments. *Science*, **284**(80), 493–495.

Sibony J. (1979). Traitement physico-chimique des nitrates et de l'amoniaque pour la production d'eau potable. TSM l'Eau, **6**, 355–359.

Sikora L. J. and Keeney D. R. (1976). Evaluation of a sulfur *Thiobacillus denitrificans* nitrate removal system. *Journal of Environmental Quality*, **5**(3), 298–302.

Soares M. I. M., Belkin S. and Abeliovich A. (1988). Biological groundwater denitrification: laboratory studies. *Water Science and Technology*, **20**(3), 189–195.

Stein L. Y. and Klotz M. G. (2016). The nitrogen cycle. *Current Biology*, **26**(3), R94–R98. https://doi.org/10.1016/j.cub.2015.12.021

Sublette K. L., Kolhatkar R. and Raterman K. (1998). Technological aspects of the microbial treatment of sulfide-rich wastewaters: a case study. *Biodegradation*, **9**, 259–271.

Sweerts J. P. R. A., De Beer D., Nielsen L. P., Verdouw H., Van Den Heuvel J. C., Cohen Y. and Cappenberg T. E. (1990). Denitrification by sulphur oxidizing *Beggiatoa* spp. mats on freshwater sediments. *Nature*, **344**, 762–763.

Tarre S., Armon R., Shelef G. and Green M. (1995). Effects of water characteristics on granular sludge formation in a USB reactor for denitrification of drinking water. *Water Science and Technology*, **30**(9), 141–147.

Taylor B. F. and Hoare D. S. (1971). Thiobacillus denitrificans as an obligate chemolithotroph – II. Cell suspension and enzymic studies. *Archives of Microbiology*, **80**, 262–276.

Taylor B. F. (1989). Thermotolerance of adenylsulfate reductase from *Thiobacillus denitrificans*. *FEMS Microbiology Letters*, **59**, 351–354.

Trouvé C., Chazal P. M., Guéroux B. and Sauvaître N. (1998a) Denitrication by new strains of *Thiobacillus denitrificans* under non standard physico-chemical conditions. Effect of temperature, pH and sulphur source. *Environmental Technology*, **19**, 601–610.

Trouvé C., Chazal P. M. and Guéroux B. (1998b). Influence of the nature of ferrous sulphide upon denitrification by *Thiobacillus denitrificans* at a pilot scale. *Journal Européen d'Hydrologie*, **29**(1), 43–62.

Trouvé C. and Chazal P. M. (1999). Autotrophic denitrification by *Thiobacillus denitrificans* with thiosulphate as sole energy source. Pilot scale experiments at low temperature. *Environmental Technology*, **20**(4), 387–395.

Ucar D., Yilmaz T., Di Capua F., Esposito G. and Sahinkaya E. (2020). Comparison of biogenic and chemical sulfur as electron donors for autotrophic denitrification in sulfur-fed membrane bioreactor (SMBR). *Bioresource Technology*, **299**, 122574. https://doi.org/10.1016/j.biortech.2019.122574

Van Benthum W. A. J., Van Loosdrecht M. C. M. and Heijnen J. J. (1997). Process design for nitrogen removal using nitrifying biofilm and denitrifying suspended growth in a biofilm airlift suspension reactor. *Water Science and Technology*, **36** (1), 119–128.

Van Der Hoek J.-P., Hijnen W. A. M., Van Benekomm C. A. and Mijnarends B. J. (1992). Optimization of the sulfur-limestone filtration process for nitrate removal from groundwater. *Journal of Water Supply: Research and Technology-Aqua*, **41**(4), 209–218.

Van de Graaf A. A., de Bruijn P., Robertson L. A., Jetten M. S. M. and Kuenen J. G. (1996). Autotrophic growth of anaerobic, ammonium-oxidising micro- organisms in a fluidized bed reactor. *Microbiology*, **142**, 2187–2196.

Van Kessel M., Speth D., Albertsen M., Nielsen P. H., Op den Camp H. J. M., Kartal B., Jetten M. S. M. and Lücker S. (2015). Complete nitrification by a single microorganism. *Nature*, **528**, 555–559 https://doi.org/10.1038/nature16459

Wan X., Xiao P., Zhang D., Lu P., Yao Z. and He Q. (2016).The kinetics for ammonium and nitrite oxidation under the effect of hydroxylamine. *Water Science and Technology*, **73**(5), 1067–1073. https://doi.org/10.2166/wst.2015.583

Wang H. and Qu J. (2003). Comparison of two combined bioelectrochemical and sulfur autotrophic denitrification processes for drinking water treatment. *Journal of Environmental Science and Health – Part A Toxic/Hazardous Substances and Environmental Engineering*, **38**(7), 1269–1284. https://doi.org/10.1081/ESE-120021125

Wang J. and Chu L. (2016). Biological nitrate removal from water and wastewater by solid-phase denitrification process. *Biotechnology Advances*, **34**(6), 1103–1112. https://doi.org/10.1016/j.biotechadv.2016.07.001

Wang X., Sun G. and Zhu Y. (2017). Thermodynamic energy of anaerobic microbial redox reactions couples elemental biogeochemical cycles. *Journal of Soils and Sediments*, **17**, 2831–2846. https://doi.org/10.1007/s11368-017-1767-4

Watsuntorn W., Ruangchainikom C., Rene E. R., Lens P. N. L. and Chulalaksananukul W. (2017). Hydrogen sulfide oxidation under anoxic conditions by a nitrate-reducing, sulfide-oxidizing bacterium isolated from the Mae Um Long Luang hot spring, Thailand. *International Biodeterioration and Biodegradation*, **124**, 196–205. https://doi.org/10.1016/j.ibiod.2017.06.013

Watsuntorn W., Ruangchainikom C., Rene E. R., Lens P. N. L. and Chulalaksananukul W. (2019). Comparison of sulphide and nitrate removal from synthetic wastewater by pure and mixed cultures of nitrate-reducing, sulphide-oxidizing bacteria. *Bioresource Technology*, **272**, 40–47. https://doi.org/10.1016/j.biortech.2018.09.125

Watsuntorn W., Khanongnuch R., Chulalaksananukul W., Rene E. R. and Lens P. N. L. (2020a). Resilient performance of an anoxic biotrickling filter for hydrogen sulphide removal from a biogas mimic: steady, transient state and neural network evaluation. *Journal of Cleaner Production*, **249**, 119351. https://doi.org/10.1016/j.jclepro.2019.119351

Watsuntorn W., Thammaporn K., Rene E. R., Lens P. N. L. and Chulalaksananukul W. (2020b). Draft Genome Sequence and Annotation of *Paracoccus versutus* MAL 1HM19, a Nitrate-Reducing, Sulfide-Oxidizing Bacterium. *Microbiology Resource Announcements*, **9**(10), e01419–19.

Widdel F. (1988). Microbiology and ecology of sulfate- and sulfur reducing bacteria. In: Biology of anaerobic micro-organisms, A. J. B. Zehnder (ed.), John Wiley & Sons, New York, pp. 469–586.

Wu L., Yan Z., Li J., Huang S., Li Z., Shen M. and Peng Y. (2020). Low temperature advanced nitrogen and sulfate removal from landfill leachate by nitrite-anammox and sulfate-anammox. *Environmental Pollution*, **259**, 113763. https://doi.org/10.1016/j.envpol.2019.113763

Xiushan Y., Garuti G. and Tilche A. (1993). Denitrification *with Thiobacillus denitrificans* in the ANANOX® process. *Biotechnology Letters*, **15**(5), 531–536.

Yang Z., Zhou S. and Sun Y. (2009). Start-up of simultaneous removal of ammonium and sulfate from an anaerobic ammonium oxidation (anammox) process in an anaerobic up-flow bioreactor. *Journal of Hazardous Materials*, **169**(1–3), 113–118. https://doi.org/10.1016/j.jhazmat.2009.03.067

Yang Y., Chen T., Morrison L., Gerrity S., Collins G., Porca E., Li R. and Zhan X. (2017). Nanostructured pyrrhotite supports autotrophic denitrification for simultaneous nitrogen and phosphorus removal from secondary effluents. *Chemical Engineering Journal*, **328**, 511–518. https://doi.org/10.1016/j.cej.2017.07.061

Zhang L., Zheng P., He Y. and Jin R. (2009a). Performance of sulfate-dependent anaerobic ammonium oxidation. *Science in China, Series B: Chemistry*, **52**(1), 86–92. https://doi.org/10.1007/s11426-008-0161-x

Zhang Y. C., Slomp C. P., Broers H. P., Passier H. F. and Van Cappellen P. (2009b). Denitrification coupled to pyrite oxidation and changes in groundwater quality in a shallow sandy aquifer. *Geochimica et Cosmochimica Acta*, **73**(22), 6716–6726. https://doi.org/10.1016/j.gca.2009.08.026

Zhou M., Wang W. and Chi M. (2009). Enhancement on the simultaneous removal of nitrate and organic pollutants from groundwater by a three-dimensional bio-electrochemical reactor. *Bioresource Technology*, **100**(20), 4662–4668. https://doi.org/10.1016/j.biortech.2009.05.002

Zou G., Papirio S., Lakaniemi A. M., Ahoranta S. H. and Puhakka J. A. (2016). High rate autotrophic denitrification in fluidized-bed biofilm reactors. *Chemical Engineering Journal*, **284**, 1287–1294. https://doi.org/10.1016/j.cej.2015.09.074

Chapter 14

Synthesis and application of sulfur nanoparticles

Sudeshna Saikia and Piet N. L. Lens

14.1 INTRODUCTION

Elemental sulfur (S^0) is one of the key intermediates in the biogeochemical cycle of sulfur (S). S^0 has a plethora of technological applications, such as in the production of sulfuric acid, chemical fibres, plastics, enamels, antimicrobial agents, insecticides, fumigants and metal glass cements, in the manufacture of dyes, fertilisers and gun-powder, and the vulcanisation of rubber (Chaudhuri & Paria, 2010; Deshpande *et al.*, 2008; Guo *et al.*, 2006; Xie *et al.*, 2009). In addition to oxidation of sulfide or polysulfide in the S cycle, also disproportionation of thiosulfate in acidic conditions results in the production of elemental sulfur (Boyd & Druschel, 2013). Depending on the conditions, the size of the S^0 particles may remain below 100 nm, and they are then known as sulfur nanoparticles (SNPs) (Tripathi *et al.*, 2018).

SNPs have garnered special research interest in the environmental sector and clinical therapeutics (Roy & Goswami, 2013). These novel applications of SNPs are attributed to their unique and improved physico-chemical properties, compared to larger S^0 particles. Among the semiconductor nanomaterials, the huge demand for SNPs is attributed to their multifaceted roles as antioxidant (Kasote *et al.*, 2015), antimicrobial agent (Choudhury *et al.*, 2013), anticancer agent (Shankar & Rhim, 2018a) as well as heavy metals adsorption material (Ghanemi *et al.*, 2011). In addition, there has been renewed interest in SNP

DOI: 10.2166/9781789060966_0445

doping in modification of carbon nanotubes, lithium/sulfur battery cathodes, infrared thermal imaging technology, temperature sensing probes and self-healing materials (Barkauskas *et al.*, 2007; Soleimani *et al.*, 2013; Zheng *et al.*, 2006). Thus, the potential application of SNPs necessitates the demand for their synthesis with precision in particle size distribution and surface properties.

In SNPs, sulfur atoms combine together by covalent bonds forming chains or rings (Häkkinen, 2012; Chapter 2). To date, several physico-chemical and biological techniques have been used for synthesis of SNPs. The synthesis yields monoclinic (β) and orthorhombic (α) SNPs with the most common allotropic form S_8 (Guo *et al.*, 2006). Figure 14.1 shows the monoclinic and orthorhombic forms of SNPs. Characterisation of the SNPs revealed particle size distribution for β-SNPs and α-SNPs are 20–50 and 10–50 nm, respectively (Roy & Goswami, 2013). SNPs have a varying shape, ranging from spherical and cylindrical shapes to tubes, wires and rods (Urakaev *et al.*, 2018).

The mechanism and kinetics of the formation and growth of SNPs suggests that nucleation and growth play a significant role in controlling particle size and particle size distribution in SNPs (Chaudhuri & Paria, 2010). Therefore, surfactant assisted routes have been introduced to synthesise SNPs and have precise control over the size distribution (Chaudhuri & Paria, 2011). Compared to other methods, biological methods are promising due to their low operational cost, high efficiency, suitability for low concentrations and for their sustainable and environment friendly approach (Tripathi *et al.*, 2018).

Considering the huge application of SNPs and the fact that these are size-dependent, there has been a constant search for specific methods to control the size and research on the effect of operating parameters to maintain the size

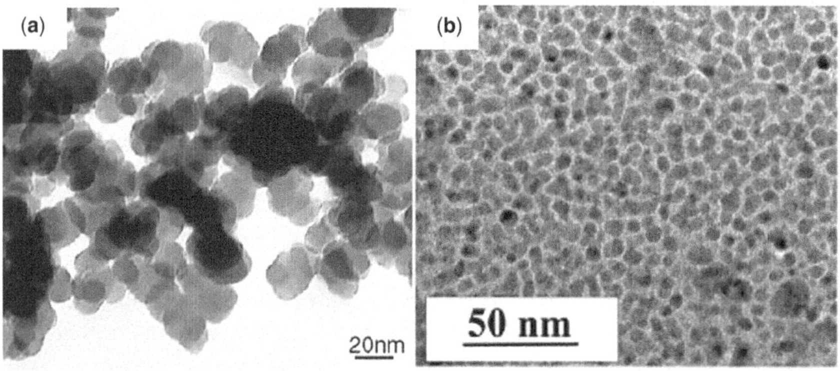

Figure 14.1 Transmission electron microscopy (TEM) image of a) monoclinic SNPs (β-SNPs) and b) orthorhombic SNPs (α-SNPs) (Guo *et al.*, 2006; Tripathi *et al.*, 2018).

distribution (Chaudhuri & Paria, 2010). This chapter provides a comprehensive topical review of the different physico-chemical and biological methods for synthesis of SNPs. It also includes the characterisation of SNPs, such as the effect of synthesis parameters (type of acid and surfactant) for the purpose of obtaining a controlled size of the SNPs. Various applications of SNPs have been overviewed as well.

14.2 SYNTHESIS OF SULFUR NANOPARTICLES (SNPs)

14.2.1 Physico-chemical methods

14.2.1.1 Mechano-chemical synthesis of SNPs

Mechanical milling, eventually assisted by high-power ultrasonic treatment, is a robust and energy efficient synthesis method to generate nanoparticles (NPs) with varying shapes and dimensionalities (Xing *et al.*, 2013). The principle of the milling process is the generation of frictional force due to the collision between the surfaces of the materials which results in increasing temperature, pressure and internal energy. Both colloid and ball milling have been investigated for the production of SNPs. A colloid mill is a type of rotor-stator grinder often used to prepare NPs in suspension through high level mechanical shear (Fan *et al.*, 2009). In ball milling, dispersions pass through a canister of tiny balls present in the grinding machine. The flow of the dispersion between the tiny balls generates friction and thus induces a grinding mechanism (Dhand *et al.*, 2015). Despite their extensive use, challenges such as polydispersity particle size, agglomeration, oxidation and reduced anisotropy limit the application of conventional milling processes (Saravanan *et al.*, 2009).

In order to improve the SNP production by mechanical milling, a surfactant assisted route has been developed: colloid milling (Turganbay *et al.*, 2012). In this method cetyltrimethylammonium bromide (CTAB), sulfanol, sodium lignin sulfonate, sodium salt of polynaphthalene sulfonic acid or water soluble polymer sodium carboxymethylcellulose (NaCMC) are added as surface active agents to reduce the surface tension at the solid/liquid interface and thus facilitate the formation of a stable suspension of sulfur. As a result, the NPs maintain shape homogeneity at the end of the process. Turganbay *et al.* (2013) further improved the yield of SNP production by colloid milling of crystalline sulfur by ultrasonic dispersion of the surfactant solution.

The ball milling method for the production of SNPs is performed through mechanical activation (MA) with a cage rotation rate regulated in the range of 100–600 rpm and an apparatus made of hardened stainless steel with a number of steel balls (Urakaev *et al.*, 2016). The dispersion medium used in this mechano-chemical synthesis consists of sodium thiosulfate pentahydrate/succinic acid system with S^0 as final product of the reaction:

$$Na_2S_2O_3 \cdot 5\,H_2O + y\,C_4H_6O_4 + z\,Na_2SO_3 \leftrightarrow (z+1)Na_2SO_3 + y\,C_4H_6O_4 + 5\,H_2O + S_n{}^0 \quad (14.1)$$

Succinic acid ($C_4H_6O_4$) and sodium thiosulfate (Na_2SO_3) act as catalyst and diluent, respectively. The stoichiometric coefficients in reaction 14.1 for the catalyst and diluent were $y = 1$ and $z = 19.6$, respectively. The key parameter of the mechanical activation (MA) process in this ball milling is the relative rate of interaction of grinding bodies, i.e. balls and cylinder walls. Nanosulfur (S_n^0) is formed due to the decomposition of the unstable intermediate, thiosulfuric acid ($H_2S_2O_3$) according to the following reaction:

$$S_2O_3{}^{2-} + 2H^+ \leftrightarrow H_2S_2O_3 \rightarrow H_2SO_3 + S \quad (14.2)$$

$$S + S \leftrightarrow S_2 + S \ldots \leftrightarrow S_{n-1} + S \rightarrow S_n \quad (14.3)$$

Here, S_n denotes SNPs in reaction 14.3. Thus, the SNPs obtained from MA are spherical and 75 nm in size.

14.2.1.2 SNP production using aqueous surfactant media

The process of nanoparticle formation needs to undergo nucleation and growth of the S^0 to obtain their equilibrium size. Nucleation is a process whereby small stable particles (with a critical radii) are formed in the liquid called nuclei. This leads to crystal growth, where the size of the particles increases due to deposition of atoms on nuclei from the bulk liquid or because of collision between the nuclei. Nucleation and growth play a crucial role in particle size, particle size distribution and nature of the S^0 phase.

The first concept of nucleation and growth was proposed by LaMer and Kenyon (1947), who explored the mechanism for the formation of nanocrystals or colloids from a homogeneous super-saturated aqueous solution. LaMer studied the synthesis of sulfur sols from the acid catalysed reaction of sodium thiosulfate and observed nucleation is a very fast process, whereas the growth is the rate limiting step. According to the process, in an acidic solution, sodium thiosulfate undergoes a disproportionation reaction to produce SNPs according to the following:

$$Na_2S_2O_3 + 2HCl \rightarrow 2NaCl + SO_2 + S\downarrow + H_2O \quad (14.4)$$

Eventually, the overall size of the colloidal sulfur was noticed to be dependent on the crystal growth and agglomeration. In addition to the diffusion rate and reaction rate, growth depends on several parameters such as temperature, reactant concentration, super-saturation, time, surface energy and hydrodynamic condition.

Taking these parameters into account, the use of surfactants has been introduced as they can regulate the growth of the particles by controlling these key parameters (Chaudhuri & Paria, 2010). Different organic (oxalic, formic, malic) and inorganic (hydrochloric, sulfuric, nitric) acids were used to investigate the effect on the SNPs particle size (Chaudhuri & Paria, 2010). Organic acids were best with small particle size (Chaudhuri & Paria, 2010) and sharp particle size distribution. This is attributed to the lower nucleation rate and more negative zeta potential values of the SNPs produced in the presence of organic acids. Similarly, the effect of different surfactants (sodium dodecylbenzene sulfonate (SDBS), CTAB and TX-100) was evaluated in the same study. The anionic surfactant SDBS was found to be highly effective for obtaining a uniform size of NP, whereas the smallest size SNPs was obtained by using CTAB (Chaudhuri & Paria, 2010). In addition to SDBS, alumina powder can be used to facilitate efficient separation of SNPs from the suspension (Ghanemi *et al.*, 2011).

Natural biopolymers such as chitosan and water soluble polymers such as polyethylene glycol (PEG) and poly-N-vinylpyrrolidone (PVP) can be used for stabilising the nanoparticles (Urakaev *et al.*, 2018). PEGylation is the simplest and cheapest method which makes the surface of the SNPs hydrophilic and induces a steric barrier by anchoring long and mobile PEG chains. Thermogravimetric analysis revealed that PEGylated SNPs are stable even at 300°C (Chaudhuri & Paria, 2010). PEG 200 and 400 are safe biocompatible polymers that confer lesser toxicity compared to other ionic surfactants. They not only hydrophilise the surface of the SNPs but also make their distribution even in a controlled manner in any polar solvent, such as water, agar or broth media. On the contrary, other surfactants or stabilising agents, e.g. a conjugate of Span 80 and Tween 80 are highly metastable in nature (Roy & Goswami, 2013). The low interfacial free energy of PEG in aqueous media can impart steric hindrance and hence control the size of nanoparticles (Otsuka *et al.*, 2003). Therefore, the use of these biocompatible polymers is most preferable since they facilitate the sustained release of SNPs in colloidal media and the potency remains intact for a longer time.

14.2.1.3 SNP production by the reverse microemulsion technique

Microemulsions have been extensively used for synthesis of monodisperse nanoparticles with a narrow size distribution (Schwuger *et al.*, 1995). A microemulsion consists of three components: a polar phase (e.g. water), a non-polar phase (e.g. oil) and a surfactant and co-surfactant (Dhand *et al.*, 2015). The surfactant molecules adsorb on the surface of the microemulsion droplets, thus reducing the interfacial energy and providing a steric barrier between the polar and the non-polar phase to prevent agglomeration, thereby controlling the size of the particles throughout the process (Solanki & Murthy, 2011). The co-surfactant further reduces the interfacial tension between the microemulsions and thus promotes formulation homogeneity and stability (Li *et al.*, 2019).

Depending on the proportion, orientation and hydrophile-lypophile balance (HLB) value of the surfactant used, two types of microemulsions are considered:

oil-in-water microemulsions (o/w) and water-in-oil (w/o) microemulsions (Malik *et al.*, 2012). The latter are also referred to as reverse microemulsions since the hydrophilic group of the surfactant is oriented inward creating an aqueous/polar (water) core, whereas the hydrophobic moieties point outward to the non-polar (oil) organic medium. Consequently, the steric barrier provided by the surfactant restricts the growth of precipitated particles and impedes intergrain coagulation (Solanki & Murthy, 2011). Thus produced nanoparticles have better characteristics of microscopic homogeneity, minimum surface energy, and isotropic and thermodynamic stability (Soleimani *et al.*, 2013). The reverse microemulsion is the most common, particularly for the synthesis of SNPs.

The production of SNP makes use of two microemulsions (microemulsion I and microemulsion II) that contain two different reactants (Figure 14.2): one of which contains the precursor (sodium polysulfide, Na_2S_x) and the other harbouring the precipitating agent (hydrochloric acid, HCl) (Guo *et al.*, 2006). Stable microemulsions are prepared by using water as polar/aqueous phase, theolin as non-polar/oil phase, a mixture of Span80 and Tween80 as surfactant and butanol as co-surfactant (Table 14.1). Upon mixing of the two microemulsions in appropriate ratios, the collision and fusion of micelles occur,

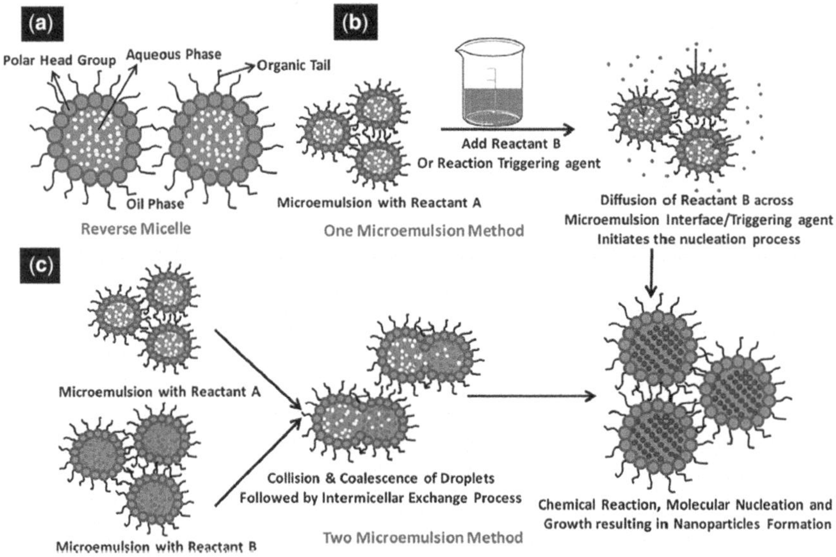

Figure 14.2 Overview of synthesis of nanoparticles by reverse microemulsion technique: (a) structure of the reverse micelle, (b) the reactions with the two reactants (For SNP, Reactant A: Na_2S_x and Reactant B: HCl) and (c) the processes (collision, exchange, nucleation, growth) required for the production of nanoparticles (reused with permission from Dhand *et al.*, Copyright (2015) RSC Advances).

Table 14.1 Overview of physicochemical synthesis methods of sulfur nanoparticles.

Method	Aqueous Phase	Oil Phase	Surface Modifier/ Surfactant/ Co-Surfactant	Reaction Time (Minute)	pH	Temperature (°C)	Allotropes	Particle Size (nm)	Reference
Reverse microemulsions	Sodium polysulfide (microemulsion I) hydrochloric acid (microemulsion II)	Theolin; cyclohexane	Butanol, Span80 and Tween80/Triton X-100 (TX-100)	30	7–7.5	Room temperature, (28 ± 0.5)	Cyclo-S8 (β-sulfur)	20; 22	Guo *et al.* (2006); Soleimani *et al.* (2013)
	Hydrogen sulfide, $FeCl_3$–malic acid chelate	Cyclohexane	TX-100, n-hexanol	15	7–7.5	Room temperature	Cyclo-S8 (α-sulfur)	10	Deshpande *et al.* (2008)
Aqueous surfactant media	Sodium thiosulfate, oxalic acid, hydrochloric acid, sulfuric acid, nitric acid	N/A	TX-100, sodium dodecyl sulfate (SDS), cetyltrimethyl ammonium bromide (CTAB)	N/A	6.4–6.5	Room temperature	α-sulfur	30	Chaudhuri & Paria (2010)
	Sodium thiosulfate, hydrochloric acid	N/A	Alumina powder, sodium dodecylbenzene sulfonate (SDBS)	20	8.5	Room temperature	α-sulfur	45–55	Ghanemi *et al.* (2011)
	Sodium thiosulfate, oxalic acid	N/A	CTAB	40	6.4–6.5	Room temperature	N/A	40 ± 10	Baskar *et al.* (2015)
	Sodium thiosulfate, sulfuric acid	N/A	Polyethyleneglycol (PEG), poly-N-vinylpyrrolidone (PVP)	N/A	N/A	60–70	β-sulfur	1–2500	Abuyeva *et al.* (2018)
	Sodium thiosulfate, hydrochloric acid	N/A	Chitosan	40	7	25	Cyclo-S8	1–30	Shankar *et al.* (2018b)
	Sublimed sulfur, anhydrous ethanol	N/A	Cysteine	30		N/A	catena-Sx; α-sulfur	50–100	Xie *et al.* (2009)
	Sublimed sulfur	N/A	PEG-200	30		Room temperature	Cyclo-S8	500	Liu *et al.* (2016)

(*Continued*)

Table 14.1 Overview of physicochemical synthesis methods of sulfur nanoparticles (*Continued*).

Method	Aqueous Phase	Oil Phase	Surface Modifier/ Surfactant/ Co-Surfactant	Reaction Time (Minute)	pH	Temperature (°C)	Allotropes	Particle Size (nm)	Reference
	Sulfur powder, sodium sulfide	N/A	PEG-400	240	10	Room temperature	Cyclo-S8	47	Ghoraba & Soleymanpour (2017)
Mechano-chemical	Crystalline sulfur, aerosol	N/A	CTAB, sulfanol, sodium lignin sulfonate, sodium salt of polynaphthalene sulfonic acid	N/A	N/A	Room temperature	Cyclo-S8	1300–2000	Turganbay *et al.* (2012)
	Crystalline sulfur		CTAB, SDBS, TX-100	>30	N/A	Room temperature	β-sulfur	1000–1500	Turganbay *et al.* (2013)
Electrochemical	Sodium thiosulfate, platinum and gold electrodes	N/A	N/A	60	N/A	22 ± 1	Cyclo-S8	35–65	Shamsipur *et al.* (2011)
Core-linked block copolymer micelles	Sodium polysulfide	N/A	4-bromophenyl vinyl sulfide N-isopropyl acrylamide (amphiphilic block copolymer); 1,2,3-trichloropropane, 1,2-dibromoethane, 2,5-dibromothiophene, and 1,4-dibromobenzene (cross-linkers)	240	N/A	30	Cyclo-S8	57–153	Nakabayashi *et al.* (2017)

followed by the exchange of reactants in the same micellar compartment. Once the critical number of molecules (N_{crit}) is obtained inside the micelle, nucleation initiates and results in SNPs formation (Dhand *et al.*, 2015). The chemical reaction that occurs when the microemulsion I and microemulsion II are mixed together is:

$$Na_2S_x + 2HCl \rightarrow 2NaCl + H_2S + (x - 1)S \qquad (14.5)$$

The product (S) formed in reaction 14.5 is insoluble and thus precipitates as light yellow powder. The SNPs obtained are characterised as spherical, higher purity monoclinic particles with an average size of 20 nm (Guo *et al.*, 2006). A similar method with different proportional ratios of the reactants and using cyclohexane as the oil phase produced monoclinic SNPs of about 22 nm (Soleimani *et al.*, 2013). Various control parameters such as type of oil phase, water-to-surfactant molar ratio, type and concentration of reactants (e.g. reducing agent or surfactant) have been observed to play a significant role in influencing the size and stability of SNPs.

Due to the generation of H_2S fumes (that are toxic and corrosive, see Chapter 11) in this process, SNP synthesis by the reverse microemulsion technique was made less hazardous by catalytic conversion of H_2S gas using iron chelates (Figure 14.3; Table 14.1). Iron chelates are generally prepared by combining iron

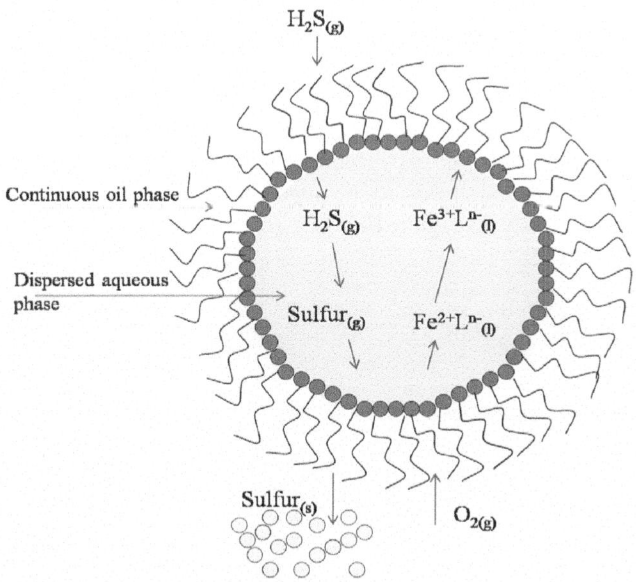

Figure 14.3 Schematic representation of synthesis of SNPs in reverse microemulsion system (reused with permission from Deshpande *et al.*, Copyright (2008) Nanoscale Research Letters).

salts and chelating agents in definite ratios. Fe^{3+} is used as a regenerable oxidant for desulfurisation processes in association with several chelating agents, such as ethylenediaminetetraacetic acid (EDTA), nitrilotriacetic acid (NTA) or diethylenetriamine pentaacetic acid (DTPA) (Demmink *et al.*, 2002). Due to the higher toxicity and lower biodegradation rate of EDTA, NTA and DTPA, the use of carboxylic acids (i.e. malic acid) is proposed as chelating agent as these possess good chelating property with a relative faster biodegradation rate (Deshpande *et al.*, 2008). The iron chelate used was synthesised by dissolving ferric chloride ($FeCl_3$) as iron salt and malic acid as chelating agent (Deshpande *et al.*, 2008). The components for the reverse microemulsion technique were thus $FeCl_3$–malic acid (1:3 mole ratio) as aqueous phase, cyclohexane as oil phase and a mixture of Triton X-100 (TX-100) and n-hexanol as surfactant and co-surfactant, respectively (Deshpande *et al.*, 2008).

Simultaneous oxidation and reduction reactions take place for converting H_2S to elemental sulfur (S^0) and ferric chloride to ferrous chloride inside the aqueous micelle (Figure 14.3). The formed S^0 undergoes nucleation to produce SNPs. The first part of the reaction follows the conventional gas desulfurisation principle where H_2S is dissolved into the liquid (reaction 14.6) and then the synthesis of SNPs proceeds through catalytic conversion using iron-chelates according reaction 14.7 (Deshpande *et al.*, 2008):

$$H_2S + H_2O_{(l)} \leftrightarrow H_2S_{(l)} \tag{14.6}$$

$$H_2S_{(l)} + 2Fe^{3+}L^{n-}_{(l)} \rightarrow S^0 \downarrow + 2H^+ + 2Fe^{2+}L^{n-}_{(l)} \tag{14.7}$$

Where 'L' denotes an organic ligand and 'n' denotes its charge. In addition to the formation of SNPs, ferric chelate ($Fe^{3+}L^{n-}$) simultaneously reduced to ferrous chelate ($Fe^{2+}L^{n-}$) as per reaction 14.7 inside the micelle (Figure 14.3). $Fe^{2+}L^{n-}$ is inactive and needs to be regenerated. Therefore, pure oxygen is sparged simultaneously to regain $Fe^{3+}L^{n-}$ (the active form of the iron-chelate) through reaction 14.9:

$$\frac{1}{2}O_{2(g)} + H_2O_{(l)} \leftrightarrow \frac{1}{2}O_{2(l)} \tag{14.8}$$

$$2Fe^{2+}L^{n-}_{(l)} + \frac{1}{2}O_{2(l)} + H_2O_{(l)} \rightarrow 2Fe^{3+}L^{n-}_{(l)} + OH^- \tag{14.9}$$

Thus, a continuous cycle of reduction of H_2S and oxidation of the chelate occurs, which makes the production of SNPs highly efficient. However, there has been a complexity in the specific composition of the reactants throughout the microemulsions process, which restricts the process further use in SNP synthesis. This brings the novel attempts of controlled radical polymerisation (CRP) and electrochemical techniques for the fabrication of SNPs (Nakabayashi *et al.*, 2017; Shamsipur *et al.*, 2011).

14.2.1.4 *Synthesis of SNPs from core-linked block copolymer micelles*

Advances in precision polymer synthesis, through developments in the controlled radical polymerisation (CRP) technique, enable the synthesis of SNPs with specific functionalities through self-assembly of amphiphilic block copolymers into well-defined core-cross linked nanoparticles (Iijima *et al.*, 1999). The most common method to achieve this is the formation of polymer micelles, via solution-state self-assembly of amphiphilic block copolymers. Core cross-linking is a useful strategy to provide the structural stability to the micelle structures, because incorporation of cross-linking agents provides weak intermolecular interactions that allow selective covalent stabilisation of the core domain and introduces desired functional units to the core moiety of the micelle (O'Reilly *et al.*, 2006).

SNPs can be synthesised using core-linked block copolymer micelles of sodium polysulfide (as sulfur source derived from elemental sulfur) combined with 4-bromophenyl vinylsulfide (BPVS) and N-isopropyl acrylamide (NIPAM) as amphiphilic block copolymers which can together form a micelle in water (solvent) (Figure 14.4). The bromide (Br) group present in the hydrophobic segment of the amphiphilic block copolymer (poly-BPVS) acts as the reactive site of the resulting micelle in aqueous solution. The presence of NIPAM in the block copolymer offers thermoresponsive properties in the formed SNPs.

Four different halide containing cross-linkers were used: 1,2,3-trichloropropane, 1,2-dibromoethane, 2,5-dibromothiophene and 1,4-dibromobenzene (Nakabayashi

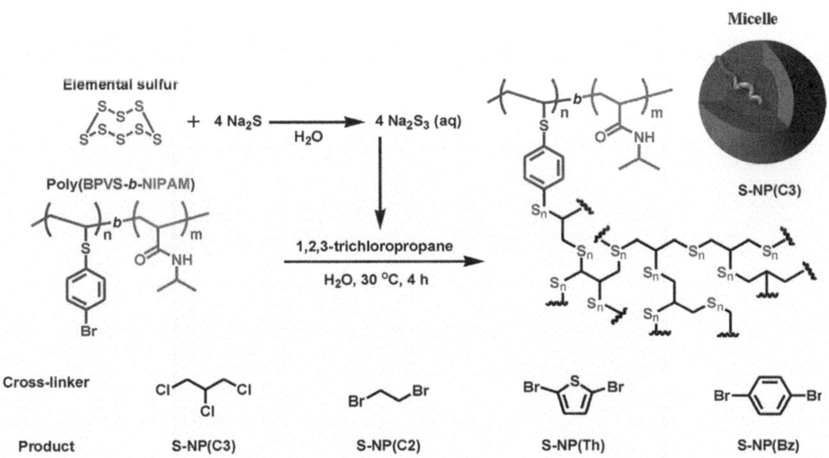

Figure 14.4 Synthesis of SNPs, S-NP(C3), S-NP(C2), S-NP(Th), and S-NP(Bz) from amphiphilic block copolymer, poly (BPVS-b-NIPAM) and the respective cross-linkers (reused with permission from Nakabayashi *et al.*, Copyright (2017) Polymer).

et al., 2017). The halide moieties present in the amphiphilic block copolymers and cross-linkers act as polymeric surfactant as the large block lengths in the copolymers provide steric stabilisation, thereby preventing agglomeration of the SNPs (Nakabayashi *et al.*, 2017). The nature of the cross-linkers is the dominant factor in controlling the sulfur content and structure of the SNPs. The cross-linking reaction with 1,2,3-trichloropropane (denoted as S-NP3 in Figure 14.4) gives a uniform stable structure of SNPs with the highest sulfur content (80%).

14.2.1.5 Electrochemical synthesis of SNPs

SNPs can also be produced through electrolysis of a sodium thiosulfate solution (Shamsipur *et al.*, 2011). The electrochemical synthesis of SNPs consists of a three-electrode cell, Ag/AgCl/(KCl) as the reference electrode (RE), a Pt wire as counter electrode (CE) and an Au wire as the working electrode (WE). This process is carried out in two steps: 1) reduction of thiosulfate ($S_2O_3^{2-}$) to sulfide (S^{2-}) and 2) oxidation of the sulfide (S^{2-}) to elemental sulfur nanoparticles (S^0/SNPs) (Figure 14.5). The pH plays an important role in the second step and a mild acidic media needs to be maintained for partial oxidation of sulfide to SNPs and to prevent the further oxidation of the sufide to sulfate (See also chapter 2 and 8). The SNPs formed through this process are in the range of 35–65 nm depending on the operational parameters (Table 14.1).

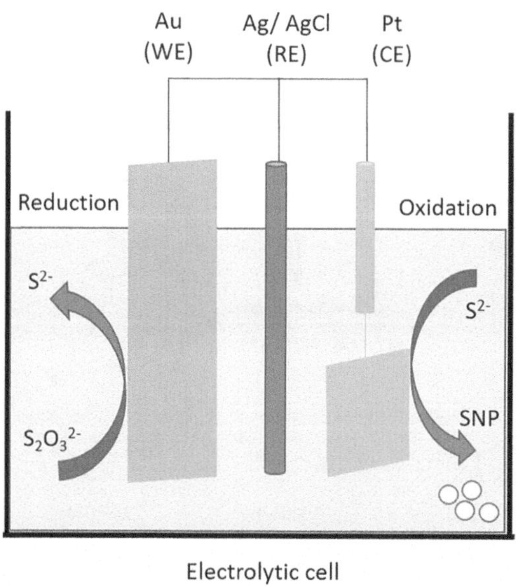

Figure 14.5 Schematic representation of electrochemical synthesis of SNPs.

14.2.2 Biological SNP production methods

It is important for nanomaterials to be produced in a sustainable and environmentally-friendly manner (Dhand *et al.*, 2015). For the green synthesis of various nanomaterials, the use of microbes and plant extracts has been proposed (Singh *et al.*, 2016). Biosynthesis of SNPs (BioSNPs) has been reported mainly with plant extracts which can be carried out at ambient temperature with minimal facilities in a cost effective manner (Table 14.2). The use of plant extracts results in shorter reaction times than microbial-mediated processes. The phenolic and other chemical compounds present in the plant extracts are reducing agents or coat the nanomaterials as capping agents and improve the SNPs reactivity and functionality (Saxena *et al.*, 2012). The most common polyphenolic compounds in plant extracts such as gallic acid, rhein, anthraquinone, and gallocatechol are well known reducing agents to attenuate reactive oxygen species (ROS) induced oxidative damage (Agbor *et al.*, 2007; Kasote *et al.*, 2015).

14.2.2.1 BioSNPs synthesis using leaf extracts

Several types of extracts from leaves of *Azadirachta indica, Catharanthus roseus, Mangifera indica, Polyalthia longifolia* and *Ficus benghalensis* have been investigated for the production of SNPs (Paralikar and Rai, 2018; Tripathi *et al.*, 2018). One of the possible mechanisms of biosynthesis of SNPs is depicted in Figure 14.6. Mixing of acids, e.g. hydrochloric, sulfuric or citric acid, with a sulfur source, e.g. thiosulfate or polysulfide solution, results in a disproportionation of the sulfur source with simultaneous reduction (S^{6+} to S^0) by phytochemicals and oxidation (S^{2-} to S^0) by acid (Tripathi *et al.*, 2018). The elemental sulfur produced then undergoes nucleation to form the SNPs. The proteins present along with the phytochemicals act as capping agents that stabilise the thus formed SNPs. The resulting polydispersed SNPs from the leaf extracts of *A. indica, C. roseus, M. indica* and *P. longifolia* have a size range of 70–80 nm, whereas in the case of *F. benghalensis* leaf extract, the SNPs have a comparatively smaller size in the range of 2–15 nm (Table 14.2).

14.2.2.2 BioSNPs synthesis using fruit extract

A novel green synthesis method for the preparation of sulfur nanoparticles from thiosulfate and hydrochloric acid by using *Albizia julibrissin* and *Punica granatum* fruit extracts was developed (Awwad *et al.*, 2015; Salem *et al.*, 2016). The surface-modified SNPs coated with compounds present in the *A. julibrissin* fruit extracts showed good long-term stability compared to the bare sulfur particles in aqueous solution with a particle size from 10 to 100 nm. The binding of amino acid residues in the protein fraction of the fruit extracts to SNPs acts as stabiliser and dispersing agent for nanoparticles and prevents the agglomeration of the SNPs. The highly crystalline SNPs with average particle size of about 20 nm were synthesised using *P. granatum* peel extracts (Salem *et al.*, 2016).

Table 14.2 Overview of biological synthesis methods of sulfur nanoparticles.

Plant Extracts or Fungal Extract	Species	Solvent	Reaction Time (Minute)	Temperature	Size (nm) and Shape	Reference
Fruit	*Albizia julibrissin*	Sodium thiosulfate and hydrochloric acid	5	Room temperature	20, spherical	Awwad *et al.* (2015)
Fruit peel	*Punica granatum*	Sodium thiosulfate and hydrochloric acid	10	Room temperature	50, spherical	Salem *et al.* (2016)
Leaf	*Azadirachta indica, Catharanthus roseus, Mangifera indica Polyalthia longifolia*	Sulfur powder, sodium sulfide, sodium polysulfide and sulfuric acid	N/A	100°C	70–80; 2–15, spherical	Paralikar and Rai (2018)
Leaf	*Ficus benghalensis*	Sodium thiosulfate and citric acid	60	Room temperature	2–15, spherical	Tripathi *et al.* (2018)
Fungi	*Colletotrichum sp.*	Sodium thiosulfate and hydrochloric acid	N/A	Room temperature	53 $\pm$ 49, spherical	Suryavanshi *et al.* (2017)

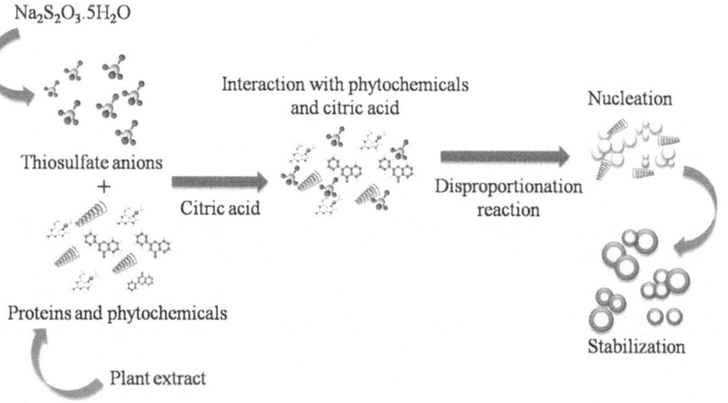

Figure 14.6 Mechanistic representation of biosynthesis of SNPs from *Ficus benghalensis* leaf extract (reused with permission from Tripathi *et al.*, Copyright (2018) RSC Advances).

14.2.2.3 BioSNPs synthesis using fungal extract

The fungal extract harvested from *Colletotrichum sp.* has been used for the synthesis of SNPs (Suryavanshi *et al.*, 2017). The process was carried out by acid catalysation of sodium thiosulfate mentioned in reaction 14.4. The fungal filtrate became yellow and turbid after the formation of SNPs. The average particle size of the formed SNPs was 53 ($\pm$49) nm. The crystalline nature of the SNPs showed resemblance with the results obtained by SNP synthesis using fruit extract of *Albizia julibrissin* (Awwad *et al.*, 2015).

14.3 CHARACTERISATION OF SNPs

The characterisation of SNPs includes measurement of shape, size and location using transmission electron microscopy (TEM) and scanning electron microscopy (SEM), surface topology with atomic force microscopy (AFM), surface modification with Fourier transform infra-red (FT-IR) spectroscopy, allotropic composition and crystallinity with X-ray diffraction (XRD), thermal stability with thermogravimetric analysis (TGA) and purity with energy-dispersive X-ray (EDX) spectroscopy (Table 14.1) (Guo *et al.*, 2006; Urakaev *et al.*, 2016; Xie *et al.*, 2009).

14.3.1 Mechanical method

The SNPs produced by mechanical ball milling have an average size bellow 100 nm (Urakaev *et al.*, 2016). Evaluation of the thermal properties of as-prepared SNPs using TGA revealed that the SNPs synthesised using mechanical ball milling have higher temperatures for complete evaporation and transformation of α-SNP to β-SNP in comparison to macro-elemental sulfur (Urakaev *et al.*, 2016). However, the temperature required to start evaporation and the polymerisation temperature were lower for SNPs produced by ball milling (Urakaev *et al.*, 2016).

14.3.2 Microemulsions based method

SNPs synthesised using the reverse microemulsions method are orthorhombic SNPs (α-SNPs) with an average diameter of about 20 nm, and have a narrow size distribution, uniform spherical shape, and high purity (Guo *et al.*, 2006). The size of the SNP is greatly influenced by the inorganic reactant in the microemulsion system, but not by the concentration of sodium polysulfide (Guo *et al.*, 2006). SNP production from H_2S gas using an Fe^{3+}–malic acid chelate microemulsion yields highly crystalline pure $\alpha-$sulfur SNPs with uniform shape and 10 nm size with a relatively narrow particle size distribution (5–15 nm) (Deshpande *et al.*, 2008).

14.3.3 Surfactant assisted method

SNPs synthesised from sodium thiosulfate using a surfactant assisted acid catalysis method are small in size (30 nm) with high crystallinity. Oxalic acid produces smaller size SNPs when compared with inorganic acids such as H_2SO_4, HNO_3 and HCl (Chaudhuri & Paria, 2010). However, use of HCl at a very low concentration resulted in the smallest sized SNP particles. The effect of different surfactants on the size of SNP was different. The use of SDBS produced more uniform-sized SNPs, whereas the smallest size SNPs (30 nm) was produced using CTAB. Furthermore, SNPs produced using SDS and SDBS along with HCl as acid catalyst had a very low zeta potential (−76.3 and −85.0 mV, respectively). Use of the surfactant CTAB for synthesis of SNPs using acid catalysis also improved crystallinity, as sharp peaks were observed in the XRD spectra of SNPs synthesised with CTAB compared with SNPs produced with other surfactants, namely TX-100, CTAB, SDBS and SDS (Chaudhuri & Paria, 2010). SNPs produced from sodium polysulfide as the inorganic sulfur reactant, cyclohexane as oil phase, a combination of Span-80 and Tween-80 as surfactant as well as 1-butanol as co-surfactant were monoclinic SNPs (β-SNPs) of 50 nm size (Soleimani *et al.*, 2013).

14.3.4 Stabilisation methods

The stabilisation of nanoparticles during their synthesis in the liquid phase can be by electrostatic, steric or electrosteric (a combination of steric and electrostatic) means or through the use of ligands. Surface modifiers such as cysteine or chitosan provide stabilisation of the SNPs through the formation of weak hydrogen bonds and forming weak bonds on the SNP surface in which S atoms acts as ligands. Cysteine coated SNP synthesised from sublimed-sulfur using ultrasonication showed a chain like aggregated structure with individual nanoparticles of 50–100 nm (Xie *et al.*, 2009). FT-IR analysis of cysteine coated SNP showed the presence of the S-H bond (2368.14 and 2340.91 cm^{-1}) that was distinct from control SNP produced without cysteine (Xie *et al.*, 2009). The disappearance of –CH or –CH_2 (1382.17 cm^{-1}) was due to the formation of S-S and S-H bonds on the SNP surface due to the presence of cysteine. Comparison between chitosan coated and non-coated SNP showed that due to the interaction with chitosan, the mean diameter of the SNPs was reduced from 45.05 (±15.55) nm to 15.80 (±7.82) nm (Shankar & Rhim, 2018a).

14.3.5 Biological methods

SNPs synthesised using different plant extracts had sizes of 89, 86, 95 and 62 nm with leaf extracts of *A. indica*, *C. roseus*, *M. indica*, and *P. longifolia*, respectively (Paralikar & Rai, 2018). In addition, the zeta potential varied depending on the leaf extract used (−15.7 mV for *A. indica*, −34.1 mV for *C. roseus*, −10.5 mV for *M. indica*, −7.12 mV for *P. longifolia*). The SNPs produced using *C. roseus* were highly stable, crystalline and amorphous. The common peaks present in both SNP and *C. roseus* leaf extract confirmed the key role played by the C-N stretch of aromatic amines and carboxylic acids, and C-O stretch of alcohols, carboxylic acids, and phenols in the SNP synthesis using this method (Paralikar & Rai, 2018).

SNPs synthesised using fungal extract of *Colletotrichum* sp. showed a maximum absorbance peak at 291 nm (Suryavanshi *et al.*, 2017). The particle size of the SNPs was 53 (±49) nm, determined using nanoparticle tracking analysis. The as-prepared SNPs had a less colloidal stability with a zeta potential value of +2.37 mV. Furthermore, the SNPs produced were crystalline in nature which was confirmed by XRD and selected area electron diffraction (SAED) combined with TEM analysis.

14.4 APPLICATION OF SNPs

14.4.1 Metal uptake by SNPs

Colloidal SNPs synthesised using the acid-catalysed reaction of thiosulfate revealed a higher uptake of mercury (Hg) than elemental sulfur (Adlim *et al.*, 2018). Immobilisation of SNPs on mini rice-husk-ash (RHA) pellet coated chitosan film resulted in a higher retention of mercury vapour up to 99.36% with a mercury uptake rate of 2.195 (±1.1) ppm Hg per gram SNPs per minute (Table 14.3).

SNPs synthesised from *F. benghalensis* leaf extract have been evaluated for their catalytic detoxification of Cr (VI) in water (Tripathi *et al.*, 2018). Various nanoparticles, e.g. Fe/Ni bimetallic nanoparticles, ZnO nanorods, and CdS nanoparticles have been explored for catalytic and photocatalytic reduction of Cr (VI) to Cr (III) (Buerge & Hug, 1998; Khataee *et al.*, 2014). However, the use of CdS, which is toxic, and photocatalysts such as TiO_2, which can work only under UV light, impose the need for new types of nanoparticles for the catalytic

Table 14.3 Application of sulfur nanoparticles for antimicrobial and cytotoxic activity.

Source of Sulfur	SNP Synthesis Method	Size of SNP	Application	Key Findings	Reference
H_2S gas	Fe^{3+}–malic acid chelate along with water-in-oil microemulsion containing cyclohexane (oil phase), Triton X-100 (surfactant) and n-hexanol (co-surfactant)	10 nm, α-SNP (average size)	Antimicrobial activity against bacteria (*Pseudomonas aeruginosa, Staphylococcus aureus*), fungi (*Aspergillus flavus, Aspergillus niger*) and yeast (*Candida albicans* NCIM 3102 and NCIM 3466 strain)	Zone inhibition against *P. aeruginosa*: 7.3 and 2.5 mm, *S. aureus*: 7.7 and 30 mm, *A. flavus*: 14 and 22.5 mm, *A. niger*: growth delayed by 48 h, *C. albicans* NCIM 3102: 8.3 and 10 mm, *C. albicans* NCIM 3466: 8.3 and 11 mm (with 30 and 250 µg/mL SNP dosage, respectively)	Deshpande *et al.* (2008)
Sodium thiosulfate	Acid catalysed method using hydrochloric acid (SNP-1) and acid catalysed and chitosan capped (SNP-2)	10–70 nm for SNP-1 and 10–30 nm for SNP-2	Antibacterial activity against *E. coli* and *S. aureus*	Minimum inhibitory concentrations against *E. coli*, 256 and 16 µg/mL for SNP-1 and SNP-2, respectively, and against *S. aureus*, 128 and 16 µg/mL for SNP-1 and SNP-2, respectively	Shankar *et al.* (2018b)
Not reported	Orthorhombic SNPs by liquid phase precipitation and PEG-400 encapsulation Monoclinic SNPs by water-in-oil microemulsion method with Span 80 and Tween 80	Orthorhombic: 10 nm size (spherical shape), and Monoclinic: 50 nm size (cylindrical shape)	Antifungal activity against *Aspergillus niger*	Minimum inhibitory concentrations against *A. niger* are 8000 mg/L (orthorhombic SNP) and 32000 mg/L (monoclinic SNP)	Roy *et al.* (2012); Choudhury *et al.* (2012)

Sodium polysulfide	Liquid-phase precipitation using formic acid and stabilisation using PEG-400 solution	20 nm (average size)	Antifungal activity against *Aspergillus niger* (MTCC-10180 and MTCC-282 strains) and *Fusarium oxysporum* (NCL-1008 and NCL-1072 strains)	Number of germinated spore of *A. niger* MTCC-10180 reduced by 22×10^4 and 27×10^4 at 125 and 500 mg/L of SNP and no spore formation at higher SNP dosage. No spore formation for *A. niger* MTCC-282 with 500, 1000 and 2000 mg/L of SNP and at 125 mg/L spore reduced by 13×10^4 in number. Spores of *F. oxysporum* NCL-1008 reduced by 6×10^4, 8×10^4 for 25 and 100 mg/L of SNP, and for 200 mg/L no spore formation. Spores of *F. oxysporum* NCL-1072 reduced by 3×10^4, 2×10^4 for 25 and 100 mg/L of SNP, and for 200 mg/L no spore formation.	Choudhury *et al.* (2011)

(*Continued*)

Table 14.3 Application of sulfur nanoparticles for antimicrobial and cytotoxic activity (*Continued*).

Source of Sulfur	SNP Synthesis Method	Size of SNP	Application	Key Findings	Reference
Ammonium polysulfide	Liquid-phase precipitation using formic acid and stabilisation using PEG-400 solution	50 nm (average size)	Antifungal activity against *Aspergillus niger* (MTCC-10180 and MTCC-282 strains) and *Fusarium oxysporum* (NCL-1008 and NCL-1072 strains)	Number of germinated spore of *A. niger* MTCC-10180 reduced by 26×10^4, 30×10^4 and 23×10^4 for 125, 500 and 1000 mg/L of SNP and no spore formation at 2000 mg/L of SNP dosage. No spore formation for *A. niger* MTCC-282 with 2000 mg/L of SNP and spores reduced by 10×10^4, 6×10^4 and 17×10^4 for 125, 500 and 1000 mg/L of SNP dosage. For *F. oxysporum* NCL-1008 no effect of SNP at 25, 50 and 200 mg/L on spore formation, at 100 mg/L dosage spore reduced by 6×10^4 in number. Spores of *F. oxysporum* NCL-1072 reduced by 5×10^4, 4×10^4 and 4×10^4 at 25, 50 and 100 mg/L of SNP, and for 200 mg/L no spore formation.	Choudhury *et al.* (2011)

Sodium thiosulfate	PEG-stabilised SNPs and custom-made non-capped SNPs commercially purchased	10 nm (PEG-stabilised SNPs) and 60 nm (custom-made non-capped SNPs)	Antifungal activity against *Escherichia coli, Klebsiella pneumoniae, Acinetobacter baumannii, Stenotrophomonas maltophilia* and *Enterobacter aerogenes*	Minimum inhibitory concentration of SNP against all the strains tested were 18.82 mg/L	Roy *et al.* (2012); Choudhury *et al.* (2012)
Sodium thiosulfate	Acid (HCl) catalysed method and chitosan coated	Not reported	Antimicrobial activity of SNP incorporated chitosan films against *Listeria monocytogenes* and *E. coli*	Complete inactivation of *L. monocytogenes* and *E. coli* within 6 h and 12 h, respectively	Shankar & Rhim (2018a)
Sodium thiosulfate	Acid catalysed method using hydrochloric acid (SNP-1) and acid catalysed and chitosan capped (SNP-2)	10–70 nm for SNP-1 and 10–30 nm for SNP-2	Cytotoxic activity against human lung carcinoma (A549), mouse colon carcinoma (CT26), human epithelial colorectal adenocarcinoma (Caco-2), and human fibroblast (CCD-986sk) cell line	Cell viability of A549 reduced by 71.8% and 40.5% with 500 μg/mL of SNP-1 and SNP-2, respectively. Cell viability of CT26 reduced by 60% with 500 μg/mL of both SNPs. Cell viability of Caco-2 were 18.7% and 32.4% with 1000 μg/mL of SNP-1 and SNP-2, respectively. Cell viability of CCD-986sk was 37.1% at 1000 μg/mL of SNP-1, but SNP-2 did not show any cytotoxicity on CCD-986sk cells.	Shankar *et al.* (2018b)

reduction of Cr (VI) to Cr (III). SNPs are such agents for this catalytic Cr (VI) reduction in anaerobic environments (Lan *et al.*, 2005). The presence of oxygen compromises the catalytic efficacy of SNPs. Tripathi *et al.* (2018) studied the catalytic detoxification of Cr (VI) from water in the presence of formic acid in an aerobic environment. The SNPs used in this study were prepared from *F. benghalensis* leaf extract. An efficiency for Cr (VI) reduction to Cr (III) of 88.7% was observed in the presence of SNPs in 80 minutes. The use of formic acid in this study released H_2, which ultimately adsorbed on the surface of the SNPs to accelerate the reduction of Cr (VI) to Cr (III) through a H_2 transfer pathway (Tripathi *et al.*, 2018).

Alumina loaded SNPs have been investigated as a potential medium for solid phase extraction (SPE) of lead (Pb), cadmium (Cd), copper (Cu) and zinc (Zn) from marine samples (Table 14.4) (Ghanemi *et al.*, 2011). The SNPs used in this study were prepared by acid-catalysed reaction of thiosulfate. Compared to the other adsorbents used in the SPE media investigated for determining the trace elements, SNPs offered a high extraction capacity and rapid extraction dynamics.

Liu *et al.* (2016) investigated the adsorption capacity of Cu by SNPs as a promising strategy for cancer therapy. Cu serves as an essential cofactor for protein kinases, e.g. in the mitogen-activated protein kinase/extracellular activating kinase (MAPK/ERK) pathway, which is constitutively activated in cancer cells to regulate cell survival and proliferation. SNPs could effectively bind Cu and form a strong surficial structure to inhibit the migration of the cancer cells (Figure 14.7). The coating of PEG is used to decrease the SNP immunogenicity and increase the adsorption capacity of Cu (Liu *et al.*, 2016). Consequently, mechanistic studies revealed that PEG coated SNPs were found as an inhibitor of the MAPK/ERK pathway as they can adsorb almost three times the amount of Cu by weight, thereby inducing apoptosis in cancer cells (Liu *et al.*, 2016).

14.4.2 Antimicrobial applications of SNPs

14.4.2.1 SNPs toxicity

The antimicrobial efficacy of SNPs to pathogenic fungi and bacteria make them a potent candidate for agricultural and biomedical applications (Roy *et al.*, 2012; Choudhury *et al.*, 2012). Various mechanisms have been proposed for their antimicrobial activities. First, depending on the size, elemental sulfur can be taken up by pathogens, where it is then intracellularly oxidised to form toxic pentathionic acid or reduced to form hydrogen sulfide. These in turn promote obstruction or oxidation of the free –SH attached forms of many essential enzymes within the cell and generate ROS stress that disrupts the mitochondrial machinery in the cell (Roy & Goswami, 2013; Owens, 1963). A second antimicrobial mechanism of SNPs can be the cross-linking interactions of proteins or lipids by the free radical of sulfur. A third mechanism is the

Table 14.4 Application of sulfur nanoparticles for heavy metal uptake.

Source of Sulfur	SNP Synthesis Method	Application	Key Findings	Reference
Sodium thiosulfate	Acid catalysed method using hydrochloric acid (SNP-1) and SDBS stabilised	SNP loaded alumina powder for Pb, Zn, Cd and Cu removal	37.8–96.1% Pb, 74.6–100% Zn, 79.5–99.5% Cd and 82.7–98% Cu recovery using SNP loaded alumina powder	Ghanemi *et al.* (2011)
Sodium thiosulfate	Acid catalysed method using HCl	SNP immobilised rice-husk-ash pellet coated chitosan film for Hg vapour uptake	Maximum 99.36% Hg uptake achieved	Adlim *et al.* (2018)
Sodium thiosulfate	Acid catalysed method using citric acid and the leaf extract of *F. benghalensis*	Catalytic reduction of Cr (VI) to Cr (III) by SNPs in the presence of formic acid; H_2 adsorption onto SNPs	Reduction efficiency of Cr (VI) to Cr (III) is 88.7% in 80 minutes	Tripathi *et al.* (2018)

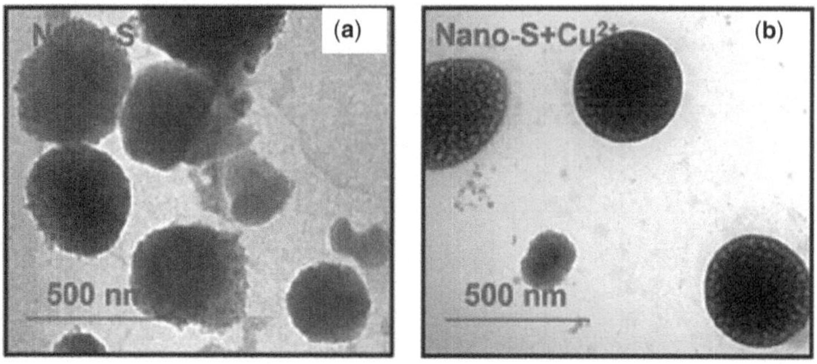

Figure 14.7 TEM micrograph of SNPs (a) before and (b) after interacting with Cu (Liu *et al.*, 2016).

interaction of SNPs with DNA, resulting in its destruction and cell death (Schneider *et al.*, 2011). Deviation in observed antimicrobial activity upon application might be attributed to differences in diffusion and bioavailability of the SNPs in different media (broth and agar) used to cultivate different organisms (Kamiya & Iijima *et al.*, 2010).

14.4.2.2 Antifungal applications of SNPs

SNPs not only limit fungal growth but also inhibit spore formation (Choudhury *et al.*, 2011). Depletion of the lipid layer due to an unusually high accumulation of saturated fatty acids could be inferred as one of the major reasons for SNPs mediated fungistasis (Roy *et al.*, 2012; Choudhury *et al.*, 2012). PEG- 400 encapsulated SNPs were tested against two known sulfur-susceptible fungi, namely *Erysiphe cichoracearum* and *Fusarium oxysporum*, and one known sulfur-resistant fungus *Aspergillus niger* (Table 14.3) (Roy & Goswami, 2013). *E. cichoracearum* and *F. oxysporum* are soil-borne pathogens, the former is responsible for vascular wilts in a wide range of woods and crops, whereas the latter causes powdery mildew in plants (Baldwin, 1950; Maraite & Meyer, 1971).

A. niger is a ubiquitous fungus most commonly coined as a food pathogen and a causative agent of invasive aspergillosis in immunocompromised patients which may lead to mortality in infants (Pfohl-Leszkowicz & Manderville, 2007). Changes in hyphal morphology and reduction in radial growth of *A. niger* and *F. oxysporum* were observed according to the concentration and particle size of SNPs (Choudhury *et al.*, 2011). Figure 14.8 illustrates the antifungal property of SNPs on hyphal morphology of *A. niger* before and after treatment. The minimum effective concentrations (MEC) found for orthorhombic SNPs were 50 ppm and 125 ppm for *F. oxysporum* and *A. niger*, respectively. From a pot culture assay, it was observed that the orthorhombic SNPs function efficiently

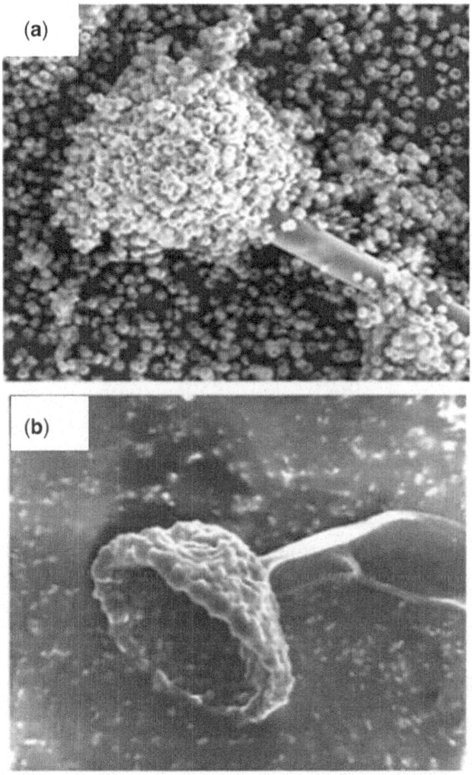

Figure 14.8 SEM micrograph of the fungal development and morphology of *A. niger* (a) after and (b) before the treatment of SNPs (Choudhury *et al.*, 2012).

against *E. cichoracearum* at a concentration of 1000 ppm and hence infected leaves showed reduced mildew contamination.

14.4.2.3 Antibacterial applications of SNPs

Antimicrobial activity against *E. coli*, *S. aureus*, *Listeria monocytogenes* was enhanced by PVP (Poly-N-vinylpyrrolidone) and chitosan coated SNPs (Shankar & Rhim, 2018a; Urakaev *et al.*, 2018). Some studies revealed antimicrobial activities of SNPs against Gram-positive but not against Gram-negative bacteria (Suleiman *et al.*, 2015). This can be explained by the fact that Gram-negative bacteria possess an outer membrane which acts as a barrier to penetrate the SNPs into the bacterial cell and may cause self-aggregation of nanoparticles (Koneman *et al.*, 2006).

SNPs have been tested against multi-drug resistant Gram-negative bacilli (GNB) (Choudhury *et al.*, 2013) due to the prevailing threat of resistance to carbapenem, a class of antibiotics (Table 14.3). The isolates of GNB taken for the study were *Escherichia coli*, *Klebsiella pneumoniae*, *Acinetobacter baumannii*, *Stenotrophomonas maltophilia* and *Enterobacter aerogenes*. They contain an enzyme called New Delhi Metallo-β-lactamase-1 (NDM-1) which is responsible for carbapenemase activity and can hydrolyse almost all hydrolysable β- lactams. The antimicrobial tests with SNPs were carried out both using the agar dilution method (ADM) and broth microdilution method (BMD). The SNPs used were coated with PEG-400 as it is a well-known biocompatible surfactant and hence induces hydrophilicity in order to permeate cells through the cell membrane. Furthermore, PEG has the ability to enhance the solubility and plasma stability of proteins which make it more potent in drug delivery (Petros & DeSimone, 2010) For all the isolates harbouring NDM-1, the ADM method revealed the antibacterial efficacy of PEGylated SNPs with a minimum inhibitory concentration (MIC) of 18.82 mg/L. However, in the BMD method, the MIC for the same SNPs was 9.41 mg/L for *A. baumannii* and *S. maltophilia* and 18.82 mg/L for the remaining isolates, which was similar to the ADM method. The difference in MICs in the two methods accounts for a varying growth pattern and the degree of susceptibility of organisms to the antimicrobial agent. Moreover, differences in the diffusion rate and bioavailability of the SNPs in different media, i.e. agar and broth, can contribute to the different response by the different organisms.

14.4.3 Medical applications of SNPs

14.4.3.1 Therapeutic application of SNP

Successful therapeutic application of SNPs depends on the SNPs administration in the host body and their pharmacokinetic parameters such as clearance rate, biological half-life and volume of distribution (Corsini *et al.*, 1999). Cumulative data obtained from in vitro dissolution of SNPs in buffer media and in vivo biodistribution properties of SNPs against a rabbit (*Oryctolagus cuniculus*) model suggests the therapeutic competency of SNPs in terms of their in vitro release kinetics and in vivo absorbance (Choudhury *et al.*, 2013). Different types of SNPs were found effective, with a sustained release pattern and periodical elimination from the body of the treated animals (Roy & Goswami, 2013).

14.4.3.2 Cytotoxicity of SNPs to cancer cells

Cytotoxicity and cell migration profiles of SNPs were estimated on various human derived cell lines, namely lung carcinoma (A549), mouse colon carcinoma (CT26), Caco-2 and fibroblast (CCD- 986sk) by Shankar and Rhim (2018a). Cell viability and cell migration assays performed in this study revealed that chitosan coated SNPs exhibited greater cytotoxicity towards tumour cells than the normal cells. Sulfur treatment is known to downregulate the epithelial growth factor receptor and upregulate the pro-apoptotic pathway, thereby inhibiting the proliferation of cancer cells (Ha *et al.*, 2013). Furthermore, chitosan showed growth inhibitory effects on cancer cells through nuclear fragmentation and chromatin condensation (Hasegawa *et al.*, 2001). The inhibition of cancer cell migration can also be attributed to the higher affinity of the cationic chitosan to negatively charged cancer cell membranes, which facilitates cell adhesion and enhances the cellular uptake of chitosan coated SNPs by endocytosis (Qi *et al.*, 2005; Tan *et al.*, 2009). Preferential toxicity and significant inhibition on cell migration in mammalian cells thus insinuate the emergence of SNPs as a viable drug or pesticide.

REFERENCES

Abuyeva, B., Burkitbayev, M., Mun, G., Uralbekov, B., Vorobyeva, N., Zharlykasimova, D. and Urakaev, F. (2018). Preparation of ointment materials based on sulfur nanoparticles in water-soluble polymers. *Materials Today: Proceedings*, **5**(11), pp. 22894–22899.

Adlim M., Zarlaida F., Khaldun I., Dewi R., Karina S. and Omar A. F. (2018). Immobilization of Sulfur from Different Precursors on Mini Rice-Husk-Ash Pellet Coated Chitosan Film and the Application for Mercury Vapor Uptake. *Indonesian Journal of Chemistry*, **19**, 386–394.

Agbor G. A., Kuate D. and Oben J. E. (2007). Medicinal plants can be good source of antioxidants: case study in Cameroon. *Pakistan Journal of Biological Sciences: PJBS*, **10**, 537–544.

Awwad A. M., Salem N. M. and Abdeen A. O. (2015). Novel approach for synthesis sulfur (S-NPs) nanoparticles using Albizia julibrissin fruits extract. *Advanced Materials Letters*, **6**, 432–435.

Baldwin M. M. (1950). Sulfur in fungicides. *Industrial and Engineering Chemistry*, **42**, 2227–2230.

Barkauskas J., Juškėnas R., Milerienė V. and Kubilius V. (2007). Effect of sulfur on the synthesis and modification of carbon nanostructures. *Materials Research Bulletin*, **42**, 1732–1739.

Baskar S., Pragati P. and Chandrababu K. (2015). Anti-microbial studies using sulphur nanoparticles on dandruff causing Malassezia Yeasts. In: Proceedings of the World Congress on Engineering, 2015 Jul 1–3, London: IAENG Publications.

Boyd E. S. and Druschel G. K. (2013). Involvement of intermediate sulfur species in biological reduction of elemental sulfur under acidic, hydrothermal conditions. *Applied and Environmental Microbiology*, **79**(6), 2061–2068.

Buerge I. J. and Hug S. J. (1998). Influence of organic ligands on chromium (VI) reduction by iron (II). *Environmental Science & Technology*, **32**(14), 2092–2099.

Chaudhuri R. G. and Paria S. (2010). Synthesis of sulfur nanoparticles in aqueous surfactant solutions. *Journal of Colloid and Interface Science*, **343**(2), pp. 439–446.

Chaudhuri R. G. and Paria S. (2011). Growth kinetics of sulfur nanoparticles in aqueous surfactant solutions. *Journal of colloid and interface science*, **354**(2), pp. 563–569.

Choudhury S. R., Ghosh M., Mandal A., Chakravorty D., Pal M., Pradhan S. and Goswami A. (2011). Surface-modified sulfur nanoparticles: an effective antifungal agent against *Aspergillus niger* and *Fusarium oxysporum*. *Applied Microbiology and Biotechnology*, **90**(2), pp. 733–743.

Choudhury S. R., Ghosh M. and Goswami A. (2012). Inhibitory effects of sulfur nanoparticles on membrane lipids of *Aspergillus niger*: a novel route of fungistasis. *Current microbiology*, **65**(1), pp. 91–97.

Choudhury S. R., Basu A., Nag T., Sengupta K., Bhowmik M. and Goswami A. (2013). Expedition of in vitro dissolution and in vivo pharmacokinetic profiling of sulfur nanoparticles based antimicrobials. *Environmental Toxicology and Pharmacology*, **36** (2), 675–679.

Corsini A., Bellosta S., Baetta R., Fumagalli R., Paoletti R. and Bernini F. (1999). New insights into the pharmacodynamic and pharmacokinetic properties of statins. *Pharmacology & Therapeutics*, **84**, 413–428.

Demmink J. F., Mehra A. N. U. R. A. G. and Beenackers A. A. C. M. (2002). Absorption of hydrogen sulfide into aqueous solutions of ferric nitrilotriacetic acid: local auto-catalytic effects. *Chemical Engineering Science*, **57**, 1723–1734.

Deshpande A. S., Khomane R. B., Vaidya B. K., Joshi R. M., Harle A. S. and Kulkarni B. D. (2008). Sulfur nanoparticles synthesis and characterization from H_2S gas, using novel biodegradable iron chelates in W/O microemulsion. *Nanoscale Research Letters*, **3**, 221.

Dhand C., Dwivedi N., Loh X. J., Ying A. N. J., Verma N. K., Beuerman R. W., Lakshminarayanan R. and Ramakrishna S. (2015). Methods and strategies for the synthesis of diverse nanoparticles and their applications: a comprehensive overview. *Rsc Advances*, **5**, 105003–105037.

Fan G., Gu Z., Yang L. and Li F. (2009). Nanocrystalline zinc ferrite photocatalysts formed using the colloid mill and hydrothermal technique. *Chemical Engineering Journal*, **155**, 534–541.

Ghanemi K., Nikpour Y., Omidvar O. and Maryamabadi A. (2011). Sulfur-nanoparticle-based method for separation and preconcentration of some heavy metals in marine samples prior to flame atomic absorption spectrometry determination. *Talanta*, **85**, 763–769.

Ghoraba, Z., Aibaghi, B. and Soleymanpour, A. (2017). Application of cation-modified sulfur nanoparticles as an efficient sorbent for separation and preconcentration of carbamazepine in biological and pharmaceutical samples prior to its determination by high-performance liquid chromatography. *Journal of Chromatography B*, **1063**, pp. 245–252.

Guo Y., Zhao J., Yang S., Yu K., Wang Z. and Zhang H. (2006). Preparation and characterization of monoclinic sulfur nanoparticles by water-in-oil microemulsions technique. *Powder Technology*, **162**, 83–86.

Ha A. W., Hong K. H., Kim H. S. and Kim W. K. (2013). Inorganic sulfur reduces cell proliferation by inhibiting of ErbB2 and ErbB3 protein and mRNA expression in MDA-MB-231 human breast cancer cells. *Nutrition Research and Practice*, **7**, 89–95.

Häkkinen H. (2012). The gold–sulfur interface at the nanoscale. *Nature Chemistry*, **4**, 443.

Hasegawa M., Yagi K., Iwakawa S. and Hirai M. (2001). Chitosan induces apoptosis via caspase-3 activation in bladder tumor cells. *Japanese Journal of Cancer Research*, **92**, 459–466.

Iijima M., Nagasaki Y., Okada T., Kato M. and Kataoka K. (1999). Core-polymerized reactive micelles from heterotelechelic amphiphilic block copolymers. *Macromolecules*, **32**, 1140–1146.

Kamiya H. and Iijima M. (2010). Surface modification and characterization for dispersion stability of inorganic nanometer-scaled particles in liquid media. *Science and Technology of Advanced Materials*, **11**, 044304.

Kasote D. M., Katyare S. S., Hegde M. V. and Bae H. (2015). Significance of antioxidant potential of plants and its relevance to therapeutic applications. *International Journal of Biological Sciences*, **11**, 982.

Khataee A., Darvishi Cheshmeh Soltani R., Hanifehpour Y., Safarpour M., Gholipour Ranjbar H. and Joo S. W. (2014). Synthesis and Characterization of Dysprosium-Doped ZnO Nanoparticles for Photocatalysis of a Textile Dye under Visible Light Irradiation. *Industrial & Engineering Chemistry Research*, **53**(5), 1924–32.

Koneman E. W., Allen S. D., Janda W. M., Schereckenberger B. C. and Winn W. C. (ed.) (2006). Koneman's Color Atlas and Textbook of Diagnostic Microbiology. Jones & Bartlett Learning, United States.

LaMer V. K. and Kenyon A. S. (1947). Kinetics of the formation of monodispersed sulfur sols from thiosulfate and acid. *Journal of Colloid Science*, **2**(2), 257–264.

Lan Y., Deng B., Kim C., Thornton E. C. and Xu H. (2005). Catalysis of elemental sulfur nanoparticles on chromium (VI) reduction by sulfide under anaerobic conditions. *Environmental Science & Technology*, **39**(7), 2087–2094.

Li L., Zhou C. H. and Xu Z. P. (2019). Self-Nanoemulsifying Drug-Delivery System and Solidified Self-Nanoemulsifying Drug-Delivery System. In: Nanocarriers for Drug Delivery, S. Mohapatra, S. Ranjan, N. Dasgupta, R. Kumar and S. Thomas (eds), Elsevier, pp, 421–449.

Liu H., Zhang Y., Zheng S., Weng Z., Ma J., Li Y. and Zheng W. (2016). Detention of copper by sulfur nanoparticles inhibits the proliferation of A375 malignant melanoma and MCF-7 breast cancer cells. *Biochemical and Biophysical Research Communications*, **477**(4), 1031–1037.

Malik M. A., Wani M. Y. and Hashim M. A. (2012). Microemulsion method: A novel route to synthesize organic and inorganic nanomaterials: 1st Nano Update. *Arabian Journal of Chemistry*, **5**, 397–417.

Maraite H. and Meyer J. A. (1971). Systemic fungitoxic action of benomyl against *Fusarium oxysporum* f. sp. melonis in vivo. *Neth. Journal of Plant Pathology*, **77**, 1–5.

Nakabayashi K., Takahashi T., Watanabe K., Lo C. T. and Mori H. (2017). Synthesis of sulfur-rich nanoparticles using self-assembly of amphiphilic block copolymer and a site-selective cross-linking reaction. *Polymer*, **126**, 188–195.

O'Reilly R. K., Hawker C. J. and Wooley K. L. (2006). Cross-linked block copolymer micelles: functional nanostructures of great potential and versatility. *Chemical Society Reviews*, **35**(11), 1068–1083.

Otsuka H., Nagasaki Y. and Kataoka K. (2003). PEGylated nanoparticles for biological and pharmaceutical applications. *Advanced Drug Delivery Reviews*, **55**(3), 403–419.

Owens R. G. (1963). Chemistry and physiology of fungicidal action. *Annual Review of Phytopathology*, **1**, 77–100.

Paralikar P. and Rai M. (2018). Bio-inspired synthesis of sulphur nanoparticles using leaf extract of four medicinal plants with special reference to their antibacterial activity. *IET Nanobiotechnology*, **12**, 25–31.

Petros R. A. and DeSimone J. M. (2010). Strategies in the design of nanoparticles for therapeutic applications. *Nature Reviews Drug Discovery*, **9**(8), 615–627.

Pfohl-Leszkowicz A. and Manderville R. A. (2007). Ochratoxin A: An overview on toxicity and carcinogenicity in animals and humans. *Molecular Nutrition & Food Research*, **51**, 61–99.

Qi L., Xu Z., Jiang X., Li Y. and Wang M. (2005). Cytotoxic activities of chitosan nanoparticles and copper-loaded nanoparticles. *Bioorganic & Medicinal Chemistry Letters*, **15**, 1397–1399.

Roy Choudhury, S., Roy, S., Goswami, A. and Basu, S. (2012). Polyethylene glycol-stabilized sulphur nanoparticles: an effective antimicrobial agent against multidrug-resistant bacteria. *Journal of Antimicrobial Chemotherapy*, **67**(5), pp. 1134–1137.

Roy Choudhury, S. and Goswami, A. (2013). Supramolecular reactive sulphur nanoparticles: a novel and efficient antimicrobial agent. *Journal of Applied Microbiology*, **114**(1), pp. 1–10.

Salem N. M., Albanna L. S. and Awwad A. M. (2016). Green synthesis of sulfur nanoparticles using Punica granatum peels and the effects on the growth of tomato by foliar spray applications. *Environmental Nanotechnology, Monitoring and Management*, **6**, 83–87.

Saravanan P., Gopalan R., Priya R., Ghosal P. and Chandrasekaran V. (2009). Textured resin-bonded Sm (Co, Fe, Cu) 5 nanostructured magnets exploiting magnetic field and surfactant-assisted milling. *Journal of Alloys and Compounds*, **477**, 322–327.

Saxena A., Tripathi R. M., Zafar F. and Singh P. (2012). Green synthesis of silver nanoparticles using aqueous solution of *Ficus benghalensis* leaf extract and characterization of their antibacterial activity. *Materials Letters*, **67**(1), 91–94.

Schneider T., Baldauf A., Ba L. A., Jamier V., Khairan K., Sarakbi M. B., Reum N., Schneider M., Röseler A., Becker K. and Burkholz T. (2011). Selective antimicrobial activity associated with sulfur nanoparticles. *Journal of Biomedical Nanotechnology*, **7**, 395–405.

Schwuger M. J., Stickdorn K. and Schomaecker R. (1995). Microemulsions in technical processes. *Chemical Reviews*, **95**(4), 849–864.

Shamsipur M., Pourmortazavi S. M., Roushani M., Kohsari I. and Hajimirsadeghi S. S. (2011). Novel approach for electrochemical preparation of sulfur nanoparticles. *Microchimica Acta*, **173**(3–4), 445–451.

Shankar S. and Rhim J. W. (2018a). Preparation of sulfur nanoparticle-incorporated antimicrobial chitosan films. *Food Hydrocolloids*, **82**, pp. 116–123.

Shankar, S., Pangeni, R., Park, J. W. and Rhim, J. W. (2018b). Preparation of sulfur nanoparticles and their antibacterial activity and cytotoxic effect. *Materials Science and Engineering C*, **92**, pp. 508–517.

Singh P., Kim Y. J., Zhang D. and Yang D. C. (2016). Biological synthesis of nanoparticles from plants and microorganisms. *Trends in Biotechnology*, **34**, 588–599.

Solanki J. N. and Murthy Z. V. P. (2011). Controlled size silver nanoparticles synthesis with water-in-oil microemulsion method: a topical review. *Industrial & Engineering Chemistry Research*, **50**, 12311–12323.

Soleimani M., Aflatouni F. and Khani A. (2013). A new and simple method for sulfur nanoparticles synthesis. *Colloid Journal*, **75**, 112–116.

Suleiman M., Al-Masri M., Al Ali A., Aref D., Hussein A., Saadeddin I. and Warad I. (2015). Synthesis of nano-sized sulfur nanoparticles and their antibacterial activities. *Journal of Materials and Environmental Science*, **6**, 513–518.

Suryavanshi P., Pandit R., Gade A., Derita M., Zachino S. and Rai M. (2017). Colletotrichum sp.-mediated synthesis of sulphur and aluminium oxide nanoparticles and its in vitro activity against selected food-borne pathogens. *LWT-Food Science and Technology*, **81**, 188–194.

Tan M. L., Choong P. F. and Dass C. R. (2009). Cancer, chitosan nanoparticles and catalytic nucleic acids. *Journal of Pharmacy and Pharmacology*, **61**, 3–12.

Tripathi R. M., Rao R. P. and Tsuzuki T. (2018). Green synthesis of sulfur nanoparticles and evaluation of their catalytic detoxification of hexavalent chromium in water. *RSC Advances*, **8**, 36345–36352.

Turganbay S., Aidarova S. B., Bekturganova N. E., Li C. S., Musabekov K. B., Kumargalieva S. S. and Toshtay K. (2012). Nanoparticles of sulfur as fungicidal products for agriculture. *Eurasian Chemico-Technological Journal*, **14**, 313–319.

Turganbay S., Aidarova S. B., Bekturganova N. E., Alimbekova G. K., Musabekov K. B. and Kumargalieva S. S. (2013). Surface-modification of sulfur nanoparticles with surfactants and application in agriculture. In: Advanced Materials Research, S. Kolisnychenko (ed.), Trans Tech Publications Ltd., pp. 475–479.

Urakaev F. K., Bulavchenko A. I., Uralbekov B. M., Massalimov I. A., Tatykayev B. B., Bolatov A. K., Dzharlykasimova D. N. and Burkitbayev M. M. (2016). Mechanochemical synthesis of colloidal sulfur particles in the $Na_2S_2O_3$–$H_2(C_4H_4O_4)$–Na_2SO_3 system. *Colloid Journal*, **78**, 210–219.

Urakaev F. K., Abuyeva B. B., Vorobyeva N. A., Mun G. A., Uralbekov B. M. and Burkitbayev M. M. (2018). Sulfur nanoparticles stabilized in the presence of water-soluble polymers. *Mendeleev Communications*, **28**, 161–163.

Xie X. Y., Zheng W. J., Bai Y. and Liu J. (2009). Cystine modified nano-sulfur and its spectral properties. *Materials Letters*, **63**, 1374–1376.

Xing T., Sunarso J., Yang W., Yin Y., Glushenkov A. M., Li L. H., Howlett P. C. and Chen Y. (2013). Ball milling: a green mechanochemical approach for synthesis of nitrogen doped carbon nanoparticles. *Nanoscale*, **5**, 7970–7976.

Zheng W., Liu Y. W., Hu X. G. and Zhang C. F. (2006). Novel nanosized adsorbing sulfur composite cathode materials for the advanced secondary lithium batteries. *Electrochimica Acta*, **51**, 1330–1335.

Chapter 15

Treatment and reuse of solid materials containing inorganic sulfur compounds

Richard Tichý, Pimluck Kijjanapanich and Piet N. L. Lens

15.1 INORGANIC SULFUR TRANSFORMATIONS IN WATER/SOLID PHASE SYSTEMS

In the environment, different types of solid phase systems can contain high concentrations of inorganic sulfurous compounds, e.g. ores, soils and sediments. The key criterion in the assessment of environmental risks of inorganic sulfurous compounds present in these materials is whether they remain bound to the solid phase or whether they are mobilized into the solution. Mobility of inorganic sulfur in natural water/solid biphasic systems is predominantly controlled by oxidative or reductive changes and precipitation constants of sulfurous compounds. In anaerobic environments, sulfur is encountered in its most reduced form, i.e. sulfidic form.

Owing to the poor solubility of most sulfidic compounds, sulfurous compounds are predominantly immobile in anaerobic environments. On the contrary, in aerobic environments, the most stable inorganic sulfur compound is sulfate, which is usually very soluble in water.

Reductive and oxidative changes of inorganic sulfurous compounds often involve an interaction with protons. Sulfur reduction is usually accompanied with alkalization of the environment, whereas sulfur oxidation often results in acidification. The general overview of the effects of reductive and oxidative

DOI: 10.2166/9781789060966_0477

changes of inorganic sulfur compounds in the environment is summarized in Table 15.1.

15.1.1 Reductive processes

The most common reductive process affecting sulfur in the environment is microbial sulfate reduction. Sulfate-reducing bacteria require an anaerobic environment ($E_h < -150$ mV) and a pH higher than 5–6. However, microbial sulfate reduction was encountered even in anoxic systems loaded with highly acidic leachates (pH < 3). This is due to the strong alkalizing capacity of the sulfate reduction, which results in the formation of micro-regions within the soil or sediment profile with conditions favorable for sulfate-reducing bacteria. Moreover, some recent studies indicate an existence of acid-tolerant sulfate-reducing bacteria. Detailed information on this subject can be found in Chapter 7.

As can be seen from the reaction stoichiometry of sulfate reduction (van Houten *et al.*, 1994), its main products are sulfide and OH^-:

$$8\,H_2 + 2\,SO_4^{2-} \rightarrow H_2S + HS^- + 5\,H_2O + 3\,OH^- \tag{15.1}$$

Sulfide itself is a rather mobile species that can escape from a system in the form of H_2S gas or soluble HS^- (Stumm & Morgan, 1981). However, sulfide reacts with numerous naturally occurring chemicals, especially cationic metals. This leads to the formation of poorly soluble metal sulfides (see Table 15.2). In this way, high amounts of reduced sulfur species are accumulated in solid materials under reductive conditions (Ivanov, 1983).

Via precipitation of sulfides, sulfur is removed from the geochemical cycling and buried in anaerobic soils or sediments. As stated above, these sulfur stocks are closely associated with stocks of cationic metals due to the formation of metal sulfides. Moreover, increased alkalinity of these systems is favorable for the formation of other (non-sulfidic) solid-phase metal stocks, e.g. hydroxides and carbonates as well as for higher sorption of metals on the surface of the soil (Ivanov, 1983; Salomons & Stigliani, 1995).

Table 15.1 Comparison of the effects of oxidative and reductive conditions on sulfur behavior in solid materials.

Conditions	Dominant Reactions	Environmental Consequences
Reductive	Sulfate reduction Release of OH^-	Alkalization Immobilization of cations (e.g. metals) Accumulation of sulfur
Oxidative	Sulfide oxidation Release of H^+	Acidification Dissolution of alumo-silicates Leaching of cations Enhanced weathering

Table 15.2 Solubility of selected metal sulfides (Ellwood *et al.*, 1992).

Metal Ion	Solubility Product ($mol \cdot L^{-1}$)
Hg^{2+}	4×10^{-54}
Cu^{2+}	8×10^{-45}
Cd^{2+}	5×10^{-29}
Zn^{2+}	10^{-20}
Mn^{2+}	1.4×10^{-14}
Fe^{2+}	10^{-20}

15.1.2 Oxidative processes

Oxidation of reduced inorganic sulfur compounds in nature proceeds via spontaneous chemical as well as microbial oxidation processes (Ivanov, 1983). Spontaneous chemical processes dominate the oxidation at neutral to alkaline pH. At acidic conditions, rates of spontaneous chemical oxidation decrease, whereas microbial oxidation rates increase. At a pH < 4, the rate of microbial oxidation of inorganic sulfur compounds is, by 5–7 orders of magnitude, higher than the rate of spontaneous chemical processes (Evangelou & Zhang, 1995; Smith *et al.*, 1988). Details about the microbial sulfur oxidation are given in Chapter 3. Here, attention is given to the facts that are specifically important for solid materials.

Owing to the acidifying character of sulfur oxidation, appropriate microbes developed a strong tolerance to the extremely low pH. Some microbes even require a low pH for their growth (Karavaiko, 1985). Among them, types with different pH-tolerance exist: some types are able to operate at pH 3–6, and are active during the initial phase of sulfur oxidation. Other bacteria are more active at pH < 3, succeeding after disappearance of the less acid-tolerant microbes (Blais *et al.*, 1993).

The most broadly known genus of the bacteria responsible for inorganic sulfurous compounds oxidation is *Thiobacillus*. Therefore, the mixed microbial communities that oxidize inorganic sulfurous compounds are called thiobacilli. The reader should bear in mind that thiobacilli is not a clearly defined set of bacterial species.

Oxidation of inorganic reduced sulfur compounds requires the presence of oxygen. However, oxidation may proceed even under anoxic conditions. Spontaneous chemical oxidation can be driven by the strong oxidizing capacity of ferric (Fe^{3+}) iron:

$$MeS + 2\,Fe^{3+} \rightarrow Me^{2+} + 2\,Fe^{2+} + S^0 \tag{15.2}$$

Here, Me stands for a cationic divalent metal, e.g. Cd, Cu, Pb or Zn.

Microbial oxidation in anoxic conditions is possible due to the capacity of certain bacteria to utilize nitrate as an electron acceptor (Mackintosh, 1978). Furthermore, *Thiobacillus ferrooxidans* is able to grow anaerobically with elemental sulfur as its electron donor and ferric iron as its terminal electron acceptor (Pronk *et al.*, 1992).

A key factor in microbial oxidation of solid inorganic reduced sulfur compounds is an adhesion of bacteria to the surface of appropriate sulfurous minerals (Bryant *et al.*, 1983; Devasia *et al.*, 1993; Karavaiko, 1985; Mirajkar *et al.*, 1997; Porro *et al.*, 1997; Solari *et al.*, 1992). Therefore, microbial sulfur oxidation is influenced by the presence of surface active agents. Addition of wetting agents increases the oxidation rate and some wetting agents are even produced by the bacteria themselves (Bryant *et al.*, 1983; Cook, 1964; Schaeffer & Umbreit, 1963). On the contrary, excessive amounts of wetting agents can inhibit the adhesion of bacteria, which results in the suppression of microbial oxidation (Ondrushka & Glombitza, 1993).

At $pH < 4$, the activity and growth of sulfur oxidizing microbes is also suppressed when organic acids are present in the system. Since these acids are not dissociated in acidic media, they are taken up by microbes as electro-neutral molecules. However, they dissociate in the cytoplasm, which often has a circumneutral pH. Acidification of cytoplasm brings higher requirements on energy supply from the cell and may eventually lead to a die-off of the bacteria (Pronk *et al.*, 1990).

The direct effect of sulfur oxidation is its release into the environment, usually in the form of sulfate. Sulfate is a rather stable and intact molecule which is soluble in water. Three exceptions to high solubility of sulfate compounds should be noted here: calcium sulfate – gypsum, lead sulfate, and jarozites (Evans, 1989). Formation of calcium and lead sulfates is illustrated in Box 15.1 by equilibrium simulations using ECOSAT ver. 4.4, a program for calculating the soil/water chemical equilibria (Wageningen University, Department of Soil Chemistry and Plant Nutrition, The Netherlands), using stability and solubility constants from Lindsay (1979). Jarozites are formed at extremely low pH (<2). Jarozite is a basic ferric sulfate with the general formula $X_3Fe(SO_4)_2(OH)_6$, where $X = K^+$, Na^+, $NH_4{}^+$, H_3O^+ (Carlson *et al.*, 1992; Karathanasis & Thompson, 1995).

As described in Chapters 2 and 3, oxidation of sulfur usually results in a production of protons and thus acidification of a system. Increased acidity results in enhanced desorption of cations from solid materials and solubilization of metals precipitated as carbonates or hydroxides. This leads to the leaching of many cationic toxic metals (Förstner, 1995).

Moreover, acidity induces a dissolution of mineral matrices, enhanced weathering, solubilization of soil alumo-silicates and leaching of aluminum (Stumm & Furrer, 1987). This effect is pronounced at $pH < 4$, as documented by Box 15.2. Leaching of aluminum from acidic soils is a world recognized environmental problem associated with toxicity to aquatic life, adverse effects on soil biota and plant roots, and exhibiting direct toxicity on animals and humans

BOX 15.1 PRECIPITATION OF CALCIUM AND LEAD SULFATE AS A FUNCTION OF pH

Figure 15.1 shows the equilibrium situations in a solution initially containing varying amounts of gypsum (calcium sulfate) and different pH. The concentrations of gypsum were selected between 12.5 mmol · L^{-1} and 42.5 mmol · L^{-1} (401.25–1364.25 mg · L^{-1} of S). The simulations also included the gaseous phase with a constant atmospheric CO_2 partial pressure. Between pH 3 and 7, no influence of pH is observed. At pH lower than 3, gypsum is solubilized, whereas at pH > 7, formation of calcium carbonate prevails.

Similar simulations were carried out to demonstrate the effects of pH on the solubility of lead sulfate. As can be seen from Figure 15.2, lead sulfate solubilization is also strongly pH dependent.

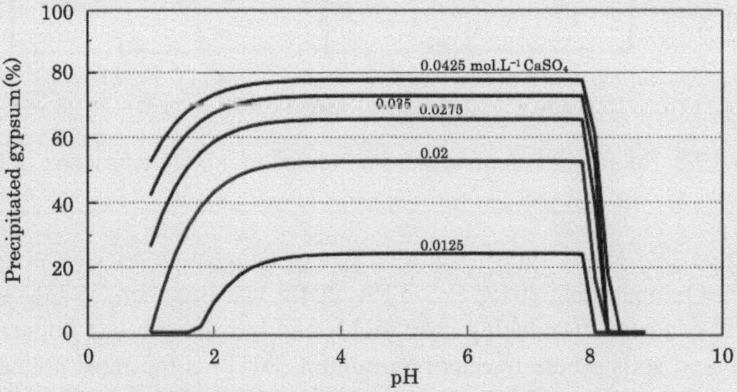

Figure 15.1 Precipitation of calcium sulfate as a function of pH.

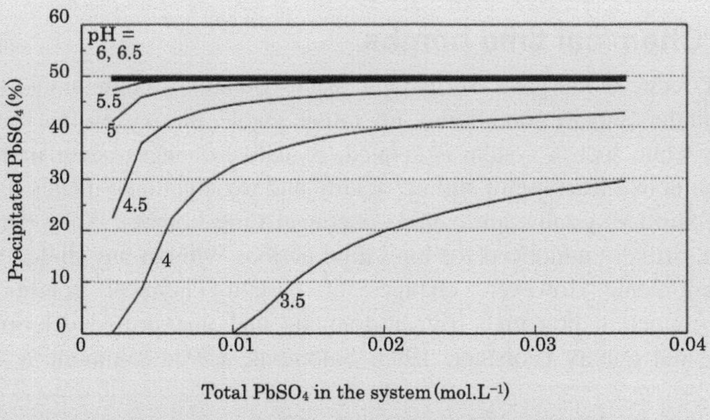

Figure 15.2 Precipitation of lead sulfate as a function of pH.

BOX 15.2 ALUMINUM RELEASE FROM SOIL MATRIX

Figure 15.3 shows the solubilization of aluminum in a soil suspension at varying pH induced by sulfuric acid. Clay, silt and sandy soils were used at a solution/soil ratio of 0.415 kg · L^{-1} (Tichý *et al.*, 1996). The arrow indicates a threshold pH value below which a strong Al solubilization can be expected.

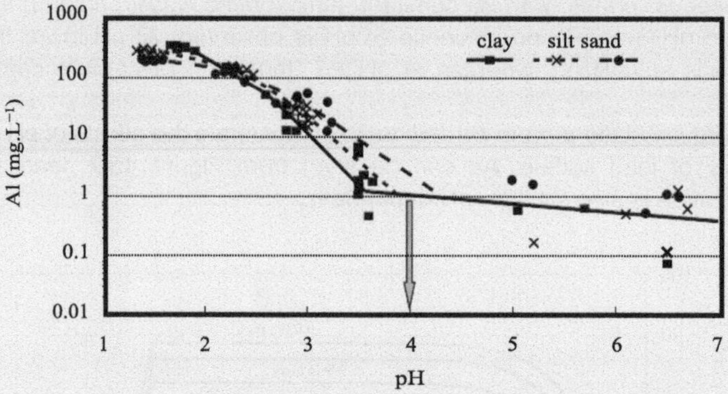

Figure 15.3 Release of aluminum from the soil matrix as a function of pH.

(Bojórquez-Quintal *et al.*, 2017; U.S. EPA, 2018; van Breemen, 1973). Aluminum is the primary factor in reducing crop yields and has the greatest impact on plant growth in acid soils where trivalent aluminum (Al^{3+}) is the most abundant form (Bojórquez-Quintal *et al.*, 2017; Kochian *et al.*, 2005; Ma *et al.*, 2002; Nogueirol *et al.*, 2015).

15.1.3 Chemical time bombs

Stocks of reduced inorganic sulfur and metal-containing compounds, that were built-up under anoxic conditions, are rather stable in oxygen-free conditions. However, when such a system is aerated, oxidative changes occur immediately, which results in a leaching of sulfate, acidity and toxic cationic metals. In popular literature, similar systems are called "chemical time bombs". They pertain in a landscape virtually unnoticed for long time periods without any direct effects on the environment. However, changes of certain chemical conditions, e.g. oxidation-reduction potential, trigger dramatic and unexpected deterioration of environmental quality (Förstner, 1995; Salomons, 1993; Salomons & Stigliani, 1995).

Triggering of the chemical time bomb effect is shown in an example of a freshwater wetland sediment in Box 15.3. The wetland sediment had been

BOX 15.3 CHANGES OF CHEMICAL PARAMETERS OF A SEDIMENT EXPOSED TO THE AIR (TICHÝ *ET AL.*, 1998A)

A wetland sediment loaded with acid mine drainage was exposed to aeration in an aqueous suspension. Figure 15.4 gives the time-courses of pH and oxidation-reduction potential (ORP) of this suspension.

Parallelly, sulfate, iron and divalent cationic metals (Cu and Zn) were released. Figure 15.5 shows the time-course of sulfate and iron release. The solubilization of other divalent cationic metals followed a monotonous increase with time (see the example of zinc).

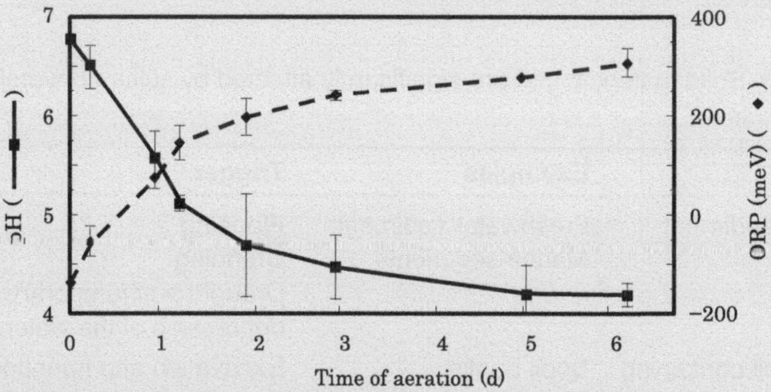

Figure 15.4 Time-course of pH and ORP.

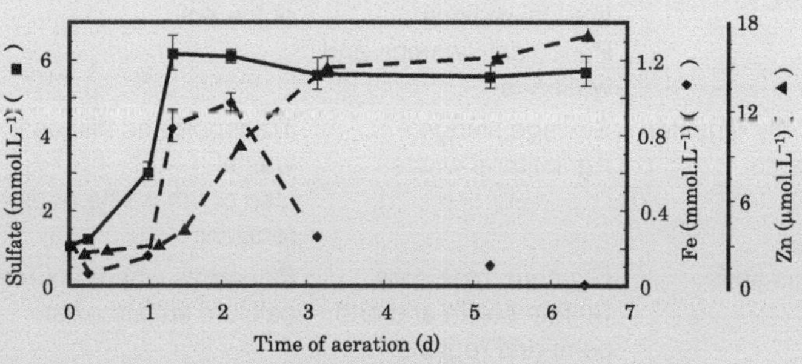

Figure 15.5 Time-course of sulfate and cationic metals release.

receiving acid mine drainage (AMD) for decades, serving as an easy and cheap pollution preventive measure. Therefore, high amounts of sulfide and metals had accumulated. Even a slight lowering of the water table resulted in a massive acidification and leaching of metals (Holmer & Storkholm 2001; Tichý, 2000).

Repeated die-off of the fish population was recorded in the recipient creek and a concrete and steel construction of a watermill downstream was heavily damaged by the corrosion (Tichý *et al.*, 1998b).

15.2 SULFUR-CONTAINING MATERIALS OF ENVIRONMENTAL CONCERN

Table 15.3 overviews different materials and systems threatened by the effects of sulfur transformations.

Table 15.3 Solid materials that are significantly affected by sulfur conversions and processes.

Material	Examples	Trigger
Aqueous sediment	Freshwater sediments Marine sediments	Flooding Dredging Draught and temporary depression of the water table
Mineral soil containing reduced sulfur species	Spoil banks Mining waste Abandoned mines	Excavation and transport Spontaneous seepage of groundwater
Periodically inundated soil	Acid sulfate soils in tropical regions Periodical swamps and wetlands	Periodical inundation and drying of the soil
Anaerobically digested organic waste	Sewage sludge Agricultural waste	Transport and disposal of the waste Use of the waste as agricultural fertilizer
Gypsiferous soils	Gypsum mine soils Sulfate soil in arid and semi-arid regions	Dry areas where sources for calcium sulfate exist
Solid wastes	Phosphogypsum Construction and demolition debris sand Flue gas desulfurization gypsum	Transport and disposal of the waste Use of the waste as agricultural fertilizer Applications where the material is placed in direct contact with the environment Landfill together with organic waste

15.2.1 Reduced inorganic sulfur containing materials

15.2.1.1 Sediments

Accumulation of substantial amounts of reduced sulfur is a well-known phenomenon encountered in all types of marine or freshwater sediments (Salomons & Stigliani, 1995). In nature, most of the sulfur in marine or fresh water originates from the weathering of sulfur-containing rocks in the catchment and from the oxidation of organic sulfur from terrestrial sources (Dornblaser *et al.*, 1994). However, at present, high contamination by sulfate in water is mostly caused by the burning of fossil fuels and discharge of sulfate-containing wastewater, including AMD (Kijjanapanich *et al.*, 2012; Mamelkina *et al.*, 2019; Ryu *et al.*, 2019) and industrial effluents (Mohammadi *et al.*, 2019; Promnuan *et al.*, 2020). Higher sulfate/sulfite atmospheric concentrations in acid rain (Bojórquez-Quintal *et al.*, 2017) and the discharge of wastewater with a high sulfate concentration affects the sulfur cycling in lakes and rivers (Kijjanapanich *et al.*, 2014c; Peiffer, 1998).

Under the water level, the sulfide produced by sulfate reduction precipitates with iron from iron phosphate precipitates present in the sediments to form iron sulfide (Lamers *et al.*, 1998; Smolders *et al.*, 2001; Smolders *et al.*, 2006; Wilfert *et al.*, 2020). This results in a release of phosphorus into the water phase and causes indirect eutrophication, which in turn, increases blooms of non-rooting species, such as algae, and turbidity of the water (Kijjanapanich *et al.*, 2014c; Smolders *et al.*, 2006).

Sediments can be exposed to oxygen when the water level decreases, when an incidental flood or storm-water event transports the sediment particles, or when the sediment is excavated (Tichý *et al.*, 1998a; Tichý & Mejstrík, 1996). The latter variant, i.e. the excavation of sediment, is often reinforced by the nautical requirements on rivers and in river estuaries, where the sedimentation is accelerated by increased salinity of the seawater. Exposure of these sediments to aerobic conditions results in a sequence of oxidative reactions, leading to the decrease of pH, increase of sulfate levels in the water and the solubilization of most cationic metals (Salomons & Stigliani, 1995). Box 15.4 gives an illustrative example of these processes.

15.2.1.2 Mining waste

Mining is a source of the most abundant waste that contains inorganic sulfurous compounds. Sulfur here is usually of geological origin and may be present in a whole range of different forms. Mining waste is not a homogenous category of material since it includes spoil, low-grade ores, mine-tailings, sedimentation lagoons, or even whole abandoned mining sites. These materials, when exposed to aeration, are oxidized, producing a significant pollution of effluents with acidity, sulfate, iron and other metals (Evangelou & Zhang, 1995; Richards *et al.*, 1993; Tichý & Mejstrík, 1996). The effluents from abandoned mines, AMD, possess serious environmental concerns worldwide. AMD is usually produced at very high quantities and for very long time periods, sometimes decades or

BOX 15.4 LONG-TERM CHEMICAL CHANGES IN A RIVERINE SEDIMENT AFTER ITS DREDGING AND PILING ABOVE THE WATER LEVEL

Maass and Miehlich (1988) documented chemical changes in a pile of riverine sediment after its dredging from the river bottom and placement at a disposal site. The time course of pH-changes and leaching of the main toxic metals (Cd and Zn) went through four different phases (Figure 15.6). The first phase was characterized by neutral pH and low concentrations of soluble Cd and Zn. Soon afterwards, oxidative processes started and the pH dropped down, resulting in accelerated metal solubilization. After this phase, the chemical conditions stabilized, reduced sulfur stocks were depleted, and natural buffering mechanisms allowed a significant pH-increase. The authors suggest that a fourth phase will appear in the future, characterized by a gradual pH-decrease and possible further release of metals (associated with long-term biochemical changes in the sediment, e.g. decrease of organic matter content).

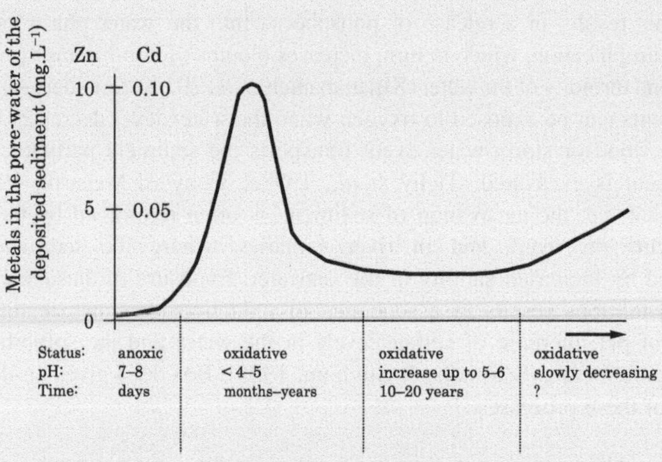

Figure 15.6 Chemical changes in riverine sediment.

centuries. AMD is characterized by a high sulfate content (up to several grams per litre), sometimes extremely low pH (<4), and elevated levels of Fe and some other metals (Kijjanapanich *et al.*, 2012). More details on the composition of AMD and its origin can be found in Chapter 7.

The characteristics of AMD depend largely upon the amounts and composition of the oxidized minerals, physical properties of the mineral surfaces, presence of electron acceptors in the water, and water flow velocity. The latter not only determines the dilution intensity, but also affects the overall chemical reactions in the system. This is illustrated in the example of pyrite oxidation: the oxidation of pyrite starts firstly with a reaction yielding sulfuric acid and ferrous iron (e.g. Equations 15.1 and 15.2). In stagnant conditions of pore-water (i.e. when the water in soil does not move), ferrous iron is oxidized at the same place, resulting in the formation of ferric precipitates and further acidification. However, movement of soil water may induce the leaching of ferrous iron out of the system. Ferrous iron may eventually enter a recipient stream or waterbody (Levy *et al.*, 1997). This effect of distant oxidation leads to orange-red (ochre) cores on solid surfaces in recipient waterbodies and distant acidification of the recipient environments.

15.2.1.3 Acid sulfate soils and periodical swamps

Certain areas experience a periodical inundation, followed by a drying phase, e.g. inundation areas in estuaries of rivers, and inundation by high flow of the sea or by some agricultural practices (e.g. rice production). When sufficient sulfate is present in the incoming water (e.g. seawater or brackish water), sulfate reduction can result in a periodic accumulation of reduced sulfur and *in situ* pyrite formation (Holmer & Storkholm, 2001). This phenomenon is pronounced in tropical regions since the temperature substantially increases reaction rates (Bloomfield & Coulter, 1973). After the water level drops down, pyrite is exposed to the air. Consequently, immediate acidification is observed *in situ* (Golez & Kyuma, 1997). In some cases, effects of distant oxidation and acidification, similar to those described in Section 15.2.1.2, have also been reported.

Acid sulfate soils are sometimes called "cat-clays" and are found in tropical and subtropical regions, comprising approximately 30% of the total area over the world (Bojórquez-Quintal *et al.*, 2017; Sade *et al.*, 2016). Similar effects are known in periodical swamps and marshes even in moderate climates (Tichý & Mejstrík, 1996; Tichý *et al.*, 1998a; Tichý *et al.*, 1998b).

15.2.1.4 Anaerobically digested sewage sludge

Surplus sludge from biological wastewater treatment is very often anaerobically digested (Dewil *et al.*, 2009; Novak *et al.*, 2006). Owing to the high content of organic matter as well as biogenic and mineral sources of sulfur, sulfate reduction occurs during sludge digestion (Kijjanapanich *et al.* 2014b; Kijjanapanich *et al.* 2014c). As a consequence, anaerobically digested sludge usually contains elevated levels of various reduced sulfur compounds. When this material comes into contact with the air, oxidative processes will start. Environmental concerns here are associated primarily with the release of toxic metals, which impose legal limitations on the sludge manipulation and disposal (Blais *et al.*, 1993; Chipasa, 2003; Couillard & Mercier, 1992; Suanon *et al.*, 2016).

15.2.2 Oxidized inorganic sulfur containing materials

15.2.2.1 Gypsiferous soils

Gypsiferous soils are those which contain significant quantities of gypsum (>10%) which may interfere with plant growth (Ernst, 2000; FAO, 1990). This type of soil covers about 94 million ha of the world's arable lands (FAO, 1993). These soils are predominantly present in dry areas (with less than 400 mm annual rainfall), where sources of calcium sulfate exist (Porta & Herrero, 1990) as well as mining areas, especially gypsum mines (Kijjanapanich *et al.*, 2014b). The agricultural utilization of gypsiferous soils is limited by the presence of gypsum resulting in low water retention capacity, shallow depth to a hardpan and vertical crusting (Alfaya *et al.*, 2009; Khresat *et al.*, 2004). This in turn reduces root penetration and causes plant cultivation problems (Smith & Robertson, 1962). It was found that the crop productivity remains low under irrigation, even with application of fertilizers or organic manures (FAO, 1990). Moreover, this type of soil has the potential to release sulfate into surface and groundwater. The physical structure such as porosity and permeability of gypsiferous soils can be improved by reducing the soil's gypsum content (Alfaya *et al.*, 2009).

15.2.2.2 Phosphogypsum

Phosphogypsum is a primary by-product of the phosphate fertilizer industry and originates from the production of phosphoric acid from phosphate rock (apatite). Each year around 100–280 million tons of gypsum by-product are generated worldwide (Tayibi *et al.*, 2009). Phosphogypsum management is one of the most serious problems currently faced by the phosphate industry. Only 15% of the worldwide production is recycled, the remaining 85% is stored without any treatment (Tayibi *et al.*, 2009).

Phosphogypsum is a moist, gray and powdery material containing residual acid, fluoride, toxic metals and radioactive compounds such as uranium and radium that may be present in the phosphate ore (Azabou *et al.*, 2005). This phosphogypsum can cause serious environmental problems including soil, water and atmosphere contamination (Jang *et al.*, 2005), due to toxic metals, especially radioactive compounds (Papageorgiou *et al.*, 2016). In addition, disposing of this gypsum in landfills may lead to exceptionally high levels of biogenic sulfide formation, resulting in, among others, odor problems (Azabou *et al.*, 2005).

15.2.2.3 Construction and demolition debris sand

Construction and demolition debris sand (CDDS) is the fine fraction (around 40–50% of the total waste) of solid waste originating from building, demolition and renovation of buildings and roads, containing high amounts of gypsum (de Vries, 2006; Montero *et al.*, 2010). The content of gypsum (by mass) in CDDS ranges from 1.5% to 9.5% (Jang & Townsend, 2001; Kijjanapanich *et al*,. 2013).

Reuse options have been proposed for CDDS, including soil amendment, alternative landfill cover, and fill material in road, embankment and construction projects. However, most of the CDDS still remains highly polluted, and the sulfate content often exceeds the prescribed limit for construction purposes (de Vries, 2006; Kijjanapanich *et al.*, 2013; Kijjanapanich *et al.*, 2014d). Therefore, CDDS needs to be washed to leach out the gypsum and other impurities from the material before use (de Vries, 2006; Kijjanapanich *et al.*, 2013; Kijjanapanich *et al.*, 2014c; Kijjanapanich *et al.*, 2014d). The presence of gypsum in CDDS may provide some benefit as a soil conditioner or nutrient source for agriculture (Jang & Townsend, 2001). However, for applications where the material is placed in direct contact with the environment, concerns have been raised by regulators regarding the chemical characteristics of the material and the potential risk to human health and the environment (Jang & Townsend, 2001; Kijjanapanich *et al.*, 2014c).

When wet landfill conditions occur, it is suggested that CDDS must be separated from other wastes, especially organic waste, and placed in a specific area of the landfill in order to prevent odor problems. This results in a rapid rise in the disposal costs of gypsum waste (Gypsum Association, 1992). The amount of gypsum going to a landfill can be reduced by separating the fine fraction from mixed construction and demolition debris. However, final disposal will require the removal of gypsum from CDDS (Montero *et al.*, 2010).

15.2.2.4 Flue gas desulfurization gypsum

Flue gas desulfurization (FGD) gypsum is a unique synthetic product derived from FGD systems at coal-based electric power plants. These systems operate by injecting absorbents such as limestone to combine with the sulfur (SO_x dissolving mainly as sulfite (SO_3^{2-}) and sulfate (SO_4^{2-}) in the scrubbing liquor) resulting in a slurry that is mostly composed of excess lime, calcium sulfite and calcium sulfate (Karnachuk *et al.*, 2002).

Some power plants can produce FGD gypsum which is nearly identical to mined natural gypsum that can be put to beneficial use (80% use in gypsum drywall products and 2% in agriculture) or sold (U.S. EPA, 2008). However, several power plants cannot produce high purity gypsum and it becomes a solid waste instead of a commercial product due to the impurities present such as fly ash and iron oxide (Panpa, 2002). FGD gypsum must then be disposed of in an approved manner into landfill sites, where it is one of the many solid waste materials which may lead to H_2S odor problems (Kijjanapanich *et al.*, 2014c).

15.3 TREATMENT STRATEGIES FOR SULFUR RICH SOLID MATERIALS

15.3.1 Prevention of sulfur pollution

Prevention of sulfur pollution is the primary measure to reduce risks associated with its accumulation in the environment (Brimblecombe *et al.* 1989; Ivanov, 1983). This involves an improved efficiency of flue-gas treatment, desulfurization of fossil fuels (Babich & Moulijn, 2003; Nuhu, 2013), or application of clean technologies in ore processing and smelting (Tichý & Mejstřík, 1996). More details about the desulfurization of fossil fuels and raw materials can be found in, respectively, Reinhoudt and Moulijn (2000) and Boon (2000).

Systematic management of mines using the whole mine life cycle can substantially reduce the production of AMD (Gray, 1997; Orava & Swider, 1996). Similarly, proper and sensitive management and policies in river basins can lead to a significant reduction of pollution, which consequently leads to lower risks associated with handling of riverine sediments (Salomons & Stigliani, 1995). Preventive administrative measures are beyond the scope of this chapter.

15.3.2 Elimination of sulfurous compounds from solid materials

Elimination of inorganic sulfurous compounds from solid materials is usually a rather costly procedure. In practice, it is applied with a primary focus on the elimination of toxic metals accompanying the sulfur.

Any potential technique for the removal of sulfur from solid materials should meet the specific requirements imposed by the fact that sulfur is present in the solid state. Principally, the removal of solid contaminants from a solid material is possible using phase or molecular separation techniques (Rulkens *et al.*, 1995, 1998).

15.3.2.1 Phase separation

Various techniques of gravity-separation and flotation are extensively applied for the enrichment of sulfur-containing non-ferrous ores (Richards *et al.*, 1993; Rulkens *et al.*, 1998; Spottiswood, 1982). Phase separation is frequently used to treat metal-polluted sediments. To our knowledge, no case of phase separation is reported for the removal of solid sulfurous compounds from waste, soil or sediment.

15.3.2.2 Molecular separation

The molecular separation, i.e. the extraction of molecules containing the sulfur atom, implies the solubilization of the pollutants followed by the selective removal of the sulfurous compounds from the process liquor (Hinsenveld, 1993). For the removal of metals, various extractants are used, e.g. mineral (HCl, H_2SO_4) or organic (lactate, formate) acids, complexing agents (ethylenediaminetetraacetic acid (EDTA), diethylenetriaminepentaacetic acid (DTPA), nitrilotriacetic acid (NTA)) or pure water. Because reduced inorganic sulfurous compounds are present mostly as solid particles, they need first to be dissolved. As mentioned above (see 15.1.2), inorganic sulfurous compounds are solubilized at oxidative conditions. Therefore, sulfur oxidation is a basic principle applied for the molecular separation of sulfurous compounds and associate toxic metals.

In the case of gypsum-containing materials, sulfates are readily dissolvable, thus pure water alone is sufficient for sulfate extraction (Kijjanapanich *et al.*, 2014c). The sulfate-containing leachate is then further treated and reused again for the sulfate extraction process of CDDS (Kijjanapanich *et al.*, 2013; Kijjanapanich *et al.*, 2014d; Kijjanapanich & Lens, 2017).

15.3.2.3 Bioleaching as a special case of molecular separation

In some cases, the oxidative process is dominated by microbial activities (growth of thiobacilli); then it is called "bioleaching" (Boon, 2000; Bruynesteyn, 1989; Karavaiko, 1985). Bioleaching is applied in practice especially in mining operations, e.g. for biohydrometallurgical mining and low-grade ore enrichment (Bruynesteyn, 1989). The process has also been applied for the treatment of sewage sludge, soils or sediments (Blais *et al.*, 1993; Iske & Glombitza, 1988; Tichý *et al.*, 1998a, b; Tyagi *et al.*, 1991).

The aim of bioleaching is to achieve pH values that are low enough to solubilize a maximum of the toxic metals. In some cases, additional elemental sulfur or ferrous iron are added to the system as substrates for bacteria (see also Chapter 7). Bioleaching has been tested in suspension reactors like Pachuca tanks, rolling reactors, or in propeller-agitated vessels (Tyagi *et al.*, 1991). A less-intensive alternative for the removal of sulfur and associated pollution from highly voluminous wastes is its treatment on heaps at the place of permanent or temporary disposal. An enhancement of the heap leaching processes, e.g. by forced aeration or addition of hydrophilic sulfurous substrates, may be a good alternative to the expensive slurry leaching processes (Tichý *et al.*, 1998a, b).

One of the few environmental applications of bioleaching has been the decontamination of anaerobically pre-treated sewage sludge and agricultural waste (Couillard & Mercier, 1994; Tyagi *et al.*, 1991). The application of anaerobic digestion for sludge containing heavy metals leads to qualitative changes in metal binding to the waste. Various bonds of metals to carboxylic or phenolic groups of organic matter, strong sorption to the cell walls, encapsulation within the vacuoles or other cellular components, chelation with complex organic molecules, and many other types of binding are replaced by sulfidic precipitates. Moreover, the structure of aerobically produced sludge is altered by the digestion, so that some forms of toxic metals can be more easily desorbed from the matrix. When the anaerobically digested sludge is aerated, the metals are solubilized in quantities that are never achieved without its anaerobic pre-treatment (Couillard & Mercier, 1994).

An example of bioleaching of metals from sewage sludge is given in Box 15.5. Here, ferrous iron is used as a substrate for bacteria. Oxidation of ferrous iron leads to an acidification of the suspension, which results in solubilization of metals. Sulfides are oxidized and get into the process liquor as well.

Meulepas *et al.* (2015) made use of the intrinsic acidification of anaerobic granular sludge to lower the pH, and thus release the heavy metals by lowering

BOX 15.5 EXAMPLE OF BIOLEACHING TECHNOLOGY APPLIED FOR THE REMOVAL OF SULFUR AND HEAVY METALS FROM POLLUTED ANAEROBICALLY PRE-TREATED SEWAGE SLUDGE (COUILLARD & MERCIER, 1994)

As shown in Figure 15.7, anaerobically digested sludge is pre-acidified and pre-aerated in a mixing vessel. Afterwards, intensive aeration and agitation is performed either in a completely stirred (CSTR) or airlift reactor. After the solid/liquid separation, the cleaned sludge is disposed of and wasted if it cannot be reused, whereas the excess process water undergoes neutralization and removal of pollutants.

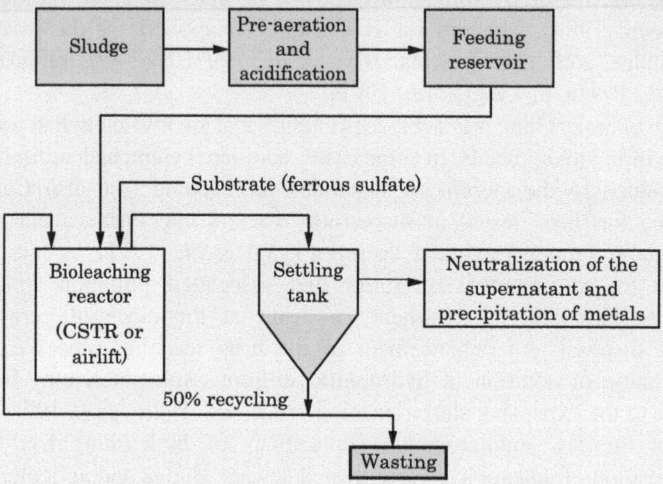

Figure 15.7 Bioleaching technology for the removal of sulfur and heavy metals.

the pH below the dissolution edge of the metal sulfides. Feeding potato peel waste to the digested sludge lowered the pH below 3, thus dissolving the heavy metals present and achieving a digestate that met the criteria for land application.

These principles are one of few feasible alternatives to handle metal polluted sewage sludge and similar types of organic waste (Tyagi *et al.*, 1993). Nowadays bioleaching techniques for heavy metal removal and recovery are also applied for metallurgical slags (Potysz *et al.*, 2016) and electronic waste, e.g. discarded printed circuit boards (Işıldar *et al.*, 2016; Işıldar *et al.*, 2017; Marra *et al.*, 2018).

15.3.3 Suppression of the sulfur oxidation

Another way to cope with the sulfur and associated pollution is to prevent its release into the environment. The text below summarizes the main strategies to prevent sulfur oxidation used in practice or proposed as feasible methods.

15.3.3.1 Inundation

For many systems, inundation can be a cost-effective and easy option. When capped with water, oxygen cannot penetrate to the sulfur compounds and the oxidation cannot start, as reported, e.g. by Schuring *et al.* (1997), for sulfur-rich hard-coal tailings. However, this method cannot always guarantee the full suppression of leaching, as bacterial leaching can proceed even in oxygen-free conditions. Moreover, spontaneous chemical solubilization can continue owing to the oxidative capacity of ferric iron (Bruynesteyn, 1989; Karavaiko, 1985; van Breemen, 1988).

15.3.3.2 Pumping of seawater

Acidophilic thiobacilli are inhibited by sodium chloride at concentrations comparable to seawater (Cameron *et al.*, 1984). Therefore, infiltration or pumping of seawater into a mine could inhibit microbial sulfide oxidation. On the other hand, some heavy metal ions, like cadmium, form stable chloride complexes. Therefore, chloride addition might considerably enhance the leaching process instead of preventing it (Salomons, 1993).

15.3.3.3 Hydrogeological isolation

Another strategy to suppress leaching is the hydrogeological isolation of the site. For this, isolation membranes or a mixture of clay and soil can be applied for preventing the vertical or horizontal water movements (Bell *et al.*, 1995). Proper treatment and restoration of the surface can lead to a reduction of polluted outflows from mine tailings or spoil banks (Richards *et al.*, 1993). Covering the deposit with soil and planting fast-growing trees is a generally applied method for the rehabilitation of large areas devastated by mining as well as for the covering of voluminous heaps of spoil worldwide (Tichý & Mejstrík, 1996).

15.3.3.4 Coating and passivation

In mining areas, lignite coal seams and overburden are the main sulfate sources which have to be covered or coated due to their high sulfate leaching potential (Paengsen, 2013). In order to reduce sulfate contamination in the mine drainage, red bed clay had to be used as cover material for coal seams and overburden. A layer of at least 30 cm of red bed clay cover was required for preventing rainwater from penetrating through to the coal pile layer. Sulfate contamination in surface runoff of all covered units was approximately 10 times less than that of the uncovered one (Paengsen, 2013).

Leaching of sulfurous compounds can also be accomplished by using specific additives which alter the physical properties of the surface of solid inorganic sulfurous compounds. Addition of fly ash and/or cement binding mixtures leads to an agglomeration of particles and a physical encapsulation of pyrite (Misra *et al.*, 1996). Belzile *et al.* (1997) reported on a possible passivation of the pyrite surface with the use of acetyl acetone, humic acids, ammonium lignosulfonates, oxalic acid or sodium silicate. The efficiency of this coating was enhanced when the pyrite was exposed to a pre-oxidation, leading to the formation of a ferric oxide layer on its surface. Evangelou (1995a) studied the possible micro-encapsulation of pyrite by a ferric phosphate layer. The coating involved a reaction mixture of H_2O_2 and phosphate. H_2O_2 oxidizes pyrite and the released Fe3+ precipitates as a ferric phosphate on the surface, thus preventing further pyrite oxidation (Georgopoulou *et al.*, 1996; Kollias *et al.*, 2015). Silicates can also be used instead of phosphate (Fytas & Bousquet, 2002). The produced ferric iron (Fe^{3+}) ions react with the silicate ions to produce ferric hydroxide-silica that precipitates on the pyrite surface producing a passive coating. This silicate coating can protect the grains of pyrite from oxidation (Fytas & Bousquet, 2002).

15.3.3.5 Use of surfactants

Addition of surface active compounds (surfactants) can prevent the attachment of bacterial cells to the sulfur-containing minerals. Thomson and Turney (1996) described the possible treatment of pyrite using various fatty acid amines, resulting in a substantial reduction of the bioleaching process. Sasaki (1997) demonstrated a possible use of oxalic, tannic and fulvic acids to prevent bacterial attachment with the best results achieved with fulvic compounds. Takeuchi and Suzuki (1997) tested cell adhesion to pyrite in the presence of potassium phosphate. Their study showed a combined effect, since potassium phosphate both inhibited the adhesion of bacteria to pyrite and kept pH at values unfavorable for bacterial growth and activity. The addition of 20 mg L^{-1} of the tenzide sodium paraffin sulfonate (E30) to the percolate of a pilot ore dump successfully inhibited the growth and activity of thiobacilli (Ondrushka & Glombitza, 1993). When the use of tenzides is considered, an estimation of the specific surface area of the sulfidic mineral is needed, because the applied tenzide concentrations should ensure full coverage of the mineral surface (Ondrushka & Glombitza, 1993).

15.3.3.6 Liming

Adding lime to a system containing inorganic sulfurous compounds is another option to suppress the leaching. It leads to the neutralization of the acidic environment. Neutral pH is unfavorable for bioleaching (Evangelou, 1995a). Moreover, neutralization of acidic conditions suppresses leaching of toxic cationic metals. In agricultural areas where waterlogging is a common practice, such as rice vegetation, anaerobic conditions occur and result in the conversion of organic matter into acetic acid and reduced sulfur compounds. This accelerates the release of heavy metals, for example cadmium, from the soil into the water, where they are adsorbed onto or taken up by the rice or crops. In this case, the pH of the soil is neutral but the addition of alkaline substances is required to react with the produced acetic acid, thereby increasing the soil pH for heavy metal immobilization (Inkham *et al.*, 2019). It was also found that limestone dust and biochar can also be used as alternative low-cost alkaline substances, which provide a more suitable soil pH for vegetation than lime (Bian *et al.*, 2014; Inkham *et al.*, 2019). However, it is usually only a temporary measure, since the lime is continuously leached out from the system. This process is also rather expensive. Cui *et al.* (2016) found that the exchangeable cadmium content increased over a long period of time due to the decreasing soil pH. Georgopoulou *et al.* (1996) even suggest that liming can be more expensive than coating with ferric phosphate.

15.3.3.7 Elimination of thiobacilli

Thiobacilli can be inactivated by specific biocides. Suppression of leaching was demonstrated using the bactericides Fluorspar and Kathon (Ondrushka & Glombitza, 1993). Fluorspar contains fluorides, which inhibit growth of *Thiobacillus* sp. at concentrations of minimum 400 mg $F^- L^{-1}$. The isothiazolone Kathon inhibits the activity of thiobacilli at concentrations of $300\ mg \cdot L^{-1}$. Application of both compounds to a pilot ore heap prevented leaching and no viable thiobacilli were found in the percolate (Ondrushka & Glombitza, 1993). However, the use of biocides cannot be recommended for full-scale applications because of the large amount of biocide required (and associated costs) and the risk of adaptation of the endogenous bacterial population.

An inhibition of growth and activity of thiobacilli fed with pyrite was also demonstrated with activated carbon (Loi *et al.*, 1993). This effect is likely caused by sequestering most of the cells from the suspension. Subsequently, microbes could not attach to the pyrite surface, and the oxidation could not initiate.

Padival *et al.* (1995) investigated the possible control of *Thiobacillus sp.* by means of microbial competition. This strategy uses excessive concentrations of nutrients to favor microbes that grow faster than thiobacilli. Experiments carried out with a mixture of thiobacilli and yeasts fed with glucose and thiosulfate have shown a substantial reduction of the activity of thiobacilli, and a significant reduction of the effluent sulfate concentrations. This method can be applied for prevention of corrosion of technological equipment. However, its large-scale application for treatment of overburdens or abandoned mines is not realistic.

15.4 TREATMENT STRATEGIES FOR SULFUR RICH SOILS

15.4.1 Treatment of acid sulfate soils

In principle, the measures to decrease acidification of acid sulfate soils are similar to those described above for sulfur rich solid materials (see 15.3.3). In practice, four basic strategies are applied, i.e. curtailing the pyrite oxidation, leaching of the reaction products, liming and self-neutralization at reductive conditions.

15.4.1.1 Curtailing the pyrite oxidation

As described in Section 15.3.3, pyrite oxidation can be suppressed by limiting the supply of oxygen or by influencing the rate of one or more of the intermediate steps in the oxidation, e.g. by decreasing the concentration of Fe^{3+}. However, in practical conditions, the only feasible prevention of oxygen entry to the system is its permanent inundation (Evangelou, 1995b; van Breemen, 1975).

15.4.1.2 Leaching of reaction products

Another treatment strategy is to remove soluble or exchangeable acidity from the soil. Leaching with fresh water is efficient to remove free sulfuric acid and partly also the Fe and Al ions that have been solubilized by the acidity. Leaching with salt or brackish water also results in the replacement of exchangeable Al by Na, Ca and Mg (van Mensvoort *et al.*, 1991). The leaching is, however, inefficient when reducing conditions prevail in the soil profile. Therefore, efficient leaching can only be done with fully aerated soil, preferably a soil exposed to an intensive drying period before the leaching (van Mensvoort *et al.*, 1991).

15.4.1.3 Liming of acid sulfate soils

Addition of lime, limestone or basalt can be applied to neutralize the acidity in acid sulfate soils, resulting in precipitation of inert Al hydroxides (Shamshuddin *et al.*, 2014). Typically, a dose of about 30 tons of $CaCO_3$ per 10,000 m^2 and per 10 cm of soil is needed (van Breemen, 1975). However, the required amount of lime also depends on the acidity of the acid soils (Shamshuddin *et al.*, 2014). The liming is usually efficient only after the water soluble acidity has been leached out. Otherwise, the huge amounts of lime required would make the process economically unfeasible. However, excessive liming would also result in a strong salinization of the site that may become unfavorable for vegetation and wildlife.

15.4.1.4 Self-neutralizing under reductive conditions

Initiation of microbial sulfate reduction in the acid sulfate soils can efficiently remove acidity and stabilize the sulfate concentration. This makes the cultivation of certain crops, e.g. rice, on acid sulfate soils a desirable ameliorative measure, because it combines both the effect of inundation and stimulation of sulfate reductive processes. However, although the toxic effects of acidity and cationic metals (especially aluminum) are suppressed at anoxic conditions, they are replaced by the toxic effects of reduced Fe^{2+}, which is one of the constraints of plant growth on acid sulfate soils (Shamshuddin *et al.*, 2014). Park *et al.* (1971) showed, however, that Fe^{2+}- toxicity can be avoided by sulfate reduction. The produced sulfide reacts with free Fe^{2+} and forms pyrite.

15.4.2 Treatment of gypsiferous soils

The quality of gypsiferous soil can be improved by reducing the soil's gypsum content, which can be achieved by *in situ* bioremediation (Kijjanapanich *et al.*, 2014b). With this technology, biological sulfate reduction was used to reduce the gypsum content of gypsiferous soil by using sulfate-reducing bacteria. Cheap organic substrates with low or no cost, such as rice husk, pig farm wastewater treatment sludge and coconut husk chips can be supplied to the soil as electron donors for the sulfate-reducing bacteria (Kijjanapanich *et al.*, 2014b). The highest sulfate removal efficiency of 59% was achieved in the soil mixed with a 40% organic mixture, corresponding to a reduction of the soil gypsum content from 25% to 7.5% (Kijjanapanich *et al.*, 2014b). For economic gains, this treated soil can be further used for agriculture and the produced sulfide can be recovered as the fertilizer elemental sulfur. However, it was also found that adding lime (0.02% of the gypsiferous soil) was required to neutralize the soil's pH due to its acidity (Kijjanapanich *et al.*, 2014b).

15.5 TREATMENT STRATEGIES FOR ACID MINE DRAINAGE

Mining waste, spoil or abandoned mines are very often difficult to control and the production of mine drainage water cannot be avoided. In these cases, proper treatment of this water is necessary. The treatment aims at the removal of (1) acidity, (2) toxic metals, (3) sulfate, and (4) hardness of the water.

Numerous technologies have been developed and are in use to treat AMD. However, many of them do not address all four above-mentioned pollutant groups in AMD. The two most important treatment principles in use are chemical treatment processes and microbial sulfate reduction.

15.5.1 Chemical treatment processes

15.5.1.1 Chemical neutralization

A first major treatment principle for AMD is neutralization. Application of alkali results in neutralization of acidity and in precipitation of toxic cations. Neutralization is most typically performed using lime. The process is usually set up in a horizontally percolated series of drains or trenches filled with the limestone (Ziemkiewicz *et al.*, 1997). A considerable negative effect of the lime application is usually an increasing hardness of the water.

Application of lime can result in a retention of sulfate as well. Sulfuric acid reacts with lime ($CaCO_3$), resulting in the formation of gypsum ($CaSO_4 \cdot 2H_2O$). However, formation of gypsum depends on pH. At pH higher than 7 (which is typical for a system saturated with lime) and in the presence of atmospheric CO_2, gypsum formation is thermodynamically unfavorable (see Box 15.1). This is overcome when limestone layers are isolated from the atmosphere (e.g. as anoxic lime drains, see Chapter 7, Section 7.3.1). The lack of free CO_2 in this system enables formation of gypsum even at high pH values. Moreover, the use of anoxic lime drains benefits from the prevalent existence of iron in its ferrous form. In this way, an undesired coating of lime particles with ferric precipitates (=armoring) is prevented.

The treated water from this technique has, nevertheless, a high sulfate concentration due to the high solubility of gypsum (2600 mg · L^{-1} in pure water at 25°C) (FAO, 1990). The latter results in a residual sulfate concentration of 1450 mg · L^{-1}, if calcium is used as the precipitant (Kijjanapanich *et al.*, 2014a).

15.5.1.2 Chemical precipitation of sulfate

15.5.1.2.1 Precipitation with barium and lead

Chemical precipitation is a well-established technology with ready availability of equipment and chemicals (U.S. EPA, 2000). Barium and lead compounds, such as barium chloride ($BaCl_2$) and lead nitrate ($Pb(NO_3)_2$), are well-known efficient chemicals for sulfate removal (Equations 15.3 and 15.4) (Benatti *et al.*, 2009; Maree *et al.*, 2004) with a sulfate removal efficiency up to 90–99% (Hlabela *et al.*, 2007; Kijjanapanich *et al.*, 2014a). The effect of the water pH on sulfate precipitation using $BaCl_2$ and $Pb(NO_3)_2$ can be negligible due to the low solubility of $BaSO_4$ and $PbSO_4$ (Benatti *et al.*, 2009; Kijjanapanich *et al.*, 2014a).

$$Ba_{(aq)}{}^{2+} + SO_{4\,(aq)}{}^{2-} \rightarrow BaSO_{4(s)} \tag{15.3}$$

$$Pb_{(aq)}{}^{2+} + SO_{4(aq)}{}^{2-} \rightarrow PbSO_{4(s)} \tag{15.4}$$

Although barium and lead compounds show good performance in sulfate precipitation, residual barium and lead which remains in the treated leachate or material after the precipitation process are toxic (Benatti *et al.*, 2009; Kijjanapanich *et al.*, 2014a; Kijjanapanich & Lens, 2017). They can result in an adverse impact on the environment if these are discharged or used without post-treatment. In addition, systems for precipitate separation and appropriate reuse or disposal of the solid phase are necessary when using chemical sulfate removal processes (Silva *et al.*, 2002). The study of Nadagouda *et al.* (2011) reports a synthesis of barium carbonate ($BaCO_3$) loaded polyvinyl chloride composites which can be used for sulfate removal. With this technique, the solubility issues of barium in treated water can be eliminated, thus reducing waste of barium carbonate and environmental issues (Nadagouda *et al.*, 2011).

15.5.1.2.2 Precipitation with calcium

Calcium compounds, such as calcium chloride ($CaCl_2$) and calcium oxide (CaO), can be cheap alternative chemicals for sulfate removal (Equation 15.5) (Benatti *et al.*, 2009; Kijjanapanich *et al.*, 2014a; Maree *et al.*, 2004), as these are less toxic than barium and lead. Moreover, gypsum precipitation can be enhanced via seeding and nucleation mechanisms (Madzivire *et al.*, 2010; Tait *et al.*, 2009). However, if calcium is used as a precipitant, a residual sulfate concentration of up to 1450 $mg \cdot L^{-1}$ of sulfate will remain due to the high solubility of calcium sulfate (gypsum).

$$Ca_{(aq)}^{2+} + SO_{4(aq)}^{2-} \rightarrow CaSO_{4(s)} \tag{15.5}$$

15.5.1.2.3 Ettringite precipitation

An alternative technology which can remove sulfate, hardness, as well as heavy metals is ettringite precipitation. With this technique, heavy metals will precipitate as metal hydroxide. Sulfate and calcium (resulting in high hardness) will precipitate as ettringite, which has a very low solubility in the mine drainage. Ettringite ($Ca_6Al_2(SO_4)_3(OH)_{12} \cdot 26H_2O$) appears as one of the most suitable minerals to be synthesized from mine drainage (Álvarez-Ayuso & Nugteren, 2005). Moreover, with precipitation as ettringite, the sulfate concentration from mine drainage can be reduced to less than 200 $mg \cdot L^{-1}$ (Dou *et al.*, 2017; Tolonen *et al.*, 2016). It was found that this technique can remove sulfate up to 99.7% from an initial sulfate concentration of 2304 $mg \cdot L^{-1}$ (Dou *et al.*, 2017).

Ettringite precipitation is not a complicated method and has a high efficiency. Ettringite precipitation can be conducted under controlled conditions such as high caustic alkalinity ($pH \approx 12$) and low sulfate concentrations (<0.1 M or <9600 $mg \cdot L^{-1}$). These conditions are the most suitable conditions to obtain ettringite as the reaction product (Álvarez-Ayuso & Nugteren, 2005). The quality of the treated water is such that it requires little further treatment and can be used for water supply. Furthermore, the precipitate from ettringite precipitation can be used as a sorbent for arsenate removal from the water phase (Tolonen *et al.*, 2016), used for the stabilization/solidification for heavy metal contaminated soils (Chrysochoou & Dermatas, 2006; Xiang *et al.*, 2018) and used as a material for concrete production.

When ettringite precipitation is used for water with high sulfate concentration, lime precipitation is used as the pre-treatment. In the pre-treatment step, part of the sulfate and heavy metals are precipitated as gypsum and metal hydroxides, respectively (Tolonen *et al.*, 2016). The precipitated gypsum is thickened and filtered. After partial sulfate removal, a smaller dosage of aluminum is needed in the ettringite precipitation step. Aluminum salt is added, leading to the precipitation of ettringite, normally using sodium aluminate or aluminum hydroxide (Tolonen *et al.*, 2016). At this step it is very important to dose with an appropriate amount of aluminum. Effluents with a high aluminum concentration should be avoided because aluminum can cause pulmonary and developmental problems in fish and benthic organisms that breath using gills (Grey, 2017). Then, the solution, which has a high caustic alkalinity and enough calcium, can be purified by carbon dioxide gas. Pure calcium carbonate is precipitated and can be separated from the resulting product water by filtration (Tolonen *et al.*, 2016).

However, aluminum chemicals are considerably more expensive than lime. Fly ash could be used as a low-cost aluminum source as an alternative way to reduce costs (Jankowski *et al.*, 2006; Mohamed *et al.*, 2003). Fly ash does not only provide aluminum for ettringite formation, but also promotes the rise of water alkalinity (Madzivire *et al.*, 2010; Wilińska & Pacewska, 2019).

15.5.1.3 Adsorption of sulfate

Adsorption separation is widely used because of its rapid and effective removal of the sulfate ion (Ao *et al.*, 2020). The adsorption method relies on electrostatic attractions which lead to anions binding onto various positively charged sites at material surfaces. This process generally occurs on the surface of minerals such as metallic oxides (Al, Fe), but also on soil components such as allophanes (typical of volcanic soils). The process may also be achieved through ligand exchange (Ao *et al.*, 2020). The adsorption of sulfate is pH dependent, and acidic conditions favor the adsorption (Ao *et al.*, 2020), which is suitable for AMD treatment.

Inorganic adsorbents, such as limestone (Silva *et al.*, 2012), modified alumino-silicate (analcime) (Runtti *et al.*, 2017), and barium exchanged geopolymers (Runtti *et al.*, 2016) can be used for sulfate adsorption with a sulfate adsorption capacity of 23.7, 13.0 and 119 $mg \cdot g^{-1}$, respectively. Recently, biochar has attracted great attention as an alternative desirable adsorbent and catalyst support due to its porous structure, low cost and environmental compatibility (Jung & Ahn, 2016). Biochar is usually prepared from agricultural wastes, such as wood, corn cob, rice straw, rice and fruit husks, and organic wastes (Ao *et al.*, 2020; Gebrewold *et al.*, 2019). However, ordinary biochar has shown poor adsorption capacity toward anionic species due to its negative surface charge (Bian *et al.*, 2014; Gebrewold *et al.*, 2019; Kameyama *et al.*, 2016). Therefore, an activation or modification process is required prior to use as adsorbent. Metal oxides such as magnesia, lanthana, zirconium and iron oxides are applied to modify biochar in order to improve its adsorption capacity for nitrate, sulfate and phosphate (Ao *et al.*, 2020; Wang *et al.*, 2016; Zhang *et al.*, 2012). A maximum sulfate adsorption capacity of 35.21 $mg \cdot g^{-1}$ was achieved when zirconium oxide modified pomelo peel biochar had been used as adsorbent (Ao *et al.*, 2020). Hardwood based activated carbon modified with polypyrrole can also be used for sulfate removal from AMD with a sulfate adsorption capacity of 44.7 $mg \cdot g^{-1}$ (Hong *et al.*, 2017).

Studies using organic materials as adsorbent for sulfate removal are relatively rare. Lignocellulosic polymers (surfactant-modified coir pith) (Namasivayam & Sureshkumar, 2007) and chitin-based shrimp shells (Moret & Rubio, 2003) have been used as sulfate adsorbent. The sulfate adsorption capacity of the lignocellulosic polymer adsorbent was found to be 8.8 mg · g^{-1} (Namasivayam & Sureshkumar, 2007). The maximum adsorption capacity of chitin-based shrimp shells reached 156 mg · g^{-1} (92%) at equilibrium pH 4.5 (Moret & Rubio, 2003).

15.5.1.4 Membrane based acid mine drainage treatment processes

There are several high-efficiency technologies that can remove contaminants such as sulfate, heavy metals and hardness from AMD. With these techniques, high quality treated water can be achieved. However, most of them have high investment and operation costs such as electrodialysis, reverse osmosis (RO) and ion exchange. Moreover, they result in a high concentration of liquid waste that needs to be handled.

Ion exchange is a high-efficiency technology which is used for removing heavy metals such as zinc and copper in AMD. The principle is based on exchanging ions of heavy metals in the water with the similar charges of the ion exchange resin in the equilibrium ratio (Brown *et al.*, 2002).

As the behavior of metals in solution is often controlled by their electrochemistry, the use of electrical technologies in the treatment of AMD has received attention from researchers and process developers. Several electrochemical techniques are often used in combination to facilitate direct deposition of a metal from a solution onto the electrodes (Brown *et al.*, 2002). However, only limited practical experience with these methods in the treatment of AMD is available. Therefore, their feasibility and economic value remain unclear.

15.5.2 Biological acid mine drainage treatment

15.5.2.1 Bioremediation of acid mine drainage

Sulfate reduction can be achieved in various technological configurations (Annachhatre & Suktrakoolvait, 2001; Kijjanapanich *et al.*, 2013; Kijjanapanich *et al.*, 2014d; Vallero *et al.*, 2005; Nevatalo *et al.*, 2010; Villa-Gomez *et al.*, 2011): (1) the upflow anaerobic sludge blanket (UASB) reactor in which biomass retention is based on the good settling characteristics of granular sludge (Lettinga *et al.*, 1980) and which is the most widely used reactor type for anaerobic wastewater treatment (Kijjanapanich & Lens, 2017), (2) the inverse fluidized bed (IFB) reactor which is based on floatable carrier material that is fluidized downward with a downflow liquid current, resulting in a promising reactor configuration for combined biological sulfate reduction (upper part of the reactor) and metal precipitate (bottom of the reactor) separation in a single unit (Kijjanapanich *et al.*, 2014d; Villa-Gomez *et al.*, 2011), (3) the anaerobic membrane bioreactor which is suitable for slow growing microorganisms, has a smaller reactor footprint and produces excellent effluent quality (Kijjanapanich *et al.*, 2014d; Lee & Kim, 2009), (4) wetlands (Eger, 1994; Tichý & Mejstrík 1996; Wildeman & Laudon, 1989), or (5) permeable reactive barrier (PRB)

(Benner *et al.*, 1999; Kijjanapanich *et al.*, 2012). The principles of using microbial sulfate reduction for AMD treatment are discussed in detail in Chapter 7.

In many installations, organic substrate, such as ethanol or lactate, is added to the system to increase the reaction rates (Kijjanapanich *et al.*, 2013; Liamleam & Annachhatre, 2007). For AMD treatment, low or no cost organic substrates, such as wood chips, compost, and sewage sludge, can also be used (Gibert *et al.*, 2004; Kijjanapanich *et al.*, 2012; Waybrant *et al.*, 1998). These organic substrates are much cheaper and less toxic compared to bulk chemicals (Kijjanapanich & Lens, 2017). Application of sulfate reduction leads to the most complete treatment of AMD: it neutralizes acidity, removes toxic metals (by precipitation as metal sulfides), removes part of the sulfate from the water and, via sulfidic precipitation, decreases the metal content in the AMD (Eger, 1994).

Among the outputs from the sulfate reduction processing of AMD is a sludge containing toxic metals and sulfides. In suspension reactors, this sludge can be (relatively) easily removed from the system and treated as a hazardous material or recycled. However, in more extensive process configurations, e.g. wetlands, this sludge remains at the treatment site. As a result, these sites accumulate high levels of reduced sulfur and toxic metals and will therefore become typical examples of chemical time bombs as described above (see 15.1.3). Owing to the pollution, such sites will require proper treatment and other pollution control measures (Gambrell, 1994). In some wetlands, proper sanitation and regeneration of the site may far exceed the costs of AMD treatment (Gaydardjiev *et al.*, 1996; Tichý *et al.*, 1998b). Unfortunately, these items are often omitted from the cost breakdown, which falsely makes these types of AMD-treatment very economical and attractive (Kadlec, 1995).

15.5.2.2 Metal and sulfur recovery from acid mine drainage

This chapter demonstrated that sulfur in solid materials has numerous effects of environmental importance. However, the amount of sulfurous compounds present is usually not the key aspect of environmental quality or deterioration. Rather, the side-effects of the sulfur cycling are of concern, especially the coupling of sulfur transformations with acidification/alkalization of the environment and with the mobilization/retention of cationic toxic metals.

The negative side-effects, like the appearance of chemical time bombs, self-acidifying soils, or AMD are evident. The treatment of sulfur pollution strongly depends on the extent of the pollution. Most of the techniques are intensive with respect to the use of chemicals, energy or equipment. However, they can only be applied for concentrated pollution in relatively small volumes of treated material (=point pollution). Sulfur is, however, mostly present in the form of diffuse pollution or as a pollution affecting large areas or volumes of waste (AMD, spoil banks and mine tailings). Here, less intensive approaches should be preferred. Extensive treatment techniques require lower inputs of chemicals or energy. However, the treatment times are appropriately prolonged. Examples of such extensive measures can be the use of natural systems, preventive measures, and whole life management of mining sites.

Transformations of sulfur in soils, sediments or solid waste can also be exploited in a positive way. Coupling of the sulfur cycle with the mobility of cationic toxic metals can be employed to control heavy metal pollution. Techniques like sulfate reduction or bioleaching have been demonstrated for pollution control of highly polluted waste streams. Sulfur cycling can also offer numerous opportunities to control diffuse pollution of the whole landscape. Amendment of sulfur at polluted sites may lead to metal mobilization, e.g. by desorption from the soil or removal from sludge, and its deposition in other compartments of the environment. Application of wetlands downstream may be a next step of this landscape management: mobilized metals with sulfide will be entrapped in the wetland sediment and accumulated there to high concentrations. After the metal concentrations are built-up in the system, the sediment can be excavated and

BOX 15.6 POSSIBLE USE OF THE MICROBIAL SULFUR CYCLE TO CONTROL HEAVY METAL POLLUTION (TICHÝ *ET AL.*, 1998A)

Figure 15.8 shows a combination of bioleaching followed by a soil/solution separation, microbial sulfate reduction producing sulfide, neutralization of acidity, precipitation of metals, and a partial microbial sulfide oxidation leading to the formation of elemental sulfur, which can be reused.

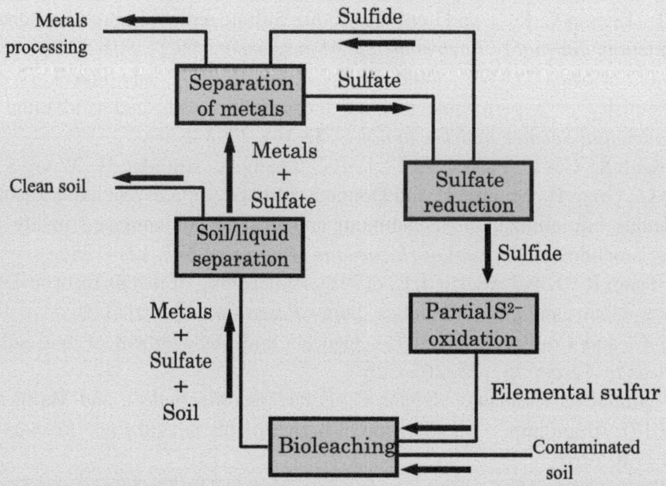

Figure 15.8 Use of the microbial sulfur cycle to control heavy metal pollution.

treated e.g. by bioleaching. In such a way, originally diffuse pollution can be converted into a concentrated and narrowly localized site, which can be periodically treated by intensive treatment systems. In the above described way, the whole microbial sulfur cycle may be employed as a tool to control heavy metal pollution in the landscape or in specific materials (see Box 15.6).

REFERENCES

Alfaya F., Cuenca-Sánchez M., Garcia-Orenes F. and Lens P. N. L. (2009). Endogenous and bioaugmented sulphate reduction in calcareous gypsiferous soils. *Environmental Technology,* **30**(12), 1305–1312.

Álvarez-Ayuso E. and Nugteren H. W. (2005). Synthesis of ettringite: a way to deal with the acid wastewaters of aluminium anodising industry. *Water Research,* **39**(1), 65–72.

Annachhatre A. P. and Suktrakoolvait S. (2001). Biological sulfate reduction using molasses as a carbon source. *Water Environment Research,* **73**(1), 118–126.

Ao H., Cao W., Hong Y., Wu J. and Wei L. (2020). Adsorption of sulfate ion from water by zirconium oxide-modified biochar derived from pomelo peel. *Science of the Total Environment,* **708**, 135092.

Azabou S., Mechichi T. and Sayadi S. (2005). Sulfate reduction from phosphogypsum using a mixed culture of sulfate-reducing bacteria. *International Biodeterioration & Biodegradation,* **56**(4), 236–242.

Babich I. V. and Moulijn J. A. (2003). Science and technology of novel processes for deep desulfurization of oil refinery streams: a review. *Fuel,* **82**(6), 607–631.

Bell A. V., Riley M. D. and Yanful E. K. (1995). Evaluation of a composite soil cover to control acid waste rock pile drainage. *CIM Bulletin,* **88**(995), 41–46.

Belzile N., Maki S., Chen Y. W. and Goldsack D. (1997). Inhibition of pyrite oxidation by surface-treatment. *Science of the Total Environment,* **196**(2), 177–186.

Benatti C. T., Tavares C. R. G. and Lenzi E. (2009). Sulfate removal from waste chemicals by precipitation. *Journal of Environmental Management,* **90**(1), 504–511.

Benner S. G., Blowes D. W., Gould W. D., Herbert R. B. and Ptacek C. J. (1999). Geochemistry of a permeable reactive barrier for metals and acid mine drainage. *Environmental Science and Technology,* **33**(16), 2793–2799.

Bian R., Joseph S., Cui L., Pan G., Li L., Liu X., Zhang A., Rutlidge H., Wong S., Chia C., Marjo C., Gong B., Munroe P. and Donne S. (2014). A three-year experiment confirms continuous immobilization of cadmium and lead in contaminated paddy field with biochar amendment. *Journal of Hazardous Materials,* **272**, 121–128.

Blais J. F., Tyagi R. D. and Auclair J. C. (1993). Bioleaching of metals from sewage sludge: microorganisms and growth kinetics. *Water Research,* **27**, 101–110.

Bloomfield C. and Coulter J. K. (1973). Genesis and management of acid sulfate soils. *Advances in Agronomy,* **25**, 265–326.

Bojórquez-Quintal E., Escalante-Magaña C., Echevarría-Machado I. and Martínez-Estévez M. (2017). Aluminum, a friend or foe of higher plants in acid soils. *Frontiers in Plant Science,* **8**, 1767.

Boon M. (2000). Chapter 5: Bioleaching of sulfide minerals. In: Environmental Technologies to Treat Sulfur Pollution: Principles and Engineering. P. N. L. Lens and L. Hulshoff Pol (eds), IWA Publishing, London, pp. 105–130.

Brimblecombe P., Hammer C., Rodhe H., Ryaboshapko A. and Boutron C. F. (1989). Human influence on the sulfur cycle. In: Evolution of the Global Biogeochemical Sulphur Cycle. SCOPE 39, P. Brimblecombe and A. Yu Lein (eds), John Wiley & Sons, Chichester, pp. 77–123.

Brown M., Barley B. and Wood H. (2002). Minewater Treatment. IWA Publishing, London.

Bruynesteyn A. (1989). Mineral biotechnology. *Journal of Biotechnology,* **11**, 1–10.

Bryant R. D., Costerton J. W. and Laishley E. J. (1983). The role of *Thiobacillus albertis* glycocalyx in the adhesion of cells to elemental sulfur. *Canadian Journal of Microbiology,* **30**, 81–90.

Cameron F., Jones M. V. and Edwards C. (1984). Effects of salinity on bacterial iron oxidation. *Current Microbiology,* **10**, 353–356.

Carlson L., Lindstrom E. B., Hallberg K. B. and Tuovinen O. H. (1992). Solid-phase products of bacterial oxidation of arsenical pyrite. *Applied and Environmental Microbiology,* **58**, 1046–1049.

Chipasa K. B. (2003). Accumulation and fate of selected heavy metals in a biological wastewater treatment system. *Waste Management,* **23**(2), 135–143.

Chrysochoou M. and Dermatas D. (2006). Evaluation of ettringite and hydrocalumite formation for heavy metal immobilization: Literature review and experimental study. *Journal of Hazardous Materials,* **136**, 20–33.

Cook T. M. (1964). Growth of *Thiobacillus thiooxidans* in shaken culture. *Journal of Bacteriology,* **88**, 620–623.

Couillard D. and Mercier G. (1992). Metallurgical residue for solubilization of metals from sewage sludge. *Journal of Environmental Engineering,* **118**, 808–813.

Couillard D. and Mercier G. (1994). An economic evaluation of biological removal of heavy metals from wastewater sludge. *Water Environment Research,* **66**, 32–39.

Cui H., Fan Y., Fang G., Zhang H., Su B. and Zhou J. (2016). Leachability, availability and bioaccessibility of Cu and Cd in a contaminated soil treated with apatite, lime and charcoal: A five-year field experiment. *Ecotoxicology and Environmental Safety,* **134P1**, 148–155.

Devasia P., Natarajan K. A., Sathyanarayana D. N. and Rao G. R. (1993). Surface chemistry of thiobacillus ferrooxidans relevant to adhesion on mineral surfaces. *Applied and Environmental Microbiology,* **59**, 4051–4055.

de Vries E. (2006). Biological Sulfate Removal from Construction and Demolition Debris Sand. Wageningen University

Dewil R., Baeyens J., Roels J. and Van De Steene B. (2009). Evolution of the total sulphur content in full-scale wastewater sludge treatment. *Environmental Engineering Science,* **26**(4), 867–872.

Dornblaser M., Giblin A. E., Fry B. and Peterson B. J. (1994). Effect of sulfate concentration in the overlying water on sulfate reduction and sulfur storage in lake sediments. *Biogeochemistry,* **24**, 129–144.

Dou W., Zhou Z., Jiang L.-M., Jiang A., Huang R., Tian X., Zhang W. and Chen D. (2017). Sulfate removal from wastewater using ettringite precipitation: Magnesium ion inhibition and process optimization. *Journal of Environmental Management,* **196**, 518–526.

Eger P. (1994). Wetland treatment for trace metal removal from mine drainage: the importance of aerobic and anaerobic processes. *Water Science and Technology,* **29**(4), 249–256.

Ellwood D. C., Hill M. J. and Watson J. H. P. (1992). Pollution control using microorganisms and magnetic separation. In: Microbial Control of Pollution. Proc. 48th Symp. Society for General Microbiology, J. C. Fry, G. M. Gadd, R. A. Herbert, C. W. Jones and I. A. Watson-Craik (eds), Cambridge University Press, Cardiff, U.K, pp. 89–112.

Ernst W. H. O. (2000). Agricultural aspects of sulfur. In: Environmental Technologies to Treat Sulfur Pollution: Principles and Engineering, P. N. L. Lens and L. Hulshoff Pol (eds), IWA Publishing, London, pp. 355–376.

Evangelou V. P. (1995a). Potential microencapsulation of pyrite by artificial inducement of ferric phosphate coatings. *Journal of Environmental Quality,* **24**(3), 535–542.

Evangelou V. P. (1995b). Pyrite Oxidation and Its Control. CRC Press, USA.

Evangelou V. P. and Zhang Y. L. (1995). Pyrite oxidation mechanisms and acid mine drainage prevention. *Critical Reviews in Environmental Science and Technology,* **25**, 141–199.

Evans L. J. (1989). Chemistry of metal retention by soils. *Environmental Science and Technology,* **23**, 1046–1056.

FAO (1990). FAO Soils Bulletin 62: Management of Gypsiferous Soils. Rome.

FAO (1993). World Soil Resources Report 66: An explanation of the FAO World Soil Resources Map at a scale of 1:25 000 000. Rome

Förstner U. (1995). Non-linear release of metals from aquatic sediments. In: Biogeodynamics of Pollutants in Soils and Sediments, W. Salomons and W. M. Stigliani (eds), Springer Verlag, Berlin, pp. 247–308.

Fytas K. and Bousquet P. (2002). Silicate micro-encapsulation of pyrite to prevent acid mine drainage. *CIM Bulletin,* **95**(1063), 96–99.

Gambrell R. P. (1994). Trace and Toxic Metals in Wetlands – A Review. *Journal of Environmental Quality,* **23**, 883–891.

Gaydardjiev S., Hadjichristova M. and Tichý R. (1996). Opportunities for using two low-cost methods for treatment of metal bearing aqueous streams. *Minerals Engineering,* **9**(9), 947–964.

Gebrewold B. D., Kijjanapanich P., Rene E. R., Lens P. N. L. and Annachhatre A. P. (2019). Fluoride removal from groundwater using chemically modified rice husk and corn cob activated carbon. *Environmental Technology,* **40**(22), 2913–2927.

Georgopoulou Z. J., Fytas K., Soto H. and Evangelou B. (1996). Feasibility and cost of creating an iron-phosphate coating on pyrrhotite to prevent oxidation. *Environmental Geology,* **28**(2), 61–69.

Gibert O., de Pablo J., Cortina J. L. and Ayora C. (2004). Chemical characterization of natural organic substrates for biological mitigation of acid mine drainage. *Water Research,* **38**, 4186–4196.

Golez N. V. and Kyuma K. (1997). Influence of pyrite oxidation and soil acidification on some essential nutrient elements. *Aquacultural Engineering,* **16**(1), 107–124.

Gray N. F. (1997). Environmental impact and remediation of acid-mine drainage – a management problem. *Environmental Geology,* **30**(1–2), 62–71.

Grey C. V. (2017). Physicochemical Treatment Options for High-Conductivity Coal Mining Runoff. Virginia Polytechnic Institute and State University, Blacksburg.

Gypsum Association. (1992). Treatment and disposal of gypsum board waste: Technical paper part II. *AWIC's Construction Dimensions*, March, **58**, 60–63.

Hinsenveld M. (1993). Physical/chemical extraction technologies. In: Demonstration of Remedial Action Technologies for Contaminated Land and Groundwater, R. F. Olfenbuttel, (ed.), NATO/CCMS, Brussel.

Hlabela P., Maree J. and Bruinsma D. (2007). Barium carbonate process for sulphate and metal removal from mine water. *Mine Water and the Environment,* **26**, 14–22.

Holmer M. and Storkholm P. (2001). Sulphate reduction and sulphur cycling in lake sediments: a review. *Freshwater Biology,* **46**, 431–451.

Hong S., Cannon F. S., Hou P., Byrne T. and Nieto-Delgado C. (2017). Adsorptive removal of sulfate from acid mine drainage by polypyrrole modified activated carbons: Effects of polypyrrole deposition protocols and activated carbon source. *Chemosphere,* **184**, 429–437.

Inkham R., Kijjanapanich V., Huttagosol P. and Kijjanapanich P. (2019). Low-cost alkaline substances for the chemical stabilization of cadmium-contaminated soils. *Journal of Environmental Management,* **250**, 109395.

Işıldar A., van de Vossenberg J., Rene E. R., van Hullebusch E. D. and Lens P. N. L. (2016). Two-step bioleaching of copper and gold from discarded printed circuit boards (PCB). *Waste Management,* **57**, 149–157.

Işıldar A., van de Vossenberg J., Rene E. R., van Hullebusch E. D. and Lens P. N. L. (2017). Chapter 8: Biorecovery of metals from electronic waste. In: Sustainable Heavy Metal Remediation Volume 2: Case Studies, E. R. Rene, E. Sahinkaya, A. Lewis and P. N. L. Lens (eds), Springer International Publishing, Switzerland, pp. 241–278.

Iske U. and Glombitza F. (1988). Moglichkeiten der nutzung biotechnologischer verfahren fur die gewinnung von metallen aus erzen, mineralen und abprodukten der chemischen industrie (Possible use of biotechnology for the extraction of metals from ores, minerals, and chemical waste). *Chemine Technologija,* **40**(11), 464–468.

Ivanov M. V. (1983). Major fluxes of the global biogeochemical cycle of sulfur. In: The Global Biogeochemical Sulfur Cycle. SCOPE 19, M. V. Ivanov and J. R. Freney (eds), John Wiley & Sons, Chichester, pp. 449–463.

Jang Y. C. and Townsend T. (2001). Sulfate leaching from recovered construction and demolition debris fines. *Advances in Environmental Research,* **5**, 203–217.

Jang M., Kang C. S. and Hyun Moon J. (2005). Estimation of ^{222}Rn release from the phosphogypsum board used in housing panels. *Journal of Environmental Radioactivity,* **80**(2), 153–160.

Jankowski J., Ward C. R., French D. and Groves S. (2006). Mobility of trace elements from selected Australian fly ashes and its potential impact on aquatic ecosystems. *Fuel,* **85**, 243–256.

Jung K.-W. and Ahn K.-H. (2016). Fabrication of porosity-enhanced MgO/biochar for removal of phosphate from aqueous solution: Application of a novel combined electrochemical modification method. *Bioresource Technology,* **200**, 1029–1032.

Kadlec R. H. (1995). Overview – Surface flow constructed wetlands. *Water Science and Technology,* **32**(3), 1–12.

Kameyama K., Miyamoto T., Iwata Y. and Shiono T. (2016). Influences of feedstock and pyrolysis temperature on the nitrate adsorption of biochar. *Soil Science and Plant Nutrition,* **62**(2), 180–184.

Karathanasis A. D. and Thompson Y. L. (1995). Mineralogy of iron precipitates in a constructed acid mine drainage wetland. *Soil Science Society of America Proceedings,* **59**, 1773–1781.

Karavaiko G. I. (1985). Microbiological Processes for the Leaching Of Metals From Ores. Centre of International Projects GKNT, Moscow.

Karnachuk, O. V., Kurochkina, S. Y., and Tuovinen, O. H. (2002). Growth of sulfate-reducing bacteria with solid-phase electron acceptors. *Applied Microbiology and Biotechnology*, **58**, 482–486.

Khresat S. A., Rawajfih Z., Buck B. and Monger H. C. (2004). Geomorphic features and soil formation of arid lands in Northeastern Jordan. *Archives of Agronomy and Soil Science*, **50**, 607–615.

Kijjanapanich P. and Lens P. N. L. (2017). Chapter 4: Biological sulphate reduction. In: Sustainable Heavy Metal Remediation Volume 2: Case studies, E. R. Rene, E. Sahinkaya, A. Lewis and P. N. L. Lens (eds), Springer International Publishing, Switzerland, pp. 115–132.

Kijjanapanich P., Pakdeerattanamint K., Lens P. N. L. and Annachhatre A. P. (2012). Organic substrates as electron donors in permeable reactive barriers for removal of heavy metals from acid mine drainage. *Environmental Technology,* **33**(23), 2635–2644.

Kijjanapanich P., Annachhatre A. P., Esposito G., van Hullebusch E. D. and Lens P. N. L. (2013). Biological sulfate removal from gypsum contaminated construction and demolition debris. *Journal of Environmental Management,* **131**, 82–91.

Kijjanapanich P., Annachhatre A. P., Esposito G. and Lens P. N. L. (2014a). Chemical sulphate removal for treatment of construction and demolition debris leachate. *Environmental Technology,* **35**(16), 1989–1996.

Kijjanapanich P., Annachhatre A. P., Esposito G. and Lens P. N. L. (2014b). Use of organic substrates as electron donors for biological sulfate reduction in gypsiferous mine soils from Nakhon Si Thammarat (Thailand). *Chemosphere,* **101**, 1–7.

Kijjanapanich P., Annachhatre A. P. and Lens P. N. L. (2014c). Biological sulfate reduction for treatment of gypsum contaminated soils, sediments and solid wastes. *Critical Reviews in Environmental Science and Technology,* **44**, 1037–1070.

Kijjanapanich P., Do A. T., Annachhatre A. P., Esposito G., Yeh D. H. and Lens P. N. L. (2014d). Biological sulfate removal from construction and demolition debris leachate: Effect of bioreactor configuration. *Journal of Hazardous Materials,* **269**, 38–44.

Kochian L. V., Piñeros M. A. and Hoekenga O. A. (2005). The physiology, genetics and molecular biology of plant aluminum resistance and toxicity. *Plant and Soil,* **274**, 175–195.

Kollias K., Mylona E., Papassiopi N. and Xenidis A. (2015). Conditions favoring the formation of iron phosphate coatings on the pyrite surface. *Desalination and Water Treatment,* **56**(5), 1274–1281.

Lamers L. P. M., Tomassen H. B. M. and Roelofs J. G. M. (1998). Sulfare-induced eutrophication and phytotoxicity in freshwater wetlands. *Environmental Science and Technology,* **32**, 199–205.

Lee M. and Kim J. (2009). Membrane autopsy to investigate $CaCO_3$ scale formation in pilot-scale, submerged membrane bioreactor treating calcium-rich wastewater. *Journal of Chemical Technology and Biotechnology,* **84**(9), 1397–1404.

Lettinga G., van Velsen A. F. M., Hobma S. W., de Zeeuw W. J. and Klapwijk A. (1980). Use of the upflow sludge blanket (USB) reactor concept for biological wastewater treatment, especially anaerobic treatment. *Biotechnology and Bioengineering,* **22**(4), 699–734.

Levy D. B., Custis K. H., Casey W. H. and Rock P. A. (1997). The aqueous geochemistry of the abandoned Spenceville copper pit, Nevada county, California. *Journal of Environmental Quality,* **26**(1), 233–243.

Liamleam W. and Annachhatre A. P. (2007). Electron donors for biological sulfate reduction. *Biotechnology Advances,* **25**(5), 452–463.

Lindsay W. L. (1979). Chemical Equilibria in Soils. John Wiley & Sons, New York.

Loi G., Mura A., Passarini N., Trois P. and Rossi G. (1993). Modification of oxidizing activity of *Thiobacillus ferroxidans* by some particulate solids. *Fuel,* **72**, 1607–1611.

Ma Q., Rengel Z. and Kuo J. (2002). Aluminium toxicity in rye (*Secale cereale*): Root growth and dynamics of cytoplasmic Ca^{2+} in intact root tips. *Annals of Botany,* **89** (2), 241–244.

Maass B. and Miehlich G. (1988). Die Wirrkung des Redoxpotentials auf die Zusammnesetzung der Porenlösung in Hafenschlicks-feldern. *Mitt Dtsch Bodekunde Ges,* **56**, 289–294.

Mackintosh M. E. (1978). Nitrogen fixation by *Thiobacillus ferrooxidans*. *Journal of General Microbiology,* **105**, 215–218.

Madzivire G., Petrik L. F., Gitari W. M., Ojumu T. V. and Balfour G. (2010). Application of coal fly ash to circumneutral mine waters for the removal of sulphates as gypsum and ettringite. *Minerals Engineering,* **23**(3), 252–257.

Mamelkina M. A., Tuunila R., Sillänpää M. and Häkkinen A. (2019). Systematic study on sulfate removal from mining waters by electrocoagulation. *Separation and Purification Technology,* **216**, 43–50.

Maree J. P., Hlabela P., Nengovhela R., Geldenhuys A. J., Mbhele N., Nevhulaudzi T. and Waamders F. B. (2004). Treatment of mine water for sulphate and metal removal using barium sulphide. *Mine Water and the Environment,* **23**, 195–203.

Marra A., Cesaro A., Rene E. R., Belgiorno V. and Lens P. N. L. (2018). Bioleaching of metals from WEEE shredding dust. *Journal of Environmental Management,* **210**, 180–190.

Meulepas R. J. W., Gonzalez-Gil G., Teshager F. M., Witharana A., Saikaly P. E. and Lens P. N. L. (2015). Anaerobic bioleaching of metals from waste activated sludge. *Science of the Total Environment,* **514**, 60–67.

Mirajkar Y. R. K., Natarajan K. A. and Somasundaran P. (1997). Growth and attachment of thiobacillus-ferrooxidans during sulfide mineral leaching. *International Journal of Mineral Processing,* **50**(3), 203–210.

Misra M., Yang K. and Mehta R. K. (1996). Application of fly-ash in the agglomeration of reactive mine tailings. *Journal of Hazardous Materials,* **51**(1–3), 181–192.

Mohamed A., Hossein M. and Hassani F. (2003). Role of fly ash and aluminum addition on ettringite formation in lime-remediated mine tailings. *Cement Concrete and Aggregates,* **25**(2), 1–10.

Mohammadi M., Mowla D., Esmaeilzadeh F. and Ghasemi Y. (2019). Enhancement of sulfate removal from the power plant wastewater using cultivation of indigenous microalgae: Stage-wise operation. *Journal of Environmental Chemical Engineering,* **7**(1), 102870.

Montero A., Tojo Y., Matsuto T., Yamada M., Asakura H. and Ono Y. (2010). Gypsum and organic matter distribution in a mixed construction and demolition waste sorting process and their possible removal from outputs. *Journal of Hazardous Materials,* **175**, 747–753.

Moret A. and Rubio J. (2003). Sulphate and molybdate ions uptake by chitin-based shrimp shells. *Minerals Engineering,* **16**(8), 715–722.

Nadagouda M. N., Pressman J., White C., Speth T. F. and McCurry D. L. (2011). Novel thermally stable poly(vinyl chloride) composites for sulfate removal. *Journal of Hazardous Materials,* **188**, 19–25.

Namasivayam C. and Sureshkumar M. V. (2007). Removal of sulfate from water and wastewater by surfactant-modified coir pith, an agricultural solid 'waste' by adsorption methodology. *Journal of Environmental Engineering and Management,* **17**(2), 129–135.

Nevatalo L. M., Makinen A. E., Kaksonen A. H. and Puhakka J. A. (2010). Biological hydrogen sulfide production in an ethanol-lactate fed fluidized-bed bioreactor. *Bioresource Technology,* **101**, 276–284.

Nogueirol R. C., Monteiro F. A. and Azevedo R. A. (2015). Tropical soils cultivated with tomato: fractionation and speciation of Al. *Environmental Monitoring and Assessment,* **187**, 160.

Novak J. T., Adams G., Chen Y. C., Erdal Z., Forbes R. H. J., Glindemann D., Hargreaves J. R., Hentz L., Higgins M. J., Murthy S. N. and Witherspoon J. (2006). Generation pattern of sulfur containing gases from anaerobically digested sludge cakes. *Water Environment Research,* **78**(8), 821–827.

Nuhu A. A. (2013). Bio-catalytic desulfurization of fossil fuels: a mini review. *Reviews in Environmental Science and Biotechnology,* **12**, 9–23.

Ondrushka J. and Glombitza F. (1993). Inhibition of natural microbiological leaching processes. In: Proc. Int. Conf. Contaminated Soil '93 May 1993, F. Berlin, G. J. Arendt, R. Annokée, W. J. Bosman and van der Brink (eds), Kluwer Academic Publishers, Dordrecht, pp. 1195–1196.

Orava D. A. and Swider R. C. (1996). Inhibiting acid-mine drainage throughout the mine life-cycle. *CIM Bulletin,* **89**(999), 52–56.

Padival N. A., Weiss J. S. and Arnold R. G. (1995). Control of Thiobacillus by means of microbial competition: implications for corrosion of concrete sewers. *Water Environment Research,* **67**, 201–205.

Paengsen K. (2013). Reduction of Sulfate Contamination in Mae Moh Mine Drainage by Using Bed Clay as Covered Material (In Thai). Chiang Mai University, Chiang Mai.

Panpa W. (2002). Production of Dental Stone from Flue-Gas Gypsum. Chulalongkorn University.

Papageorgiou F., Godelitsas A., Mertzimekis T. J., Xanthos S., Voulgaris N. and Katsantonis G. (2016). Environmental impact of phosphogypsum stockpile in remediated Schistos waste site (Piraeus, Greece) using a combination of γ-ray spectrometry with geographic information systems. *Environmental Monitoring and Assessment,* **188**(3), 133.

Park N. J., Park J. S. and Kim Y. S. (1971). Effect of lime on growth of rice and changes in pH, Eh, Fe^{2+} and Al in an acid sulfate soil. *Journal of Korean Soil Science and Fertilizer,* **4**, 167–175.

Peiffer S. (1998). Geochemical and microbial processes in sediments and at the sediment-water interface of acidic mining lakes. *Water, Air, & Soil Pollution,* **108**, 227–229.

Porro S., Ramirez S., Reche C., Curutchet G., Alonsoromanowski S. and Donati E. (1997). Bacterial attachment – its role in bioleaching processes. *Process Biochemistry,* **32**(7), 573–578.

Porta J. and Herrero J. (1990). Micromorphology and genesis of soils enriched with gypsum. In: Soil Micromorphology: A Basic and Applied Science, L. A. Douglas (ed.), Elsevier, Amsterdam, pp. 321–339.

Potysz A., Lens P. N. L., van de Vossenberg J., Rene E. R., Grybos M., Guibaud G., Kierczak J. and van Hullebusch E. D. (2016). Comparison of Cu, Zn and Fe bioleaching from Cu-metallurgical slags in the presence of Pseudomonas fluorescens and Acidithiobacillus thiooxidans. *Applied Geochemistry,* **68**, 39–52.

Promnuan K., Higuchi T., Imai T., Kongjan P., Reungsang A. and O-Thong S. (2020). Simultaneous biohythane production and sulfate removal from rubber sheet wastewater by two-stage anaerobic digestion. *International Journal of Hydrogen Energy,* **45**(1), 263–274.

Pronk J. T., Meulenberg R., Hazeu W., Bos P. and Kuenen J. G. (1990). Oxidation of reduced inorganic sulfur compounds by acidophilic thiobacilli. *FEMS Microbiology Reviews,* **75** (2–3), 293–306.

Pronk J. T., de Bruyn J. C., Bos P. and Kuenen J. G. (1992). Anaerobic growth of *Thiobacillus ferrooxidans. Applied and Environmental Microbiology,* **58**, 2227–2230.

Reinhoudt H. R. and Moulijn J. A. (2000). Chapter 4: Catalytic removal of sulfur from diesel oil by hydrotreating. In: P. N. L. Lens and L. Hulshoff Pol, (eds), Environmental Technologies to Treat Sulfur Pollution: Principles and Engineering. IWA Publishing, London, pp. 87–104.

Richards I. G., Palmer J. P. and Barratt P. A. (1993). The Reclamation of Former Coal Mines and Steelworks. Studies in Environmental Science. Elsevier, Amsterdam.

Rulkens W. H., Tichý R. and Grotenhuis J. T. C. (1995). Sites polluted with heavy metals: Current techniques for clean-up and desirable future developments. In: Heavy metals in the environment. Proc. Int. Conf., Hamburg, September 1995, R.-D. Wilken, U. Förstner and A. Knöchel (eds), CEP Consultants Ltd., Edinburgh, pp. 11–19.

Rulkens W. H., Tichý R. and Grotenhuis J. T. C. (1998). Remediation of polluted soil and sediment: Perspectives and failures. *Water Science and Technology,* **37**(8), 27–36.

Runtti H., Luukkonen T., Niskanen M., Tuomikoski S., Kangas T., Tynjälä P., Tolonen E.-T., Sarkkinen M., Kemppainen K., Rämö J. and Lassi U. (2016). Sulphate removal over barium-modified blast-furnace-slag geopolymer. *Journal of Hazardous Materials,* **317**, 373–384.

Runtti H., Tynjälä P., Tuomikoski S., Kangas T., Hu T., Rämö J. and Lassi U. (2017). Utilisation of barium-modified analcime in sulphate removal: Isotherms, kinetics and thermodynamics studies. *Journal of Water Process Engineering,* **16**, 319-328.

Ryu S., Naidu G., Hasan Johir M. A., Choi Y., Jeong S. and Vigneswaran S. (2019). Acid mine drainage treatment by integrated submerged membrane distillation–sorption system. *Chemosphere,* **218**, 955–965.

Sade H., Meriga B., Surapu V., Gadi J., Sunita M. S. L., Suravajhala P. and Kavi Kishor P. B. (2016). Toxicity and tolerance of aluminum in plants: tailoring plants to suit to acid soils. *Biometals,* **29**, 187–210.

Salomons W. H. (1993). Non-linear and delayed responses of toxic chemicals in the environment. In: Proc. Int. Conf. Contaminated Soil '93 May 1993, Berlin, F. Arendt, G. J. Annokée, R. Bosman and W. J. van der Brink (eds), Kluwer Academic Publishers, Dordrecht, pp. 225–238.

Salomons W. and Stigliani W. M. (1995). Biogeodynamics of Pollutants in Soils and Sediments. Springer Verlag, Berlin.

Sasaki K. (1997). A study on the dissolution of pyrite and its suppression in acidic environments. *Bunseki Kagaku,* **46**(7), 611–612.

Schaeffer W. I. and Umbreit W. W. (1963). Phosphotidylinositol as a wetting agent in sulfur oxidation by *Thiobacillus thiooxidans. Journal of Bacteriology,* **85**, 492–493.

Schuring J., Kolling M. and Schulz H. D. (1997). The potential formation of acid-mine drainage in pyrite-bearing hard-coal tailings under water-saturated conditions – an experimental approach. *Environmental Geology,* **31**(1–2), 59–65.

Shamshuddin J., Elisa Azura A., Shazana M. A. R. S., Fauziah C. I., Panhwar Q. A. and Naher U. A. (2014). Chapter 3: Properties and management of acid sulfate soils in Southeast Asia for sustainable cultivation of rice, oil palm, and cocoa. In: Advances in Agronomy, D. L. Sparks (ed.), Elsevier, Burlington, Vol. **124**, pp. 91–142.

Silva A. J., Varesche M. B., Foresti E. and Zaiat M. (2002). Sulphate removal from industrial wastewater using a packed-bed anaerobic reactor. *Process Biochemistry,* **37**, 927–935.

Silva A. M., Lima R. M. F. and Leão V. A. (2012). Mine water treatment with limestone for sulfate removal. *Journal of Hazardous Materials*, **221–222**, 45–55.

Smith R. and Robertson V. S. (1962). Soil and irrigation classification of shallow soils overlying gypsum beds, northern Iraq. *Journal of Soil Science,* **13**, 106–115.

Smith J. R., Luthy R. G. and Middleton A. C. (1988). Microbial ferrous iron oxidation in acidic solution. *Journal Water Pollution Control Federation,* **60**, 518–530.

Smolders A. J. P., Lamers L. P. M., Moonen M., Zwaga K. and Roelofs J. G. M. (2001). Controlling phosphate release from phosphate-enriched sediments by adding various iron compounds. *Biogeochemistry,* **54**, 219–228.

Smolders A. J. P., Lamers L. P. M., Lucassen E. C. H. E. T., van der Velde G. and Roelofs J. G. M. (2006). Internal eutrophication: How it works and what to do about it – a review. *Chemical Ecology,* **22**(2), 93–111.

Solari J. A., Huerta G., Escobar B., Vargas T., Badilla-Ohlbaum R. and Rubio J. (1992). Interfacial phenomena affecting the adhesion of *Thiobacillus ferrooxidans* to sulfide mineral surfaces. *Colloids and Surfaces,* **69**, 159–166.

Spottiswood E. G. (1982). Introduction to Mineral Processing. John Wiley & Sons Inc., New York.

Stumm W. and Furrer G. (1987). The dissolution of oxides and aluminum silicates; Examples of surface-coordination-controlled kinetics. In: Aquatic Surface Chemistry, W. Stumm (ed.), John Wiley & Sons, New York, pp. 197–219.

Stumm W. and Morgan J. J. (1981). Aquatic Chemistry. An Introduction Emphasizing Chemical Equilibria in Natural Waters. John Wiley & Sons, Inc., New York.

Suanon F., Sun Q., Mama D., Li J., Dimon B. and Yu C.-P. (2016). Effect of nanoscale zero-valent iron and magnetite (Fe_3O_4) on the fate of metals during anaerobic digestion of sludge. *Water Research,* **88**, 897–903.

Tait S., Clarke W. P., Keller J. and Batstone D. J. (2009). Removal of sulfate from high-strength wastewater by crystallisation. *Water Research,* **43**(3), 762–772.

Takeuchi T. L. and Suzuki I. (1997). Cell hydrophobicity and sulfur adhesion of *Thiobacillus thiooxidans. Applied and Environmental Microbiology,* **63**(5), 2058–2061.

Tayibi H., Choura M., Lopez F. A., Alguacil F. J. and Lopez-Delgado A. (2009). Environmental impact and management of phosphogypsum. *Journal of Environmental Management,* **90**, 2377–2386.

Thomson B. M. and Turney W. R. (1996). Minerals and mine drainage. *Water Environment Research,* **68**(4), 542–546.

Tichý R. (2000). Chapter 14: Treatment of solid materials containing inorganic sulfur compounds. In: Environmental Technologies to Treat Sulfur Pollution: Principles and Engineering, P. N. L. Lens and L. Hulshoff Pol (eds), IWA Publishing, London, pp. 329–354.

Tichý R. and Mejstrík V. (1996). Heavy metal contamination from open-pit coal-mining in Europe's Black Triangle and possible remediation. *Environmental Reviews,* **4**, 321–341.

Tichý R., Grotenhuis J. T. C., Rulkens W. H. and Nýdl V. (1996). Strategy for leaching of zinc from artificially contaminated soil. *Environmental Technology,* **17**, 1181–1182.

Tichý R., Lens P., Grotenhuis J. T. C. and Bos P. (1998a). Solid-state reduced sulfur compounds: environmental aspects and bio-remediation. *Critical Reviews in Environmental Science and Technology,* **28**(1), 1–40.

Tichý R., Rulkens W. H., Grotenhuis J. T. C., Nýdl V., Cuypers C. and Fajtl J. (1998b). Bioleaching of metals from soils or sediments. *Water Science and Technology,* **37**(8), 119–128.

Tolonen E.-T., Hu T., Rämö J. and Lassi U. (2016). The removal of sulphate from mine water by precipitation as ettringite and the utilisation of the precipitate as a sorbent for arsenate removal. *Journal of Environmental Management,* **181**, 856–862.

Tyagi R. D., Couillard D. and Tran F. T. (1991). Comparative study of bacterial leaching of metals from sewage sludge in continuous stirred tank and air-lift reactors. *Process Biochemistry,* **26**, 47–54.

Tyagi R. D., Sreekrishnan T. R., Campbell P. G. C. and Blais J. F. (1993). Kinetics of heavy metal bioleaching from sewage sludge. 2. Mathematical model. *Water Research*, **27**, 1653–1661.

U.S. EPA. (2000). Wastewater Technology Fact Sheet Chemical Precipitation EPA 832-F00-018. Washington D.C.

U.S. EPA. (2008). Agricultural Uses for Flue Gas Desulfulfurization (FGD) Gypsum EPA530-F-08-009. Washington D.C.

U.S. EPA. (2018). Fact Sheet: Final 2018 Aquatic Life Ambient Water Quality Criteria for Aluminum in Freshwaters EPA 822-F-18-003. Washington D.C.

Vallero M. V. G., Lettinga G. and Lens P. N. L. (2005). High rate sulfate reduction in a submerged anaerobic membrane bioreactor (SAMBaR) at high salinity. *Journal of Membrane Science,* **253**(1–2), 217–232.

van Breemen N. (1973). Dissolved aluminium in acid sulfate soils and in mine waters. *Soil Science Society of America Proceedings*, **37**, 694–697.

van Breemen N. (1975). Acidification and deacidification of coastal plain soils as a results of periodic flooding. *Soil Science Society of America Proceedings*, **39**, 1153–1157.

van Breemen N. (1988). Effects of seasonal redox processes involving iron on the chemistry of periodically reduced soils. In: Iron in Soils and Clay Minerals D Reidel Publ Co, Dordrecht, pp. 797–809.

van Houten R. T., Hulshoff Pol L. W. and Lettinga G. (1994). Biological sulfate reduction using gas-lift reactors fed with hydrogen and carbon dioxide as energy and carbon source. *Biotechnology and Bioengineering,* **44**, 586–594.

van Mensvoort M. E. F., Duang V. N. and van der Schans J. (1991). Improvement of acid sulphate soils by leaching with salt or brackish water. In: Rice production on acid soils of the tropics, P. Deturck and F. N. Ponnamperuna (eds), Inst. Fundamental Studies, Kandy, pp. 219–224.

Villa-Gomez D., Ababneh H., Papirio S., Rousseau D. P. L. and Lens P. N. L. (2011). Effect of sulfide concentration on the location of the metal precipitates in inversed fluidized bed reactors. *Journal of Hazardous Materials,* **192**, 200–207.

Wang Z., Shen D., Shen F. and Li T. (2016). Phosphate adsorption on lanthanum loaded biochar. *Chemosphere,* **150**, 1–7.

Waybrant K. R., Blowes D. W. and Ptacek C. J. (1998). Selection of Reactive Mixtures for Use in Permeable Reactive Walls for Treatment of Mine Drainage. *Environmental Science and Technology,* **32**(13), 1972–1979.

Wildeman T. R. and Laudon L. S. (1989). Use of wetlands for treatment of environmental problems in mining: non-coal-mining applications. In: Constructed Wetlands for Wastewater Treatment. Municipal, Industrial and Agricultural, D. A. Hammer (ed.), Lewis Publishers, Chelsea, Michigan, pp. 221–231.

Wilfert P., Meerdink J., Degaga B., Temmink H., Korving L., Witkamp G. J., Goubitz K. and van Loosdrecht M. C. M. (2020). Sulfide induced phosphate release from iron phosphates and its potential for phosphate recovery. *Water Research,* **171**, 115389.

Wilińska I. and Pacewska B. (2019). Comparative investigation of reactivity of different kinds of fly ash in alkaline media. *Journal of Thermal Analysis and Calorimetry,* **138**, 3857–3872.

Xiang L., Zhu H. and Zhou R. (2018). Effect of Ca/Al ratio on stabilization/solidification of lead-contaminated soil by ettringite. *Hindawi Advances in Civil Engineering,* **2018**, 3612962.

Zhang M., Gao B., Yao Y., Xue Y. and Inyang M. (2012). Synthesis of porous MgO-biochar nanocomposites for removal of phosphate and nitrate from aqueous solutions. *Chemical Engineering Journal,* **210**, 26–32.

Ziemkiewicz P. F., Skousen J. G., Brant D. L., Sterner P. L. and Lovett R. J. (1997). Acid-mine drainage treatment with armored limestone in open limestone channels. *Journal of Environmental Quality,* **26**(4), 1017–1024.

Index

IWA Publishing's authorised EU representative for General Product Safety Regulations is Diane D'Arras, 15 rue Duret, 75116 Paris, France, e-mail: safety@iwap.co.uk.

Printed and bound by CPI Group (UK) Ltd, Croydon, CR0 4YY
12/05/2026
02108864-0016